CRC Handbook Series in Organic Electrochemistry

Volume VI

Authors:

Louis Meites
Petr Zuman
Professors of Chemistry
Department of Chemistry
Clarkson College of Technology

Elinore B. Rupp
Research Associate in Chemistry
Department of Chemistry
Clarkson College of Technology

and

Theodore L. Fenner
Ananthakrishnan Narayanan

With financial support from the National Science Foundation

CRC Press, Inc.
Boca Raton, Florida

Library of Congress Cataloging in Publication Data

Meites, Louis.
   Handbook of organic electrochemistry.

(CRC handbook series in organic electrochemistry; v. 6).
   "List of abbreviations": p. inserted.
   Editors: v. Louis Meites, et al.
   Bibliography: v. 6, p.
   Includes index.
   1. Electrochemical analysis--Handbooks, manuals,
etc. 2. Chemistry, Organic--Handbooks, manuals, etc.
I. Meites, Louis. II. Zuman, Petr, joint author.
[DNLM: 1. Chemistry, Organic--Handbooks. 2. Electro-
chemistry--Handbooks. W1 C58K]
QD272.E4C17   547.1'37   77-24273
ISBN 0-8493-7220-8 (set)
ISBN 0-8493-7226-7 (v. 6)

This Handbook was prepared with the support of the National Science Foundation Grants No. CHE-7617435, CHE-7815284, and CHE-802635. However, any opinions, findings, conclusions, or recommendations expressed herein are those of the editors and do not necessarily reflect the view of NSF.

This book represents information obtained from authentic and highly regarded sources. Reprinted material is quoted with permission, and sources are indicated. A wide variety of references are listed. Every reasonable effort has been made to give reliable data and information, but the author and the publisher cannot assume responsibility for the validity of all materials or for the consequences of their use.

All rights reserved. This book, or any parts thereof, may not be reproduced in any form without written consent from the publisher.

Direct all inquiries to CRC Press, Inc., 2000 Corporate Blvd., N.W., Boca Raton, Florida, 33431.

© 1983 by Clarkson College of Technology

International Standard Book Number 0-8493-7220-8 (Complete Set)
International Standard Book Number 0-8493-7226-7 (Volume VI)

Library of Congress Card Number 77-24273
Printed in the United States

LIST OF ABBREVIATIONS

Handbook Series in Organic Electrochemistry
Vol. VI

INDICATOR AND WORKING ELECTRODES

Material

Cb	carbon black
Cp	carbon paste
g	graphite
gl	glassy
gp	pyrolytic graphite
gw	wax-impregnated graphite
px	epoxy resin
sr	silicone rubber
tr	optically transparent
[ ]=	plated
( )=	amalgam

Configuration

process

dp	dropping
pd	periodic displacement of solution
ro	rotating
sr	stirred
st	streaming
vi	vibrating
xf	stationary electrode, flowing solution
xx	stationary electrode and solution

shape

be	bead or sphere
bu	button
co	coil
cy	cylinder
di	disc
fl	flag or foil
gr	grid or gauze
hb	hemisphere
hd	hanging drop
me	mesh
po	pool
ri	ring
rd	ring disc
sd	sessile drop
tu	tubular
wi	wire

APPARATUS:

Cell

0	two-electrode
1	iR correction
2	three-electrode
3	separation by agar bridge
4	liquid junction, e.g., Kalousek
5	glass or ceramic diaphragm
6	salt bridge
7	plastic membrane
8	thin layer
9	stirred solution

Oxygen removal

A	inert gas
B	air
C	hydrogen
D	sulfite
E	hydrazine
F	other

Data acquisition

0	recorder
1	manual
2	oscillographic
3	digital
4	other

ALL OTHER:

A

A	area [cm^2], or (alone or as subscript) anodic
ACET	acetate
AcNHME	N-methylacetamide
ads	adsorption
AM	ammonia
ANHYD	anhydride
Ap	approximate value
a.s.	anomalous shape
Aux	auxiliary (counter) electrode
Av(n)	average of n values

B

B	poorly separated
BARB	barbital
BB	Bates and Bower
BENZ	benzoic acid
BOR	borate or borax
BPT, bp	boiling point
BR	Britton and Robinson

C

C	concentration [$mmol\ dm^{-3}$], or (alone or as superscript) cathodic
C=x-x	concentration range
c	calculated value
C_0	concentration of compound being reduced
C_R	concentration of compound being oxidized
CA	convective chronoamperometry
calc	calculated, calculation
CARB	carbonate
CC	convective chrono-coulometry
ce	chemical reaction followed by an electron-transfer step
CHA	carbon-hydrogen analysis
CHN	carbon-hydrogen-nitrogen analysis
CITR	citrate
CL	Clark and Lubs
ClACET	chloroacetic acid
corr	corrected for, or (as subscript) corrected (usually for double-layer effects)
CP	controlled-potential electrolysis
CT	convective triangular-wave voltammetry

D

D	diffusion coefficient [$10^{-6} cm^2 s^{-1}$]
d	diameter [cm]
D:D	comparison of diffusion coefficients
De	not energy-sufficient
detn	determination
DI	differential pulse polarography
dim	dimerization
DME	dropping mercury
DMF	N,N-dimethylformamide
DMSO	dimethyl sulfoxide
do	doubtful
DP	derivative pulse polarography
Dplot	Delahay plot
dpn	disproportionation
DQ	incremental charge polarography
dr	drawn out

E

E	exponential (e.g., $1E4 = 1 \times 10^4$), or potential
E_{app}	applied emf [V]
E_{disc}	potential of the disc (for a ring-disc electrode) [V]
E_i	incision potential [V]
E_j	liquid-junction potential [V]
E_{max}	potential of a maximum [V]
E_{min}	potential of a minimum [V]
E_p	peak potential [V]
$E_{p,A}$	anodic peak potential [V]
$E_{p,C}$	cathodic peak potential [V]
$E_{p/2}$	half-peak potential [V]
E_{pi}	photoionization potential [V]
E_{ref}	potential of a reference electrode [V]
E_{rf}	radio-frequency potential at which current changes sign [V]
E_{ring}	potential of the ring (for a ring-disc electrode) [V]
E_s	steric substituent constant

ALL OTHER (cont.)

E_{su}	summit potential [V]
E_t	tangent potential [V]
$E_{\pi_{max}}$	potential of peak emission (on scanning) [V]
E_λ	half-neutralization potential [V]
E_τ	switching potential [V]
$E_{\tau/4}$	quarter-transition-time potential [V]
E_o	standard potential [V]
$E_{0.22}$	0.22-transition-time potential [V]
$E_{0.85}$	0.85-peak potential [V]
$E_{\frac{1}{4}}$	quarter-wave potential [V]
$E_{\frac{1}{2}}$	half-wave potential [V]
$E_{\frac{1}{2}}^o$	half-wave potential at pH=0
$E_{\frac{1}{2}}^\uparrow$	wave merges with initial current rise
$E_{\frac{3}{4}}$	three-quarter-wave potential [V]
E_{2d}	second-harmonic null potential [V]
EC	programmed-current chronopotentiometry
EC=	electrode constant $m^{2/3} t^{1/6}$ [$mg^{2/3} s^{-1/2}$]
ec	electron-transfer step followed by a chemical reaction
ece	electron-transfer step followed by a chemical reaction, then a second electron-transfer step
ecl	electrochemiluminescence
eec	two electron-transfer steps followed by a chemical reaction
ECM	electrocapillary maximum
EE	current-step chronopotentiometry
EF	ac chronopotentiometry
EK	alternating-voltage chronopotentiometry
EL	chronopotentiometry with linear current sweep
El	electrode
Elog	(alone) $dE/d\log[i/(i_l-i)]$ [mV] (followed by any symbol except i^a or τ) see Description and Remarks column
Elogi^a	$dE/d\log[i^a/(i_l-i)]$ [mV]
Elogv	E_p vs. log v
Elogτ	$dE/d\log[t^{\frac{1}{2}}/(\tau^{\frac{1}{2}}-t^{\frac{1}{2}})]$ [mV]
Elogω	$E_{\frac{1}{2}}$ vs. log ω
EM	derivative chronopotentiometry
eqn	equation
ER	chronopotentiometry
ESR	electron-spin resonance
EV	chronopotentiometry with superimposed ac
EW	controlled-potential electrogravimetry
EY	cyclic chronopotentiometry

F

F	from a figure
$\underline{F}$	faraday
F( )	given non-linear functional dependence
f	frequency of a periodic signal (Hz)
f( )	functional dependence, not given
Fc	ferrocene-ferricinium ion
FORM	formate
FP	demodulation polarography

G

GC	gas-phase chromatography
GEL	gelatin
GLC	gas-liquid chromatography
GLYC	glycine

H

H	reduction of hydrogen ion
h	height of the mercury column [cm]
HL	high-level faradaic rectification
HMDE	hanging mercury-drop
HMP	hexamethylphosphoramide
HOMO	highest occupied molecular orbital
HQ	hydroquinone

I

I	diffusion current constant [$\mu A\ s^{1/2} mg^{2/3} m\underline{M}^{-1}$]
i	current [μA]
i_A	anodic current [μA]
i_a	adsorption-controlled current
I_{ac}	diffusion current constant for ac polarography [$mmho\ m\underline{M}^{-1}$]
i_C	cathodic current [μA]
i_c	capacity controlled current
i_{cat}	catalytically controlled current
i_d	diffusion-controlled current
i_{disc}	current at the disc of a ring-disc electrode [μA]
i_k	kinetically controlled current
i_l	limiting current [μA]
$i_{l,A}$	anodic limiting current [μA]
$i_{l,C}$	cathodic limiting current [μA]
i_{max}	current at the maximum [μA]
i_{min}	current at the minimum [μA]
i_p	peak current [μA]
$i_{p,A}$	anodic peak current [μA]
$i_{p,C}$	cathodic peak current [μA]
i_{ring}	current at the ring of a ring-disc electrode [μA]
i_{su}	summit current [μA]
IB	amperometry with two indicator electrodes
ID	differential amperometry
i.d.	inside diameter
I:I	comparison of diffusion current constants
i:i	comparison of (limiting) currents
IL	chronoamperometry with linear potential sweep
Ilk	from Ilkovič equation
IM	chronoamperometry with non-linear potential sweep
IO	amperometry
IP	Izmailov and Pivneva
IQ	stirred-pool-electrode chronocoulometry
IR	chronoamperometry
iR	product of current and resistance
IRS	infrared spectroscopy
IS	stirred-pool-electrode-chronoamperometry
i-$\underline{t}$	potentiostatic measurement
IU	double potential-step chronoamperometry
IV	polarographic chronoamperometry

J

j	current density [$\mu A\ cm^{-2}$]
j:j	comparison of current densities
j_l	limiting current density [$\mu A\ cm^{-2}$]
j_p	peak current density [$\mu A\ cm^{-2}$]
$j\tau^{\frac{1}{2}}/C$	chronopotentiometric constant [$\mu A\ s^{1/2}\ cm^{-2} m\underline{M}^{-1}$]

K

k	heterogeneous rate constant for electron transfer [$cm\ s^{-1}$], or constant of proportionality
k_{app}	apparent rate constant
$k_{A,h}$	anodic heterogeneous rate constant [$cm\ s^{-1}$]
$k_{A,h}^o$	the value of $k_{A,h}$ at 0 V vs. NHE [$cm\ s^{-1}$]
$k_{C,h}$	cathodic heterogeneous rate constant [$cm\ s^{-1}$]
$k_{C,h}^o$	the value of $k_{C,h}$ at 0 V vs. NHE
k_h	heterogeneous rate constant [$cm\ s^{-1}$]
k_h^o	heterogeneous rate constant at 0 V vs. NHE
$k_{s,h}$	the common value of $k_{A,h}$ and $k_{C,h}$ at the standard potential
$k_{s,h}^{app}$	apparent value of $k_{s,h}$
k_1, k_2	numerical constants
Kplot	Koutecký plot

L

L	light excluded
l	thickness of thin-layer electrode gap [μm]
LLC	liquid-liquid chromatography
lp	limited proof
LSC	liquid-solid chromatography
LSQ	least-squares value

ALL OTHER (cont.)

LUMO	lowest unoccupied molecular orbital	PD	derivative polarography	$r_{r,o}$	outer radius of ring [cm]
		PE	pretreated electrode	redn	reduction
		PF	differential polarography	rms	root mean square
		PG	square-wave polarography	R_4NCE	$Hg\|Hg_2Cl_2(\underline{s}), R_4NCl(\underline{c})$ electrode (followed by identity of R and value of $\underline{c}$)

M

M	merges with final current rise	PH	higher harmonic ac polarography	RP	radio-frequency polarography
m	rate of flow of Hg [mg s^{-1}]	PHEN	phenol	rxn	reaction
MAS,MS	mass spectrometry	PHOS	phosphate		
MB	McIlvaine	PHTH	phthalate		
Mc( )	mistake corrected (column)	PI	alternating-voltage polarography		

S

MeCN	acetonitrile	PK	Kalousek polarography		
MG	1,2-dimethoxyethane (monoglyme)	pK_a	-log(acidic dissociation constant)	S	merging with succeeding wave
MM	modulation polarography	pK'	pH at inflection point of polarographic dissociation curve	satd	saturated
Mn	minimum			SCE	saturated calomel electrode (with KCl unless otherwise specified)
MPT,mp	melting point	PL	intermodulation polarography	Se	energy-sufficient
MSE	mercury-mercurous sulfate electrode	Pl	plateau	Sr	surface reaction
MWT,mw	molecular weight	PM	multisweep polarography	St	slope of tangent $E_{\frac{1}{2}}$
Mx	maximum	PO	oscillopolarography	sttd	stated
		Po	postwave	SULF	sulfate
		pos	positive	SULN	sulfolane

N

		PP	pulse polarography	supp. elect.	supporting electrolyte
n	number of electrons	ppt	precipitate, precipitation	sur	surface reaction
n_a	number of electrons transferred through rate-determining step of a cathodic process	PQ	polarographic coulometry		
		PR	cyclic triangular-wave polarography		

T

n_{app}	an apparent number of electrons transferred	Pr	prewave	T	temperature [°C]
		PrCN	butyronitrile	t	drop time [s]
n_b	number of electrons transferred through rate-determining step of an anodic process	PROC	propylene carbonate	$\underline{t}$	time [s]
		PRW	Prideaux and Ward	t_s	switching (reversal) time
		PS	staircase polarography		
		PT	Tast polarography	t(c)	controlled drop time [s]
N_o	collection efficiency (ring-disc electrode)	PV	total ac polarography	t(oc)	drop time on open circuit [s]
		PVC	poly(vinyl chloride)		
NBS	National Bureau of Standards pH scale	PVI	in-phase ac polarography	t(sc)	drop time on short circuit [s]
		PVQ	quadrature polarography		
NCE	normal calomel electrode	PW	single-sweep polarography	Tafel	Tafel slope (dE/dlogi)
neg	negative			Tc	temperature coefficient of the limiting current [% deg^{-1}]
NHE	normal hydrogen electrode	PX	double-tone polarography		
NMR	nuclear magnetic resonance	PY	polarography		
N-S	αn_a or $k_{s,h}$ calculated from peak separation (Nicholson-Shain equation)	PYR	pyridine	THF	tetrahydrofuran
				TI	thin-layer chronoamperometry

Q

				TLC	thin-layer chromatography
ns	not stated			TLE	thin-layer electrode
		Q	in characteristic potential column, incision quotient; elsewhere, quantity of charge [μC]	TLQE	thin-layer coulometry
				Tomeš	$E_{\frac{3}{4}} - E_{\frac{1}{4}}$ (mV)

O

				TQ	thin-layer chronocoulometry
0	no wave observed	QE	controlled-potential coulometry	TRIS	tris(hydroxymethyl)aminomethane
o	organic	QI	quasi-reversible		
(oc)	open circuit	QP	controlled-current coulometry without a reagent precursor	TX100	Triton X-100
oh =	number of hydroxide ions consumed through rate-determining step				
		Qp	integrated charge corresponding to the faradaic current [μC]		

U

oxidn	oxidation	QR	chronocoulometry	(u)	arbitrary unit
		QT	controlled-current coulometry with a reagent precursor	UB	universal buffer
				UVS,UV	ultraviolet spectroscopy

P

		QU	double potential-step chronocoulometry		
P	merges with previous wave	QW	electrogravimetry		
p	preliminary				

V

p=	number of protons consumed through rate-determining step			V	volts
				v	scan rate [mV s^{-1}]
PA	triangular-wave polarography			V_f	rate of flow through a tubular electrode [cm^3 s^{-1}]

R

PB	higher harmonic ac polarography with phase-sensitive rectification	R	reversible	VA	triangular-wave voltammetry
		r	(alone) reliable, or (with a subscript) radius	VC	current-scanning voltammetry
PC	current-scanning polarography	r_d	radius of disc [cm]	VD	derivative voltammetry
		$r_{r,i}$	inner radius of ring [cm]	VF	differential voltammetry
				VIS	visible spectroscopy

ALL OTHER (cont.)

VM	multisweep voltammetry
VO	oscillovoltammetry
VR	cyclic triangular-wave voltammetry
Vr	volume reaction
VV	ac voltammetry
VY	hydrodynamic voltammetry

W

W	well defined
WGE	wax-impregnated graphite

X

X	ill defined
XC	infrequently used unit for concentration
XE	infrequently used characteristic potential
XEl	infrequently used indicator, working, or reference electrode
XEs	infrequently used pilot ion or reference substance
Xeqn	infrequently cited equation
XP	infrequently used experimental parameter
XR	infrequently used response constant
XS	infrequently used symmetry parameter
XT	infrequently used technique

Each of the preceding nine symbols is defined in the Description and Remarks column opposite the code number for which it is used.

Y

Y	addmittance (mmho)

Symbols:

→	becomes or range of variation or up to
↠	extrapolated to
⇌̸	irreversible
≠	not equal to
↑	(in Characteristic Potential column) wave merges with initial current rise (in Description and Remarks column) increase(s)
↓	decrease(s)
∝̸	not proportional
∅	quantum yield efficiency
□	square wave
∧	triangular wave
/	ramp
(?)	questionable

α	cathodic charge-transfer coefficient
β	Hückel integral energy unit
β_a	anodic charge-transfer coefficient
Γ	surface excess [10^{-10} mol cm^{-2}]
ΔE	pulse amplitude [mV]
Δe	amplitude of sinusoidal potential [mV]
ΔE_p	$E_{p,A} - E_{p,C}$
$\Delta E_{p/2}$	half-peak width
$\Delta E_{su/2}$	half-summit width
Δi	amplitude (peak-to-peak) of alternating current [μA]
$\Delta(1/\lambda)$	change in wavenumber [nm^{-1}]
η	relative viscosity
λ	wavelength [nm]
λ_{max}	wavelength of maximum absorbance [nm]
μ	ionic strength [mol dm^{-3}]
ν	frequency (= c/λ)
π	intensity of luminescence
π^*	excited π-orbital
π_{max}	maximum intensity of luminescence
ρ	Hammett reaction constant
ρ^*	Taft reaction constant
Σ	(except in equations) secondary source
σ, σ_x	Hammett substituent constant
σ^n	substituent constant for aromatic rings
σ^*, σ^*_x	Taft substituent constant
τ	transition time [s]
τ_A	anodic transition time [s]
τ_C	cathodic transition time [s]
τ_d	diffusion-controlled transition time [s]
τ_f	forward transition time [s]
τ_r	reverse transition time [s]
Φ_2	potential difference across the diffuse double layer
φ	phase angle
φ^0	zero-charge potential (-ECM)
χ	for $A \underset{k_{-1}}{\overset{k_1}{\rightleftarrows}} B$, $\chi = 2k_1(k_1 + k_{-1})^{\frac{1}{2}} t^{\frac{1}{2}}/k_{-1}$
ψ^0	ψ potential in double layer
ω	rotation rate (s^{-1})

PREFACE

More than 60 years have gone by since Professor Jaroslav Heyrovský conducted the preliminary investigations that led to the development of polarography. In that time, nurtured by the efforts of a host of brilliant and dedicated scientists, it has grown from promising infancy to vigorous maturity. Mathematicians, electrochemists, analytical chemists, hydrodynamicists, biochemists, clinical chemists and analysts, inorganic chemists, physical organic and synthetic organic chemists, and others have contributed to its development and have conferred a wide spectrum of abilities on it. Analysts use it for detection, identification, and determination; organic chemists use it for studying the rates and equilibria of organic reactions, elucidating their mechanisms, making structural assignments, and designing or improving synthetic procedures; inorganic chemists use it for deducing the compositions of complexes and studying the thermodynamics and kinetics of their formations and dissociations; clinicians use it for diagnosis and for following the treatment of diseases. Many other kinds of uses have already been found for it, and many more can be foreseen for the future. Without doubt it is one of the most powerful and most widely applicable techniques of scientific measurement and investigation.

As polarography itself has grown, other techniques related to it have been revived or invented and have shared in its growth. Some of these, like ac polarography and oscillopolarography, involve changes of the natures of the exciting signal and the measured response. Some, like voltammetry, involve different kinds of mass-transfer processes and different indicator electrodes. Others, like controlled-potential electrolysis and coulometry, involve the same fundamental relationships among current, potential, composition, and time, but employ them in different ways and for different purposes.

The development and growth of this family of interrelated techniques has been recorded in a large number of publications. About 125 research papers dealing with polarography had been published by 1930, about 1000 by 1940, about 3500 by 1950, and about 10,000 by 1960. Professor Heyrovský's annual *Bibliography of Publications Dealing with the Polarographic Method* included a total of 17,306 citations through 1964, while the *Bibliografica Polarografica* published by the Consiglio Nazionale delle Richerche lists 21,798 through 1967. No more recent reliable figure is available, but an estimate of 60,000 through 1982 can hardly be very far from the mark.

Ready access to the information previously gathered is of course essential to further progress throughout the development of any field of science. The analytical chemist and the clinical chemist need data on the behaviors of the substances they wish to determine and of other substances that might accompany these so that they can design and use analytical methods that yield accurate and reliable results. Data on the behaviors of individual substances at different pH-values, ionic strengths, and temperatures and in different buffers and solvents are needed in correlating these with their chemical and biochemical properties; data obtained with different techniques and indicator electrodes are needed in clarifying electrochemical phenomena; data on the behaviors of different but related substances provide not only structure-reactivity correlations but also correlations with many other kinds of data from different fields, ranging from ultraviolet infrared, and nuclear magnetic resonance spectroscopy to the abilities of polycyclic hydrocarbons to act as semiconductors; data on the behaviors of starting materials often enable the synthetic electrochemist to select conditions that will give the best possible yield of the purest possible product in the least possible time. Data compilations can also indicate which compounds or groups of compounds need further study, aid in the selection of systems that will repay study by newly devised techniques, and guide efforts to achieve uniformity in reporting the results of research.

A number of authors, including two of ourselves, strove to achieve these goals in compilations of electroanalytical data published between about 1950 and 1965. These differed widely in scope, length, and arrangement, and in the amounts of critical judgment they embodied. Some were limited to inorganic substances and others to organic ones. Some were intended only as representative samples of the available information; others were attempts to present it much more fully. Some were limited to half-wave potentials, to polarography alone, to data obtained in aqueous media, or in other ways; others, especially the more recent ones, attempted greater diversity. We shall not evaluate them in detail, for some of them were our own, all served a useful and important purpose, and we have striven to learn from the merits and defects of each. It is more appropriate to lament their increasing age, which renders them all less satisfactory and less useful than they were when they were published. The simple fact is that the volume of data available is already so large and is growing so rapidly that their collection, evaluation, and selection is no longer a feasible task for any single electrochemist.

That the need for such work was becoming more acute as the possibility of doing it decreased has been a matter of concern to Commission V.5 on Electroanalytical Chemistry of the International Union of Pure and Applied Chemistry for almost 20 years. A subcommission charge with responsibility for electroanalytical data and nomenclature, established as early as 1963 and having first one and then another of us as chairman, attempted for years to encourage, advise, and cooperate with everyone engaged in data compilation in this area. During that time there were several groups, in as many different countries, that seriously considered engaging in such work, but the magnitudes of the task and of the expenditures of money and time required by work on a worthwhile scale prevented some of these from actually beginning work and rendered significant achievement impossible for those who made the attempt.

When two of us came together on the staff of Clarkson College of Technology in 1970, we had published 12 different compilations of polarographic data: one of us had been involved in this work since 1950 and the other since 1955. In our common concern for continuing it we were fortunate enough to be able to secure financial support from the National Library of Medicine of the National Institutes of Health. That support enabled us to publish, in 1974, a volume containing data on 2015 organic compounds having 1 to 11 carbon atoms per molecule.

In mid-1974, support for the work was taken over by the National Science Foundation, and a new publisher had to be selected for reasons beyond our control. The first two volumes of the present series were published in 1977. They included the data contained in the 1974 volume, with various changes and corrections, and also gave data on 1848 new compounds having 12 or more carbon atoms per molecule. The third volume was published in 1978 and contained data on 1309 compounds, including 228 that had appeared in the first two volumes and for which additional data (obtained by different techniques, under different conditions, or in different solvents and supporting electrolytes) were given in Volume III. The fourth volume, published in 1980, followed the same pattern as Volume III. It contained data on 731 compounds, of which 190 had also appeared in Volumes I thorugh III. Volume V contained 528 compounds, of which 196 had also appeared in Volumes I through IV. The first five volumes thus dealt with a total of 5817 compounds.

This volume extends the series still further. It follows the same pattern as Volumes III, IV, and V, and includes 580 compounds. There are newer or additional data for 129 compounds that also appeared in previous volumes.

Although these six volumes contain the largest collection of organic electrochemical data ever made, it is essential to describe their limitations because they include only a fraction of the data that have appeared in the original literature. We have excluded electrochemical techniques, such as conductometry, high-frequency conductometry, and dielectrometry*, in which neither the electrical double layer nor any electrode reaction need be considered. These furnish information so widely different from, and so rarely used in conjunction with, that provided by polarography and its congeners that combining them would be difficult and of little use. We have also excluded a number of other techniques, such as electrography and potentiometric, amperometric, and other titration techniques that, although of great analytical utility, do not in general provide fundamental data of lasting importance. Finally, we have excluded potentiometry itself to avoid duplicating the efforts of others who have prepared or are preparing compilations of standard and formal potentials. These exclusions leave a group of approximately 65 closely interrelated techniques, which include polarography, voltammetry, amperometry, controlled-potential coulometry, chronopotentiometry, chronoamperometry, and chronocoulometry, and a host of variants on these such as ac and differential pulse polarography, triangular-wave voltammetry, and potential-step chronoamperometry and chronocoulometry. Table VIII lists the techniques that were used to obtain the data that appear in this volume.

This series of volumes is devoted to data on the behaviors of organic substances and organometallic compounds in which the metal is bound to a carbon atom. The importance of porphyrin and its derivatives to scientists in health-related fields led us to include hemes and related compounds in these volumes, and we also included some substitution-inert complexes in which the ligand is electroactive. A companion series, entitled *CRC Handbook Series of Inorganic Electrochemistry,* deals with inorganic substances, including the complexes of metal ions with organic ligands that are not electroactive, and data on complex compounds included in this volume are cross-referenced there.

Volumes I and II of the *CRC Handbook Series of Organic Electrochemistry* were still further restricted to information published in the 12-year period from 1960 through 1971. Volumes III, IV, and V contained additional material from the same period, and also covered the later period beginning in 1972 and extending into 1976. Volume VI follows the same pattern, covering additional information published between 1960 and 1978. Subsequent volumes will extend this coverage in both directions.

Within these restrictions the coverage of the literature in this volume and its predecessors is comprehensive but by no means complete. According to the estimates we gave above, there must have been roughly 30,000 to 35,000 publications in the field between the beginning of 1960 and the end of 1978, and we have not been able to inspect all of these. We have never relied on secondary sources such as *Chemical Abstracts,* and our coverage of the original publications has naturally been influenced by the availability

---

* The names used in this compilation are taken from Meites, L., Nürnberg, H. W., and Zuman, P., The classification and nomenclature of electroanalytical techniques, *Pure and Applied Chemistry,* **45**, 81, 1976.

of journals and reprints. Although we have occasionally allowed our desire to record the fact that a compound has been found to be amenable to investigation by polarographic and related techniques, and our conviction that further study of its behavior would be worthwhile, to persuade us to include preliminary or dubious results (all of which we have scrupulously attempted to identify as such), we have rejected many more such results than we have included. Our attempt has been to select the best of the available data, the most useful, and the ones most likely to withstand the test of time.

To this end we and our collaborators have repeatedly scrutinized every bit of information recorded here. We have separately evaluated the conformity of the apparatus and experimental techniques to the best modern standards, the accuracy of the handling of the data obtained, the certainty with which products and intermediates were identified, the strength of the evidence given for each suggested reaction course or mechanism, and every other point that the published description of each research allowed us to check. In an effort to minimize inconsistency we have compared the data obtained under different conditions, by different techniques, and for different but related compounds. We have assayed alternative courses and mechanisms proposed by different authors and for related compounds. Despite all these precautions it would of course be idle to pretend that the result is wholly free from errors or contradictions. Some of these are in the original literature and have survived our best attempts to detect and expunge them; others have been introduced in transcription and typing. We can only hope that the merits and utility of this collection are sufficient to induce its users to forgive its defects, and urge those who find errors in the pages that follow to call them to our attention so that others may be apprised of them.

There is, however, one sort of defect for which we must disclaim responsibility: this is the result of the woeful lack of standardization in reporting data that runs rampant through the original literature and that we have been powerless to correct. In some circumstances this is trivial enough to be merely unaesthetic; in others it is an ineradicable blot on the meaningfulness or utility of the work that the data represent. In polarography some authors deduce ratios of n-values for processes that give rise to successive waves on a polarogram by comparing the heights of the waves, while others apply a correction for the dependence of $m^{2/3}t^{1/6}$ on potential by comparing diffusion current constants instead. In Table I, which constitutes the major portion of this volume, these two practices are represented by the entries "i:i" and "I:I", respectively, in the 21st column. Though the second of them is unarguably superior for diffusion-controlled waves, it would be so unlikely that the first would lead to incorrect values of n that the universal adoption of the better expedient would merely improve the beauty and legibility of a tabulation like this one. At the other extreme, there are still authors who report potentials that are referred to mercury-pool and other unpoised "reference" electrodes even though every reputable textbook or monograph published in the field for at least 3 decades has inveighed against this archaic practice. There are actually even some who fail to describe the compositions of the solvents, supporting electrolytes, or electrodes they have used. Such defects becloud the significance of the data reported and render them far less useful than they might have been for identification, devising analytical methods, and many other purposes. In the present state of the art it has not been possible to exclude all of the data that are thus or otherwise blemished, but we strongly hope that these volumes will help to promote standardization and uniformity so that such blemishes will be less numerous in the raw material gathered for future ones.

We said above that this work owes its existence to the financial support we have received. That of the National Science Foundation is acknowledged in official terms on the title page. But in addition we are glad to have this opportunity to record our personal gratitude to Drs. A. F. Findeis, and Janet Osteryoung, William E. Ohnesorge, and Henry N. Blount at the Foundation: their sympathy with our aspirations and patience with our problems have given us moral support that has meant as much to us as the Foundation's financial support.

This work would also have been impossible without the advice, encouragement, and cooperation that we have received from many of our electrochemical colleagues. First among these are the ones whose names appear on the title page. Dr. Elinore B. Rupp joined the project early in 1975 and played a leading role in every stage of the preparation of this volume. Dr. Theodore L. Fenner joined us in 1974, and his care, diligence, and accuracy as an editor have contributed much to its quality. Dr. Ananthakrishnan Narayanan worked with us from mid-1976 through 1980; his responsibility has been chiefly to the parallel *CRC Handbook Series in Inorganic Electrochemistry,* of which Volume I was published in 1980 and Volume II in 1981; Volumes III and IV are currently in the production stage. He has also contributed substantially to the present volume. We also owe a continuing debt of gratitude to those who have worked on the project in prior years. These include Dr. William J. Scott, who shared in its birth pangs, and Dr. Bruce H. Campbell and Dr. Alex M. Kardos, who helped to prepare Volumes I and II and to develop most of the procedures employed in preparing this volume. It is a pleasure to be able to express our thanks to all of these for the talents, devotion, and enthusiasm that they brought to tasks that are both tedious and difficult. The bulk of this volume is a reproduction of a typescript prepared by Mrs. Helen Tyler. Her contributions to its appearance and legibility and to its accuracy and reliability are visible whereas her care, dedication, and patient good cheer are not, but all have been equally indispensable throughout our work. We are grateful to Professor Ernest I. Becker for his continued generous advice and help with problems of organic nomenclature. Dr. John J. Rupp assisted us in checking the ACS Registry Numbers, which appear in the present volume as a further aid to identifying the compounds listed. We have also had the benefit of consultation and help from our fellow members of Commission V.5 on Electroanalytical Chemistry of IUPAC, whose encouragement, support, and experience in data compilation in many diverse fields have been invaluable to us. Finally, it is a pleasure to acknowledge the advice and assistance that we have received at various stages from many eminent leaders of the electrochemical community.

Louis Meites
Petr Zuman
Elinore B. Rupp
Potsdam, New York
October, 1982

TABLE OF CONTENTS

General Introduction.. 1

Table I.   Electrochemical Data.. 5

Table II.   Structural Formulas..335

Table III.   Courses and Mechanisms of Half-Reactions..357

Table IV.   Compounds Included in Table I...431

Table V.   Functional-Group Index..461

Table VI.   Chemical Abstracts Service Registry Numbers..501

Table VII.   Index of Solvents Employed..507

Table VIII. Index of Techniques Employed...513

Table IX.   Index of Indicator Electrodes Employed..519

Table X.   Key to Literature Citations...523

Table XI.   Author Index...531

Corrigenda 537

GENERAL INTRODUCTION

We have attempted in these volumes to give as much information as possible in as little space as possible. The information is arranged in ways that are often arbitrary, and it is presented with the aid of symbols and abbreviations whose meanings are often not obvious. This General Introduction, and the introductions to the individual tables that follow, are provided in the hope that they will enable the reader to make full use of the tables in the shortest possible time.

This volume of the *CRC Handbook Series in Organic Electrochemistry* comprises 11 tables. Of these, the longest and most important is Table I, which gives information about the electrochemical behaviors of 580 organic and organometallic compounds and about the experimental techniques and conditions employed in studying them. The other ten tables serve to supplement, cross-index, and interpret the contents of Table I.

Despite the physical bulk of this volume and its predecessors in this series and the stringency with which we applied the criteria, described in the Preface, for restricting their scopes, ruthless abbreviation and condensation have been necessary in compressing the information that is given here into the space available for it. The list of abbreviations and symbols is therefore essential to the correct decoding and interpretation of entries that would be unintelligible without it. One copy of this list appears inside the front and back covers of this volume, and there is a separate copy of it on unbound sheets that can be removed and used to follow the lines across the wide pages of Table I. In preparing this volume we added a few abbreviations to the lists that appeared in Volumes I through V. Hence the current list can be used in conjunction with those volumes, whereas the earlier lists will occasionally fail to permit the complete interpretation of an entry appearing in this volume.

Each table is preceded by a description of its contents, its organization, and the manner in which it can be used. The present introductory discussion gives a more general description of the volume as a whole, stressing the purpose of each table and the ways in which different tables are related and supplement each other.

Table I contains the purely electrochemical data that have been obtained by polarography and the other techniques listed in Table VIII. There are so many columns that they extend across two facing pages. The division between pages is very nearly such that the left-hand pages identify the compounds for which data are given, the techniques by which the data were obtained, and the electrodes used, and also describe the solvent and supporting electrolyte, apparatus, and experimental conditions; the right-hand pages give the data and other information obtained and provide cross-references to additional information contained in other tables.

The compounds listed in Table I are arranged according to their empirical formulas, and this of course provides one mode of access to the information on any particular compound. Empirical formulas are written in the sequence C, H, and all other atoms in the alphabetical order of their symbols. These formulas are then arranged alphabetically and according to increasing number of atoms; thus, for example, $C_5H_4O_6$ precedes C_5H_8O, and this in turn precedes $C_6H_4O_2$. To permit the rapid location of any empirical formula, the empirical formula of the first compound that appears on each pair of facing pages is given in the upper left-hand corner of the left-hand page, and that of the last compound appearing on that pair of pages is given in the upper right-hand corner of the right-hand page. By virtue of these facts, formamidine disulfide ($[NH_2C(:NH)S-]_2$ or $C_2H_6N_4S_2$) can be located quickly on the pair of pages that contain information for compounds having formulas between CH_3Cl_3Sn and C_2H_6O, and then further identified as the fourth compound on this pair of pages by means of either the empirical formulas listed in the second column or the names listed in the third. Different compounds having the same empirical formulas are listed in the alphabetical order of the names assigned to them.

Another mode of access to the same information is provided by Table IV. This is an index of names and a few common synonyms. It contains the entry "Formamidine disulfide, FA03". The code number FA03 appears in both the first and last columns of Table I. Each code number consists of two letters followed by two digits; in Table I the code numbers appear in the order FA00, FA01, ..., FA99, FB00, FB01, ..., FF28. The code numbers serve not only to abbreviate cross-references between tables and in the tightly packed columns of Table I but also to distinguish between different compounds that have identical empirical formulas. The first letter of the code number identifies the volume of this series: the letters A and B refer to compounds given in Table I of Volumes I and II, C to those in Volume III, D to those in Volume IV, E to those in Volume V, and F to those in the present volume. References to the *CRC Handbook Series in Inorganic Electrochemistry* can be recognized by virtue of the fact that they consist of two digits followed by two letters.

Table V is another index that makes it possible to locate compounds of interest in Table I: it classifies the compounds according to the electroactive (or possibly electroactive) functional groups they contain. One frequently wants to examine data on heterocyclic compounds, on carboxylic acids, or on another more or less specific class of substances, and the purpose of Table V is to simplify searches of this sort. For example, folic acid appears among the carboxylic acids, amides, amines, heterocyclic compounds, and hydroxy compounds in Table V, and inspection of the compounds listed under this heading would quickly reveal any other closely related compound that appears in this volume. To minimize false leads in such searches, this index is selective: trichloro(methyl)stannane does not appear among the compounds containing halogen atoms, because its behavior, so far as it is known, is not representative of those compounds. On the other hand, 2-propyn-1-ol would appear both among the hydroxy compounds and among the alkyne compounds because both the hydroxy group and the triple bond are known (or can be assumed) to be involved in its electrochemical reactions.

Table VI lists the compounds under American Chemical Society Registry Numbers, arranged sequentially. These Numbers, which also appear in Column 3 of Table I, should enable the user to find a compound in Table I without having to be concerned with nomenclature.

Table VII is a list of the solvents that appear in the individual entries of Table I. It permits comparing data on different substances in a single solvent or solvent mixture. The typical entry

Benzonitrile, FB25,FC54,FF14,FF96

provides rapid and easy access to all of the information that this volume includes about electrochemistry in this solvent. Water is omitted from this index because an entry for purely aqueous solutions would have contained 73 citations and would therefore have been far too long to be of any use.

Table VIII is an index to the techniques employed. For the sake of brevity classical dc polarography is excluded from it, as the dropping mercury electrode is from Table IX, but for each of the other 65-odd techniques that are included in these volumes an entry like

Chronoamperometry(IR), FE22,FE23,FE24,FF74

provides not only access to all of the chronoamperometric information that these volumes contain, but also an indication of how much chronoamperometry has contributed to our knowledge of the electrochemical properties of organic substances during the years that these volumes cover. Since Table VIII lists some techniques for which no citations are given, it may be useful to state the policy we adopted in dealing with data obtained by rarely used techniques. There were a few of these that were rejected because they, or the interpretations based on them, disagreed radically with other information, but such cases were very few and we much more often found ourselves deciding to include an entry because it was obtained by a technique little represented in the data file. Such a selection procedure tends to yield overestimates of the utility and importance of rarely used techniques. The overwhelming preponderance of classical dc polarography is reflected by the omission of an entry for it from Table VIII because that would have consumed more space than we thought reasonable. If one had been included, its length would, if anything, have understated the predominating importance that dc polarography still has in obtaining fundamental electrochemical information.

Table IX is an index to the indicator electrodes employed. It provides access to all of the data obtained with mercury-pool, carbon-paste, rotating disc, and each of the numerous other electrode materials and configurations represented in Table I, with the single exception of the dropping mercury electrode, for which there would have been no less that 353 citations, and which is therefore omitted here as water is omitted from Table VII and polarography is omitted from Table VIII.

Tables IV to IX are thus indexes to Table I. Tables II, III, X, and XI, on the other hand, are supplements to Table I, providing additional information and used with Table I as a starting point. Inspection of these Tables is often necessary for the proper use of Table I. Table II gives, in the order of their code numbers, the structural formulas that could not conveniently be compressed into line form for inclusion in Column 4 of Table I. Table III gives equations that represent the courses and mechanisms of half-reactions, together with values of the rate and equilibrium constants for many of the homogeneous chemical steps involved in these. The equations are given in numerical order and keyed to entries in the fifteenth column (headed "C/M") of Table I; the appearance of a number in that column of Table I in the entry for any compound signifies that information about the course or mechanism of the half-reaction is given opposite the same number in Table III. Table X provides full literature references and is used together with the condensed citations that appear in the fourteenth column (headed "Ref.") of Table I. For each such citation, Table X gives the name of the journal, the volume number, the year of publication, the page number, and the names of the authors. Table XI is an author index which gives, in alphabetical order, the names of all of the authors listed in Table X and, for each, a reference to each citation of his work in Table X. Finally, there is a list of corrigenda giving all of the previously uncorrected errors in prior volumes of which we are aware.

This volume is dedicated to all those whose names appear in Table XI, in recognition of the labor that they have expended on assembling and bringing to its present state the enormous body of knowledge and understanding that is summarized here. Much remains to be discovered, and much of our understanding is still very fragmentary. We hand the volume over to its users in the hope that it will both ease and help to guide their work in the future, as well as the work of those who will follow them.

TABLE I.
ELECTROCHEMICAL DATA

This table consists of 28 columns divided between two facing pages. Here we shall first describe the contents, significance, and arrangement of these columns.

Columns 1 and 28: "Code No."

The contents of these columns are identical so that the user can more easily follow the lines of this very wide table across two pages. The columns give the four-character code number (see explanation given in the General Introduction) assigned to each compound and also indicate whether or not more information about a compound will be found on the following pair of pages.

In many instances the data for a single compound had to be divided between two, and occasionally even among three, pairs of facing pages. To avoid overlooking part of an entry thus divided, the user should observe the following points:

1a. When the last line is blank in Columns 1 and 29, the entry does *not* continue onto the next pair of pages, whereas
1b. An entry that *is* continued overleaf is identified by the symbol "CONT" on the last line in both Column 1 and 28.
2a. When the index in the upper left-hand corner of a left-hand page consists solely of a code number and an empirical formula (for example, "FA08, C_3H_4O") the first entry on that page is *not* continued from the preceding pair of pages, whereas
2b. An entry that *is* being continued from the preceding pair of pages is identified by the appearance of "(CONT.)" between the code number and the empirical formula in the upper left-hand corner of the left-hand page (for example, "FA04 (CONT.) C_2H_6O").

For a number of these compounds additional data will be found in an earlier volume of this series. Each compound that also appears in an earlier volume is identified by the appearance, in both Column 1 and Column 28, of the code number assigned to it in each previous volume in which it appeared, but with small letters in place of the capitals used elsewhere. For acrylonitrile the entry in these columns

FA07
ab34
ca23
da11

shows that additional data appear in Volume I under the code number AB34, and in Volume III under the code number CA23, and in Volume IV under the code number DA11. As was stated in the General Introduction, the first letter of a code number identifies the volume to which reference is made. The second letter and the two digits of the code number have only an ordinal significance.

Column 2: "Empirical Formula"

The empirical formulas in this column are given in the customary way, with carbon first, hydrogen second, and then the other elements in the alphabetical order of their chemical symbols, as described in the General Introduction. Deuterated compounds appear immediately after the corresponding protonated ones.

Column 3: "Name and C.A. Number"

This column gives the name and, below it, the Chemical Abstracts Service Registry Number, introduced by the abbreviation C.A. for *Chemical Abstracts,* where available. Some of the names are the simplest and shortest trivial names for the compounds they denote, but most follow IUPAC or *Chemical Abstracts* nomenclature. The Registry Number is given only once for each compound; when an entry continues onto the following pair of pages the name of the compound is repeated but the Registry Number is not. Some compounds have not been assigned Registry Numbers, either because their structures are unknown or because Registry Numbers have not yet been published by *Chemical Abstracts.*

Wiswesser Line Notation (WLN) designations of the compounds were given in Volumes I through IV, but the Registry Numbers appear to be both more easily used and more widely accepted.

Column 4: "Structural Formula"

Only line structural formulas are given here. For many of these compounds, however, the structural formula cannot conveniently be given in line form, and is therefore shown in Table II. The entry "Table II" in Column 4 means that the structural formula may be found in Table II of this volume opposite the appropriate code number taken from Column 1. To avoid duplicating structural formulas in different volumes, the entry "Table II-2", which appears in Column 4, provides a cross-reference to Table II in Volume II, while the entry "Table II-5" is a cross-reference to Table II in Volume V. For the compound for which the entries in Columns 1 and 3 are

FA33 1,2,3,5-tetrafluorobenzene
cb39

the reference "Table II-3" means that the structural formula may be found in Table II of Volume III with the code number (CB39) given for 1,2,3,5-tetrafluorobenzene in Column 1.

Column 5: "Solvent"

The symbols used in this column to denote non-aqueous solvents are given in the list of abbreviations. The entry "H_2O" denoted an aqueous solution; an entry like "MeCN" denotes a solution in the nominally pure (and anhydrous) solvent given; in an entry like "MeCN 50" the number denotes the percentage (by volume, unless otherwise stated) of the specified non-aqueous solvent in a solvent mixture, of which the balance is understood to be water. An occasional entry like "EtOH(aq)" reflects our inability to deduce the composition of the solvent mixture from the published information given; such entries appear only when the importance of the data seemed to us to override the laxity of the reporting. Of the entries

MeOH 50 and EtOH 50
C_6H_6 50 Me_2CO 25

the left-hand one denotes a nominally anhydrous mixture containing equal parts by volume of methanol and benzene. The right-hand one denotes a mixture containing 50% (V/V) of ethanol, 25% (V/V) of acetone, and 25% (V/V) of water; when the sum of the percentages given for the constituents of any solvent mixture is less than 100, the balance is always understood to be water.

Aqueous solutions are listed first, then alcohols, and finally other solvents in the order in which they appear on the list of abbreviations. Entries for mixtures of a non-aqueous solvent with water immediately precede those for the corresponding anhydrous solvent. A few data that seemed worth including were taken from papers that did not fully describe the composition of the solvent; for these data the symbol "ns" (for "not stated") is given in this column.

Column 6: "Technique"

The technique is identified by a two-letter symbol that can usually be interpreted by means of the list of abbreviations. Polarography ("PY") appears first in the data for any solvent, and the other techniques follow in the alphabetical order of the symbols assigned to them. Some of the information in this volume has been obtained by techniques as yet so rarely used that we have hesitated to assign symbols to them until further evidence of their utility appears. Each such technique is identified by the letters "XT" in Column 6, and Column 27 always gives its name on the same line in the form "XT = second-derivative chronoamperometry with linear potential sweep". In assigning such names we have followed the principles embodied in the 1975 IUPAC recommendations regarding the nomenclature of electroanalytical techniques (*Pure and Applied Chemistry*, **45**, 81, 1976).

Column 7: "Medium"

This column describes the composition of the supporting electrolyte. Concentrations are given in moles per cubic decimeter wherever possible: the entry "KOH 0.1" denoted 0.1 M potassium hydroxide. Britton-Robinson, McIlvaine, and other commonly used buffers, both mixed and simple, are identified by means of abbreviations. The word "buffer" means that the solution was said to be buffered at the pH quoted in Column 9, but that no information was given about the composition of the buffer employed. Maximum suppressors are identified in this column and their concentrations are given in weight/volume % (grams per 100 cm^3).

Column 8: "μ, M"

This gives the ionic strength of the solution, in moles per cubic decimeter, except where this is fully specified by the entry in Column 7. A dash in this column means that no information about the ionic strength is available in addition to that given in Column 7.

Column 9: "pH"

Values of the pH are generally given to the nearest 0.1 unit. An entry like "5.7→6.2" would signify that the pH-value changed over the range given during the course of a measurement; one like "5—7" means that all of the data in the columns that follow are independent of pH over the stated range. The entries for any given medium are usually arranged in order of increasing pH. A dash in this column means that no information about the acidity of the solution is available in addition to that in Column 7.

Column 10: "T, °C"

This column gives the temperature at which the data were obtained and, where possible, the precision with which it was maintained. A dash in this column means that no information about the temperature of the solution was given in the original publication.

Column 11: "Electrodes"

The indicator or working electrode is given first; then, after a double solidus or virgule ("//"), the reference electrode. In three-electrode configurations the nature of the auxiliary or counter electrode is given in Column 13.

New materials and configurations of indicator electrodes continue to appear and the resulting proliferation of the arbitrary symbols (e.g., "RPDE" for a rotating platinum-disc electrode) used in prior volumes has led us to design a new scheme for identifying them. The pertinent symbols appear under "INDICATOR AND WORKING ELECTRODES" at the beginning of the list of abbreviations. The material is first identified by either a chemical symbol or an arbitrary abbreviation taken from this list: "Pt" identifies a platinum electrode, "g" a graphite electrode, and "Cp" one made from carbon paste. Plated or coated electrodes are identified by giving first the symbol or abbreviation of the base material, then that of the coating in *square brackets*: "Pt[Hg]" identifies a mercury-plated platinum electrode. Amalgam electrodes, and others in which a liquid metal serves as the solvent for another metal, are identified, in the usual way, by giving first the symbol of the dissolved metal, then that of the solvent in *parentheses*: "K(Hg)" denotes a potassium amalgam. Alloy electrodes are identified by dashes between the symbols of their constituents, of which at least one is followed, wherever possible, by the weight percentage of that constituent, so that "Pt-Rh10" represents a platinum-rhodium alloy containing 10% of rhodium. Then, after a colon, the configuration is described by one or two two-letter abbreviations, of which the first gives the process (e.g., "ro" for rotating and "vi" for vibrating) while the second gives the shape (e.g., "be" for bead or sphere and "di" for disc). Thus a rotating gold-disc electrode would be identified by the symbol "Au:rodi" and a vibrating mercury-plated platinum-bead electrode would be identified by the symbol "Pt[Hg]: vibe." The abbreviation "ns" (for "not stated") is often used to describe electrodes of which full specifications were not given in the original source, such as a "rotating platinum electrode", which is transcribed as "Pt:rons" to indicate that no information about its shape was provided, while a platinum-rhodium electrode of unstated composition would appear as "Pt-Rh(ns)".

Reference electrodes are divided into two groups. One comprises the saturated calomel electrode, its variants (such as the "lithium S.C.E.", $Hg/Hg_2Cl_2(s)$, LiCl(s), and others of the same ilk), and the normal hydrogen electrode. These are almost invariably prepared with water, so that their use with a non-aqueous solution entails a liquid-junction potential between the non-aqueous solution of the compound being studied and the aqueous solution in the reference electrode. Some workers have sought to circumvent this by preparing similar electrodes in the same solvent or solvent mixture as the compound being studied; when this has been done, the symbol "(o)" (for "organic") follows the abbreviation that would denote the ordinary aqueous form of the reference electrode.

The other group comprises silver-silver halide electrodes, mercury pools, metal-metal-ion electrodes, and others normally prepared in the solvent used for the compound being studied (and often, often, indeed, employed as internal "reference" electrodes). For such an electrode, the abbreviation alone signifies that the solvent was the same throughout the cell, while the symbol "(aq)" (for water) following the abbreviation signifies that the reference electrode was prepared with water and used as an external reference electrode.

Some authors have used "pilot ions", whose half-wave potentials in non-aqueous solvents have been reliably established in prior work by themselves or others, to provide internal reference potentials and minimize problems like those that arise from variations of liquid-junction potentials with time. Potassium ion, rubidium ion, ferrocene, ferricinium ion, and several others have been used. This practice gives rise to entries of the form

$$DME//Hg//$$
$$K^+$$

which denotes the use of a dropping mercury electrode as the indicator electrode, of a mercury pool as the reference electrode, and of potassium ion as the added pilot ion.

The symbol "XEl" is used in this column for two different purposes. One is to denote an indicator electrode that is too rarely used to justify assigning it an abbreviation. The other is to represent a reference electrode of which the full description would not fit conveniently into the very limited space available in this Column and Column 18. Regardless of the purpose, every electrode represented by "XEl" in these columns is fully described on the same line in column 27. Similarly, the symbol "XEs" is used occasionally (as in the entries for code numbers 50CD-69CD) to denote an unusual pilot ion or standard substance, and is also defined on the same line in Column 27 wherever it occurs.

Column 12: "App."

Some salient features of the apparatus are summarized in this column. The basic code is a three-position alphanumeric, consisting of one or more digits that describe the cell, one or more letters that indicate whether and how dissolved oxygen was removed, and a final digit that identifies the technique of data acquisition. By means of the list of abbreviations (in which the abbreviations used in this column appear immediately following the symbols for electrodes), the typical entry "OAO" may be decoded to find that a two-electrode cell was used, that deaeration was accomplished with a stream of inert gas, and that the data were recorded on a pen-and-ink recorder. Inert gases are not identified because that would, in our judgment, consume more space than it would be worth, but they do not include hydrogen because hydrogen is sometimes not inert. More elaborate entries, such as "23CD3", are occasionally necessary; this one denotes the use of a three-electrode cell with an agar bridge, deaeration by both hydrogen and sulfite, and digital data-acquisition. When insufficient information was given to permit completing any part of this three-position code, a dash always appears in the appropriate position, as in the entry "0-0". This permits easy differentiation between, for example, "--0" and "0--".

Column 13: "Experimental Parameters"

Some of the information given in this column, such as the concentration C of the compound studied (which is always given in millimoles per cubic decimeter), would be significant no matter what technique was employed; some of it is governed by the nature of that technique. For polarographic data we give, wherever possible, values of the drop time t (in seconds), the rate of flow of mercury m (in milligrams per second), and the height h (in centimeters) of the column of mercury above the capillary tip.

The drop time for a dropping electrode with enforced mechanical detachment is denoted by the symbol "t(c)"; in the absence of the "(c)" (for "controlled") the value of t, like that of m, should be regarded only as an approximation, for there is little agreement about the conditions under which it is measured and the ways in which it is reported. An author chiefly interested in mass-transfer phenomena is apt to measure t at a potential on the plateau of the wave, one chiefly interested in charge-transfer phenomena is apt to measure it at the half-wave potential, and one studying a number of different compounds having different half-wave potentials is apt to measure it at the potential of the electrocapillary maximum (open circuit) or at the potential of the reference electrode employed (short circuit). Similarly, the value of m may be measured at the same potential as that of t, but often it is measured at the potential of the electrocapillary maximum while t is measured at a different potential. Sometimes m is measured with the tip of the capillary immersed in mercury. In view of this diversity, exact specification would have been difficult and wasteful of space, and the difficulty would have been compounded in dealing with papers that report values of t and m without specifying how they were measured. Consequently we have usually foregone any attempt at greater exactness; the user who needs more information must seek it in the original literature. The value of h serves as a crude but useful indication of whether the capillary was a conventional one or not: an abnormally high or low value would signify that there was something unusual about the capillary, and the original literature should again be consulted for details.

Sometimes the value of $m^{2/3}t^{1/6}$ (abbreviated "EC") is given, always in $mg^{2/3}s^{-1/2}$, when this was stated in the paper cited in lieu of the value of t or m or both. What was said above about the difficulty of quoting, and often even of deducing, the potential at which a value of m or t was measured applies equally to the potential at which $m^{2/3}t^{1/6}$ was evaluated.

For other techniques other kinds of data are given instead. For chronopotentiometry these include the area of the indicator electrode and the current or current density; for stationary-electrode voltammetry they include the area of the indicator electrode and the scan rate; for cyclic voltammetry they include the starting and reversal potentials and the area of the indicator electrode, and so on. The list of abbreviations must always be consulted regarding the units of the quantities given in this column. In the space that was available for these purposes it is quite impossible to give a full description of the experimental conditions, but an attempt has been made to give an accurate idea of their nature.

This column is also used to give the identity of the auxiliary or counter electrode when a three-electrode configuration was used, and to indicate whether the indicator or working electrode was subjected to chemical, mechanical, or electrolytic pretreatment.

The symbol "ns" (for "not stated") is often used in this column to mean that information was not available about the concentration of the compound studied ("C = ns") or about an important experimental parameter.

Column 14: "Ref."

Each entry in this column is a nine-character alphanumeric of which two letters and three digits appear on the first line while the remaining four digits appear on the second line. The first two letters denote the journal; these abbreviations do not appear in the list of abbreviations, but may be deciphered with the aid of Table X. The next three digits give the volume number, packed to three digits by the addition of leading zeros if necessary. The four digits on the second line give the page number, again with leading zeros added if necessary to yield a total of four digits. Since the letters "AA" denote *"Analytica Chimica Acta"*, the typical entries

AA072 and AA080
0169   0017

would refer to papers published in *Anal. Chim. Acta,* **72,** 169 and *Anal. Chim. Acta,* **80,** 17, respectively. The year is not given here but may be obtained by consulting Table X, which gives the names of the authors as well. In a reference to a journal that does not employ volume numbers, the two leading digits "19" are dropped from the number of the year, and the two remaining digits (e.g., 66 for a paper published in 1966) are followed by a letter. This is "O" if, like the *Bulletin de la Société Chimique de France*, the journal has no subsections; otherwise it denoted the subsection of a journal like the *Journal of the Chemical Society (London)*. Thus the entries

JL66B   and BF63O
0103   2252

would denote *J. Chem. Soc., Sec. B,* 103(1966) and *Bull. Soc. Chim. Fr.,* 2252 (1963), respectively.

To the shame chiefly of the authors, but also to some extent of the referees, editors, and journals involved, we have found a number of cases in which nearly, and sometimes exactly, identical papers have been published in two or three different journals. This practice has been repeatedly criticized even in the formerly common situation where publication in a local or regional journal in the language of a small country was followed by publication in an international journal having a much wider circulation and in a more widely known language. Repeated publication of the same material in journals of comparable circulation and in the same language seems to us to defraud the scientific community. In such cases we have simply suppressed all of the citations save one; we prefer the risk of being criticized for having covered the literature a little less thoroughly than we have actually done to that of allowing the offenders to profit from actions that we must hope to have been ill-judged rather than something more reprehensible.

Columns 15 through 26

Information that belongs in any of these columns but that was not given in the original paper is represented by a dash on the first line of an entry to aid the user in following separate entries across the page.

Column 15: "C/M"

A number that appears in this column is a cross-reference to Table III. Opposite the same number in that table, equations are given that describe the course or mechanism of the half-reaction. Table III may also contain other information, such as the rate constants of homogeneous chemical steps in the overall mechanism; its introduction should be consulted for further details.

The numbers that appear in Column 15 are not in numerical order because virtually identical equations may often be written for the reductions or oxidations of a number of compounds that have widely different empirical formulas and are therefore widely scattered in Table I. Cross-references in Table III list the code numbers of other compounds to which the same mechanism, or a closely related one, is applicable. For further details see the introduction to Table III.

Columns 16—18: "Characteristic Potential"

This term denotes a potential whose nature depends on the technique used. Typical characteristic potentials are the half-wave potential in polarography, the quarter-transition-time potential in chronopotentiometry, and the peak or half-peak potential in linear-sweep (stationary-electrode) voltammetry. Regardless of its nature, the characteristic potential always depends on the identity of the electroactive substance, on the kinetics or thermodynamics of the electron-transfer process, and of course on the experimental conditions; for any particular technique and under any completely defined set of experimental conditions the value of any characteristic potential is a reproducible property of the electroactive substance.

Column 16 gives the symbol of the characteristic potential whose numerical value (in volts) appears in Column 17. Column 18 identifies the reference electrode to which that value is referred; this is not necessarily the same as the reference electrode used for the experimental measurements, which was identified in Column 11.

As is also true of the techniques employed, some of the characteristic potentials appearing here have not yet been assigned individual symbols. Each of these is identified by the letters "XE" in Column 16, and Column 27 always defines it on the same line in the form "XE = potential where $d^2i/dE^2 = 0$".

Some investigators report values of the half-wave potential obtained by triangular-wave voltammetry and other techniques that do not yield experimental values of the half-wave potential. These values are of dubious significance and are identified by the symbol "$E_{1/2}(?)$" in Column 16.

Columns 19 and 20: "Response Const."

The generic term, "response constant" may denote either the measured value of the independent variable under the experimental conditions employed or some function of that value: as with the characteristic potential, the nature of the response constant depends on the technique used, and in addition it depends on the behavior of the system being studied. For a diffusion-controlled process the preferred polarographic response constant is the diffusion current constant $i_d/Cm^{2/3}t^{1/6}$, but the ratio i_d/C is often given instead when a value of $m^{2/3}t^{1/6}$ could not be obtained from the original, and even values of the limiting current i_ℓ alone are sometimes quoted. Values of i_d/C and i_ℓ provide at least some indication of the relative heights of different waves. Wherever possible, values of i_ℓ are accompanied (in Column 27) by an indication of whether the current is controlled by diffusion or some other process.

Values of i_ℓ, i_ℓ/C, and I pertain to individual waves unless these symbols are preceded by the symbol Σ: a value of Σi_l, for example, represents the total limiting current of the wave, measured at a potential on its plateau and including the contributions of all prior waves. In general the symbol Σ denotes that the wave in question is better defined than the preceding one, so that the sum of the currents is more precise than the usual incremental value would be.

For diffusion-controlled processes the preferred response constants in chronopotentiometry and stationary-electrode voltammetry are $i\tau^{1/2}/AC$ and $i_p/ACv^{1/2}$, respectively, given here in terms of current densities $j\tau^{1/2}/C$ and $j_p/Cv^{1/2}$ for compactness. We have attempted to give all currents in microamperes, concentrations in millimoles per cubic decimeter, and areas in square centimeters, but as currents are sometimes reported in arbitrary units such as millimeters of recorder deflection we have occasionally had to quote these, using the symbol "(u)" to denote their limited significance.

For processes that are not diffusion-controlled, and even for diffusion-controlled processes when some techniques are used, closed-form descriptions of the preferred response constant may not be available, and this is another reason why currents, ratios of current or current density to concentration, and other incomplete response "constants" appear in these columns. There are even some authors who have reported diffusion coefficients in place of the response constants obtained experimentally; the justification for doing this is meager, but we have sometimes been forced to quote these values for want of anything with a more solid basis.

Column 19 identifies the response constant whose value is given in Column 20. Units are given in the list of abbreviations: a polarographic diffusion current constant identified by an "I" in Column 19 is always in $\mu A\ mmol^{-1}dm^3mg^{-2/3}s^{1/2}$, a calculated diffusion coefficient identified by a "D" in Column 19 is always 10^6 times the reported value in cm^2s^{-1}, and so on. Some rarely used response constants have not been assigned individual symbols but are identified by the letters "XR" in Column 19 and defined on the same line in Column 27.

Columns 21 and 22: "n"

Usually the value of n appearing in Column 22 is the total number of electrons involved in the overall half-reaction that consumes one molecule or ion of the electroactive substance, and the abbreviation appearing in Column 21 identifies the technique by which that value was obtained. Values obtained by controlled-potential coulometry ("QE" in Column 21), however, generally pertain to the experimentally determined ratio of the number of faradays consumed to the number of moles of electroactive substance taken or consumed; in the original literature this ratio is often denoted by a symbol like "n_{app}" to emphasize its origin. Nonintegral values of n obtained by controlled-potential coulometry and other techniques reflect the occurrence of coupled chemical reactions that produce or consume electroactive material during a measurement.

Columns 23—25: "Electrokinetic Data"

Information about charge-transfer kinetics is given in these columns in various ways whose diversity again reflects the lack of standardization in the literature. Some authors calculate and report values of the symmetry parameter αn_a (usually abbreviated here as αn), the heterogeneous rate constant at 0 V vs. N.H.E. ($k_{f,h}$) or at the standard or formal potential of the couple ($k_{s,h}$), or of other parameters similar to these; others report values of more directly accessible quantities, such as (in polarography) the slope of a plot of E vs. $\log[i/(i_d - i)]$ or $E_{3/4} - E_{1/4}$. Some calculate and report the number of hydrogen or hydroxide ions consumed in the rate-determining electron-transfer step and the steps that precede it; some report the slope of a plot of the characteristic potential against the pH. Column 23 gives the symbol of the parameter whose value is given in Column 24, while Column 25 identifies the experimental data from which that value was deduced.

Column 26: "Products and Identification"

This column indicates what final products were obtained and what techniques were used to isolate and identify them. Products are usually identified by names; only for unusual structures (e.g., some radical ions) are their formulas given. The percent yield of the product, when available, follows its name. Techniques used for isolation and identification are indicated by abbreviations given in the list of abbreviations. Additional informtion, such as the lifetime of a radical, is sometimes included.

Column 27: "Description and Remarks"

This column, in which abbreviations are very dense, indicates whether the wave or other signal is cathodic (reduction) or anodic (oxidation), whether it is well- or ill-defined, whether or not it shows a maximum or some other reproducible deviation from the idealized simple form, whether the response is controlled by mass-transfer or some other process such as adsorption, and gives other information not provided in the preceding columns, including definitions of the special symbols "XT", "XEl" and "XEs", "XE", and "XI" where these appear in Columns 6, 11, 16, or 19 respectively. Correlation of measured quantities, such as the half-wave potential, with structural parameters (such as the exchange integral β, the Hammett substituent constant σ, or the Taft polar constant σ^*) are often indicated here. Relationships between measured quantities, such as the current, and experimental variables, such as the concentration, are indicated together with the limits of their validities. Sometimes the nature of the electrochemical process is indicated by entries such as "redn. of H^+" (which means that the process consists of the direct or catalytic reduction of hydrogen ion rather than the reduction of the organic species) or "redn. of maleic acid", which for a maleate salt means that it is not the organic cation but the maleate anion that is responsible for the behavior observed. In addition, this column gives, very briefly, our assessment of the reliability of the data and interpretation. Usually we have confined ourselves to stating whether we believe the information and conclusions to be reliable and reasonably accurate and precise, of limited precision, or in need of further confirmation, but occasionally we have indicated doubts concerning individual components of the entry. The symbol "(?)" is used to signify that we doubt the correctness of the information it follows, while the symbol "(!)" usually means either that the information reported cannot be validly obtained in the manner specified, as for example when the slope of a "log plot" is adduced as evidence for an overall n-value, or that we are dubious of its significance.

When data reported in one of the papers abstracted for this volume were thought to be inferior to data reported in a previous volume for the same compound, only the name of the compound and the experimental condition are given, and a cross-reference (in the form "see AG07") to the earlier data appears in this column.

The second part of the table appears on pages 328 to 333. Its arrangement is generally similar to that of the first part, but there are some differences that arise from the special nature of experiments on electrochemiluminescence (see "Columns 1 and 28: 'Code No.' " above). Since two different compounds are often involved in experiments on electrochemiluminescence, cross-references are more complex here than in the main portion of this table. Thus, the second entry in this column on page 328 is

FF82
fb87
ff29

"FF82" is the code number asssigned to this combination of electroactive substances: 10-methylphenothiazine as the one that is oxidized (which is denoted as "R" and identified in Column 2), and tris(2,2'-bipyridine)ruthenium(2+) diperchlorate as the one that is reduced (which is denoted as "O" and identified in Column 3). The "ff29" on the third line pertains to the second compound and signifies that electrochemical data on the behavior of O may be found under code number FF29 in the present volume. On consulting code number FB87 the user will find a further reference to additional data on 10-methylphenothiazine under code number CH91 in Volume II and code number DE04 in Volume IV. If these volumes had contained data on R but not on O, or if they had contained no data on R but had data on O, the entry would have been

FF82		FF82
fb87	or	—
—		ff29

respectively.

Column 2: "Compound oxidized (=R) and empirical formula"

For the compound that is oxidized in the experiment, this column gives the information contained in Columns 2 and 3 ("Empirical Formula" and Name) of the first part of this table.

Column 3: "Compound reduced (=O) and empirical formula"

This column gives the same information for the compound that is reduced in the experiment. The structural formulas of both compounds may be obtained through the cross-references provided in Columns 1 and 23.

Column 4: "Solvent"

See "Column 5: 'Solvent' " above.

Column 5: "Medium" through Column 13: "C/M"

"See Column 7: 'Medium' " through "Column 15: 'C/M' " above.

Columns 14—18: "Excitation"

These columns give information about the electrical signal employed. Column 14 shows whether a square wave ("☐"), a square wave superimpoed on a ramp ("☐ + /"), or a triangular wave ("∧") was used, and Column 15 gives the frequency, in hertz, of the square or triangular component. Columns 16 and 17 give the more positive (E_1) and the moe negative (E_2) limits, respectively, of the range of potentials employed, and Column 18 specifies the reference electrode or value to which E_1 and E_2 are referred.

Columns 19—21: "Emission"

These columns give information about the luminescence obtained. Column 19 identifies the species responsible for it, if this is known, using symbols like "1R*" to denote an excited singlet species of R (the compound oxidized). Column 20 tells whether the process is energy-sufficient ("S") or deficient ("De"), and Column 21 gives information on the wavelength of maximum emission, the excitation frequency at which the intensity of emission is a maximum, the quantum efficiency, and other quantities of interest.

Column 22: "Remarks"

This column contains an "O" if luminescence was not observed under the conditions stated, gives other information not contained in the preceding columns, and includes our assessment of the reliability of the data and interpretation.

FA00 CH₃Cl₃Sn

Code No.	Empirical Formula	Name and C.A. Number	Structural Formula	Solvent	Tech.	Medium		μ, M	pH	T, °C	Electrodes	App.	Experimental Parameters
FA00	CH₃Cl₃Sn	trichloro(methyl)-stannane C.A. 993-16-8	CH₃SnCl₃	MeOH 40 C₆H₆ 60	PY	NaClO₄	0.4	-	-	-	DME// Ag/AgCl	---	C=1
FA01	CH₄O	methanol C.A. 67-56-1	CH₃OH	H₂O	VA	H₂SO₄	0.5	-	-	-	Pt: xxns// SCE	35-0	0→1.55-0 V, C=1E3,v=590, A=1
						NaOH	0.5	-	-	25±2	Pd: xxns// Hg/HgO	---	C=2.5,v=78
FA02	C₂H₅Cl₃Sn	trichloro(ethyl)-stannane C.A. 1066-57-5	C₂H₅SnCl₃	MeOH 40 C₆H₆ 60	PY	LiCl	0.4	-	-	-	DME//SCE	---	C=0.4
FA03	C₂H₆N₄S₂	formamidine disulfide C.A. 3256-06-2	[NH₂C(:NH)S]₂	H₂O	PY	H₂SO₄	3	-	-	25	DME//SCE?	0-0	C=0.25, m=1.08, t=1.77 or 3.15,h=25
													C=0.5
													C=1.0
													C=2.5
						NaClO₄ HCl	0.1 ns	1.05					C=1
								1.90					
								2.00					
								3.10					
FA04 CONT	C₂H₆O	ethanol C.A. 64-17-5	CH₃CH₂OH	H₂O	VA	H₂SO₄	0.5	-	-	-	Pt: xxns// SCE	35-0	0→1.55→0V, C=50,v=238, A=1

TABLE I. Electrochemical Data

C_2H_6O (CONT.) FA04

| Ref. | C/M | Charact. | Potential | | Response | Const. | | n | Electrokinetic Data | | | Products and | Description and | Code |
		Value	Value	vs.		Value	Tech.		Parameter	Value	From	Identification	Remarks	No.
EA021 0395	421 a	$E_{\frac{1}{2}}$	-0.29F	Ag/AgCl	i_ℓ	1.1F	QE (-0.5V) sttd	1.74 2	-	-	-	starting material 16.6%, dichloro(di-methyl)stannane 38.1%, tin 45.3%	C, i_k, Mx in 0.4M LiCl, r	FA00
			-0.66F			6.8F		-					C	
			-1.22F			6.3F	QE (-1.3V) sttd	2.50 3				trichloro(methyl)-stannane 27.5%, di-chloro(dimethyl)-stannane 15.2%, tin 57.3%	C, Mx in 0.4M LiCl	
EA020 0323	-	E_p	0→0.4	SCE	i_p	-	-	-	-	-	-	-	$A, C, i_p \downarrow$ as $C \uparrow$ due to supp elect, p	FA01
			0.69F			3.2E3F							A	
			0.91F			5.5E3F							A	
			-			-							A, M	
			0.58F			3.5E3F							A, on scan from 1.55 to 0 V	
EA019 0063	-	E_p	-0.1F	Hg/HgO	i_p	2F	-	-	-	-	-	-	A, p	
EA021 0395	421 a	$E_{\frac{1}{2}}$	0.28F	Ag/AgCl	i_ℓ	1.9F	QE (0.50V)	1.65	-	-	-	starting material 14.8%, dichloro(di-ethyl)stannane 34%, tin 50.7%	C, Mx, no Mx in 0.1M $NaClO_4$, author claims 3 waves, p	FA02
			0.90F			2.5F	QE (1.40V)	2.86				trichloro(ethyl)-stannane 48%, di-chloro(diethyl)-stannane 13.5%, tin 38.5%	C, Mx	
EA020 0427	-	$E_{\frac{1}{2}}$	0.0Ap	SCE?	i_ℓ	0.8F	-	-	-	-	-	-	C, p	FA03
			-0.06F			1.7F	-	-	-	-	-	-	C, p	
			-0.07F			3.4F	-	-	-	-	-	-	C, p	
			-0.08F			9.4F	-	-	Elog	60 and 17	sttd	-	C, p	
			0.00F			2.77F	-	-	α	46 0.6		-	$C, \not=, i_d, p$	
			-0.27F			0.62F							C, i_d	
			-0.58			0.62F							C, i_d	
			0.00F			2.54F	-	-		-			C, p	
			-0.30F			0.62F							C	
			-0.62F			0.69F							C	
			0.00F			2.23F	-	-		-	-		C, p	
			-0.32F			0.92F							C	
			-0.62F			0.69F							C	
			0.00F			2.08F	-	-		-	-		C, p	
			-0.38F			0.92F							C	
			-0.67F			0.81F							C	
EA020 0323	-	E_p	0.92F	SCE	i_p	8E2F	-	-	-	-	-	-	A, p	FA04
			1.34F			1.3E3F							A, $i_p = k_1 + k_2 v$ for $v > 350$	
			0.74F			1.4E3F							C	
			0.56F			3.1E3F							A, on scan from 1.55 to 0 V	CONT

FA04 (CONT.) C_2H_6O

Code No.	Empirical Formula	Name and C.A. Number	Structural Formula	Solvent	Tech.	Medium		μ, M	pH	T, °C	Electrodes	App.	Experimental Parameters
FA04	C_2H_6O	ethanol	CH_3CH_2OH	H_2O	VR	H_2SO_4	0.5	-	-	-	Pt: xxns// SCE	35-0	$0 \to 1.55 \to 0$ V, C=500, v=120
FA05	C_2H_7N	ethylamine C.A. 75-04-7	$CH_3CH_2NH_2$	H_2O?	VY	KOH	1	-	-	-	Ni/ Ni(OH)$_2$: rons// SCE	---	C=166, thickness of Ni(OH)$_2$=80 mC cm^{-2}, A= 4200, ω=20, PE
FA06	$C_3H_3BrO_3$	3-bromo-2-oxo-propanoic acid C.A. 1113-59-3	$BrCH_2COCOOH$	H_2O	PY	H_2SO_4	-	-	0.4	-	DME//SCE	2--	C=1, m=8.8, t=4.5, h=50, EC=5.44
						MB	-		2.3				
									2.9				
									3.5				
									4.2				
									5.5				
									6.5				
									8.6				
				EtOH 20	PY	H_2SO_4	-	-	1.0	-	DME//SCE	2--	C=1, m=8.8, t=4.5, h=50, EC=5.44
CONT		ethanol	CH_3CH_2OH						2.0				

TABLE I. Electrochemical Data   17

$C_3H_3BrO_3$ (CONT.) FA06

Ref.	C/M	Charact. Potential		Response Const.		n Tech.	Electrokinetic Data			Products and Identification	Description and Remarks	Code No.
		Value	vs.		Value		Parameter	Value	From			
EA020 0323	-	E_p	0.56F	SCE	i_p	1.5E4F	-	-	-	-	-	A, on first scan from 0 to 1.55 V only, p	FA04
			0.86F			5E3F						A, on first scan from 0 to 1.55 V only	
			1.33F			1.4E4F						A, on first and subsequent scans from 0 to 1.55 V	
			0.6F			2.6E4F						A, on first and subsequent scans from 1.55 to 0 V	
			0.95F			8E3F						A, on second and subsequent scans from 0 to 1.55 V	
EA018 0923	-	$E_{\frac{1}{2}}$	0.3ApF	SCE	j_ℓ	4E4ApF	-	-	-	-	CP → MeCN(68%)	A, p	FA05
EA020 0369	421 b	$E_{\frac{1}{2}}$	-0.10F	SCE	i_ℓ	1.1F	-	-	$dE_{\frac{1}{2}}/dpH$	0	sttd	-	$C, i_k, E_{\frac{1}{2}} \neq f(pH)$ for pH < 2.3, $i_\ell = kh^{0.3}$, Tc=5, r	FA06
	421 c		-0.10F			1.13F	-	-		60		-	C, i_k, r	
			-			2.1F	-	-		60		-	C, i_k, r	
			-0.20F			2.1F	-	-		60		2-oxopropanoic acid, from $E_{\frac{1}{2}}, 2$	$C, i_k, i_\ell = kh^{0.3}$, Tc=5, r	
			-1.03F			-	-	-		70	calc		$C, i_k, i_\ell = kh^{0.3}$, Tc=5, redn of 2-oxopropanoic acid	
			-0.27F			2.1F	-	-		0	sttd	2-oxopropanoic acid, from $E_{\frac{1}{2}}, 2$	C, i_k, r	
			-1.07F			-	-	-		70	calc		C, i_k, redn of 2-oxopropanoic acid	
			-0.30F			2.1F	QE (-0.7V)	2.0		0	sttd	2-oxopropanoic acid, (semicarbazone, PY, UVS)	C, i_k, r	
			-1.17F			1.6F	-	-		70	calc	-	C, i_k, redn of 2-oxopropanoic acid	
			-0.30F			2.1F	-	-		0	sttd	2-oxopropanoic acid, from $E_{\frac{1}{2}}, 2, E_{\frac{1}{2}}, 3$	$C, i_\ell \uparrow$ as pH ↑, r	
			-1.22F			1.3F	-	-		70?			$C, i_\ell \uparrow$ as pH ↑, redn of 2-oxopropanoic acid, for pH > 7	
			-1.53F			-	-	-		0			C, redn of 2-oxopropanoate ion	
			-0.30F			3.6F	-	-		0		2-oxopropanoate ion, from $E_{\frac{1}{2}}, 3$	$C, i_\ell \uparrow$ as pH ↑, r	
			-1.50F			2.4F	-	-		0			$C, i_\ell \uparrow$ as pH ↑, redn of 2-oxopropanoate ion	
			-0.30F			-	-	-		0		2-oxopropanoate ion, from $E_{\frac{1}{2}}, 3$	$C, i_\ell \uparrow$ as pH ↑, compound decomposes if pH > 9, r	
			-1.53F			-	-	-		0			$C, i_\ell \uparrow$ as pH ↑, redn of 2-oxopropanoate ion	
EA020 0369	421 bc	$E_{\frac{1}{2}}$	-0.08F	SCE	i_ℓ	1.0F	$i : i_d$	0.35F	$dE_{\frac{1}{2}}/dpH$	0	calc	-	$C, i_\ell = kC$ for C < 0.5, Tc < 0 for T=15-25, $i_\ell = kh^{0.66}, r$	
			-0.08F			1.1F		0.28F		0		-	C, r	

CONT

FA06 (CONT.) $C_3H_3BrO_3$

Code No.	Empirical Formula	Name and C.A. Number	Structural Formula	Solvent	Tech.	Medium		μ, M	pH	T, °C	Electrodes	App.	Experimental Parameters
FA06	$C_3H_3BrO_3$	3-bromo-2-oxo-propanoic acid	$BrCH_2COCOOH$	EtOH 20	PY	MB	–	–	2.8	–	DME//SCE	2--	C=1,m=8.8, t=4.5,h=50, EC=5.44
									3.4				
									4.0				
									5.0				
									6.3				
									7.5				
									9.7				
					PR	MB	–	–	5.5	–	DME//SCE	2-2	C=1,t=5, v=1000
FA07 ab34 ca23 da11	C_3H_3N	acrylonitrile C.A. 107-13-1	$CH_2:CHCN$	MeCN 90	PY	Et_4NClO_4 0.1		–	–	25	DME// Ag/ $AgClO_4$ 0.01	12AO	C=1.9,Pt Aux
				MeCN 99	PY	Et_4NClO_4 0.1		–	–	25	DME// Ag/ $AgClO_4$ 0.01	12AO	C=1.9,Pt Aux
				MeCN 99.95	PY	Et_4NClO_4 0.1		–	–	25	DME// Ag/ $AgClO_4$ 0.01	12AO	C=1.9,Pt Aux
				DMF 90	PY	Et_4NClO_4 0.1		–	–	25	DME//SCE	12AO	C=2,Pt Aux
				DMF 95	PY	Et_4NClO_4 0.1		–	–	25	DME//SCE	12AO	C=2,Pt Aux
				DMF 99	PY	Et_4NClO_4 0.1		–	–	25	DME//SCE	12AO	C=2,Pt Aux
				DMF 99.9	PY	Et_4NClO_4 0.1		–	–	25	DME//SCE	12AO	C=2,Pt Aux

TABLE I. Electrochemical Data

C_3H_3N FA07

Ref.	C/M	Charact. Potential		Response Const.		n Tech.		Electrokinetic Data			Products and Identification	Description and Remarks	Code No.
		Value	vs.		Value			Parameter	Value	From			
EA020 0369	421 c	$E_{\frac{1}{2}}$	-0.08F	SCE	i_ℓ	1.5	$i:i_d$	0.39F	$dE_{\frac{1}{2}}/dpH$	0	calc	-	C,r	FA06
			-0.33F			-		-		140			C	
			-			2.1F		0.54F		0		2-oxopropanoic acid, from $E_{\frac{1}{2}},2$	C,r	
			-			0.8F		0.2F		100			C,redn of 2-oxopropanoic acid	
			-			1.1F		0.29F		52			C,redn of 2-oxopropanoate ion	
			-0.08F			1.7F		0.44F		0		2-oxopropanoic acid, from $E_{\frac{1}{2}},2,E_{\frac{1}{2}},3$	C,r	
			-0.50F			1.5F		0.39F		100			C,redn of 2-oxopropanoic acid	
			-1.13F			1.0F		0.26F		52			C,redn of 2-oxopropanoate ion	
			-0.08F			1.2F		0.31F		0		QE,n=2→2-oxopropanoate ion;Br⁻	C,r	
			-0.580			2.5F		0.65F		0		-	C,i_ℓ= kC for C < 4, i_ℓ=kh$^{0.5}$,Tc=5.7	
			-1.17F			0.9F		0.23F		52		-	C,redn of 2-oxopropanoic acid	
			-0.08F			1.0F		0.26F		0		-	C,r	
			-0.58			2.9F		0.76F		0			C	
			-1.25F			}1.3F		0.34F		52			C,0 for pH > 7	
			-1.54F							0			C	
			-0.08F			1.0F		0.26F		0		-	C,r	
			-0.58			2.9F		0.76F		0			C	
			-1.54F			2.7F		0.70F		0			C	
			-0.08F			-		-		0		-	C,r	
			-0.58							0			C	
			-1.54F							0			C	
EA020 0369	421 c	E_p	-0.17F	SCE	i_p	1F	-	-	-	-	-	-	C,i_p ↓ on subsequent scans,p	
			-0.56F			2F	-	-	-	-	-	-	C	
JE042 0189	50s	$E_{\frac{1}{2}}$	-2.27F	Ag/ AgClO$_4$ 0.01	i_ℓ	7.4F	QE	1.9	-	-	-	-	C,r	FA07 ab34 ca23 da11
JE042 0189	50s	$E_{\frac{1}{2}}$	-2.37F	Ag/ AgClO$_4$ 0.01	i_ℓ	10F	-	-	-	-	-	-	C,r	
JE042 0189	50s	$E_{\frac{1}{2}}$	-2.3F	Ag/ AgClO$_4$ 0.01	i_ℓ	3.4F	$i:i_d$, QE	0.80 0.3	-	-	-	-	C,r	
			-2.47F			2.2F	$i:i_d$	0.4					C	
JE042 0189	50s	$E_{\frac{1}{2}}$	-2.1F	SCE	i_ℓ	10.9F	-	-	-	-	-	-	C,r	
JE042 0189	50s	$E_{\frac{1}{2}}$	-2.1F	SCE	i_ℓ	12.5F	-	-	-	-	-	-	C,r	
JE042 0189	50s	$E_{\frac{1}{2}}$	-2.0F	SCE	i_ℓ	7.4F	-	-	-	-	-	-	C,r	
			-2.28F			2.7F							C	
JE042 0189	50s	$E_{\frac{1}{2}}$	-1.96F	SCE	i_ℓ	3.8F	$i:i_d$	0.7	-	-	-	-	C,r	
			-2.16F			1.1F		0.17					C	

FA08 C$_3$H$_4$O

Code No.	Empirical Formula	Name and C.A. Number	Structural Formula	Solvent	Tech.	Medium		μ, M	pH	T, °C	Electrodes	App.	Experimental Parameters
FA08	C$_3$H$_4$O	2-propyn-1-ol C.A. 107-19-7	HC⋮CCH$_2$OH	DMF	PY	Bu$_4$NI	0.1	-	-	-	DME// SCE(o)	2-0	C=2
FA09 ab82 ca39 ea33	C$_3$H$_6$O	acetone C.A. 67-64-1	(CH$_3$)$_2$CO	H$_2$O	VA	H$_2$SO$_4$	1	-	-	25±0.05	HMDE// NHE	25--	-0.6→-1.1→ -0.6V, C=860, A=0.03, v=200
FA10 ab83	C$_3$H$_6$O	2-propen-1-ol C.A. 107-18-6	CH$_2$:CHCH$_2$OH	MeCN	IL	Bu$_4$NBF$_4$ H$_2$O	0.15 (5-10)E-3	-	-	-	Pt: rodi// Ag/ AgClO$_4$ 0.01	25A0	C=ns, Pt Aux
					VA	Bu$_4$NBF$_4$ H$_2$O	0.15 (5-10)E-3	-	-	-	glC: xxdi// Ag/ AgClO$_4$ 0.01	25A0	C=ns, v=200, Pt Aux
											Pt: xxdi// Ag/ AgClO$_4$ 0.01		
				DMF	PY	Bu$_4$NI	0.1	-	-	-	-	-	C=2
FA11	C$_3$H$_8$O	1-propanol C.A. 71-23-8	CH$_3$CH$_2$CH$_2$OH	H$_2$O	VA	H$_2$SO$_4$	0.5	-	-	-	Pt: xxns// SCE	35-0	0→1.55→0V, C=55, v=238, A=1
													C=2E3, v=120
FA12	C$_3$H$_8$O	2-propanol C.A. 67-63-0	(CH$_3$)$_2$CHOH	H$_2$O	VA	H$_2$SO$_4$	0.5	-	-	-	Pt: xxns// SCE	35-0	C=55, v=238, A=1
													C=500, v=200, A=1
													C=2000, v=560, A=1
													C=4000, v=700, A=1
FA13	C$_3$H$_9$ClSn	chloro(trimethyl)- stannane C.A. 1066-45-1	(CH$_3$)$_3$SnCl	MeOH 40 C$_6$H$_6$ 60	PY	LiCl Tween 80	ns	-	-	-	DME// Ag/AgCl	---	ns

TABLE I. Electrochemical Data

C_3H_9ClSn FA13

Ref.	C/M	Charact. Potential		Response Const.		n Tech.	n	Electrokinetic Data			Products and Identification	Description and Remarks	Code No.
		Value	vs.		Value			Parameter	Value	From			
EA019 0629	-	-	-	-	-	-	-	-	-	-	-	O,p	FA08
EA019 0555	402 b	E_p	-	i_p	-	Tafel	1	Tafel	0.12	sttd	CP (-1.2 V) → 2-propanol 71%, GLC	C, sweep does not include i_p, r	FA09 ab82 ca39
		1.04F	NHE		3.57E4							A, i_p ↑ as C ↑	ea33
EA019 0565	442 a	-	-	-	-	Levich	2	-	-	-	CP(2.4±0.2 V) → acrolein 30%, GLC, yield ↓ as [H_2O] ↑; polymer	A, $i_\ell = k\omega^{\frac{1}{2}}$, i_d, (D assumed=21), p	FA10 ab83
EA019 0565	442 a	$E_{p/2}$ 2.95	Ag/ AgClO$_4$ 0.01	-	-	-	-	Tafel	190-165	C= 0.014-0.458M	-	A, p	
		-							-			C, O	
		2.65		-	-	-	-	-	-	-	-	A, p	
												C, O	
EA019 0629	-	-	-	-	-	-	-	-	-	-	-	O, p	
EA020 0323	-	E_p	0 → 0.4F	SCE	i_p	-	-	-	-	-	-	-	A, i_p ↓ if C ↑, due to adsorbed H_2, p	FA11
		1.04F			0.6E3F							A, $i_p = k_1 + k_2v$	
		1.36F			1.0E3F							A, $i_p = k_1 + k_2v$	
		0.75F			1.0E3F							C, H	
		0.56F			1.3E3F							A, on scan from 1.55 to 0 V	
		1.01F			1.7E3F							A, p	
		1.33F			8.5E3F							A	
		0.67F			9.6E3F							A, on scan from 1.55 to 0 V	
EA020 0323	-	E_p	0.56F	SCE	i_p	330F	-	-	-	-	-	-	A, $i_p = k_1 + k_2v$ for v > 100	FA12
		1.41F			470F	-	-	-	-	-	-	A, $i_p = k_1 + k_2v$ for v > 450	
		-			2300F	-	-	-	-	-	-	A, $i_p = k_1 + k_2v$ for v > 200, $i_p = f(C)$, p	
		-			5400F	-	-	-	-	-	-	A, $i_p = k_1 + k_2v$ for v > 560; p	
		-			8000F	-	-	-	-	-	-	A, $i_p = k_1 + k_2v$ for v > 700, $i_p \ne f(C)$, p	
EA021 0395	240 c	$E_{\frac{1}{2}}$	-1.100	Ag/AgCl	i(u)	1	-	-	-	-	-	CP → hexamethyldi-stannane	C, i_d, p	FA13
		-1.650			1			α	0.4	sttd	-	C, ≠, i_d	

FA14 $C_4H_2N_2$

Code No.	Empirical Formula	Name and C.A. Number	Structural Formula	Solvent	Tech.	Medium	μ, M	pH	T, °C	Electrodes	App.	Experimental Parameters
FA14 ac31 ca51	$C_4H_2N_2$	trans-2-buten-dinitrile C.A. 764-42-1	trans-NCCH:CHCN	MeCN 85	PY	Et_4NClO_4 0.1	-	-	25	DME// Ag/ $AgClO_4$ 0.01	12A0	C=3, Pt Aux
				MeCN 90	PY	Et_4NClO_4 0.1	-	-	25	DME// Ag/ $AgClO_4$ 0.01	12A0	C=3, Pt Aux
					PW	Et_4NClO_4 0.1	-	-	25	DME// Ag/ $AgClO_4$ 0.01	12A2	C=0.31-3.68, Pt Aux
				MeCN 99	PY	Et_4NClO_4 0.1	-	-	25	DME// Ag/ $AgClO_4$ 0.01	12A0	C=3, Pt Aux
				MeCN 99.95	PY	Et_4NClO_4 0.1	-	-	25	DME// Ag/ $AgClO_4$ 0.01	12A0	C=3, Pt Aux
FA15	$C_4H_6O_3S$	3-methylthio-2-oxo-propanoic acid C.A. 18542-43-3	$CH_3SCH_2COCOOH$	EtOH 3	QE	BR	-	4.5 and 10.5	-	Hg: nspo// SCE	259A -	
FA16	$C_4H_8O_2S$	methyl 3-mercapto-propanoate C.A. 2935-90-2	$H_3COOCCH_2CH_2SH$	EtOH 25	PY	CL $NaClO_4$ 0.1 TX100 0.002%	-	9.37	25.0 ±0.1	DME//SCE	-A1	C=2.0, m=2.96, t=2.39, h=38.3, EC=2.35
FA17 ad32	$C_4H_{10}Cl_2Sn$	dichloro(diethyl)-stannane C.A. 866-55-7	$(C_2H_5)_2SnCl_2$	MeOH 40 C_6H_6 60	PY	LiCl 0.4 Tween 80 ns	-	-	-	DME// Ag/AgCl	---	C=0.8
FA18 ad41	$C_4H_{10}O$	1-butanol C.A. 71-36-3	$CH_3(CH_2)_3OH$	H_2O	VA	H_2SO_4 0.5	-	-	-	Pt: xxns// SCE	35-0	0→1.55→0V, C=50, v=238, A=1
FA19 ad42	$C_4H_{10}O$	2-butanol C.A. 78-92-2	$CH_3CH_2CHOHCH_3$	H_2O	VA	H_2SO_4 0.5	-	-	-	Pt: xxns// SCE	35-0	0→1.55→0 V, C=50, v=1169, A=1

TABLE I. Electrochemical Data

$C_4H_{10}O$ FA19

Ref.	C/M	Charact. Potential		Response Const.		n		Electrokinetic Data			Products and Identification	Description and Remarks	Code No.
		Value	vs.		Value	Tech.		Parameter	Value	From			
JE042 0189	50s	$E_{\frac{1}{2}}$	-1.46F	Ag/AgClO$_4$ 0.01	i_ℓ	2.4F	-	-	-	-	-	-	C,r	FA14 ac31 ca51
			-2.21F			2F							C	
JE042 0189	50s	$E_{\frac{1}{2}}$	-1.54F	Ag/AgClO$_4$ 0.01	i_ℓ	2.8F	QE	0.94	-	-	-	-	C,r	
			-2.34F			3.2F							C	
JE042 0189	50s	-	-	-	-	-	-	-	$dE_p/d\log v$	20±1	for v < 1E3	-	C, first wave only, r	
										0	for v > 1E3			
									$dE_p/d\log C$	21	sttd			
		-								-			C?	
JE042 0189	50s	$E_{\frac{1}{2}}$	-1.6F	Ag/AgClO$_4$ 0.01	i_ℓ	2F	-	-	-	-	-	-	C,r	
			-2.5F			3F							C, $E_{\frac{1}{2}}$ and i_ℓ (Ap)	
JE042 0189	50s	$E_{\frac{1}{2}}$	-1.6F	Ag/AgClO$_4$ 0.01	i_ℓ	2F	$i:i_d$ QE	0.5 0.6	-	-	-	-	C,r	
			-			-	-						C?	
EA018 0691	53d e	-	-	-	-	-	-	-	-	-	-	-	C,r	FA15
EA018 0933	-	$E_{\frac{1}{2}}$	-0.552	SCE	I	1.02	-	-	α $k^0_{A,h}$	0.48 1.13E-4	Elogt	CP (-0.4 V, pH 9.37)→ H$_3$COOCCH$_2$CH$_2$SHg, ($E_{\frac{1}{2}}$); Hg^{+2}?(ppt with H$_2$S)	A,$\neq$(Elog), i_ℓ=0.39h$^{\frac{1}{2}}$, i_ℓ=kC, Tc=2.05, i_d, i_ℓ=f[EtOH] for [EtOH] < 40%, pK'_a = 9.95, D=2.90	FA16
EA021 0395	-	$E_{\frac{1}{2}}$	-	Ag/AgCl	i_ℓ	-	-	-	$dE_{\frac{1}{2}}/d\log C$	0	sttd	-	C, i_a, Mx	FA17 ad32
			-1.03F			4.3F				-			C, i_d	
			-1.60F			1.0F							C, M	
EA020 0323	-	E_p	1.03F	SCE	i_p	1E3F	-	-	-	-	-	-	A, X, S, p	FA18 ad41
			1.25F			1E3F							A, P, i_p=$k_1 + k_2 v$ for v > 350	
			0.57F			1E3F							C	
			0.53F			1E3F							A, on scan from 1.55 to 0 V	
EA020 0323	-	E_p	0.6F	SCE	i_p	4.2E2F	-	-	-	-	-	-	A, X, p	FA19 ad42
			1.4F			1.1E3F							A, i_p=$k_1 + k_2 v$ for v > 200	

FA20 $C_5H_4O_6$

Code No.	Empirical Formula	Name and C.A. Number	Structural Formula	Solvent	Tech.	Medium		μ, M	pH	T, °C	Electrodes	App.	Experimental Parameters
FA20	$C_5H_4O_6$	2,2,4,5-tetra-hydroxy-4-cyclo-pentene-1,3-dione C.A. 63056-19-9	Table II	H_2O	PY	buffer	ns	-	0	-	DME//SCE	---	ns
									1.0				
									3.0				
FA21	$C_5H_7BrO_3$	ethyl 3-bromo-2-oxopropanoate C.A. 70-23-5	$BrCH_2COCOOCH_2$-CH_3	H_2O	PY	H_2SO_4	ns	-	0.6	-	DME//SCE	2--	C=1,m=8.8, t=4.5,h=50, EC=5.44
						MB	-		2.1				
									3.0				
									4.3				
									5.5				
									7.1F				
									8.4				
				EtOH 20	PY	H_2SO_4	ns	-	1	-	DME//SCE	2--	C=1,m=8.8, t=4.5,h=50, EC=5.44
						MB	-		3.3				
									4				
									5				
									9.7				

TABLE I. Electrochemical Data   25

$C_5H_7BrO_3$ FA21

Ref.	C/M	Charact. Potential		Response Const.		n		Electrokinetic Data			Products and Identification	Description and Remarks	Code No.
		Value	vs.	Value		Tech.		Parameter	Value	From			
EA021 0913	-	$E_{1/2}$	-0.230	SCE	-	-	-	-	-	-	-	pinacol	C, pK_{a_1}=0.5, pK_{a_2}=1.50, r	FA20
			-0.300		-	-	sttd	0.8-0.9	-	-	-	pinacol	C,r	
			-0.480		-	-	-	-	-	-	-	1,2,4-trihydroxy-3,5-dioxocyclo-pentene	C,r	
EA020 0369	283 e	$E_{1/2}$	-0.10F	SCE	i_ℓ	0.8F	$i:i_d$	0.3F	$dE_{1/2}/dpH$	0	sttd	-	C, $i_\ell=kh^{0.1}$, Tc=4.5, i_k, r	FA21
			-0.10F			0.8F		0.3F		0		ethyl 2-oxopropano-ate from i_2	C,r	
			-0.70F			-		-		50			C, i_k, $i_\ell=kh^{0.35}$, Tc=5.5, redn of ethyl 2-oxopropano-ate	
			-0.10F			1.0F		0.3F		0		ethyl 2-oxopropano-ate from i_2	C,r	
			-0.77F			0.7F		-		50			C, i_k, redn of ethyl 2-oxopropanoate	
			-			0.1F		-		-			C, i_k, redn of ethyl 2-oxopropanoate	
			-0.10F			1.1F		0.3		0		ethyl 2-oxopropano-ate from i_2	C,r	
			-0.82F			0.6F		0.2F		50			C, i_k, redn of ethyl 2-oxopropanoate	
			-1.05			0.4F		0.2F		67			C, i_k, redn of ethyl 2-oxopropanoate	
			-0.10F			1.2F	QE (-0.6V)	2		0		ethyl 2-oxopropano-ate from i_2, Br^-	C,r	
			-0.87F			0.6F	QE (-1.1V)	4		50			C, i_k, redn of 2-oxopropanoic acid	
			-1.13F			0.4F		-		67			C, i_k, redn of ethyl 2-oxopropanoate	
			-0.10F			1.0F		-		0		-	C,r	
			-			0.6F				-			C, i_k	
			-			0.4F							C, i_k	
			-0.10F			-		-		0		-	C, compound decomposes if pH > 9, r	
EA020 0369	283 e	$E_{1/2}$	-0.05F	SCE	i_ℓ	0.9F	$i:i_d$	0.2F	$dE_{1/2}/dpH$	0	calc	-	C, i_ℓ = kC for C < 1 and pH < 2.4, $i_\ell=kh^{0.8}$, Tc=5 < 0 for T=25-30, r	
			-0.05F			0.96F		0.3F		0		-	C,r	
			-0.970			-		-		-			C	
			-0.05F			0.9F		0.3F		0		-	C,r	
			-			0.75F		0.2F		-			C	
			-0.05F			0.8F		0.2F		0		-	C, i_ℓ = kC for C < 0.5 at pH=5.5, r	
			-0.96F			0.7F		0.2F		36			C	
			-0.05F			0.8F		0.2F		0		-	C,r	
			-1.13F			0.7F		0.2F		36			C	

FA22 C_5H_8O

Code No.	Empirical Formula	Name and C.A. Number	Structural Formula	Solvent	Tech.	Medium		μ, M	pH	T, °C	Electrodes	App.	Experimental Parameters
FA22	C_5H_8O	2-methyl-3-butyn-2-ol C.A. 115-19-5	$(CH_3)_2C(OH)C\vdots CH$	DMF	PY	Bu_4NI	0.1	-	-	-	DME// SCE(o)	2-O	C=2
FA23 ea88	C_5H_8O	3-methyl-3-buten-2-one C.A. 814-78-8	$H_2C:C(CH_3)COCH_3$	MeOH 25	PY	UB	ns	-	1.0	-	DME// Hg/ Hg_2Cl_2, LiCl 1.0	2AO	C=ns, m=4.9, t=1.9
									3.5				
									4.6				
									6.0				
									8.0				
									9.3				
									10.5				
									13				
				MeOH	PY	$LiClO_4$	0.1	-	-	-	DME// Ag/AgCl, LiCl	2AO	C=1, m=4.9, t=1.9
						$LiClO_4$ $C_6H_5SO_3H$	0.1 4E-4						
						$LiClO_4$ $C_6H_5SO_3H$	0.1 8E-4						
						$LiClO_4$ $C_6H_5SO_3H$	0.1 1.5E-3						
						$LiClO_4$ $C_6H_5SO_3H$	0.1 2.0E-3						
				MeCN 99.95	PY	Et_4NClO_4	0.1	-	-	25	DME//SCE	12AO	C=2, Pt Aux
					PW	Et_4NClO_4	0.1	-	-	25	DME// Ag/ $AgClO_4$ 0.01	12AO	C=0.31, Pt Aux
													C=3.1
				DMF 99.9	PY	Et_4NClO_4	0.1	-	-	25	DME//SCE	12AO	C=2.14, Pt Aux

C_5H_8O FA23

Ref.	C/M	Charact. Potential		Response Const.		n Tech.	n	Electrokinetic Data			Products and Identification	Description and Remarks	Code No.
		Value	vs.		Value			Parameter	Value	From			
EA019 0629	-	-	-	-	-	-	-	-	-	-	-	o,p	FA22
JE035 0381	-	$E_{\frac{1}{2}}$	-1.02F	Hg/ Hg_2Cl_2, LiCl 1.0	-	-	sttd	1	$dE_{\frac{1}{2}}/dpH$	69F	-	-	C,r	FA23 ea88
		-1.18F		-	-	-	-		69F		-	C,r	
		-1.27F		-	-	-	-		69F			C,r	
		-1.55F							0F			C	
		-1.38F		-	-	-	-		69F			C,r	
		-1.56F							0F			C	
		-1.56F		-	-	-	-	-	-	-	-	C,r	
		-1.65F		-	-	-	-		-			C,r	
		-1.67F		-	-	-	-		0F			C,r	
		-1.67F		-	-	-	-		0F			C,r	
JE035 0381	-	$E_{\frac{1}{2}}$	-1.56F	Ag/AgCl, LiCl	i_ℓ	6.0F	-	-	-	-	-	-	C,S,p	
		-1.8F			1.2F							C,P	
		-0.92F		i_ℓ	2.0F	-	-	-	-	-	-	C,p	
		-1.56F			6.0F							C,S	
		-1.80F		?								C,P	
		-0.92F			3.6F	-	-	-	-	-	-	C,p	
		-1.67F			4.0F							C,2 merging waves	
		-0.90F			4.4F	-	-	-	-	-	-	C,p	
		-1.23F			1.9F							C	
		-1.86F			0.9F							C	
		-0.9F			4.4F	-	-	-	-	-	-	C,p	
		-1.23F			4.4F							C	
		-			-							C,O	
JE042 0189	50s	$E_{\frac{1}{2}}$	-1.9	SCE	-	-	i:i	1	-	-	-	-	C,r	
		-						-				c(?)	
JE042 0189	50s	-	-	-	-	-	-	-	$dE_p/dlogv$	22	for v < 1E6		C,first wave only,r	
										0	for v > 1E6			
		-								-			C?	
		-								22	for v < 2E5		C,r	
										0	for v > 2E5			
									$dE_p/dlogC$	24	sttd			
		-								-			C?	
JE042 0189	50s	$E_{\frac{1}{2}}$	-1.9	SCE	-	-	i:i	1	-	-	-	-	C,r	
								-					c(?)	

FA24 $C_5H_8O_3$

Code No.	Empirical Formula	Name and C.A. Number	Structural Formula	Solvent	Tech.	Medium	μ, M	pH	T, °C	Electrodes	App.	Experimental Parameters
FA24	$C_5H_8O_3$	ethyl 2-oxopropanoate C.A. 617-35-6	$CH_3COCOOCH_2CH_3$	EtOH 20	PY	MB	-	3.0	-	DME//SCE	2--	m=8.8,C=ns, t=4.5,h=50, EC=5.44
								6.5				
FA25	$C_5H_8O_3S$	3-(ethylthio)-2-oxopropanoic acid C.A. 51033-95-5	$CH_3CH_2SCH_2CO-COOH$	EtOH 3	PY	H_2SO_4	1	-1.0	25	DME//SCE	3--	C=1,m=6.4, t=3.0,h=55, EC=4.15
								0.15				
						BR	-	2.10				
								5.00				
								6.40				
								7.35				
								10.50				
								13.60				
					QE	BR	-	4.5	-	Hg: srpo// SCE	259A -	C=10, E_{app}= -1.20 V
								10.5				E_{app}= -1.50 V
FA26	$C_5H_9BrO_3S$	3-dimethylsulfonio-2-oxopropanoic acid bromide C.A. 51033-97-7	$(CH_3)_2\overset{+}{S}CH_2CO-COOH\ Br^-$	H_2O	PY	H_2SO_4	1	-1.6	25	DME//SCE	3--	C=1,m=6.4, t=3.0,h=55, EC=4.15
								-1.0				
								0				
								0.7				
								1.1				
						BR	-	2.05				
								5.95				
								6.8				
								7.5				
								7.7				
								7.95				
FA24 CONT		ethyl 2-oxopropanoate	$CH_3COCOOCH_2CH_3$					8.4				

TABLE I. Electrochemical Data

$C_5H_9BrO_3S$ (CONT.)   FA26

Ref.	C/M	Charact. Potential Value	vs.	Response	Const. Value	n Tech.	Electrokinetic Data Parameter	Value	From	Products and Identification	Description and Remarks	Code No.
EA020 0369	65c k	$E_{\frac{1}{2}}$ -0.800	SCE	-	-	-	-	-	-	-	-	C,r	FA24
		-1.050		-	-	-	-	-	-	-	C,r	
EA018 0691	53d e	$E_{\frac{1}{2}}$ -0.3F	SCE	-	-	-	-	$dE_{\frac{1}{2}}/dpH$	0	sttd	-	C,$E_{\frac{1}{2}} \neq f(pH)$ for pH < -0.5, i_k, Tc=8, r	FA25
		-0.35F		i_ℓ	3.3F	-	-		70		-	C, $dE_{\frac{1}{2}}/dpH$ for pH= -0.5 to 2.3, i_k, Tc=8, r	
		-0.53F			1.9F	-	-		70		-	C, i_k, Tc=8, r	
		-0.98F			5.2F	-	-		130		-	C, $dE_{\frac{1}{2}}/dpH$ for pH= 2.3 to 5, i_k, Tc=8, r	
		-1.07F			4.0F	-	-		70		-	C, $dE_{\frac{1}{2}}/dpH$ for pH > 6, i_k, Tc=8, r	
		-1.35F			2.7F	-	-		0		-	C, $dE_{\frac{1}{2}}/dpH$ for pH > 6	
		-			1.5F	-	-		70		-	C, r	
		-1.32F			5.8F				0		-	C	
		-1.33F			8.3F	-	-		0		-	C, r	
		-1.38F			4.3F	-	-		0		-	C, i_k, $i_\ell \downarrow$ as pH $\uparrow$, r	
EA018 0691	53d e	-	-	-	-	QE	2	-	-	-	3-ethylthio-2-hydroxypropanoic acid,CHA;ethylthiol 7.5%,PY,UVS	C,r	
		-		-	-		2	-	-	-	3-ethylthio-2-hydroxypropanoic acid,CHA;ethylthiol 7.5%,PY,UVS	C,r	
EA018 0691	421 b	-	-	i_ℓ	2.9F	-	-	-	-	-	-	C,r	FA26
		-			2.3F	-	-	-	-	-	-	C,r	
		$E_{\frac{1}{2}}$ -0.33F	SCE		1.3F	-	-	$dE_{\frac{1}{2}}/dpH$	0F	-	-	C,$E_{\frac{1}{2}} \neq f(pH)$ for pH < 0.7, i_k, Tc=10, r	
	421 c	-0.34F			1.4F	-	-		65F		-	C, i_k, Tc=10, r	
		-			1.64F	-	-		65F		-	C, i_k, r	
		-0.46F			2.1F	-	-		65F		-	C, i_k, Tc=10, r	
		-0.73F			2.3F	-	-		65F		-	C, i_k, Tc=10, r	
		-			3.43F	-	-		65F		-	C, i_k, Tc=10, r	
		-0.83F			4.7F	-	-		0F		-	C, $dE_{\frac{1}{2}}/dpH$ for pH = 7.5-8.3, i_k, Tc=10, r	
		-			4.8F	-	-		0F		-	C, i_k, Tc=10, r	
		-0.83F			5.6F	-	-		0F		-	C, i_k, Tc=10, r	
		-0.83F			3.8F	-	-		60F		-	C, i_k, Tc=10, r	

CONT

FA26 (CONT.) $C_5H_9BrO_3S$

Code No.	Empirical Formula	Name and C.A. Number	Structural Formula	Solvent	Tech.	Medium		μ, M	pH	T, °C	Electrodes	App.	Experimental Parameters
FA26	$C_5H_9BrO_3S$	3-dimethylsulfonio-2-oxopropanoic acid bromide	$(CH_3)_2{}^+$-$CH_2COCOOH\ Br^-$	H_2O	PY	BR	-	1	9.40	25	DME//SCE	3--	C=1,m=6.4, t=3.0,h=55, EC=4.15
					QE	BR	-	-	3	-	Hg: nsns// SCE	25A	C=10, E_{app}= -0.70 V
									8				C=100, E_{app}= -1.20 V
FA27	$C_5H_{12}O$	2-methyl-2-butanol C.A. 75-85-4	$CH_3CH_2C(CH_3)_2OH$	H_2O	VA	H_2SO_4	0.5	-	-	-	Pt: xxns// SCE	35-O	0→1.55→0 V, C=55,v=238, A=1
FA28	$C_5H_{12}O$	1-pentanol C.A. 71-41-0	$CH_3(CH_2)_4OH$	H_2O	VA	H_2SO_4	0.5	-	-	-	Pt: xxns// SCE	35-O	0→1.55→0 V, C=50,v=238
FA29	$C_6Cl_5NO_2$	pentachloronitrobenzene C.A. 82-68-8	$C_6Cl_5NO_2$	MeOH ?%	DI	BR	-	0.12	2.0	-	DME//SCE	2AO	C=ns,m=1.32, t=4.6,h=76, t(c)=1, ΔE=100,Pt Aux
									3				
									4.8				
									6.4				
									7.0				
									11.5				
FA30	C_6F_6	hexafluorobenzene C.A. 392-56-3	C_6F_6	$HFSO_3$	VA	HOAc	0.1	-	-	20	Pt: xxwi// Pd/H_2	25-O	C=ns,A=0.6, v=100,Pt Aux
FA31 ai43	C_6N_4	tetracyanoethylene C.A. 670-54-2	$(NC)_2C:C(CN)_2$	MeCN	VA	$LiClO_4$	0.1	-	-	25	gw: xxdi// SCE	25AO	C=ns,0.9→-0.5→0.9 V, d=0.62,v=67, Pt Aux
FA32 cb35	C_6HF_5	pentafluorobenzene C.A. 363-72-4	C_6HF_5	$HFSO_3$	VA	HOAc	0.1	-	-	20	Pt: xxwi// Pd/H_2	25-O	C=ns,A=0.6, v=100,Pt Aux
FA33 cb39	$C_6H_2F_4$	1,2,3,5-tetrafluorobenzene C.A. 2367-82-0	Table II-3	$HFSO_3$	VA	HOAc	0.1	-	-	20	Pt: xxwi// Pd/H_2	25-O	C=ns,A=0.6, v=100,Pt Aux

TABLE I. Electrochemical Data

$C_6H_2F_4$ FA33

Ref.	C/M	Charact. Potential		Response	Const.	n Tech.	n	Electrokinetic Data			Products and Identification	Description and Remarks	Code No.
		Value	vs.		Value			Parameter	Value	From			
EA018 0691	421 c	$E_{\frac{1}{2}}$	0.87F	SCE	i_ℓ	0.8F	-	-	$dE_{\frac{1}{2}}/dpH$	60F	-	-	$C,i_k,Tc=10,r$	FA26
EA018 0691	421 c	-	-	-	-	-	QE	2	-	-	-	2-oxopropanoic acid, PY,(semicarbazone, PY,UVS)	C,r	
		-		-			2	-	-	-	2-oxopropanoic acid, PY,(semicarbazone, PY,UVS)	C,r		
EA020 0323	-	-	-	-	-	-	-	-	-	-	-	-	A,O,p	FA27
EA020 0323	-	E_p	-	SCE	i_p	-	-	-	-	-	-	-	A,Pr,p	FA28
			1.30F			2E3F							A,$i_p = k_1 + k_2 v$ for $v > 450$	
			0.70F			1E3F							C	
			0.54F			1E3F							A,on scan from 1.55 to 0 V	
AA096 0335	124 j	E_{su}	-0.03F	SCE	-	-	-	-	dE_{su}/dpH	68F	-	-	$C,i_{su}=kC$ for $C=10^{-5}$ to 10^{-2}, redn of $-NO_2$,r	FA29
			-0.51F							120F			C,redn of -NHOH	
			-0.09F							68F			$C,i_{su}=kC$ for $C = 10^{-5}$ to 10^{-2}, redn of $-NO_2$,r	
			-0.62F							120F			C,redn of -NHOH, O for pH > 3.5	
			-0.22F		-	-	-			94F		-	C,pK=5.0,redn of $-NO_2$,r	
	-		-0.37F		-	-	-			94F		-	C,pK_1'=7.0,redn of $-NO_2$,r	
			-0.42F		-	-	-			0F		-	C,redn of $-NO_2$,r	
	124 h		-0.42F		-	-	-		$\Delta E_{su/2}$	0F 53F		-	$C,W,i_a,i_{su}=kC$ for $C=10^{-5}$ to 10^{-2}, redn of $-NO_2$,r	
EA018 0331	-	$E_{p/2}$	1.32	Pd/H_2	i_p/C	59	-	1	ΔE_p	60	sttd	CP(in presence of SbF_5)→radical cation (stable),ESR	A,R,r	FA30
			-			-							C	
EA021 0973	50t	-	-	-	$i_{p,A}/i_{p,C}$	0.79	-	-	-	-	-		$C,i_{p,A}/i_{p,C} \to 0.57$ as $v \to 6.7, D=19.1$,r	FA31 ai43
			-			-							A	
EA018 0331	-	$E_{p/2}$	1.13	Pd/H_2	i_p/C	56	sttd	1	ΔE_p	60	sttd	radical cation (half-life = a few minutes)	A,R,r	FA32 cb35
			-			-							C	
EA018 0331	-	$E_{p/2}$	1.08	Pd/H_2	i_p/C	58	sttd	1	ΔE_p	65	sttd	radical cation (half-life = a few minutes)	A,R,r	FA33 cb39
			-			-							C	

FA34 $C_6H_2O_6$

Code No.	Empirical Formula	Name and C.A. Number	Structural Formula	Solvent	Tech.	Medium		μ, M	pH	T, °C	Electrodes	App.	Experimental Parameters
FA34	$C_6H_2O_6$	1,2-dihydroxy-1-cyclohexen-3,4,5,6-tetraone C.A. 63183-44-8	Table II	H_2O	PY	H_2SO_4	ns	-	-1.6	25	DME//SCE	2A0	C=1,m=2.92, t=2.7,h=50
									-0.7				
									0.4				
									2.0				
					MB	0.3		3.0				
									3.6				
									4.2				
									5				
									7.0				
									8.0				
					CARB	0.3		9.3				
					PR	H_2SO_4	ns	-	0.3	25	DME//SCE	2A2	0→-0.7→ 0.3→-0.7V, C=1,m=2.92, t=2.7,h=50, v=1E3
						BR	0.3		4.2				0→-0.9→ 0.1→-0.9V
									6.85				0→-0.6→ 0→0.6V
									8.3				0→-0.7→ 0→0.7V
FA35	$C_6H_3F_3$	1,3,5-trifluorobenzene C.A. 372-38-3	Table II	$HFSO_3$	VA	HOAc	0.1	-	-	20	Pt: xxwi// Pd/H_2	25-0	C=ns,A=0.6, v=100, Pt Aux

TABLE I. Electrochemical Data

$C_6H_3F_3$ FA35

Ref.	C/M	Charact. Potential		Response Const.		n Tech.	n	Electrokinetic Data			Products and Identification	Description and Remarks	Code No.
		Value	vs.		Value			Parameter	Value	From			
EA020 0951	423 a	$E_{\frac{1}{2}}$	-0.33F	SCE	i_ℓ	-	sttd	4	$dE_{\frac{1}{2}}/dpH$	0	sttd	-	C,r	FA34
			-0.33F			-		-		100		-	C, $dE_{\frac{1}{2}}/dpH$ for pH= -0.7 to 2.0, r	
			-0.44F			10.5F		4	Elog $dE_{\frac{1}{2}}/dpH$	125 100		hexahydroxybenzene	C,⇌,pKa_1=3.9,pKa_2=4.2,r	
			-0.86F			0.3F		-		80		-	C	
			-0.60F			8.6F	-	-		30		-	C,pK'=1.75,r	
			-0.98			1.7F				80			C	
	423 b		-0.10F			0.3F	sttd	2		25		-	C,⇌,r	
			-0.62F			7.8F		2		30			C	
			-1.08F			1.8F	-			80			C	
			-0.11F			0.8F	-	-		25		-	C,r	
			-0.65F			7.3F				30			C	
			-1.12F			1.2F				80			C	
			-0.12F			3.0F		⎫	Elog $dE_{\frac{1}{2}}/dpH$	62 90		QE(-0.05 V) → hexa-hydroxybenzene(n= 3.4) 80%, yield ↓ as C ↑	C,r	
			-0.68F			5.2F		⎬ 3		210		-	C	
			-			0.3F		⎭		-			C	
			-0.20F			8.8F	-	-		90		CP→dimer 79%, yield ↓ as pH ↑ or ↓	C,r	
			-0.88F			1.7F				210		-	C	
			-0.36			10.7F	sttd	4	Elog $dE_{\frac{1}{2}}/dpH$	30 90		-	C,i_ℓ=kh°,Tc=7,i_k,r,	
			-0.36F			5.5F		2	Elog $dE_{\frac{1}{2}}/dpH$	31 60			C,r	
			-0.50F			5.5F		2		113			C	
			-0.46F			5.5F		2		60			C,r	
			-0.64F			5.5F		2		113			C	
EA020 0951	423 a	E_p	-0.540	SCE	i_p	7.3F	-	-	-	-	-	-	C,r	
			0.140			7.3F							A	
			0.120			2.7							C,0 on first scan A	
	423 b		-0.115			8.8F	-	-	-	-	-	-	C,r	
			-0.780			3.2F							C	
			-0.085			11.2							A	
			-0.340			17F	-	-		-	-	-	C,r	
			-0.380			8F							C	
			-0.310			27F							A	
			-0.43F			13F	-	-		-	-	-	C,r	
			-0.57F			12F							C	
			-0.50F			13F							A	
			-0.390			11F							A	
EA018 0331	-	$E_{p/2}$	1.07	Pd/H_2	i_p/C	66	-	-	-	-	-	radical cation (half-life <1E-3s)	A,ece,r	FA35
			1.51			91							A	

FA36 $C_6H_4N_2O_4$

Code No.	Empirical Formula	Name and C.A. Number	Structural Formula	Solvent	Tech.	Medium	μ, M	pH	T, °C	Electrodes	App.	Experimental Parameters
FA36 af99 eb12	$C_6H_4N_2O_4$	1,2-dinitrobenzene C.A. 528-29-0	$2\text{-}O_2NC_6H_4NO_2$	MeCN	XT	Et_4NClO_4 0.1	-	-	-30 ±1	HMDE// Ag/ AgClO$_4$ 0.01	2-2	C=ns,v= (1.5-5.5)E3
									0±1			
									20± 0.5			
				DMF 98	XT	Et_4NClO_4 0.1	-	-	0	HMDE// Ag/AgI, Bu$_4$NI 0.1, Et$_4$NClO$_4$ 0.1	2-2	C=ns,v= (1.5-5.5)E3
				DMF	XT	Et_4NClO_4 0.1	-	-	0±1	HMDE// SCE	2-2	C=ns,v= (1.5-5.5)E3
									30± 0.5			
									50			
FA37 ag00 da84 eb13	$C_6H_4N_2O_4$	1,3-dinitrobenzene C.A. 99-65-0	$3\text{-}O_2NC_6H_4NO_2$	MeCN	XT	Et_4NClO_4 0.1	-	-	-30 ±1	HMDE// Ag/ AgClO$_4$ 0.1	2-2	C=ns,v= (1.5-5.5)E3
									0±1			
									20± 0.5			
				DMF 98	XT	Et_4NClO_4 0.1	-	-	0	HMDE// Ag/AgI, Bu$_4$NI 0.1, Et$_4$NClO$_4$ 0.1	2-2	C=ns,v= (1.5-5.5)E3
CONT												

TABLE I. Electrochemical Data   35

$C_6H_4N_2O_4$ (CONT.) FA37

| Ref. | C/M | Charact. Potential | | Response Const. | n | Electrokinetic Data | | | Products and | Description and | Code |
		Value	vs.	Value	Tech.	Parameter	Value	From	Identification	Remarks	No.
JE047 0115	-	XE	0.166 ±0.002	Ag/ AgClO$_4$ 0.01	-	-	-	-	-	-	-	-	C,XT=semiintegral chronoamperometry with linear potential sweep,XE= difference between potentials for the two steps,p C	FA36 af99 eb12
			0.229 ±0.002			-		-		-	-	-	C,p C
			0.257 ±0.002			-		-		-	-	-	C,p C
JE057 0027	-	XE	0.225	Ag/AgI, Bu$_4$NI 0.1, Et$_4$NClO$_4$ 0.1	-	-	-	-	-	-	-	-	C,XT=semiintegral chronoamperometry with linear potential sweep,XE= difference between potentials for the two steps,see reference for dependence of XE on [Li$^+$],[H$_2$O], and T, p C	
JE047 0115	-	XE	0.346 ±0.002	SCE	-	-	-	-	-	-	-	-	C,XT=semiintegral chronoamperometry with linear potential sweep,XE= difference between potentials for the two steps,p C	
			0.369 ±0.002			-		-		-	-	-	C,p C
			0.389 ±0.002			-		-		-	-	-	C,p C
JE047 0115	-	XE	0.281 ±0.002	Ag/ AgClO$_4$ 0.1	-	-	-	-	-	-	-	-	C,XT=semiintegral chronoamperometry with linear potential sweep,XE= difference between potentials for the two steps,p C	FA37 ag00 da84 eb13
			0.316 ±0.002			-		-		-	-	-	C,p C
			0.342 ±0.002			-		-		-	-	-	C,p C
JE057 0027	-	XE	0.338	Ag/AgI, Bu$_4$NI 0.1, Et$_4$NClO$_4$ 0.1	-	-	-	-	-	-	-	-	C,XT=semiintegral chronoamperometry with linear potential sweep,XE= difference between potentials for the two steps,see reference for dependence of XE on [Li$^+$],[H$_2$O], and T, p C	
														CONT

FA37 (CONT.) $C_6H_4N_2O_4$

Code No.	Empirical Formula	Name and C.A. Number	Structural Formula	Solvent	Tech.	Medium	μ, M	pH	T, °C	Electrodes	App.	Experimental Parameters
FA37 ag00 da84 eb13	$C_6H_4N_2O_4$	1,3-dinitrobenzene	$3\text{-}O_2NC_6H_4NO_2$	DMF	XT	Et_4NClO_4 0.1	-	-	0±1	HMDE// SCE	2-2	C=ns, v= (1.5-5.5)E3
									30± 0.5			
									50± 0.5			
FA38 ag01 eb14	$C_6H_4N_2O_4$	1,4-dinitrobenzene C.A. 100-25-4	$4\text{-}O_2NC_6H_4NO_2$	MeCN	XT	Et_4NClO_4 0.1	-	-	-30 ±1	HMDE// Ag/ AgClO_4 0.01	2-2	C=ns, v= (1.5-5.5)E3
									0±1			
									20± 0.5			
				DMF 98	XT	Et_4NClO_4 0.1	-	-	0	HMDE// Ag/AgI, Bu_4NI 0.1, Et_4NClO_4 0.1	2-2	C=ns, v= (1.5-5.5)E3
				DMF	XT	Et_4NClO_4 0.1	-	-	0±1	HMDE// SCE	2-2	C=ns, v= (1.5-5.5)E3
									20± 0.5			
									60± 0.5			
FA39 ag08 cb70 da86 eb22 CONT	$C_6H_4O_2$	1,4-benzoquinone C.A. 106-51-4	Table II	MeCN	VR	$NaClO_4$ 0.1 $HClO_4$ 4E-3	-	-	-	Pt: xxbu// SCE(?)	05A-	1.3→-0.6→ 1.3→-0.6 V, C=2, v=156
						$NaClO_4$ 0.1 $HClO_4$ 4E-3 2,6-lutidine 1.6E-3						
						$NaClO_4$ 0.1 $HClO_4$ 4E-3 2,6-lutidine 3.2E-3						

TABLE I. Electrochemical Data

$C_6H_4O_2$ (CONT.) FA39

| Ref. | C/M | Charact. Potential | | Response Const. | | n | | Electrokinetic Data | | | Products and | Description and | Code |
		Value	vs.		Value	Tech.		Parameter	Value	From	Identification	Remarks	No.
JE047 0115	-	XE	0.408 ±0.002	SCE	-	-	-	-	-	-	-	-	C,XT=semiintegral chronoamperometry with linear potential sweep,XE= difference between potentials for the two steps,p C	FA37 ag00 da84 eb13
			0.434 ±0.002 -		-	-	-	-	-	-	-	-	C,p C	
			0.451 ±0.002 -		-	-	-	-	-	-	-	-	C,p C	
JE047 0115	-	XE	0.140 ±0.002 -	Ag/ AgClO$_4$ 0.01	-	-	-	-	-	-	-	-	C,XT=semiintegral chronoamperometry with linear potential sweep,XE= difference between potentials for the two steps,p C	FA38 ag01 eb14
			0.186 ±0.002 -		-	-	-	-	-	-	-	-	C,p C	
			0.212 ±0.002 -		-	-	-	-	-	-	-	-	C,p C	
JE057 0027	-	XE	0.202 -	Ag/AgI, Bu$_4$NI 0.1, Et$_4$NClO$_4$ 0.1	-	-	-	-	-	-	-	-	C,XT=semiintegral chronoamperometry with linear potential sweep,XE= difference between potentials for the two steps,see reference for dependence of XE on [Li$^+$],[H$_2$O], and T, p C	
JE047 0115	-	XE	0.306 ±0.002 -	SCE	-	-	-	-	-	-	-	-	C,XT=semiintegral chronoamperometry with linear potential sweep,XE= difference between potentials for the two steps,p C	
			0.321 ±0.002 -		-	-	-	-	-	-	-	-	C,p C	
			0.355 ±0.002 -		-	-	-	-	-	-	-	-	C,p C	
EA018 0519	119 j	E$_p$	1.06F	SCE	i$_p$(u)	9F	-	-	-	-	-	-	A,p	FA39 ag08 cb70 da86 eb22
			0.24F			15F							C	
			0.00F			2F							C	
			1.05F			9F							A,p	
			0.29F			13F							C	
			-0.30F			3F							C	
			0.74F			3F	-	-	-	-	-	-	A,p	
			1.03F			3F							A	
			0.31F			6F							C	
			-0.30F			12F							C	
														CONT

FA39 (CONT.) $C_6H_4O_2$

Code No.	Empirical Formula	Name and C.A. Number	Structural Formula	Solvent	Tech.	Medium		μ, M	pH	T, °C	Electrodes	App.	Experimental Parameters
FA39 ag08 cb70 da86 eb22	$C_6H_4O_2$	1,4-benzoquinone	Table II	MeCN	VR	$NaClO_4$ $HClO_4$ 2,6-lutidine	0.1 4E-3 4.0E-3	–	–	–	Pt: xxbu// SCE(?)	05A-	1.3→-0.6→ 1.3→0.6V, C=2,v=156
FA40 ag12	$C_6H_4O_6$	tetrahydroxy-1,4-benzoquinone C.A. 319-89-1	Table II-2	H_2O	PY	H_2SO_4	ns	–	-0.5 0.4	25	DME//SCE	2AO	C=0.25,m= 2.92,t=2.7, h=50
						MB	0.3		2.0 5.0 6.6				
						CARB	0.3		9.00 11.2				
FA41 ag22	C_6H_5Cl	chlorobenzene C.A. 108-90-7	C_6H_5Cl	HF	VY	KF	0.1	–	–	0	Pt: rodi// Cu/CuF_2, KF 0.2	2-0	C=ns,A=5E-3, v=100,ω=1, Pt Aux
FA42 ag54 da94 eb33	$C_6H_5NO_3$	4-nitrophenol C.A. 100-02-7	$4-HOC_6H_4NO_2$	MeOH ?%	DI	BR	–	0.12	2.2 4.0 4.9 6.6 7.4 9.0 10.0	–	DME//SCE	2AO	C=ns,m=1.32, t=4.6,h=76, ΔE=100,t(c)= 1,Pt Aux
FA43 ag66 da96	C_6H_6	benzene C.A. 71-43-2	C_6H_6	HF	VY	KF	0.1	–	–	0	Pt: rodi// Cu/CuF_2, KF 0.2	2-0	C=ns,A=5E-3, v=100,ω=1, Pt Aux
FA44 ag85 cb79 db04 eb39 CONT	$C_6H_6N_2O$	3-carbamoylpyridine C.A. 98-92-0	Table II-2	MeCN	PY	Et_4NClO_4	0.1	–	–	–	DME//SCE //Rb+	35AFO	m=1.0,t=4.0, h=70

TABLE I. Electrochemical Data

$C_6H_6N_2O$ (CONT.) FA44

Ref.	C/M	Charact. Potential Value	vs.	Response Const. Value		n	Tech.	Electrokinetic Data Parameter	Value	From	Products and Identification	Description and Remarks	Code No.
EA018 0519	119 j	E_p 0.74F	SCE	$i_p(u)$	5F	-	-	-	-	-	-	A,p	FA39 ag08 cb70 da86 eb22
		1.00F			1F							A	
		0.35F			2.5F							C	
		-0.29F			25F							C	
EA020 0951	423 a	$E_{\frac{1}{2}}$ 0.10F	SCE	-	-	-	-	$dE_{\frac{1}{2}}/dpH$	0	sttd	hexahydroxybenzene	C,r	FA40 ag12
		0.10F		-	-	-	-		60		-	C,r	
		-0.41F							113			C	
		0.18F		$i_\ell(u)$	1		2		115		rhodizonic acid	A,r	
		0.0F			}1		}2		60		hexahydroxybenzene	C	
		-0.59F							30		-	C	
	423 b	-0.15F		-	-	-	-		85		-	A,r	
		-0.19F							90			C	
		-0.68							30			C,O for pH > 5	
		-0.30F		-	-	-	-		60		-	A,r	
		-0.33F							115			C	
		-0.45F		-	-	-	-		60		-	A,r	
		-0.61F							90			C	
		-0.59F		-	-	-	-		60		-	A,r	
		-0.81F							90			C	
JE054 0232	-	$E_{\frac{1}{2}}$ 1.342	NHE	-	-	-	-	Elog	51	sttd	-	A,p	FA41 ag22
AA096 0335	123 e	E_{su} -0.22F	SCE	-	-	sttd	6	dE_{su}/dpH	67F	-	-	C,redn of $-NO_2$,r	FA42 ag22 da94 eb33
		-0.34		-	-		6		67F		-	C,pK_1=4.6,redn of $-NO_2$,r	
	123 i	-0.42F		-	-		4		100F		-	C,redn of $-NO_2$,r	
		-0.59F		-	-		4		100F		-	C,pK_2=7.0,redn of $-NO_2$,r	
	-	-0.63F		-	-		-		138F		-	C,redn of $-NO_2$,r	
		-0.85F		-	-		-		138F		-	C,redn of $-NO_2$,r	
		-0.90F		-	-		-		0F		-	C,i_{su}=kC for C=10^{-5} to 10^{-2},redn of $-NO_2$,r	
JE054 0232	-	$E_{\frac{1}{2}}$ 1.775	NHE	-	-	-	-	-	-	-	-	A,p	FA43 ag66 da96
JA095 5482	277 f	$E_{\frac{1}{2}}$ -2.0	SCE	-	-	Tomeš	1	Tomeš	46	sttd	radical anion	C,i_d,redn of pyridine ring,r	FA44 ag85 cb79 db04 eb39
		-2.45					1		-			C,i_d,redn of pyridine ring	

CONT

FA44 (CONT.) $C_6H_6N_2O$

Code No.	Empirical Formula	Name and C.A. Number	Structural Formula	Solvent	Tech.	Medium	μ, M	pH	T, °C	Electrodes	App.	Experimental Parameters
FA44 ag85 cb79 db04 eb39	$C_6H_6N_2O$	3-carbamoylpyridine	Table II-2	DMSO	PY	Et_4NClO_4 0.1	-	-	-	DME//SCE //Rb+	35AF0	C=1.01, m=1.02, t=5.0, h=70
												C=2.1
												C=1.01
					VA	Et_4NClO_4 0.1 HQ 8.2E-3	-	-	-	HMDE// SCE//Rb+	35AF0	C=ns, 0→-3? →0V, v-100
												0→-2.2→0V, v=480
												v=1E3
						Et_4NClO_4 0.1 HQ 1E-3						0→-2.5→0V, C=1.01, v=100
						Et_4NClO_4 0.1 HQ 8.2E-3						
FA45 ah06 cb85 db07 eb47	$C_6H_6O_2$	hydroquinone C.A. 123-31-9	$4-HOC_6H_4OH$	MeCN	VA	$NaClO_4$ 0.1	-	-	-	Pt: xxbu// SCE(?)	05A-	0→1.2→ -0.8→0.7V, C=2.0, v=156
						$NaClO_4$ 0.1 2,6-lutidine 4E-4						
						$NaClO_4$ 0.1 2,6-lutidine 1.2E-3						
						$NaClO_4$ 0.1 2,6-lutidine 2.0E-3						
					VY	$NaClO_4$ 0.2	-	-	-	Pt: rodi// SCE(?)	05A-	C=1.0, ω=25-500
						$NaClO_4$ 0.2 $HClO_4$ 1E-2						
						$NaClO_4$ 0.2 2,4-lutidine 1E-2						

TABLE I. Electrochemical Data

$C_6H_6O_2$ FA45

Ref.	C/M	Charact. Potential		Response Const.		n Tech.	n	Electrokinetic Data			Products and Identification	Description and Remarks	Code No.
		Value	vs.		Value			Parameter	Value	From			
JA095 5482	277 f	$E_{\frac{1}{2}}$	-1.94F	SCE	i_ℓ	1.0F	QE	1	-	-	-	6,6'-dimer of reduced nicotinamide, PY,UVS,addition of 3mM HQ destroys product	C,R,i_d,r	FA44 ag85 cb79 db04 eb39
			-2.01		I	1.23	Tomeš	1	Tomeš	48-50	sttd	radical anion	$C,R,i_\ell \propto h^{\frac{1}{2}},i_d,$ redn of pyridine ring,r	
			-2.50			-		1	-				$C,i_\ell \propto h^{\frac{1}{2}},i_d,$ redn of pyridine ring	
			-1.89F		i_ℓ	2.0F	-	-	-	-	-	-	C,R,r	
JA095 5482	277 g	E_p	-2.10	SCE	$i_p/v^{\frac{1}{2}}$	23.7	-	-	-	-	-	-	C,⊬,r	
			-2.60			-							C,⊬	
			-0.58			2.37							A,⊬	
			-2.11			23.2	-	-	-	-	-	-	C,r	
			-0.58			5.8							A	
			-2.11			22.0							C,r	
			-0.58			8.0							A	
			-1.97F		i_p	5.4F	-	-	-	-	-	-	C,r	
			-						$dE_{\frac{1}{2}}/d[HQ]$	+	sttd		C, new wave E=more neg than -2.5 V,0 if $[H^+]$ < 5E-13,⊬	
			-0.17			-							A	
			-1.94F			5.7F	-	-	-	-	-	-	C,⊬,r	
			-			-							C	
			-0.14F			-							A	
EA018 0519	119 j	E_p	1.06F	SCE	$i_p(u)$	27F	-	-	$E_p-E_{p/2}$	105	sttd	-	A,⊬,r	FA45 ah06 cb85 db07 eb47
			0.24F			5F							C	
			0.83F			3F	-	-	-	-	-	-	A,r	
			1.05F			24F							A	
			0.22F			5F							C	
			-0.30F			2F							C	
			0.61F			9F	-	-	-	-	-	-	A,r	
			1.03F			17F							A	
			0.27F			2F							C	
			-0.33F			4F							C	
			0.67			12.5F	-	-	-	-	-	-	A,r	
			1.05F			8F							A	
			0.36F			0.5F							C	
			-0.32F			7F							C	
EA018 0519	119 j	-	-	-	$i_\ell/C\omega^{\frac{1}{2}}$	53.6± 0.3	-	-	-	-	-	-	A,r	
			-	-	-	53.7± 0.3	-	-	-	-	-	-	A,r	
			-	-	-	54.0± 0.8	-	-	-	-	-	-	A,r	

FA46 $C_6H_8N_2O$

Code No.	Empirical Formula	Name and C.A. Number	Structural Formula	Solvent	Tech.	Medium		μ, M	pH	T, °C	Electrodes	App.	Experimental Parameters
FA46	$C_6H_8N_2O$	2,4-diaminophenol C.A. 95-86-3	Table II	H_2O	VA	H_2SO_4	2.5	-	-	25	Cp: nsns// SCE	2-O	C=1, A=0.18, v=13, Pt Aux
						H_2SO_4	1						
						H_2SO_4	0.5						
FA47 db27	$C_6H_{10}O$	3-methyl-3-penten-2-one C.A. 565-62-8	$CH_3-CH:C(CH_3)-COCH_3$	MeCN 99.95	PY	Et_4NClO_4	0.1	-	-	25	DME//SCE	12A0	C=2.24, Pt Aux
				MeCN	PW	Et_4NClO_4	0.1	-	-	25	DME// Ag/ $AgClO_4$ 0.01	12A2	C=0.4, Pt Aux
													C=1.33
FA48 ah74 eb71	$C_6H_{10}O$	4-methyl-3-penten-2-one C.A. 141-79-7	$(CH_3)_2C:CHCOCH_3$	MeCN 99.95	PW	Et_4NClO_4	0.1	-	-	25	DME// Ag/ $AgClO_4$ 0.01	12A2	C=0.28, Pt Aux
													C=2.8
				DMF 99	PW	Et_4NClO_4	0.1	-	-	25	DME//SCE	12A2	C=3.9, Pt Aux
				DMF 99.9	PY	Et_4NClO_4	0.1	-	-	25	DME//SCE	12A0	C=2.2, Pt Aux
						Et_4NClO_4 PHEN	0.1 0.01						
						Et_4NClO_4 PHEN	0.1 0.05						
					PW	Et_4NClO_4	0.1	-	-	25	DME//SCE	12A2	C=3.9

TABLE I. Electrochemical Data

$C_6H_{10}O$ FA48

Ref.	C/M	Charact. Potential		Response Const.		n		Electrokinetic Data			Products and Identification	Description and Remarks	Code No.
		Value	vs.		Value	Tech.		Parameter	Value	From			
JE043 0397	–	E_p	0.635	SCE	i_p	21.7	QE	2.05	–	–	–	–	A,r	FA46
			0.27			–		–					c	
			0.630			23.8		2.12	–	–	–	–	A,r	
			0.22			–		–					c	
			0.586			23.5		2.09	–	–	–	–	A,r	
			0.23			–		–					c	
JE042 0189	50s	$E_{\frac{1}{2}}$	-2.08	SCE	–	–	i:i	1	–	–	–	–	C,data for first wave only,r	FA47 db27
			–										c(?)	
JE042 0189	50s	–	–	–	–	–	–	$dE_p/dlogv$	20	for v < 3E5	–	C,r	
									0	for v > 3E5			
			–						–			c(?)	
		–	–	–	–	–	–		20	for v < 1E6	–	C	
									0	for v > 1E6			
								$dE_p/dlogC$	17	sttd			
			–						–			c?	
JE042 0189	50s	–	–	–	–	–	–	$dE_p/dlogv$	19	for v < 2.5 E3	–	C,r	FA48 ah74 eb71
									0	for v > 2.5E3			
			–						–			c(?)	
		–	–	–	–	–	–		20	for v < 1.8E4	–	C,r	
									0	for v > 1.8E4			
								$dE_p/dlogC$	17	sttd			
			–						–			c(?)	
JE042 0189	50r	–	–	–	–	–	–	$dE_p/dlogv$	20	for v < 3.8E4	–	C,data for first wave only,r	
									0	for v > 3.8E4			
			–						–			c(?)	
JE042 0189	50r	$E_{\frac{1}{2}}$	-2.04F	SCE	i_l	1.4F	i:i	1	–	–	–	C,data for first wave only,r	
			–			–							c(?)	
			-1.9F			1.5F		–	–	–	–	C,r	
			-2.2F			1.4F							c	
			–			–							c(?)	
			-1.82F			1.2F		–	–	–	–	C,r	
			-2.16F			1.5F							c	
			–			–							c(?)	
JE042 0189		–	–	–	–	–	–	$dE_p/dlogC$	21	for v > 3.6E4	–	C,r	
									0	for v < 3.6E4			
			–						–			c(?)	

FA49 $C_6H_{10}O_2$

Code No.	Empirical Formula	Name and C.A. Number	Structural Formula	Solvent	Tech.	Medium		μ, M	pH	T, °C	Electrodes	App.	Experimental Parameters
FA49 ah77 cc12	$C_6H_{10}O_2$	ethyl-<u>trans</u>-2-butenoate C.A. 623-70-1	$CH_3CH:CHCOO-C_2H_5$	MeCN 99.95	PW	Et_4NClO_4	0.1	-	-	25	DME// Ag/ $AgClO_4$ 0.01	12A2	C=0.21-2.28, Pt Aux
				DMF 99.9	PY	Et_4NClO_4	0.1	-	-	25	DME//SCE	12A0	C=3.7, Pt Aux
						Et_4NClO_4 PHEN	0.1 0.01						
					PW	Et_4NClO_4	0.1	-	-	25	DME//SCE	12A2	C=4.4, Pt Aux
FA50	$C_6H_{14}O$	1-hexanol C.A. 111-27-3	$CH_3(CH_2)_5OH$	H_2O	VA	H_2SO_4	0.5	-	-	-	Pt: xxns// SCE	35-0	$0 \to 1.55 \to 0V$, C=50,v=238, A=1
FA51	$C_6H_{15}ClSn$	chloro(triethyl)-stannane C.A. 994-31-0	$(C_2H_5)SnCl$	MeOH 40 C_6H_6 60	PY	LiCl Tween 80	0.4 ns	-	-	-	DME// Ag/AgCl	---	C=ns
FA52	$C_7H_3F_5$	methylpentafluoro-benzene C.A. 771-56-2	$CH_3C_6F_5$	$HFSO_3$	VA	HOAc	0.1	-	-	-40	Pt: xxwi// Pd/H_2	25-0	$0.75 \to 1.50 \to 0.75V$, C=ns, A=0.6, v=100, Pt Aux
										-9			
										20			
FA53 aj81 db68	$C_7H_6O_2$	benzoic acid C.A. 65-85-0	C_6H_5COOH	H_2O	CP	PHOS KCl	ns ns	-	6.2	20±2	Pb: rons// SCE	---	C=ns, E_{app} = -1.95 V
					IL	KCl HCl	0.1 -	-	2.48	20±2	Pb: xxwi// SCE	3-02	C=8, v=220
									2.66				
									2.86				
									3.22				

TABLE I. Electrochemical Data

$C_7H_6O_2$ FA53

Ref.	C/M	Charact. Potential		Response Const.		n Tech.	n	Electrokinetic Data			Products and Identification	Description and Remarks	Code No.
		Value	vs.		Value			Parameter	Value	From			
JE042 0189	50s	-	-	-	-	-	-	$dE_p/dlogv$	22	for $v<$ 5E5	-	C, data for first wave only, r	FA49 ah77 cc12
									0	for $v>$ 5E5			
								$dE_p/dlogC$	23	sttd			
		-			-				-			C?	
JE042 0189	50s	$E_{\frac{1}{2}}$	-2.26F	SCE	i_ℓ	11F	i:i	1	-	-	-	-	C, data for first wave only, r
		-			-		-					C(?)	
		-2.11F			12F		-		-	-	-		C, r
		-2.33F			8.7F							C	
JE042 0189	50s	-	-	-	-	-	-	$dE_p/dlogv$	19	for $v<$ 4E5	-	C, r	
									0	for $v>$ 4E5			
		-							-			C(?)	
EA020 0323	-	E_p	1.45F	SCE	i_p	3.4E3F	-	-	-	-	-	A, $i_p = k_1 + k_2 v$ for $v > 450$, p	FA50
			0.63F			6.6E2F						A, on scan from 1.55 to 0 V	
EA021 0395	240 c	$E_{\frac{1}{2}}$	-1.100	Ag/AgCl	$i_\ell(u)$	1	-	-	-	-	CP → hexaethyldi-stannane	C, i_d, p	FA51
			-1.650			1		α	0.4	sttd	-	C, ≠, i_d	
EA018 0331	-	$E_{p/2}$	1.05F	-	i_p	50	sttd	1	-	-	-	-	A, R, r	FA52
			1.09F			50							C, R	
		E_p	0.97F	Pd/H_2		80F	-	-	-	-	-	A, R, r	
			1.28F			11F						A, R	
			1.34F			28F						C, R	
			1.03F			39F						C, R	
		$E_{p/2}$	0.97	-	i_p/C	52		1	ΔE_p	60	sttd for $v >$ 100	radical cation (half-life 0.1 sAp)	A, R, r
			1.28			15		1		70	sttd	-	A
			-			-		-		-			C
			-			-		-		-			C
EA019 0049	439 a	-	-	-	-	-	-	-	-	-	-	no aldehyde or alcohol	C, p	FA53 aj81 db68
EA019 0049	439 a	E_p	-1.59F	SCE	i_p	2.8F	-	-	-	-	-	-	C, p
			-1.68F			-							C
			-1.60F			-							C, Pr, p
			-1.69F			2.3F							C
			-1.69F			1.8F	-	-	-	-	-		C, p
			-1.69F			1.5F	-	-	-	-	-		C, E_p → more neg as C ↑ and v ↑, p

FA54 C_7H_7Br

Code No.	Empirical Formula	Name and C.A. Number	Structural Formula	Solvent	Tech.	Medium	μ, M	pH	T, °C	Electrodes	App.	Experimental Parameters
FA54 aj95	C_7H_7Br	benzyl bromide C.A. 100-39-0	$C_6H_5CH_2Br$	MeCN	CP	ns	-	-	25	Al: nsns// Ag/Ag$^+$ 0.01, Et$_4$NBr 0.2	-	-
FA55 ak45	C_7H_8	toluene C.A. 108-88-3	$C_6H_5CH_3$	HF	VY	KF 0.1	-	-	0	Pt: rodi// Cu/CuF$_2$, KF 0.2	2-O	C=1,A=0.07, v=100,ω=1, PE,Pt Aux
FA56 ec29	C_7H_8O	benzyl alcohol C.A. 100-51-6	$C_6H_5CH_2OH$	DMF	PY	Bu$_4$NI 0.1	-	-	-	DME// SCE(o)	2-O	C=2
FA57 ak74	$C_7H_9ClN_2O$	1-methylpyridinium-3-carbamoyl chloride C.A. 1005-24-9	Table II	MeCN	PY	Et$_4$NClO$_4$ 0.1	-	-	-	DME// SCE//Rb$^+$	35AFO	C=0.54,m=1.0, t=4.0,h=70
						Et$_4$NClO$_4$ 0.1 BENZ 7.9E-4						
						Et$_4$NClO$_4$ 0.1 BENZ 2.74E-3						
						Et$_4$NClO$_4$ 0.1 BENZ 5.42E-3						
					PV	Et$_4$NClO$_4$ 0.1	-	-	-	DME// SCE//Rb$^+$	35AFO	C=ns,50 Hz, Δe=10
						Et$_4$NClO$_4$ 0.1 BENZ 2.5E-3						C=1
					VA	Et$_4$NClO$_4$ 0.1	-	-	-	glC: xxns// SCE//Rb$^+$	35AFO	0.35 → -1.3 → 0.35 V, C=0.39,v= 100-2.5E-3
				DMSO	PY	Et$_4$NClO$_4$ 0.1	-	-	-	DME// SCE//Rb$^+$	35AFO	C=ns,m=1.02, t=5.0,h=70
					VA	Et$_4$NClO$_4$ 0.1	-	-	-	HMDE// SCE//Rb$^+$	35AFO	C=ns,v=100
												v=200
												v=1E3
												v=2.5E3
						Et$_4$NClO$_4$ 0.1 BENZ 3E-3						C=1

$C_7H_9ClN_2O$ FA57

Ref.	C/M	Charact. Potential		Response	Const. Value	n Tech.		Electrokinetic Data			Products and Identification	Description and Remarks	Code No.
		Value	vs.					Parameter	Value	From			
JE035 0013	-	-	-	-	-	-	-	-	-	-	-	see also AJ95	FA54 aj95
JE054 0232	-	$E_{\frac{1}{2}}$	1.175	NHE	i_ℓ	236.6F	-	-	Elog	85	-	-	A,W,p	FA55 ak45
EA019 0629	-	-	-	-	-	-	-	-	-	-	-	-	O,p	FA56 ec29
JA095 5482	42e	$E_{\frac{1}{2}}$	-0.20	SCE	-	-	-	-	-	-	-	-	A,oxidn of Hg	FA57 ak74
			-1.04		I	3.3			Tomeš	46	sttd	neutral radical;QE in MeCN or DMSO, n=1 → 6,6-dimer of reduced compound, yellow,UVS	$C,R,i_\ell \propto h^{\frac{1}{2}},i_d$, redn of pyridine ring,r	
	42f		-1.04			2.7	-	-	-	-	-	-	C,r	
			-1.79			0.8							C,≠	
			-1.04			2.8	-	-	-	-	-	-	C,r	
			-1.79			1.7							C,≠	
			-1.04			2.6	-	-	-	-	-	-	C,r	
			-1.79			2.8							C,≠	
JA095 5482	42e	E_{su}	-1.17	SCE	-	-	sttd	1	$\Delta E_{su/2}$	00	sttd	-	C,W,R,r	
	42f		-1.17		$i_{su}(u)$	1	-	-	-	-	-	-	C,W,r	
			-2.15			1							C	
JA095 5482	42e	E_p	-1.08	SCE	-	-	-	-	-	-	-	-	$C,i_p/v^{\frac{1}{2}} \neq f(v)$,r	
			-0.05										A	
JA095 5482	42e	$E_{\frac{1}{2}}$	-0.18	SCE	-	-	-	-	-	-	-	-	A,oxidn of Hg	
			-1.01		I	1.4			Tomeš	45	sttd	neutral radical	$C,R,i_\ell \propto h^{\frac{1}{2}},i_d$, redn of pyridine ring,r	
JA095 5482	42e	E_p	-0.29	SCE	$i_p/v^{\frac{1}{2}}$	3.5	-	-	-	-	-	-	A,≠,oxid of prod of first redn,r	
			-1.09			14.2	-	-	$E_p-E_{p/2}$	45	sttd		C,≠	
			-0.29			4.4							A,r	
			-1.09			12.9							C	
			-0.29			7.5	-	-	-	-	-	-	A,r	
			-1.10			15.0							C	
			-1.10			15.2	-	-	-	-	-	-	C, R for v > 1E4,r	
	43f		-			-	-	-	$E_p-E_{p/2}$	45	sttd	-	C,r	
			-			-				56-68			C	

FA58 C7H9NO

Code No.	Empirical Formula	Name and C.A. Number	Structural Formula	Solvent	Tech.	Medium		μ, M	pH	T, °C	Electrodes	App.	Experimental Parameters
FA58 ak87 cc82	C7H9NO	2-anisidine C.A. 90-04-0	2-CH$_3$OC$_6$H$_4$NH$_2$	H$_2$O	VY	H$_2$SO$_4$	0.1	-	-	-	g:xftu// SCE	--0	C=0.1, ℓ=1.2, r$_i$=0.1, V$_f$=0.17
FA59 ak88	C7H9NO	3-anisidine C.A. 536-90-3	3-CH$_3$OC$_6$H$_4$NH$_2$	H$_2$O	VY	H$_2$SO$_4$	0.1	-	-	-	g:xftu// SCE	--0	C=0.1, ℓ=1.2, r$_i$=0.1, V$_f$=0.17
FA60 ak89	C7H9NO	4-anisidine C.A. 104-94-9	4-CH$_3$OC$_6$H$_4$NH$_2$	H$_2$O	VY	H$_2$SO$_4$	0.1	-	-	-	g:xftu// SCE	--0	C=0.1, ℓ=1.2, r$_i$=0.1, V$_f$=0.17
FA61	C7H12O3S	ethyl 3-ethylthio-2-oxopropanoate C.A. 51033-96-6	CH$_3$CH$_2$SCH$_2$CO-COOCH$_2$CH$_3$	EtOH 3	PY	H$_2$SO$_4$	-	1	0.5	25	DME//SCE	3--	C=1, m=6.4, t=3.0, h=55, EC=4.15
						BR	-	1	2.1				
									5.00				
									7.35				
									9.1				
FA62	C7H13BrO3S	ethyl 3-dimethylsulfonio-2-oxopropanoate bromide C.A. 51033-98-8	(CH$_3$)$_2$$\overset{+}{S}CH_2$CO-COOCH$_2CH_3$ Br$^-$	H$_2$O	PY	H$_2$SO$_4$	-	1	0	25	DME//SCE	3--	C=1, m=6.4, t=3.0, h=55, EC=4.15
						BR	-	1	2.05				
									4				
									5.95				
									7.95				
FA63	C8H6N2	1,5-naphthyridine C.A. 254-79-5	Table II	H$_2$O	PA or VA	HClO$_4$	ns	-	-2	25	HMDE or glC: nsns// SCE	2AO	C=ns, v=1(?), Pt Aux
													v=13.6(?)
									1.5				C=0.5-2, v=2.5E3, Pt Aux
													v=4.2E3, Pt Aux
													v=1.1E4, Pt Aux

$C_8H_6N_2$ FA63

Ref.	C/M	Charact. Value	Potential vs.	Response	Const. Value	Tech.	n	Electrokinetic Data Parameter	Value	From	Products and Identification	Description and Remarks	Code No.
EA021 1085	–	$E_{\frac{1}{2}}$	0.740	SCE	i_ℓ	19.6	–	–	–	–	–	–	A,p	FA58 ak87 cc82
EA021 1085	–	$E_{\frac{1}{2}}$	0.860	SCE	i_ℓ	20.6	–	–	–	–	–	A,p	FA59 ak88
EA021 1085	–	$E_{\frac{1}{2}}$	0.672	SCE	i_ℓ	10.8	–	–	–	–	–	A,p	FA60 ak89
			0.938			10.0							A	
EA018 0691	65c k	$E_{\frac{1}{2}}$	−0.35F	SCE	i_ℓ	2.0F	sttd (i:i_d?)	0.4F	$dE_{\frac{1}{2}}/dpH$	65	sttd	–	C,i_k,Tc=8,r	FA61
			−0.50F			0.7F		0.1F		65		–	C,i_k,Tc=8,r	
			−0.67F			0.7F		0.1F		65		–	C,i_k,Tc=8,r	
			−0.80F			0.7F		0.1F		65		–	C,i_k,Tc=8,r	
			−0.90F			–		–		65		–	C,r	
EA018 0691	283 ac	$E_{\frac{1}{2}}$	−0.33F	SCE	–	–	–	–	$dE_{\frac{1}{2}}/dpH$	0F	–	–	C,i_k,Tc=10,r	FA62
			−0.35			0.5F	sttd (i:i_d?)	0.1F		0F		–	C,i_k,Tc=10,r	
			−0.35F			–		–		55F		–	C,i_k,Tc=10,r	
			−0.46F			2.0F		0.4F		55F		–	C,i_k,Tc=10,r	
			−1.2F			1.3F		0.3F		–		–	C	
			−0.50F			0.3F		0.1F		55F		–	C,i_k,Tc=10,r	
			−0.90F			0.95F		0.2F		–		–	C	
EA021 0421	–	E_p	−0.29F	SCE	–	–	–	–	$dE_p/d\log v$	59	sttd	–	C,p	FA63
			0.36F		–	–	–	–		–		–	C,p	
		–	–	–	$i_{p,A}/i_{p,C}$	0.20±0.02F	–	–	$dE_{\frac{1}{2}}/dpH$	59	sttd	–	A,$dE_{\frac{1}{2}}/dpH$=120 for pH > 2.9,p	
			–			–		–		–		–	C	
			–			0.32±0.03	–	–		–		–	A,p	
			–			–						–	C	
			–			0.52±0.03F	–	–		–		–	A,p	
			–			–							C	

FA64 $C_8H_6N_2$

Code No.	Empirical Formula	Name and C.A. Number	Structural Formula	Solvent	Tech.	Medium		μ, M	pH	T, °C	Electrodes	App.	Experimental Parameters
FA64 a188 cd24	$C_8H_6N_2$	quinoxaline C.A. 91-19-0		MeCN	VA	Et_4NClO_4	ns	-	-	-	Pt: xxbu// SCE	2-0	C=ns, A=0.22
FA65 am01 cd27 dc15	$C_8H_6O_3$	phenylglyoxalic acid C.A. 611-73-4	$C_6H_5COCOOH$	H_2O	PY	H_2SO_4	0.5	-	-	25	DME//SCE	2--	C=ns, m=1.59, t=5.2
						ClH_2CCO_2H	0.5		2.8				
						ACET	0.5		4.8				C=0.5
						ACET	0.5	-	-	25	DME//SCE	2--	C=11.1, m=1.41, t=6.32, h=65
						ACET strychnine 1.81E-5	0.5	-	4.8	25	DME//SCE	2--	C=0.5, m=1.59, t=5.2
						ACET strychnine 2.94E-5	0.5						
						ACET strychnine 1.67E-5	0.5						
						ACET strychnine 1.09E-4	0.5	-	-	25	DME//SCE		C=11.1, m=1.41, t=6.32, h=65
						BR KNO_3 0.2 NH_4OAc 0.5		-	6.58	25	DME//SCE	2--	C=0.5, m=1.59, t=5.2
						BR KNO_3 0.2 NH_4OAc 0.2			6.64				
						BR KNO_3 0.2 NH_4OAc 5E-3							
						BR KNO_3 0.2			6.70				
						BR KNO_3 0.2 ephedrine chlorhydrate 3.3E-2			-				
						BOR	ns		8.5				
						BOR morphine 1.2E-4	ns						
CONT													

$C_8H_6O_3$ (CONT.) FA65

Ref.	C/M	Charact. Potential Value	vs.	Response Const.	Value	n Tech.	Electrokinetic Data Parameter	Value	From	Products and Identification	Description and Remarks	Code No.
JE049 0111	-	$E_{p/2}$ — 2.19	SCE	$j_p/Cv^{\frac{1}{2}}$	10.89 8.63	- -	- -	- -	- -	- -	C, redn of quinoxaline, p A	FA64 a188 cd24
EA021 0407	-	$E_{\frac{1}{2}}$ -0.35	SCE	-	-	- -	-	-	-	-	C, p	FA65 am01 cd27 dc15
		-0.55 ±0.2		-	-	- -	-	-	-	-	C, p	
		-0.65		i_ℓ	3.6F	- -	-	-	-	-	C; $E_{\frac{1}{2}}$ → more neg on adding nicotine, strychnine(i-t curve shows adsorption), brucine, quinine, emetine, quinidine, cinchonidine or cinchonine; $E_{\frac{1}{2}}$ unaffected by codiene, sparteine, morphine, ephedrine, or amphetamine, p	
EA019 0865	-	$E_{\frac{1}{2}}$ -1.04F	SCE	i_ℓ	130F	- -	-	-	-	-	C, p	
EA021 0407	-	$E_{\frac{1}{2}}$ -0.63F	SCE	i_ℓ	3.6F	- -	-	-	-	-	C, p	
		-0.65F			3.6F	- -	-	-	-	-	C, p	
		-0.72F			3.6F	- -	-	-	-	-	C, p	
EA019 0865	-	$E_{\frac{1}{2}}$ -0.89F	SCE	i_ℓ	135F	- -	-	-	-	CP→R(-)-C_6H_5CHOH COOH(19% optical activity)	C, optical activity of products ↑ at low C, low T, low i; p	
EA021 0407	-	$E_{\frac{1}{2}}$ -0.84F	SCE	i_ℓ	3.5	- -	-	-	-	-	C, p	
		-0.87F -1.19F			3.6F 1.19F	- -	-	-	-	-	C, p C	
		0.89F 1.21F			3.3F 0.9F	- -	-	-	-	-	C, p C	
		-0.93F -1.24F			3.0F 1.3F	- -	-	-	-	-	C, p C	
		-0.84F		-	-	- -	-	-	-	-	C, p	
		-1.2		-	-	- -	-	-	-	-	C, p	
		-1.1		-	-	- -	-	-	-	-	C, p	

CONT

FA65 (CONT.) $C_8H_6O_3$

Code No.	Empirical Formula	Name and C.A. Number	Structural Formula	Solvent	Tech.	Medium	μ, M	pH	T, °C	Electrodes	App.	Experimental Parameters
FA65 am01 cd27 dc15	$C_8H_6O_3$	phenylglyoxalic acid	$C_6H_5COCOOH$	H_2O	PY	ACET,AM 0.5	-	9.2	25	DME//SCE	2--	C=0.5, m=1.59, t=5.2
						ACET,AM 0.5 strychnine 5E-4						
FA66 am45 cd37 dc24	C_8H_8	1,3,5,7-cyclooctatetraene C.A. 629-20-9	Table II-2	MeCN	PY	Et_4NClO_4 ns	-	-	0	DME//SCE (NaCl, aq)	26A0	C=ns, m=2.19, h=65, EC=2.05, Pt Aux
									22			
					VR	Et_4NClO_4 ns	-	-	22	HMDE// SCE (NaCl, aq)	26A0	-0.8 → -2.1 → -0.8 → -1.5 V, C=ns, v=200, Pt Aux
				DMF	PY	Et_4NClO_4 ns	-	-	22	DME//SCE (NaCl, aq)	26A0	C=ns, m=2.19, h=65, EC=2.05, Pt Aux
FA67 am68 cd41	C_8H_8O	acetophenone C.A. 98-86-2	$C_6H_5COCH_3$	MeOH 36	PY	NaOAc 0.64 HOAc 0.74	-	4.8	-	DME// Hg/ Hg_2SO_4, K_2SO_4 (satd, aq)	---	C=2.5, t=32, v=3
				MeOH	PY	Me_4NCl 0.1 TX100 0.05%	-	-	25	DME// Ag/AgCl, KCl satd	015-	C=1
						S-$C_6H_5CH_2$- $CH(CH_3)NH_2$- CH_3^+ Cl^- 0.1 TX100 0.05%						
						S-$C_6H_5CH_2CH$- $(CH_3)NH$- $(CH_3)_2^+$ Cl^- 0.1 TX100 0.05%						
						S-$C_6H_5CH_2CH$- $(CH_3)N$- $(CH_3)_3^+$ I^- 0.1 TX100 0.05%						
						1R,2S-C_6H_5CH- $OHCH(CH_3)$- $NH_2CH_3^+$ Cl^- 0.1 TX100 0.05%						
CONT												

TABLE I. Electrochemical Data

C_8H_8O (CONT.) FA67

Ref.	C/M	Charact. Potential		Response Const.		n		Electrokinetic Data			Products and Identification	Description and Remarks	Code No.
		Value	vs.		Value	Tech.		Parameter	Value	From			
EA021 0407	-	$E_{\frac{1}{2}}$	-1.06	SCE	-	-	-	-	-	-	-	-	C,p	FA65 am01 cd27 dc15
			-1.09		-	-	-	-	-	-	-	-	C,p	
JE056 0409	-	$E_{\frac{1}{2}}$	-1.74	SCE(NaCl, aq)	-	-	-	-	-	-	-	-	C,p	FA66 am45 cd37 dc24
			-1.86		-	-	-	-	-	-	-	-	C	
			-1.76		-	-	sttd	1	St	2.50	sttd	-	$C, i_\ell = kh^{\frac{1}{2}}, i_d, p$	
			-1.89					1				dianion	$C, i_\ell = kh^{\frac{1}{2}}, i_d$	
JE056 0409	-	E_p	-1.8F	SCE(NaCl, aq)	i_p	28F	-	-	-	-	-	-	C,p	
			-1.94			18F							C	
			-			small							A, related to cathodic peak at -1.94 V	
			-1.74F			5F							A, related to cathodic peak at -1.8 V	
			-1.06F			15F							A, observed only when sweep includes both cathodic peaks	
			-1.13F			8F							C,0 on first scan, related to anodic peak at -1.74 V	
JE056 0409	-	$E_{\frac{1}{2}}$	-1.64	SCE(NaCl, aq)	I	1.57	1:1	1	St	1.11	sttd	-	$C, i_\ell = kh^{\frac{1}{2}}, i_d, p$	
			-1.80			1.33		1				dianion	$C, i_\ell = kh^{\frac{1}{2}}, i_d$	
EA018 0433	-	$E_{\frac{1}{2}}$	-1.70F	Hg/ Hg_2SO_4, K_2SO_4 (satd, aq)	i_ℓ	30F	-	-	-	-	-	-	C,p	FA67 am68 cd41
EA019 0611	65g	$E_{\frac{1}{2}}$	-1.58	Ag/AgCl, KCl satd	-	-	-	-	-	-	-	methylphenylcarbinol 41%, 2,3-diphenyl-butanediol 30%	C,r	
			-1.45		-	-	-	-	-	-	-	methylphenylcarbinol 38%(R-1.3%), 2,3-diphenyl-2,3-butanediol 36%	C,r	
			-1.38		-	-	-	-	-	-	-	methylphenylcarbinol 38%(S-1.4%), 2,3-diphenyl-2,3-butanediol 30%	C,M,r	
			-1.59		-	-	-	-	-	-	-	methylphenylcarbinol 22%(S-2.0%), 2,3-diphenyl-2,3-butanediol 58%	C,r	
			-1.43		-	-	-	-	-	-	-	methylphenylcarbinol 44%(R-4.2%), 2,3-diphenyl-2,3-butanediol 41%	C,r	

CONT

FA67 (CONT.) C_8H_8O

Code No.	Empirical Formula	Name and C.A. Number	Structural Formula	Solvent	Tech.	Medium	μ, M	pH	T, °C	Electrodes	App.	Experimental Parameters
FA67 am68 cd41	C_8H_8O	acetophenone	$C_6H_5COCH_3$	MeOH	PY	1R,2S-C_6H_5CH-OHCH(CH_3)N-$H_2CH_3^+$ Br^- 0.1 TX100 0.05%	–	–	25	DME// Ag/AgCl, KCl satd	O15- –	C=1
						1R,2S-C_6H_5CH-OHCH(CH_3)NH-$(CH_3)_2^+$ Cl^- 0.1 TX100 0.05%						
						1R,2S-C_6H_5CH-OHCH(CH_3)N-$(CH_3)_3^+$ Cl^- 0.1 TX100 0.05%						
						1R,2S-C_6H_5CH-OHCH(CH_3)N-$(CH_3)(C_2H_5)_2^+$ I^- 0.1 TX100 0.05%						
						S-$C_6H_5CH_2$CH-(CH_2OH)N-$(CH_3)_3^+$ I^- 0.1 TX100 0.05%						
					CP	R-$C_6H_5CH_2$CH-CH(CH_3)NH_2-CH_3^+ Cl^- 0.1 TX100 0.05%	–	–	–	DME// Ag/AgCl, KCl satd	O15- –	C=1
						1R,2S-C_6H_5CH-OHCH(CH_3)NH_2-CH_3^+ Cl^- 0.1 TX100 0.05%						
						1S,2S-C_6H_5CH-OHCH(CH_3)NH_2-CH_3^+ Cl^- 0.1 TX100 0.05%						
						1S,2R-C_6H_5CH-OHCH(CH_3)NH_2-CH_3^+ Cl^- 0.1 TX100 0.05%						
						1S,2R-C_6H_5CH-OHCH(CH_3)NH-$(CH_3)_2^+$ Cl^- 0.1 TX100 0.05%						
						1S,2R-C_6H_5CH-OHCH(CH_3)N-$(CH_3)_3^+$ Cl^- 0.1 TX100 0.05%						
						1S,2S-C_6H_5CH-OHCH(CH_3)N-$(CH_3)_3^+$ Cl^- 0.1 TX100 0.05%						
FA68	$C_8H_9HgNO_2$	4-aminophenylmercury acetate C.A. 6283-24-5	4-$H_2NC_6H_4Hg^+$ $C_2H_3O_2^-$	H_2O	PW	NaOH 0.1	–	–	25	DME//SCE	2-2	C=0.0075, m=0.293, v=450? v=1000? v=4894 v=1E4

TABLE I. Electrochemical Data

$C_8H_9HgNO_2$ FA68

Ref.	C/M	Charact.	Potential		Response Const.	n		Electrokinetic Data			Products and Identification	Description and Remarks	Code No.
			Value	vs.	Value	Tech.		Parameter	Value	From			
EA019 0611	65g	$E_{\frac{1}{2}}$	-1.46	Ag/AgCl, KCl satd	-	-	-	-	-	-	-	methylphenylcarbinol 38%(R-4.4%),2,3-diphenyl-2,3-butanediol 49%	C,r	FA67 am68 cd41
			-1.44		-	-	-	-	-	-	-	methylphenylcarbinol 38%(S-2.4%);2,3-diphenyl-2,3-butanediol 33%	C,M,r	
			-1.58		-	-	-	-	-	-	-	methylphenylcarbinol 38%(S-8.4%);2,3-diphenyl-2,3-butanediol 30%	C,r	
			-1.56		-	-	-	-	-	-	-	methylphenylcarbinol 16%;2,3-diphenyl-2,3-butanediol 58%	C,r	
			-1.59		-	-	-	-	-	-	-	methylphenylcarbinol 41%(R-1.5%);2,3-diphenyl-2,3-butanediol 33%	C,r	
EA019 0611	65g	$E_{\frac{1}{2}}$	-	Ag/AgCl, KCl satd	-	-	-	-	-	-	-	methylphenylcarbinol 41%(S-1.2%);2,3-diphenyl-2,3-butanediol 33%	C,r	
			-		-	-	-	-	-	-	-	methylphenylcarbinol 33%(optical purity= 4.2%);2,3-diphenyl-2,3-butanediol 48%	C,optical purity ↓ as T ↓ below 20° or ↑ above 40°	
			-		-	-	-	-	-	-	-	methylphenylcarbinol 38%(S-7.8%);2,3-diphenyl-2,3-butanediol 33%	C,r	
			-		-	-	-	-	-	-	-	methylphenylcarbinol 38%(S-4.5%);2,3-diphenyl-2,3-butanediol 41%	C,r	
			-		-	-	-	-	-	-	-	methylphenylcarbinol 38%(R-3.0%);2,3-diphenyl-2,3-butanediol 30%	C,r	
			-		-	-	-	-	-	-	-	methylphenylcarbinol 33%(R-7.6%);2,3-diphenyl-2,3-butanediol 41%	C,r	
			-		-	-	-	-	-	-	-	methylphenylcarbinol 49%,2,3-diphenyl-2,3-butanediol 25%	C,r	
EA018 0237	8g	E_p	-0.433F	SCE	-	-	-	-	$dE_p/d\log v$	30	sttd	-	A,r	FA68
			-0.450F		-	-	-	-		30		-	A,r	
			-0.504F		-	-	-	-		30		-	A,r	
			-0.507F		-	-	-	-		0		-	A,r	

Code No.	Empirical Formula	Name and C.A. Number	Structural Formula	Solvent	Tech.	Medium		μ, M	pH	T, °C	Electrodes	App.	Experimental Parameters
FA69 ao07 cd87	$C_8H_{12}O_4$	diethyl fumarate C.A. 623-91-6	$C_2H_5OOCCH:CH-COOC_2H_5$	MeCN 96	PY	Et_4NClO_4	0.1	-	-	25	DME// Ag/ $AgClO_4$ 0.01	12A0	C=1.5, Pt Aux
				MeCN 99	PY	Et_4NClO_4	0.1	-	-	25	DME// Ag/ $AgClO_4$ 0.01	12A0	C=1.5, Pt Aux
				MeCN 99.95	PY	Et_4NClO_4	0.1	-	-	25	DME// Ag/ $AgClO_4$ 0.01	12A0	C=1.5, Pt Aux
FA70	$C_8H_{18}NO$	di-tert-butyl-nitroxide C.A. 2406-25-9	$[(H_3C)_3C]_2NO$	MeCN	QE	$LiClO_4$	0.5	-	-	30	Pt: nsgz// Ag/ $AgNO_3$ 0.1	259A0	C=0.125-5, E_{app}=0.83
					VA	$LiClO_4$	0.5	-	-	30	Pt: xxwi// Ag/ $AgNO_3$ 0.1	25A0	0→0.4→0 V, C=5, v=258, Pt Aux, PE
FA71	$C_8H_{20}ClNS$	2-[bis(1-methyl-ethyl)amino]ethane-thiol hydrochloride C.A. 41480-75-5	$HSCH_2CH_2NH-[CH(CH_3)_2]_2^+ Cl^-$	H_2O	PY	HCl LiCl	0.1 0.1	-	-	-	DME//SCE	OA1	C=1.0, m=1.58, t=3.24
						CL LiCl	0.1		2.60				
									3.0				
									3.80				
						ACET KCl	ns 0.1		4.50				
						ACET KNO_3	ns 0.1						
						ACET $KClO_4$	ns 0.1						
						ACET LiCl	ns 0.1						
CONT													

$C_8H_{20}ClNS$ (CONT.) FA71

Ref.	C/M	Charact. Potential Value	vs.	Response Const. Value		Tech.	n	Electrokinetic Data Parameter	Value	From	Products and Identification	Description and Remarks	Code No.
JE042 0189	50s	$E_{\frac{1}{2}}$ -1.26F	Ag/ AgClO$_4$ 0.01	i_ℓ	9.2F	-	-	-	-	-	-	C,r	FA69 ao07 cd87
		-1.44F			2.4F							C	
JE042 0189	50s	$E_{\frac{1}{2}}$ -1.30F	Ag/ AgClO$_4$ 0.01	i_ℓ	5.6F	QE	0.65	-	-	-	-	C,r	
		-1.48F			6.4F		-					C	
JE042 0189	50s	$E_{\frac{1}{2}}$ -1.20F	Ag/ AgClO$_4$ 0.01	i_ℓ	3.6F	i:i QE	0.6 0.5	-	-	-	-	C,r	
		-1.84F			0.8F	-						C,Mn	
EA018 0241	424 a	-	-	-	-	QE	0.3 ± 0.15	$k_{A,h}$	1.33 E-2	i-t curves	cation,ESR	A,ec,r	FA70
EA018 0241	424 a	E_p 0.595F	SCE	i_p	19.2F	sttd	1	-	-	-	-	A,≠ C,0	
EA018 0327	141 e	$E_{\frac{1}{2}}$ -0.14F	SCE	i_ℓ	0.3F	-	-	-	-	-	-	A,Pr,r	FA71
		0.1F			2.1F	sttd	1					A	
		-0.18F			0.3F	-		$dE_{\frac{1}{2}}/dpH$	95	calc	-	A,Pr,i_ℓ=kC for C≤1.0,r	
		0.0F			-		1		70			A,M,R,$E_{\frac{1}{2}}$≠f(C),i_ℓ= $kh^{\frac{1}{2}}$,i_ℓ=kC,i_d,Tc= 1.75	
		-0.21F			0.3F	-			95			A,Pr,i_ℓ = kC for C ≤ 1,r	
		-0.03F			2.1F		1		70			A,R,$E_{\frac{1}{2}}$≠f(C),i_ℓ=$kh^{\frac{1}{2}}$, i_ℓ=kC,i_d,Tc=1.75	
		-0.28F			0.3	-			95			A,Pr,i_ℓ = kC for C ≤ 1,r	
		-0.05F			2.0F		1		70			A,R,$E_{\frac{1}{2}}$≠f(C),i_ℓ=$kh^{\frac{1}{2}}$, i_ℓ=kC,i_d,Tc=1.75	
		-0.315			0.40	-		-			-	A,Pr,r	
		-0.120			2.00							A,R,W	
		-0.340			0.30							A,Pr,r	
		-0.110			2.20							A,R,W	
		-0.325			0.40	-		-			-	A,Pr,r	
		-0.105			2.10							A,R,W	
		-0.320			0.3	-		-			QE at 0.1 V → Hg compound (precipitated with H$_2$S)	A,Pr,Γ=3.98,r	
		-0.115			2.5							A,R,W	

CONT

FA71 (CONT.) $C_8H_{20}ClNS$

Code No.	Empirical Formula	Name and C.A. Number	Structural Formula	Solvent	Tech.	Medium		μ, M	pH	T, °C	Electrodes	App.	Experimental Parameters
FA71	$C_8H_{20}ClNS$	2-[bis(1-methyl-ethyl)amino]ethane-thiol hydrochloride	$HSCH_2CH_2NH-[CH(CH_3)_2]_2^+ Cl^-$	H_2O	PY	ACET NaCl	ns 0.1	–	4.50	–	DME//SCE	OA1	C=1.0, m=1.58, t=3.24
						ACET $NaClO_4$	ns 0.1						
						ACET Me_4NBr	ns 0.1						
						CL LiCl	0.1		4.8				
						ACET LiCl	ns 0.1		6.00				
						CL LiCl	0.1		6.80				
									7.60				
									9.20				
									10.20				
									11.18				
									12.60				
FA71	$C_8H_{20}ClNS$	2-[bis(1-methyl-ethyl)amino]ethane-thiol hydrochloride	$HSCH_2CH_2NH-[CH(CH_3)_2]_2^+ Cl^-$			NaOH LiCl	0.1 0.1		–				

TABLE I. Electrochemical Data

$C_8H_{20}ClNS$ FA71

Ref.	C/M	Charact. Potential Value	vs.	Response Const. Value		n Tech.	Electrokinetic Data Parameter	Value	From	Products and Identification	Description and Remarks	Code No.
EA018 0327	141 e	$E_{\frac{1}{2}}$ −0.315	SCE	i_ℓ	0.40	−	−	−	−	−	A,Pr,r	FA71
		−0.115			2.00						A,R,W	
		−0.340			0.35	−		−		−	A,Pr,r	
		−0.110			2.15						A,R,W	
		−0.315			0.7	−		−		−	A,Pr,r	
		−0.138			1.10						A,⇌	
		0.02			1.40						A,⇌	
		−0.35F			0.4F	−	$dE_{\frac{1}{2}}/dpH$	95	calc	−	A,Pr,i_ℓ = kC for C ≤ 1,r	
		−0.14F			2.0F	1	Elog	70 60	sttd	−	A,R,$E_{\frac{1}{2}}\neq f(C)$,$i_\ell=kh^{\frac{1}{2}}$, $i_\ell=kC$,i_d,Tc=1.75	
		−0.41F			0.4	−	−	−	−	−	A,Pr,r	
		−0.2F			1.7						A,R	
		−0.51F			0.4F	−	$dE_{\frac{1}{2}}/dpH$	95	calc	−	A,Pr,i_ℓ = kC for C ≤ 1,r	
		−0.3F			1.8F	1		70			A,R,$E_{\frac{1}{2}}\neq f(C)$,$i_\ell=kh^{\frac{1}{2}}$, $i_\ell=kC$,i_d,Tc=1.75	
		−0.75F			0.4F	−		95		−	A,Pr,i_ℓ = kC for C ≤ 1,r	
		−0.35F			4.0F	1		70			A,R,$E_{\frac{1}{2}}\neq f(C)$,$i_\ell=kh^{\frac{1}{2}}$, $i_\ell=kC$,i_d,Tc=1.75	
		−0.81F			0.4F	−		95			A,Pr,i_ℓ = kC for C ≤ 1.0,r	
		−0.44			4.0F	1		70			A,R,$E_{\frac{1}{2}}\neq f(C)$,$i_\ell=kh^{\frac{1}{2}}$, $i_\ell=kC$,i_d,Tc=1.75	
		−0.9F			0.4F	−		95		−	A,Pr,i_ℓ = kC for C ≤ 1.0,r	
		−0.53F			4.0F	1		70			A,R,$E_{\frac{1}{2}}\neq f(C)$,$i_\ell=kh^{\frac{1}{2}}$, $i_\ell=kC$,i_d,Tc=1.75, pK=10.60	
		−0.89F			0.4F	−		−		−	A,Pr,i_ℓ = kC for C ≤ 1.0,r	
		−0.55F			3.6F	1		0	sttd		A,R,$E_{\frac{1}{2}}\neq f(C)$,$i_\ell=kh^{\frac{1}{2}}$, $i_\ell=kC$,i_d,Tc=1.75	
		−0.91F			0.4F	−		−		−	A,Pr,i_ℓ = kC for C ≤ 1,r	
		−0.55F			4.0F	1		0			A,R,$E_{\frac{1}{2}}\neq f(C)$,$i_\ell=kh^{\frac{1}{2}}$, $i_\ell=kC$,i_d,Tc=1.75	
		$E_{\frac{1}{2}}$ −0.9F			0.6	−	−	−	−	−	A,Pr,r	
		−0.55F			4.0F						A	

FA72 $C_9H_4O_3$

Code No.	Empirical Formula	Name and C.A. Number	Structural Formula	Solvent	Tech.	Medium		μ, M	pH	T, °C	Electrodes	App.	Experimental Parameters
FA72 ao58 ec92	$C_9H_4O_3$	ninhydrin C.A. 485-47-2	Table II-2	H_2O	PY	buffer	ns	-	0	-	DME//SCE	---	ns
									1.0				
									3.0				
								-	1.3	-	DME//SCE	2-0	C=0.1
									5.5				
									7.8				
									13.2				
									3.05				C=2
									5.10				
									8.0				
									10				
						buffer Tween 80	ns 0.5%		10				
					PA	buffer	ns	-	5.05	-	DME//SCE	2-0	0→-0.7→ 0.2V,C=5, v=1E3
						MB	0.3	-	6.0	22	DME//SCE	2-2	-0.1→ -1.5→-0.7V, C=1,h=50, v=1E3
					PR	MB	0.3	-	2.9	22	DME//SCE	2-2	-0.1→ -0.9→-0.1 → -0.9V, C=1,h=50, v=1E3
													-0.1→ -1.1→-0.1 → -1.1V
FA72 CONT	$C_9H_4O_3$	ninhydrin C.A. 485-47-2	Table II-2	H_2O	PY	buffer	ns	-	10	-			-0.2→ -0.7→-0.2 →-0.7V,C=5, v=1E3

TABLE I. Electrochemical Data

$C_9H_4O_3$ (CONT.) FA72

Ref.	C/M	Charact. Potential		Response Const.		n Tech.		Electrokinetic Data			Products and Identification	Description and Remarks	Code No.
		Value	vs.		Value			Parameter	Value	From			
EA021 0913	381 a	$E_{\frac{1}{2}}$	0.150	SCE	-	-	-	-	-	-	-	-	C,p	FA72 ao58 ec92
			-0.650										C	
	381 b		0.100		-	-	-	-	-	-	-	-	C,p	
			-0.700										C	
			0.0		-	-	-	-	-	-	-	-	C,p	
			-0.790										C	
EA020 0981 (EA020 0965)	381 b	$E_{\frac{1}{2}}$	0.01F	SCE	-	-	-	-	$dE_{\frac{1}{2}}/dpH$	36F	-	-	C,Pr,only wave studied,p	
			-							-			C(?)	
			-0.14F		-	-	-	-		26F		-	C,Pr,p	
			-							-			C(?)	
			-0.22F		-	-	-	-		63F		-	C,Pr,p	
			-							-			C(?)	
	381 c		-0.56F		-	-	-	-		63F		-	C,Pr,p	
			-							-			C(?)	
	381 b		~0.02F		i_ℓ	0.37F	sttd	0.1	-	-	-	-	C,Mx,r	
			-0.13F			1.75F		0.4	$dE_{\frac{1}{2}}/dlogC$	60	-	-	C, i_ℓ ↑ as pH ↑,r	
			-0.20F			0.88F	-	-	-	-	-	-	C, i_ℓ ↑ as pH ↑,r	
			-0.43F			0.81F							C,Mx	
	381 c		-0.32F			3.9F	-	-	-	-	-	-	C, i_ℓ ↑ as pH ↑ for pH < 11,r	
			-0.47F			2.8F							C,Mx	
			-0.32F			3.9F	-	-	-	-	-	-	C,r	
EA020 0981	381 b	E_p	-0.2F	SCE	i_p	4F	-	-	-	-	-	-	A,r	
			0.0F			7.3F							A	
EA020 0965	381 b	$E_{\frac{1}{2}}$	-0.930	SCE	i_ℓ	7.0F	-	-	-	-	-	-	C,⇌,r	
			-1.190			6.1F							C,⇌	
EA020 0965	381 b	E_p	-0.830	SCE	i_p	7.5F	-	-	-	-	-	-	C,⇌,r	
			-0.830			7.3F	-	-	-	-	-	-	C,⇌,r	
			-1.010			4.0F							C,⇌	
			-0.180			3.3F							A,R	
			-0.200			3.3F							C,R,0 on first scan	
EA020 0981	381 c		-0.26F			43	-	-	-	-	-	-	A,r	

CONT

FA72 (CONT.) $C_9H_4O_3$

Code No.	Empirical Formula	Name and C.A. Number	Structural Formula	Solvent	Tech.	Medium		μ, M	pH	T, °C	Electrodes	App.	Experimental Parameters
FA72 ao58 ec92	$C_9H_4O_3$	ninhydrin		H_2O	PR	buffer	ns	-	10	-	DME//SCE	2-2	-0.2→-0.7→ -0.2→-0.7 V, C=ns, v=3E4
FA73	$C_9H_6O_3$	2,3-dihydroxy-1H-inden-1-one C.A. 59130-97-1	Table II	H_2O	PY	H_2SO_4	ns	-	0	22	DME//SCE	2-0	C=1, h=50
						MB	0.3		5				
									5.5				
									6.3				
									7.0				
									7.7				
									8.3				
						CARB	0.3		9.0				
									10.5				
						NaOH	ns		13.3				
						buffer	ns	-	2.7	-	DME//SCE	2-0	C=0.1
									3.0				C=1

Code No.	Empirical Formula	Name and C.A. Number	Structural Formula	Solvent	Tech.	Medium		μ, M	pH	T, °C	Electrodes	App.	Experimental Parameters
FA72 CONT	$C_9H_4O_3$	ninhydrin		H_2O		buffer	ns						

TABLE I. Electrochemical Data

$C_9H_8O_3$ (CONT.) FA73

Ref.	C/M	Charact. Potential Value	vs.	Response Const. Value		n Tech.	Electrokinetic Data Parameter	Value	From	Products and Identification	Description and Remarks	Code No.
EA020 0981	381 c	E_p -0.26F -0.38F	SCE	i_p	43F 16F	- -	-	-	-	-	A,r C,0 on first scan	FA72 ao58 ec92
EA020 0973	381 e	$E_{\frac{1}{2}}$ -0.67F	SCE	i_ℓ	5.5F	sttd 2	$dE_{\frac{1}{2}}/dpH$	44	sttd	CP (-0.70 V)→2,3-di-hydroxyindan-1-one, UVS,PY	C,$E_{\frac{1}{2}}$= -0.660- 0.044pH for pH < 5, r	FA73
		-0.95F			5.5F		2		44		1,2,3-trihydroxy-indane	C,$E_{\frac{1}{2}}$= -0.950- 0.044pH for pH < 7.5	
	381 f	-0.88F -1.21F			5.5F 5.3F	-		44 44		-	C,r C	
		-0.91F -1.00F			4.3F 1.2F	-		44 95		-	C,r C,$E_{\frac{1}{2}}$= -0.480- 0.095pH for pH=5-7.5	
		-1.23F			5.5F			44			C	
		-0.94F -1.06F -1.27F			2.3F 3.1F 5.5	1Ap 1Ap		44 95 44		-	C,r C C	
		-0.97F -1.09F -			1.1F 3.5F 5.7F	-		44 95 44		-	C,r C C	
		-1.00F -1.12F -1.28F -1.34F			0.3F 1.9F 17.9F	-		44 95 - 34		-	C,r C C C,$E_{\frac{1}{2}}$= -1.070- 0.034pH for pH= 7.5-10.5	
		-1.36F			9.6F	-		34		-	C,r	
	381 g	-1.38F			9.8F	3.91		34		CP (pH=9.5,E -1.70V) → 1,2,3-trihydroxy-indane,UVS	C,r	
		-1.42F			9.8F	-		45			C,$E_{\frac{1}{2}}$= -0.820- 0.045pH for pH ≥ 10.5,r	
		-1.58F			9.8F	-		45		-	C,r	
EA020 0981	381 e	$E_{\frac{1}{2}}$ 0.01F	SCE	-	-	- -	Tomeš $dE_{\frac{1}{2}}/dpH$	40 46 F	sttd -	-	A,0 for pH < 1,i_ℓ ↑ as pH ↑,$dE_{\frac{1}{2}}/dpH$ for pH=2.7-5.5	
		-						-			A,0 for pH < 1,i_ℓ ↑ as pH ↑	
		0.07F		-	-	QE 1		50F		pinacol	A,0 for pH < 1, $dE_{\frac{1}{2}}/dpH$ for pH= 3.0-6.3,i_ℓ ↑ as pH ↑, r	
		-0.10F						55F			A,0 for pH < 1, $dE_{\frac{1}{2}}/dpH$ for pH= 3.0-5.5,i_ℓ ↑ as pH ↑	CONT

FA73 (CONT.) $C_9H_6O_3$

Code No.	Empirical Formula	Name and C.A. Number	Structural Formula	Solvent	Tech.	Medium		μ, M	pH	T, °C	Electrodes	App.	Experimental Parameters
FA73	$C_9H_6O_3$	2,3-dihydroxy-1H-inden-1-one	Table II	H_2O	PY	buffer	ns	–	5.0	–	DME//SCE	2-0	C=1
									5.3				C=0.007
													C=0.15
													C=1
													C=4.2
									5.5				C=0.1
									5.6				C=1
									6.3				
									7.8				C=0.1
									8				C=1
									13.2				C=0.1
									13.5				C=1
					PA	buffer	ns	–	5.40	–	DME//SCE	2-2	-0.5 → -0.2 → -0.5 V, C=1, v=1E3
													0 → -0.5 → 0 V
					PW	buffer	ns	–	5.40	–	DME//SCE	2-2	-0.5 → -0.1 V, C=1, v=1E3
	$C_9H_6O_4$ (see $C_9H_4O_3$)												
FA74 ao77 ce06	C_9H_7N	cinnamonitrile C.A. 4360-47-8	$C_6H_5CH:CHCN$	MeCN 90	PY	Et_4NClO_4	0.1	–	–	25	DME// Ag/ AgClO$_4$ 0.01	12A0	C=2.9, Pt Aux
				MeCN 95	PY	Et_4NClO_4	0.1	–	–	25	DME// Ag/ AgClO$_4$ 0.01	12A0	C=2.9, Pt Aux
CONT													

TABLE I. Electrochemical Data

C_9H_7N (CONT.) FA74

Ref.	C/M	Charact.	Potential Value	vs.	Response Const.	Value	Tech.	n	Electrokinetic Data Parameter	Value	From	Products and Identification	Description and Remarks	Code No.
EA020 0981	381 f	$E_{\frac{1}{2}}$	-0.06F	SCE	-	-	-	-	$dE_{\frac{1}{2}}/dpH$	50F	-	-	A,r	FA73
			-0.21F							55F			A	
			-0.109F		-	-	-	-	$dE_{\frac{1}{2}}/dlogC$	0F	-	-	A,r	
			-0.156F							-			A, $dE_{\frac{1}{2}}/dlogC \neq k$	
			-0.109F		-	-	-	-		48F	-	-	A,r	
			-0.186F							-			A	
			-0.078F		-	-	-	-	$dE_{\frac{1}{2}}/dpH$	48F	-	-	A,r	
										50F				
			-0.219F							55F			A	
			-0.050F		-	-	-	-	$dE_{\frac{1}{2}}/dlogC$	48F	-	-	A,r	
			-0.230F							-			A	
			-0.14F		-	-	-	-	$dE_{\frac{1}{2}}/dpH$	35F	-	-	A, $dE_{\frac{1}{2}}/dpH$ for pH 5.5-7.8	
			-0.09F		-	-	sttd	1		50F	-	-	A,r	
			-0.23F					1		13F			A, $dE_{\frac{1}{2}}/dpH$ for pH=5.6-8.0	
			-0.15F		-	-	-	-		50F	-	-	A,r	
			-0.23F							13F			A	
			-0.24F		-	-	-	-		65F	-	-	A,r	
			-0.26F		-	-	QE	2		49F		ninhydrin	A,Mx, $dE_{\frac{1}{2}}/dpH$ for pH > 8, r	
	381 g		-0.59F		-	-	-	-		65F	-	-	A,r	
			-0.53F		-	-	sttd	2		49F		-	A,r	
EA020 0981	381 f	E_p	-0.21F	SCE	i_p	2F	-	-	-	-	-	-	A,r	
			-0.29F			2F							C	
			-0.21F			2F	-	-	-	-	-	-	A,r	
			-0.08F			4F							A	
			-			-							C,0	
EA020 0981	381 f	E_p	-0.17F	SCE	i_p	2.7F	-	-	-	-	-	-	A,r	
			-0.0F			4.0F							A	
JE042 0189	50s	$E_{\frac{1}{2}}$	-2.24F	Ag/ AgClO$_4$ 0.01	i_ℓ	20.8F	-	-	-	-	-	-	C,r	FA74 ao77 ce06
JE042 0189	50s	$E_{\frac{1}{2}}$	-2.24F	Ag/ AgClO$_4$ 0.01	i_ℓ	19.2F	-	-	-	-	-	-	C,r	
			-2.56F			4.8F							C	CONT

FA74 (CONT.) C_9H_7N

Code No.	Empirical Formula	Name and C.A. Number	Structural Formula	Solvent	Tech.	Medium	μ, M	pH	T, °C	Electrodes	App.	Experimental Parameters
FA74 ao77 ce06	C_9H_7N	cinnamonitrile	$C_6H_5CH:CHCN$	MeCN 99	PW	Et_4NClO_4 0.1	-	-	25	DME// Ag/ AgClO$_4$ 0.01	12A2	C=0.49, Pt Aux
												C=4.94
				MeCN 99.95	PY	Et_4NClO_4 0.1	-	-	25	DME// Ag/ AgClO$_4$ 0.01	12AO	C=2.9, Pt Aux
					PW	Et_4NClO_4 0.1	-	-	25	DME// Ag/ AgClO$_4$ 0.01	12AO	C=2.01, Pt Aux
												C=4.94
FA75 ao78	C_9H_7N	isoquinoline C.A. 119-65-3	Table II-2	MeCN	IL	Et_4NClO_4 ns	-	-	-	Pt: xxbu// SCE	2-O	C=ns, A=0.22
FA76 ao79 ec93	C_9H_7N	quinoline C.A. 91-22-5	Table II-2	MeCN	IL	Et_4NClO_4 ns	-	-	-	Pt: xxbu// SCE	2-O	C=ns, A=0.22
FA77 ao81 ec94	C_9H_7NO	8-hydroxyquinoline C.A. 148-24-3	Table II-2	H_2O?	PY	AM 0.2 Co^{2+} 4E-3	-	-	25± 0.1	DME//SCE	2AO	C=0.01, EC=2.25
					DP	AM 0.2 Co^{2+} 4E-3	-	-	-	DME//SCE	2AO	C=ns, EC=2.25

TABLE I. Electrochemical Data

C_9H_7NO FA77

Ref.	C/M	Charact. Potential		Response Const.		n Tech.		Electrokinetic Data			Products and Identification	Description and Remarks	Code No.
		Value	vs.		Value			Parameter	Value	From			
JE042 0189	50s	-	-	-	-	-	-	$dE_p/d\log v$	19	for $v <$ 9E2	-	C,data for first wave only,r	FA74 ao77 ce06
									0	for $v >$ 9E2			
		-							-			C(?)	
		-	-	-	-	-	-	$dE_p/d\log v$	20	for $v <$ 9E3	-	C,$dE_p/d\log[H_2O]$ =12 for $[H_2O]$=0.05- 1%(V/V),r	
									0	for $v >$ 9E3			
								$dE_p/d\log C$	21	sttd			
								$dE_p/d\log[H_2O]$	12	sttd			
		-							-			C(?)	
JE042 0189	50s	$E_{\frac{1}{2}}$	-2.2F	Ag/ AgClO$_4$ 0.01	i_ℓ	10.4F	QE i:i	0.81	-	-	-	C,r	
			-2.72F			11.2F		1				C	
JE042 0189	50s	-	-	-	-	-	-	$dE_p/d\log v$	19	for $v <$ 8E2	-	C,data for first wave only,r	
									0	for $v >$ 8E2			
		-							-			C(?)	
		-	-	-	-	-	-		18	for $v <$ 2.6E3	-	C,r	
									0	for $v >$ 2.6E3			
								$dE_p/d\log C$	22	sttd			
		-							-			C(?)	
JE049 0111	-	$E_{p/2}$	1.84	SCE	-	-	-	-	-	-	-	A,p	FA75 ao78
JE049 0111	-	$E_{p/2}$	1.97	SCE	-	-	-	-	-	-	-	A,p	FA76 ao79 ec93
AA096 0353	-	E_{max}	-1.25	SCE	-	-	-	-	-	-	-	C,Mx due to Co^{2+}, $i_{max}\downarrow$ as C $\uparrow$ and $\rightarrow$ 0 for C > 0.5,p	FA77
		$E_{\frac{1}{2}}$	-1.45									C	
		E_{max}	-1.59									C,Mx,i_{cat} or $i_{max}\uparrow$ as C $\uparrow$,$i_{max}\propto$ h	
AA096 0353	-	E_p	-1.145	SCE	-	-	-	-	-	-	-	C,p	
			-1.305									C	
			-1.53									C	

FA78 C$_9$H$_8$O

Code No.	Empirical Formula	Name and C.A. Number	Structural Formula	Solvent	Tech.	Medium	μ, M	pH	T, °C	Electrodes	App.	Experimental Parameters
FA78	C$_9$H$_8$O	1-phenyl-2-propyn-1-ol C.A. 4187-87-5	C$_6$H$_5$CHOHC⋮CH	DMF	PY	Bu$_4$NI 0.1	-	-	-	DME// SCE(o)	2-0	C=2
FA79	C$_9$H$_{10}$O	p-methoxystyrene C.A. 637-69-4	4-CH$_3$OC$_6$H$_4$CH:CH$_2$	MeCN	VY	Et$_4$NClO$_4$ 0.1	-	-	0	Pt: rodi// SCE	5-0	C=0.4, r_d=0.25, v=25, ω=9
						Et$_4$NClO$_4$ 0.1 PYR 3E-4						
						Et$_4$NClO$_4$ 0.1 PYR 6E-4						
						Et$_4$NClO$_4$ 0.1 PYR 4E-3						
						Et$_4$NClO$_4$ 0.1 PYR 2E-2						
						Et$_4$NClO$_4$ 0.1			-	Pt: rord// SCE		C=0.4, E_{disc}=1.47, $r_{r,i}$=0.275, $r_{r,o}$=0.335, v=25, ω=9, N_o=0.33±0.1
						Et$_4$NClO$_4$ 0.1 PYR 5E-5						
						Et$_4$NClO$_4$ 0.1 PYR 1E-4						
						Et$_4$NClO$_4$ 0.1 PYR 1.25E-4						
						Et$_4$NClO$_4$ 0.1 PYR 3E-4						
						Et$_4$NClO$_4$ 0.1 PYR 2E-2						
FA80	C$_9$H$_{10}$O	3-phenyl-2-propen-1-ol C.A. 104-54-1	C$_6$H$_5$CH:CHCH$_2$OH	DMF	PY	Bu$_4$NI 0.1	-	-	-	DME// SCE(o)	2-0	C=2
FA81 ap32 ce31	C$_9$H$_{10}$O	propiophenone C.A. 93-55-0	C$_6$H$_5$COCH$_2$CH$_3$	MeOH	PY	Me$_4$NCl 0.1 TX100 0.05	-	-	25	DME// Ag/AgCl, KCl satd	015-	C=1
						S-C$_6$H$_5$CH$_2$- CH(CH$_3$)N- (CH$_3$)$^+$Cl$^-$ 0.1 TX100 0.05						
CONT						1R,2S-C$_6$H$_5$CH- OHCH(CH$_3$NH$_2$- CH$_3$$^+Cl^-$ 0.1 TX100 0.05						

TABLE I. Electrochemical Data

$C_9H_{10}O$ (CONT.) FA81

Ref.	C/M	Charact. Potential		Response Const.		n Tech.		Electrokinetic Data			Products and Identification	Description and Remarks	Code No.
		Value	vs.		Value			Parameter	Value	From			
EA019 0629	33c	$E_{\frac{1}{2}}$	-2.74	SCE(o)	-	-	-	-	-	-	-	-	C,r	FA78
EA019 0681	-	$E_{\frac{1}{2}}$	1.38F	SCE	i_ℓ	6.7F	-	-	-	-	-	-	A,r	FA79
			1.63F			-							A,M	
			1.40F			7.9F	-	-	-	-	-	-	A,r	
			1.66F			-							A,M	
			1.43F			11.5F	-	-	-	-	-	-	A,r	
			1.7F			-							A,M	
			1.43F			13.6F	-	-	-	-	-	-	A,r	
			1.43F			13.9F	-	-	-	-	-	-	A,r	
		$E_{\frac{1}{2}}$, ring	0.19F			16.1F	-	-	-	-	-	-	C,r	
			0.19F			8.3F	-	-	-	-	-	-	C,r	
			-0.4F			6.8F	-	-	-	-	-	-	C	
			-			1.7F	-	-	-	-	-	-	C,X,r	
			-0.4F			13.7F	-	-	-	-	-	-	C	
			-0.4F			16.1F	-	-	-	-	-	-	C,r	
			-0.4F			12.7F	-	-	-	-	-	-	C,r	
			-1.23F			3.9F	-	-	-	-	-	-	C,M	
			-1.0F			10.0F	-	-	-	-	-	-	C,r	
			-1.2F			11.7F	-	-	-	-	-	-	C,M	
EA019 0629	33b	$E_{\frac{1}{2}}$	-2.57	SCE(o)	-	-	QE (-2.7V)	1.85	-	-	-	1-phenyl-1-propene 70%,3-phenyl-1-propanol 30%	C,r	FA80
			-2.80					-					C	
EA019 0611	65y	$E_{\frac{1}{2}}$	-1.60	Ag/AgCl, KCl satd	-	-	-	-	-	-	-	ethylphenylcarbinol 51%,3,4-diphenyl-3,4-hexanediol 25%	C,r	FA81 ap32 ce31
			-1.60		-	-	-	-	-	-	-	ethylphenylcarbinol 39% (S-2.3%);3,4-diphenyl-3,4-hexanediol 42%	C,r	
			-1.51		-	-	-	-	-	-	-	ethylphenylcarbinol 42% (R-1.8%);3,4-diphenyl-3,4-hexanediol 37%	C,r	

CONT

FA81 (CONT.) $C_9H_{10}O$

Code No.	Empirical Formula	Name and C.A. Number	Structural Formula	Solvent	Tech.	Medium	μ, M	pH	T, °C	Electrodes	App.	Experimental Parameters
FA81 ap32 ce31	$C_9H_{10}O$	propiophenone	$C_6H_5COCH_2CH_3$	MeOH	PY	1R,2S-C_6H_5CH-OHCH(CH_3)N-$(CH_3)_3^+Cl^-$ 0.1 TX100 0.05	-	-	25	DME// Ag/AgCl, KCl satd	O15-	C=1
FA82 ap63	C_9H_{12}	1,3,5-trimethyl-benzene C.A. 108-67-8	Table II	HF	VY	KF 0.1	-	-	0	glC: rodi// Cu/CuF_2 KF 0.2	2-O	C=1, A=0.07, v=100, ω=1, Pt Aux
FA83 ce59	$C_9H_{14}O$	3,5,5-trimethyl-2-cyclohexen-1-one C.A. 78-59-1	Table II-3	MeCN 95	PY	Et_4NClO_4 0.1	-	-	25	DME// Ag/ $AgClO_4$ 0.01	12AO	C=2.0, Pt Aux
				MeCN 99.95	PY	Et_4NClO_4 0.1	-	-	25	DME// Ag/ $AgClO_4$ 0.01	12AO	C=2, Pt Aux
					PW	Et_4NClO_4 0.1	-	-	25	DME// Ag/ $AgClO_4$ 0.01	12A2	C=1.85, Pt Aux
												C=2.03
				DMF 99	PW	Et_4NClO_4 0.1	-	-	25	DME// Ag/ $AgClO_4$ 0.01	12A2	C=2.63, Pt Aux
				DMF 99.9	PW	Et_4NClO_4 0.1	-	-	25	DME// Ag/ $AgClO_4$ 0.01	12A2	C=2.63, Pt Aux
FA84 CONT	$C_9H_{16}NO_2$	2,2,6,6-tetramethyl-4-oxo-1-piperidinyl-oxy C.A. 2896-70-0	Table II	MeCN	PY	Et_4NClO_4 0.1	-	-	-	DME// Ag/ $AgClO_4$ 0.01, Et_4NClO_4 0.1	2AO	C=ns, Pt Aux
					VA	Et_4NClO_4 0.1	-	-	-	glC: xxdi// Ag/ $AgClO_4$ 0.01, Et_4NClO_4 0.1	2A-	C=ns, d=1.3, v=25, Pt Aux

$C_9H_{16}NO_2$ (CONT.) FA84

Ref.	C/M	Charact. Potential Value	vs.	Response Value	Const.	n	Tech.	Electrokinetic Data Parameter	Value	From	Products and Identification	Description and Remarks	Code No.
EA019 0611	65y	$E_{\frac{1}{2}}$	-1.59	Ag/AgCl, KCl satd	-	-	-	-	-	-	-	ethylphenylcarbinol 46% (S-7.2%); 3,4-diphenyl-3,4-hexanediol 32%	C,r	FA81 ap32 ce31
JE054 0232	-	$E_{\frac{1}{2}}$	0.915	NHE	i_ℓ	77F	-	-	Elog	55	sttd	-	A,p	FA82 ap63
JE042 0189	50s	$E_{\frac{1}{2}}$	-2.16F	Ag/AgClO$_4$ 0.01	i_ℓ	6.7F	-	-	-	-	-	-	C, data for first wave only, r	FA83 ce59
			-		-					-			C(?)	
JE042 0189	50s	$E_{\frac{1}{2}}$	-2.42F	Ag/AgClO$_4$ 0.01	i_ℓ	6.2F	i:i	1	-	-	-	-	C, data for first wave only, r	
			-		-					-			C(?)	
JE042 0189	50s	-	-	-	-	-	-	-	$dE_p/d\log v$	19	for v < 2.1E3	-	C,r	
										0	for v > 2.1E3			
			-							-			C(?)	
			-		-		-	-		22	for v < 2.1E4	-	C,r	
										0	for v > 2.1E4			
									$dE_p/d\log C$	20(?)	sttd			
			-							-			C(?)	
JE042 0189	50r	-	-	-	-	-	-	-	$dE_p/d\log v$	20	for v < 3.2E4	-	C, data for first wave only, r	
										0	for v > 3.2E4			
			-							-			C(?)	
JE042 0189	50r	-	-	-	-	-	-	-		21	for v < 2.5E4	-	C, data for first wave only, r	
										0	for v > 2.5E4			
			-							-			C(?)	
EA020 0469	424 b	$E_{\frac{1}{2}}$	-1.65	Ag/AgClO$_4$ 0.01, Et$_4$NClO$_4$ 0.1	-	-	sttd	1	-	-	-	-	C,p	FA84
			-					1					C	
EA020 0469	424 b	-	-	-	-	-	-	-	$E_p-E_{p/2}$ ΔE_p	75 220	sttd	-	C,r	

FA84 (CONT.) C₉H₁₆NO₂

Code No.	Empirical Formula	Name and C.A. Number	Structural Formula	Solvent	Tech.	Medium	μ, M	pH	T, °C	Electrodes	App.	Experimental Parameters
FA84	C₉H₁₆NO₂	2,2,6,6-tetramethyl-4-oxo-1-piperidinyl-oxy	Table II	MeCN	VA	Et₄NClO₄ 0.1	-	-	-	Pt: xxdi // Ag/AgClO₄ 0.01, Et₄NClO₄ 0.1	2A-	C=ns,v=25, Pt Aux
					VY	Et₄NClO₄ 0.1	-	-	-	glC: rodi // Ag/AgClO₄ 0.01, Et₄NClO₄ 0.1	2AO	C=ns,ω=10, d=1.3, Pt Aux
										Pt: rodi // Ag/AgClO₄ 0.01, Et₄NClO₄ 0.1	2AO	C=ns,ω=10, Pt Aux
FA85	C₉H₁₈NO	2,2,6,6-tetramethyl-piperidinenitroxide cation C.A. 45842-10-2		MeCN	VA	LiClO₄ 0.5	-	-	30	Pt: xxdi // Ag/AgNO₃ 0.1	25AO	0.6→0.1→0.6V,C=2, v=36
						LiClO₄ 0.5 HCl 1E-3						C=2?,v=7
												C=2?,v=100
						LiClO₄ 0.5 HCl 4.6E-3						C=2?,v=7
												C=2,v=100
						LiClO₄ 0.5 HCl 0.021						C=2?,v=7
												C=2?,v=100
						LiClO₄ 0.5 HCl 0.06						C=2?,v=7
												C=2?,v=100

$C_9H_{18}NO$ FA85

Ref.	C/M	Charact. Potential		Response Const.		n Tech.		Electrokinetic Data			Products and Identification	Description and Remarks	Code No.
		Value	vs.		Value			Parameter	Value	From			
EA020 0469	424 b	–	–	–	–	–	–	$E_p-E_{p/2}$	60 180	sttd	–	A,R,p C	FA84
								ΔE_p	950				
EA020 0469	424 b	$E_{\frac{1}{2}}$ -1.61	Ag/ AgClO$_4$ 0.01, Et$_4$NClO$_4$ 0.1	–	–	sttd	1	–	–	–	–	C,R,p	
		–					1					C	
		$E_{\frac{1}{2}}$ 0.37		–	–	sttd	1	–	–	–	oxoammonium cation	A,R,p	
		-1.78		i_ℓ/C	6.3		1					C	
												C	
EA018 0615	50a	E_p 0.25F	Ag/AgNO$_3$ 0.1	i_p	38F	–	–	–	–	–	protonated 2,2,6,6-tetramethylpiperdinenitroxide radical, ESR	C,R,r	FA85
		0.32F			22F							A,R	
		0.271F		–	–	–	–	$dE_p/d\log v$	30	for v < 3E4	RH$^{\pm}$, ESR	C,r	
									0	for v > 3E4			
		0.253F		–	–	–	–	–	–	–	–	C,r	
		0.283F		–	–	–	–	$dE_p/d\log v$	29	for v < 5E4	–	C,r	
									0	for v > 5E4			
		0.253F		–	–	–	–	–	–	–	–	C,r	
		0.290F		–	–	–	–	$dE_p/d\log v$	30	for v < 1E5	–	C,r	
									0	for v > 1E5			
		0.260F		–	–	–	–	–	–	–	–	C, $dE_p/d\log v \neq k$ for v > 26, r	
		0.297F		–	–	–	–	$dE_p/d\log v$	28	for v < 5E5	–	C,r	
									0	for v > 5E5			
		0.265F		–	–	–	–		28		–	C,r	
EA020 0469	424 b	–	–	–	–	–	–	$E_p-E_{p/2}$	60 180	sttd			
								ΔE_p	950				

FA86 $C_9H_{18}NO$

Code No.	Empirical Formula	Name and C.A. Number	Structural Formula	Solvent	Tech.	Medium		μ, M	pH	T, °C	Electrodes	App.	Experimental Parameters
FA86	$C_9H_{18}NO$	2,2,6,6-tetramethyl-1-piperidinyloxy C.A. 2564-83-2	Table II	MeCN	PY	Et_4NClO_4	0.1	-	-	-	DME// Ag/ $AgClO_4$ 0.01, Et_4NClO_4 0.1	2AO	C=ns, Pt Aux
					IL	$LiClO_4$	0.5	-	-	30	Pt: xxwi// Ag/ $AgNO_3$ 0.1	25AO	-0.1(30s)→ 0.7(30s)→ -0.1V, C=5, v=30.1, Pt Aux PE
													v=92
													v=470.6
													v=1110
					QE	$LiClO_4$	0.5	-	-	30	Pt: nsgz// Ag/ $AgNO_3$ 0.1	25AO	A=15(80 mesh), E_{app}= 0.73 V, Pt Aux
					VA	Et_4NClO_4	0.1	-	-	-	glC: xxdi// Ag/ $AgClO_4$ 0.01, Et_4NClO_4 0.1	2A-	C=ns, v=25, Pt Aux
											Pt: xxdi// Ag/ $AgClO_4$ 0.01, Et_4NClO_4 0.1		
					VR	$LiClO_4$	0.5	-	-	30	Pt: xxdi// Ag/ $AgNO_3$ 0.1	2A-	0→1.7→0 →1.7V, C=2, v=193, Pt Aux
					VY	Et_4NClO_4	0.1				glC: rodi// Ag/ $AgClO_4$ 0.01, Et_4NClO_4 0.1	2AO	C=ns, ω=10, d=1.3, Pt Aux
											Pt: rodi// Ag/ $AgClO_4$ 0.01, Et_4NClO_4 0.1		
													-2.2V(30s)→ pos
FA86		2,2,6,6-tetramethyl-1-piperidinyloxy C.A. 2564-83-2	Table II	MeCN	PY	Et_4NClO_4							

TABLE I. Electrochemical Data

$C_9H_{18}NO$ FA86

Ref.	C/M	Charact. Potential		Response Const.		Tech.	n	Electrokinetic Data			Products and Identification	Description and Remarks	Code No.
		Value	vs.		Value			Parameter	Value	From			
EA020 0469	424 b	$E_{\frac{1}{2}}$ -1.76	Ag/ AgClO$_4$ 0.01, Et$_4$NClO$_4$ 0.1	-	-	sttd	1	-	-	-	-	C,p	FA86
EA018 0241	424 b	E_p 0.652	SCE	$i_p/v^{\frac{1}{2}}$	1.64	-	-	ΔE_p	65	sttd	-	A,R,i_d,r	
		0.587			1.69							C,R	
		0.652			1.41	-	-		65		-	A,r	
		0.587			1.48							C	
		0.662			1.34	-	-		75		-	A,r	
		0.587			1.45							C	
		0.662			1.35	-	-		-	-	-	A,r	
		0.592			1.42							C	
EA018 0241	424 b	-	-	-	-	QE	1.04	-	-	-	cation, ESR(no signal)	A,r	
EA020 0469	424 b	-	-	-	-	-	-	$E_p - E_{p/2}$	70	sttd	-	C,r	
		-	-	-	-	-	-		60		-	A,r	
		-							90			C	
EA019 0593	-	E_p 0.32F	Ag/AgNO$_3$ 0.1	i_p	131F	-	-	-	-	-	-	A,R,r	
		0.23F			130F							C,R	
EA020 0469	424 b	$E_{\frac{1}{2}}$ 0.32	Ag/AgClO$_4$ 0.1, Et$_4$NClO$_4$ 0.1	-	-	sttd	1	-	-	-	oxoammonium cation	A,R,r	
		-1.74					1					C,$\neq$	
		0.32	Ag/AgClO$_4$ 0.01, Et$_4$NClO$_4$ 0.1	i_ℓ/C	-	sttd	1	-	-	-	oxoammonium cation	A,R,r	
		-1.96			6.4		1					C	
		-1.78			-		-				-	A,p	

FA87 C₁₀F₈

Code No.	Empirical Formula	Name and C.A. Number	Structural Formula	Solvent	Tech.	Medium	μ, M	pH	T, °C	Electrodes	App.	Experimental Parameters
FA87 ed20	C₁₀F₈	octafluoronaphthalene C.A. 313-72-4	Table II-5	HFSO₃	VA	HOAc	0.1	-	-80 20	Pt: xxwi// Pd/H₂	25-0	C=ns,A=0.6, v=100,Pt Aux
FA88	C₁₀H₆N₂	3-cyanoquinoline C.A. 34846-64-5	Table II	EtOH 95	IL	NH₄OAc	1	-	-	DME//SCE	---	C=1.0,m=1.00, t(c)=1.00
					PT	NH₄OAc	1	-	25± 0.2	DME//SCE	2A-	C=1.0,m=1.00, t(c)=1.00
FA89	C₁₀H₈N₂	2,2'-bipyridine C.A. 366-18-7	Table II-2	MeCN	VA	Bu₄NBF₄	0.1	-	-	Pt: xxdi//Ag	2F-	0→-2.6→0V, C=ns,Pt Aux 0→1.9→0V
FA90	C₁₀H₈O₄	ethyleneglycol phthalate C.A. 29382-68-1	Table II	EtOH 40	PY	BR	-	7 8.7 9.9	20	DME//SCE	04AO	C=ns,m=1.77, t=5,h=60
FA91 ar15	C₁₀H₉N	3-methylisoquinoline C.A. 1125-80-0	Table II-2	MeCN	IL	Et₄NClO₄	ns	-	-	Pt: xxpu//SCE	2-0	C=ns,A=0.22
FA92	C₁₀H₁₀F₃NO₂	α-methoxy-α-(3'-trifluoromethylphenyl)acetaldoxime C.A. 64756-76-9	3-F₃CC₆H₄CH(OCH₃)CH:NOH	MeOH 10	PY	BR	-	4	-	DME//SCE	2AO	C=ns, m=1.73, t=2.03,h=50, v=2,Pt Aux
					DI	BR	-	4 7.0	-	DME//SCE	2AO	C=0.001-0.1, m=1.73, t=0.3,h=50, ΔE=100,v=2, Pt Aux C=ns

$C_{10}H_{10}F_3NO_2$ FA92

Ref.	C/M	Charact. Potential		Response Const.		n Tech.	Electrokinetic Data			Products and Identification	Description and Remarks	Code No.
		Value	vs.		Value		Parameter	Value	From			
EA018 0331	–	$E_{p/2}$	0.65F	Pd/H_2	i_p	20F	–	–	–	–	–	A,R,r	FA87 ed20
		1.50F			14F	–					A	
		1.90F			14F	QE 1					A,R	
		1.92F			17F	1					C,R	
		0.69F			8F	–					C,R	
		0.55		i_p/C	42	1	ΔE_p	60	–	radical cation (stable),ESR	A,R,r	
		1.31			107	3		–		–	A	
		1.80			85	–					A	
		0.6F			–						C	
EA021 0753	–	–	–	–	–	sttd 1	$dE_p/dlogv$	25	sttd		C,R	FA88
							$dE_p/dlogC$	28				
		–				1		–			C,⊬	
EA021 0753	–	$E_{\frac{1}{2}}$	-1.09F	SCE	I	1.79	QE (-1.3V) 1.0	$Elogi^{\frac{2}{3}}$ $dE_{\frac{1}{2}}/dlogC$ $dE_{\frac{1}{2}}/dlogt$	59F 28F 22F	–	CP→orange solution 68%, yellow precipitate 20%, both are dimers, MW, MAS, IRS, UVS	C,R	
		-1.47F				1.79	QE (-1.6V) 1.0		61F 31F 23F		1,4-dihydro-3-cyano-quinoline, yellow 60%, MPT, NMR, IRS	C,⊬	
JA095 6582	–	E_p	-2.13	SCE	–	–	–	–	–	–	–	C,r	FA89
		–			–	–	–	–	–	–	–	A,O,r	
C0028 1985	438 a	–	–	–	i_ℓ/i_d	0.75F	sttd 4	–	–	–	–	C,p	FA90
		–				0.90F	4	–	–	–	–	C	
		$E_{\frac{1}{2}}$	-1.83	SCE		1.00F	4	–	–	–	–	C,$\Delta E_{\frac{1}{2}}=0.39\sigma^*$,p	
JE049 0111	–	$E_{p/2}$	1.67	SCE	–	–	–	–	–	–	–	A,p	FA91 ar15
AA092 0353	–	–	–	–	–	–	–	αn_a p=	0.375 1	sttd	α-methoxy-α-(3-trifluoromethylphenyl)-ethylamine and N-hydroxy-α-methoxy-α-(3-trifluoromethylphenyl)ethylamine; TLC, GLC	C,W,X for pH>5,i_d,p	FA92
		–							–			A,O	
AA090 0353	–	E_{su}	-1.16	SCE	–	–	–	–	–	–	–	C, $i_{su}=k_1+k_2C$ for C=0.001-0.050,p	
			-1.30		–	–	–	–	–	–	–	C, redn of <u>anti</u> form, p	
			-1.45									C, redn of <u>syn</u> form	

FA93 $C_{10}H_{10}Fe$

Code No.	Empirical Formula	Name and C.A. Number	Structural Formula	Solvent	Tech.	Medium	μ, M	pH	T, °C	Electrodes	App.	Experimental Parameters
FA93 ar36 cf30 dd14 ed33	$C_{10}H_{10}Fe$	ferrocene C.A. 102-54-5	$(\pi\text{-}C_5H_5)_2Fe$	H_2O	PVQ	Et_4NClO_4 0.1	-	-	-	Pt: xxbu// SCE	013-0	C=0.05, A=0.20, f=80, PE
					VA	Et_4NClO_4 0.1	-	-	-	Pt: xxbu// SCE	013-0	C=0.05, A=0.20, v=10, PE
												v=1000
				EtOH	PVQ	Et_4NClO_4 0.02	-	-	-	Pt: xxbu// Ag/ AgClO$_4$ 0.01, Et$_4$NClO$_4$ 0.01	013-0	C=0.5-1, A=0.20, f=80, PE
					VA	Et_4NClO_4 0.02	-	-	-	Pt: xxbu// Ag/ AgClO$_4$ 0.01, Et$_4$NClO$_4$ 0.01	013-0	C=0.5-1, A=0.20, v=10, PE
												v=1000
				MeCN	PVQ	Et_4NClO_4 0.02	-	-	-	Pt: xxbu// Ag/ AgClO$_4$ 0.01, Et$_4$NClO$_4$ 0.01	013-0	C=1-10, A=0.20, f=80, PE
					VA	Et_4NClO_4 0.02	-	-	-	Pt: xxbu// Ag/ AgClO$_4$ 0.01, Et$_4$NClO$_4$ 0.01	013-0	C=1-10, A=0.20, v=10, PE
												v=1000
				DMF	PVQ	Et_4NClO_4 0.02	-	-	-	Pt: xxbu// Ag/ AgClO$_4$ 0.01, Et$_4$NClO$_4$ 0.01	013-0	C=1-10, A=0.20, f=80, PE
					VA	Et_4NClO_4 0.02	-	-	-	Pt: xxbu// Ag/ AgClO$_4$ 0.01, Et$_4$NClO$_4$ 0.01	013-0	C=1-10, A=0.20, v=10, PE
												v=1000
				DMSO	PVQ	Et_4NClO_4 0.02	-	-	-	Pt: xxbu// Ag/ AgClO$_4$ 0.01, Et$_4$NClO$_4$ 0.01	013-0	C=0.2-10, A=0.20, f=80, PE
CONT												

$C_{10}H_{10}Fe$ (CONT.) FA93

Ref.	C/M	Charact. Potential Value	vs.	Response Const. Value	n Tech.		Electrokinetic Data Parameter	Value	From	Products and Identification	Description and Remarks	Code No.
EA018 0975	-	E_{su} 0.195 ±0.005	SCE	-	-	-	-	$\Delta E_{su/2}$	125	sttd	-	A, slightly distorted, i_a, r	FA93 ar36 cf30 dd14 ed33
EA018 0975	-	E_p 0.194 ±0.005	SCE	-	-	-	-	ΔE_p k	60 0.01	N-S	-	A, R, r	
	-	-	-	-	-	-	-	ΔE_p	70	sttd	-	A, r	
EA018 0975	-	E_{su} -0.094 ±0.005	Ag/ AgClO$_4$ 0.01, Et$_4$NClO$_4$ 0.01	-	-	-	-	$\Delta E_{su/2}$	95	sttd	-	A, r	
EA018 0975	-	E_p -0.062 ±0.005	Ag/ AgClO$_4$ 0.01, Et$_4$NClO$_4$ 0.01	-	-	-	-	ΔE_p k	67 0.016	sttd N-S	-	A, $i_{p,A}/v^{\frac{1}{2}}$ ↑ as v ↓, r	
	-	-	-	$i_{p,C}/i_{p,A}$ → 1	-	-	-	ΔE_p	110	sttd	-	A, r	
EA018 0975	-	E_{su} 0.092 ±0.005	Ag/ AgClO$_4$ 0.01, Et$_4$NClO$_4$ 0.01	-	-	-	-	$\Delta E_{su/2}$	90	sttd	-	A, r	
EA018 0975	-	E_p 0.120 ±0.005	Ag/ AgClO$_4$ 0.01, Et$_4$NClO$_4$ 0.01	-	-	-	-	ΔE_p k	70 0.044	sttd N-S	-	A, $i_{p,A}/v^{\frac{1}{2}}$ ↑ as v ↓, r	
	-	-	-	$i_{p,C}/i_{p,A}$ → 1	-	-	-	ΔE_p	80	sttd	-	A, r	
EA018 0975	-	E_{su} 0.034 ±0.005	Ag/ AgClO$_4$ 0.01, Et$_4$NClO$_4$ 0.01	-	-	-	-	$\Delta E_{su/2}$	95	sttd	-	A, r	
EA018 0975	-	E_p 0.058 ±0.005	Ag/ AgClO$_4$ 0.01, Et$_4$NClO$_4$ 0.01	-	-	-	-	ΔE_p k	60 0.033	sttd	-	A, $i_{p,A}/v^{\frac{1}{2}}$ ↑ as v ↓, r	
	-	-	-	$i_{p,C}/i_{p,A}$ → 1	-	-	-	ΔE_p	100	sttd	-	A, r	
EA018 0975	-	E_{su} 0.180 ±0.005	Ag/ AgClO$_4$ 0.01, Et$_4$NClO$_4$ 0.01	-	-	-	-	$\Delta E_{su/2}$	95	sttd	-	A, r	CONT

FA93 (CONT.) $C_{10}H_{10}Fe$

Code No.	Empirical Formula	Name and C.A. Number	Structural Formula	Solvent	Tech.	Medium		μ, M	pH	T, °C	Electrodes	App.	Experimental Parameters
FA93 ar36 cf30 dd14 ed33	$C_{10}H_{10}Fe$	ferrocene	$(\pi-C_5H_5)_2Fe$	DMSO	VA	Et_4NClO_4	0.02	-	-	-	Pt: xxbu// Ag/ $AgClO_4$ 0.01, Et_4NClO_4 0.01	013-0	C=0.2-10, A=0.20, v=10, PE
													v=1000
FA94	$C_{10}H_{10}O$	2-phenyl-3-butyn-2-ol C.A. 127-66-2	Table II	DMF	PY	Bu_4NI PHEN	0.1	-	-	-	DME// SCE(o)	2-0	C=2
FA95	$C_{10}H_{10}O$	3-methoxy-3-phenyl-1-propyne C.A. 50874-13-0	$C_6H_5CH(OCH_3)C\!:\!CH$	DMF	PY	Bu_4NI	0.1	-	-	-	DME// Ag/AgI, I^- 0.1	3-0	C=1, Pt Aux
					QE	Bu_4NI PHEN	0.1 0.01	-	-	-	Hg?// Ag/AgI, I^- 0.1	3-0	C=1, E_{app} = -2.3 V, Pt Aux
FA96 ar51	$C_{10}H_{10}O$	4-phenyl-3-buten-2-one C.A. 122-57-6	$C_6H_5CH\!:\!CHCOCH_3$	EtOH	PW	Bu_4NOH	0.1	-	-	25	DME// Ag/ $AgClO_4$ 0.01	12A2	C=0.24, Pt Aux
						PHEN $Bu_4NOC_6H_5$	ns 0.1						C=1.3
				MeCN 95	PY	Et_4NClO_4	0.1	-	-	25	DME// Ag/ $AgClO_4$ 0.01	12A0	C=2.7, Pt Aux
					PW	Et_4NClO_4	0.1	-	-	25	DME// Ag/ $AgClO_4$ 0.01	12A2	C=4.86, Pt Aux
				MeCN 99	PY	Et_4NClO_4	0.1	-	-	25	DME// Ag/ $AgClO_4$ 0.01	12A0	C=2.7, Pt Aux

CONT

TABLE I. Electrochemical Data

$C_{10}H_{10}O$ (CONT.) FA96

Ref.	C/M	Charact. Potential Value	vs.	Response	Const. Value	Tech.	n	Electrokinetic Data Parameter	Value	From	Products and Identification	Description and Remarks	Code No.
EA018 0975	-	E_p	0.200 ±0.005	Ag/ AgClO$_4$ 0.01, Et$_4$NClO$_4$ 0.01	-	-	-	-	ΔE_p k	65 0.038	sttd N-S	-	A, $i_{p,A}/v^{1/2}$ ↑ as v ↓, r	FA93 ar36 cf30 dd14 ed33
	-	-	-	$i_{p,C}/i_{p,A}$	→ 1	-	-	ΔE_p	100	sttd	-	A,r	
EA019 0629	33c	$E_{1/2}$	-2.82	SCE(o)	-	-	QE (-2.88V)	5.2	-	-	-	3-phenylbutane 90%	C,r	FA94
EA020 0853	258 b	$E_{1/2}$	-2.27	Ag/AgI, I$^-$ 0.1	i_ℓ	6.5	QE (-2.3V)	1.5	-	-	-	1-methoxy-1-phenyl-propene (70% cis, 30% trans)	C,r	FA95
			-2.44			6.5	-		-	-	-		C	
EA020 0853	258 b	-	-	-	-	-	QE	2.25	-	-	-	1-methoxy-1-phenyl-propene (50% cis, 50% trans)	C,r	
JE042 0189	50s	-	-	-	-	-	-	-	$dE_p/d\log v$	17	for v < 4.5E3	-	C, data for first wave only, r	FA96 ar51
										0	for v > 4.5E3			
									$dE_p/d\log C$	20	sttd			
		-	-	-	-	-	-	-	-	-			C(?)	
		-	-	-	-	-	-	-	$dE_p/d\log v$	19	for v < 1.7E4	-	C,r	
										0	for v > 1.7E4			
									$dE_p/d\log C$	20.0	sttd			
		-	-	-	-	-	-	-	-	-			C(?)	
JE042 0189	50s	$E_{1/2}$	-1.72F	Ag/ AgClO$_4$ 0.01	i_ℓ	4F	-	-	-	-	-	-	C,r	
			-2.04F			3.6F							C	
			-2.2F			1F							C,Mx	
JE042 0189	50s	-	-	-	-	-	-	-	$dE_p/d\log v$	18	for v < 7E5	-	C, $dE_p/d\log[H_2O]$=17 for [H$_2$O]=0.5-5%(V/V), data for first wave only, r	
										0	for v > 7E5			
									$dE_p/ d\log[H_2O]$	17	sttd			
		-	-	-	-	-	-	-	-	-			C(?)	
JE042 0189	50s	$E_{1/2}$	-1.82	Ag/ AgClO$_4$ 0.01	i_ℓ	3.2F	-	-	-	-	-	-	C,r	
			-2.1F			3.6F							C	
			-2.3F			small							C	CONT

FA96 (CONT.) $C_{10}H_{10}O$

Code No.	Empirical Formula	Name and C.A. Number	Structural Formula	Solvent	Tech.	Medium	μ, M	pH	T, °C	Electrodes	App.	Experimental Parameters
FA96 ar51	$C_{10}H_{10}O$	4-phenyl-3-buten-2-one	$C_6H_5CH:CHCOCH_3$	MeCN 99.5	PW	Et_4NClO_4 0.1	-	-	25	DME// Ag/ AgClO$_4$ 0.01	12A2	C=0.24, Pt Aux
												C=2.4, Pt Aux
												C=4.86, Pt Aux
				MeCN 99.95	PY	Et_4NClO_4 0.1	-	-	25	DME// Ag/ AgClO$_4$ 0.01	12A0	C=2.7, Pt Aux
					PW	Et_4NClO_4 0.1	-	-	25	DME// Ag/ AgClO$_4$ 0.01	12A2	C=0.25, Pt Aux
												C=2.47, Pt Aux
												C=4.86, Pt Aux
CONT	$C_{10}H_{10}O$	4-phenyl-3-buten-2-one	$C_6H_5CH:CHCOCH_3$	MeCN 99.5	PW	Et_4NClO_4 0.1			25	DME//		

TABLE I. Electrochemical Data

$C_{10}H_{10}O$ (CONT.) FA96

Ref.	C/M	Charact. Potential		Response Const.		n		Electrokinetic Data			Products and Identification	Description and Remarks	Code No.
		Value	vs.	Value		Tech.		Parameter	Value	From			
JE042 0189	50s	-	-	-	-	-	-	$dE_p/dlogv$	19	for v < 6.5E3	-	$C, dE_p/dlog[H_2O]=17$ for $[H_2O]=0.5-5\%(V/V)$, data for first wave only, r	FA96 ar51
									0	for v > 6.5E3			
								$dE_p/dlogC$	19	sttd			
								$dE_p/dlog[H_2O]$	17				
		-							-			C(?)	
		-	-	-	-	-	-	$dE_p/dlogv$	19	for v < 6.5E4	-	C, r	
									0	for v > 6.5E4			
								$dE_p/dlogC$	19	sttd			
								$dE_p/dlog[H_2O]$	17				
		-							-			C(?)	
		-	-	-	-	-	-	$dE_p/dlogv$	21	for v < 9E4	-	C, r	
									0	for v > 9E4			
								$dE_p/dlogC$	19	sttd			
		-							-			C(?)	
JE042 0189	50s	$E_{\frac{1}{2}}$	-1.94F	Ag/AgClO$_4$ 0.01	i_l	3.2F	i:i	0.91	-	-	-	-	C, r
			-2.3F			3.8F		1Ap					C
JE042 0189	50s	-	-	-	-	-	-	$dE_p/dlogv$	20	for v < 9E2	-	C, data for first wave only, r	
									0	for v > 9E2			
								$dE_p/dlogC$	20	sttd			
		-							-			C(?)	
		-	-	-	-	-	-	$dE_p/dlogv$	20	for v < 7E3	-	C, r	
									0	for v > 7E3			
								$dE_p/dlogC$	20	sttd			
		-							-			C(?)	
		-	-	-	-	-	-	$dE_p/dlogv$	21	for v < 1.1E4	-	C, r	
									0	for v > 1.1E4			
								$dE_p/dlogC$	20	sttd			
		-							-			C(?)	
									$dE_p/dlogv$	19	for			CONT

FA96 (CONT.) $C_{10}H_{10}O$ (1)

Code No.	Empirical Formula	Name and C.A. Number	Structural Formula	Solvent	Tech.	Medium	μ, M	pH	T, °C	Electrodes	App.	Experimental Parameters
FA96 ar51	$C_{10}H_{10}O$	4-phenyl-3-buten-2-one	$C_6H_5CH:CHCOCH_3$	DMF 90-99.9	PW	Et_4NClO_4 0.1	-	-	25	DME// Ag/ AgClO$_4$ 0.01	12A2	C=5.3, Pt Aux
FA97 ar52 ed34	$C_{10}H_{10}O$	1-tetralone C.A. 529-34-0	Table II-2	MeOH	PY	Me_4NCl 0.1 TX100 0.05	-	-	25	DME// Ag/AgCl, KCl satd	O15-	C=1
						1R,2S-C_6H_5CH-OHCH(CH$_3$)N-(CH$_3$)$_3^+$Cl$^-$ 0.1 TX100 0.05						
					CP	Me_4NCl 0.1 buffer TX100 0.05	-	7Ap	25	DME// Ag/AgCl, KCl satd	O15-	C=1
						1R,1S-C_6H_5CH-OHCH(CH$_3$)N-(CH$_3$)$_3^+$Cl$^-$ buffer 0.1 TX100 0.05						
FA98 ar55 dd16	$C_{10}H_{10}O_2$	methyl cinnamate C.A. 103-26-4	$C_6H_5CH:CHCOOCH_3$	MeCN 90	PY	Et_4NClO_4 0.1	-	-	25	DME// Ag/ AgClO$_4$ 0.01	12A0	C=1.8, Pt Aux
				MeCN 95	PY	Et_4NClO_4 0.1	-	-	25	DME// Ag/ AgClO$_4$ 0.01	12A0	C=1.8, Pt Aux
				MeCN 98	PW	Et_4NClO_4 0.1	-	-	25	DME// Ag/ AgClO$_4$ 0.01	12A2	C=5.04, Pt Aux
				MeCN 99	PW	Et_4NClO_4 0.1	-	-	25	DME// Ag/ AgClO$_4$ 0.01	12A2	C=0.18, Pt Aux
												C=1.98, Pt Aux
CONT												

TABLE I. Electrochemical Data

$C_{10}H_{10}O_2$ (CONT.) FA98

Ref.	C/M	Charact. Potential		Response Const.		n Tech.	Electrokinetic Data			Products and Identification	Description and Remarks	Code No.
		Value	vs.		Value		Parameter	Value	From			
JE042 0189	50r	-	-	-	-	-	-	$dE_p/dlogv$	21	for $v <$ 2E4	-	C, data for first wave only, r	FA96 ar51
									0	for $v >$ 2E4			
		-							-			C(?)	
EA019 0611	65y	$E_{\frac{1}{2}}$	-1.59	Ag/AgCl, KCl satd	-	-	-	-	-	-	carbinol 29%, pinacol 50%	C, r	FA97 ar52 ed34
			-1.59		-				-	-	carbinol 47%(S-6.6%); pinacol 36%	C, r	
EA019 0611	65y	-	-	-	-	-	-	-	-	-	carbinol 34%, pinacol 43%	C, r	
		-							-	-	carbinol 43%(S-5.5%); pinacol 39%	C, r	
JE042 0189	50s	$E_{\frac{1}{2}}$	-1.9F	Ag/AgClO$_4$ 0.01	i_ℓ	2.8F	-	-	-	-	-	C, r	FA98 ar55 dd16
			-2.08F			2.8F						C	
JE042 0189	50s	$E_{\frac{1}{2}}$	-2.06F	Ag/AgClO$_4$ 0.01	i_ℓ	3.8F	-	-	-	-	-	C, r	
			-2.28F			3.4F						C	
JE042 0189	50s	-	-	-	-	-	-	$dE_p/dlogv$	23	for $v <$ 2.1E4	-	C, $dE_p/dlog[H_2O]$=20 for $[H_2O]$=1-2%(V/V), data for first wave only, r	
									0	for $v >$ 2.1E4			
								$dE_p/dlog[H_2O]$	20	sttd			
		-							-			C(?)	
JE042 0189	50s	-	-	-	-	-	-	$dE_p/dlogv$	19	for $v <$ 3E2	-	C, $dE_p/dlog[H_2O]$=6 for $[H_2O]$=0.05-1%(V/V), data for first wave only, r	
									0	for $v >$ 3E2			
								$dE_p/dlogC$	21	sttd			
								$dE_p/dlog[H_2O]$	6	sttd			
		-							-			C(?)	
		-	-	-	-	-	-	$dE_p/dlogv$	22	for $v <$ 3.7E3	-	C, r	
									0	for $v >$ 3.7E3			
		-							-			C(?)	

CONT

FA98 (CONT.) $C_{10}H_{10}O_2$

Code No.	Empirical Formula	Name and C.A. Number	Structural Formula	Solvent	Tech.	Medium	μ, M	pH	T, °C	Electrodes	App.	Experimental Parameters
FA98 ar55 dd16	$C_{10}H_{10}O_2$	methyl cinnamate	$C_6H_5CH\text{:}CHCOOCH_3$	MeCN 99	PW	Et_4NClO_4 0.1	-	-	25	DME// Ag/ AgClO$_4$ 0.01	12A2	C=5.04, Pt Aux
				MeCN 99.95	PY	Et_4NClO_4 0.1	-	-	25	DME// Ag/ AgClO$_4$ 0.01	12A0	C=1.8, Pt Aux
					PW	Et_4NClO_4 0.1	-	-	25	DME// Ag/ AgClO$_4$ 0.01	12A2	C=5.04, Pt Aux
				DMF 80	PY	Et_4NClO_4 0.1	-	-	25	DME//SCE	12A0	C=3.7, Pt Aux
					PW	Et_4NClO_4 0.1	-	-	25	DME// Ag/ AgClO$_4$ 0.01	12A2	C=4.7, Pt Aux
				DMF 95	PW	Et_4NClO_4 0.1	-	-	25	DME// Ag/ AgClO$_4$ 0.01	12A2	C=4.7, Pt Aux
				DMF 98	PY	Et_4NClO_4 0.1	-	-	25	DME//SCE	12A0	C=3.7, Pt Aux
				DMF 99.9	PY	Et_4NClO_4 0.1	-	-	25	DME//SCE	12A0	C=3.7, Pt Aux
FA99	$C_{10}H_{10}O_3S$	3-[(4-methylphenyl)-thio]-2-oxopropanoic acid C.A. 35699-38-8	$4\text{-}CH_3C_6H_4SCH_2\text{-}COCOOH$	EtOH 5	PY	H_2SO_4	1	0.06	25	DME//SCE	3--	C=1, m=6.4, t=3.0, h=55, EC=4.15
								0.6				
								1.45				
						BR	1	3.10				
FA99 CONT												

$C_{10}H_{10}O_3S$ (CONT.) FA99

Ref.	C/M	Charact. Potential		Response Const.		n Tech.		Electrokinetic Data			Products and Identification	Description and Remarks	Code No.
		Value	vs.		Value			Parameter	Value	From			
JE042 0189	50s	-	-	-	-	-	-	$dE_p/dlogv$	23	for v < 1E4	-	C,r	FA98 ar55 dd16
									0	for v > 1E4			
									-			C(?)	
JE042 0189	50r	$E_{\frac{1}{2}}$	-2.16	Ag/ AgClO$_4$ 0.01	i_ℓ	2.8F	i:i	1	-	-	-	-	C,r
			-2.5F			1.5F		1					C
JE042 0189	50r	-	-	-	-	-	-	$dE_p/dlogv$	22	for v < 4E3	-	C, data for first wave only,r	
									0	for v > 4E3			
			-									C(?)	
JE042 0189	50s	$E_{\frac{1}{2}}$	-1.82F	SCE	i_ℓ	3F	-	-	-	-	-	C,r	
			-2.0F			2.8F							C
JE042 0189	50s	-	-	-	-	-	-	$dE_p/dlogv$	22	for v < 3E3	-	C,$dE_p/dlog[H_2O]$=20 for $[H_2O]$=5-20%(V/V),data for first wave only,r	
									0	for v > 3E3			
								$dE_p/dlog[H_2O]$	20	sttd			
			-						-			C(?)	
JE042 0189	50s	-	-	-	-	-	-	$dE_p/dlogv$	18	for v < 7E2	-	C, data for first wave only,r	
									0	for v > 7E2			
			-						-			C(?)	
JE042 0189	50s	$E_{\frac{1}{2}}$	-1.86F	SCE	i_ℓ	4F	-	-	-	-	-	C,r	
			-2.24F			4F							C
JE042 0189	50s	$E_{\frac{1}{2}}$	-1.84F	SCE	i_ℓ	3.8F	i:i QE	1 1	-	-	-	-	C,R,r
			-2.34F			2.6F		-					C
EA018 0691	421 b	$E_{\frac{1}{2}}$	-0.45F	SCE	i_ℓ	2.6	sttd (i:i$_d$?)	0.6F	$dE_{\frac{1}{2}}/dpH$	0F	-	-	C,i_k,Tc=9,$dE_{\frac{1}{2}}/dpH$ for pH < 0.6	FA99
	421 c		-0.44F			1.7F		0.4F		65F		-	C,$dE_{\frac{1}{2}}/dpH$ for pH=0.6-2.3,i_k,Tc=9, r	
			-0.52F			2.3F		0.6F		65F		-	C,i_k,Tc=9,r	
			-0.68F			5.7F		1.4F		115F		-	C,$dE_{\frac{1}{2}}/dpH$ for pH=2.3-5.9,i_k,Tc=9, r	

CONT

FA99 (CONT.) $C_{10}H_{10}O_3S$

Code No.	Empirical Formula	Name and C.A. Number	Structural Formula	Solvent	Tech.	Medium		μ, M	pH	T, °C	Electrodes	App.	Experimental Parameters
FA99	$C_{20}H_{20}O_3S$	3-[(4-methylphenyl)-thio]-2-oxopropanoic acid	$4\text{-}CH_3C_6H_4SCH_2\text{-}COCOOH$	EtOH 5	PY	BR	–	1	4.6	25	DME//SCE	3--	C=1,m=6.4, t=3.0,h=55, EC=4.15
									7.06				
									8.0				
									12				
									13.5				
FB00	$C_{10}H_{10}O_4$	dimethyl phthalate C.A. 131-11-3	$C_6H_4(COOCH_3)_2$	EtOH 40	PY	BR	–	–	6.3	30	DME//SCE	O4AO	C=ns, m=1.77, t=5,h=60
										40			
										50			
					BR NaCl	– 0.1		6.4	20			
						BR	–		7.3	30			
										50			
					BOR NaCl	0.08 0.1		7.6	20			C=1.25
									8.25				
									9.1				
									9.3				

$C_{10}H_{10}O_4$ FB00

Ref.	C/M	Charact. Potential		Response	Const. Value	Tech.	n	Electrokinetic Data			Products and Identification	Description and Remarks	Code No.
		Value	vs.					Parameter	Value	From			
EA018 0691	421 c	$E_{\frac{1}{2}}$	-0.95F	SCE	i_ℓ	6.0F	i:i, QE	2	$dE_{\frac{1}{2}}/dpH$	115F	-	QE(pH=4.5,-1.05 V)→ 2-oxopropanoic acid,PY;4-methyl-phenylthiol,UVS,PY	C,i_d,Tc=2,r	FA99
			-1.2F			0.7F	sttd (i:i_d?)	0.2		-			C	
			-1.16F			6.0F		2		60F		-	$C,dE_{\frac{1}{2}}/dpH$ for pH > 5.9,i_d,Tc=2,r	
			-1.35F			0.6F		0.2		0F			$C,dE_{\frac{1}{2}}/dpH$ for pH=7.5-12.3	
			-1.68F			-		-		0F			$C,dE_{\frac{1}{2}}/dpH$ for pH=7.5-12.3	
			-1.29F			4.0F		1F		60F		-	C,0 for pH > 8,r	
			-1.32F			2.4F		0.6F		0F			C	
			-1.65F			1.2F		0.3F		0F			C	
			-1.34F			-		-		50F		QE/CP(pH=11,-1.40 V) → n=2, 2-oxopropanoic acid,PY;4-methylphenylthiol, UVS,PY	$C,dE_{\frac{1}{2}}/dpH$ for pH > 12.3,Tc= -2, i_a,r	
			-1.63F			-		-		56F			$C,dE_{\frac{1}{2}}/dpH$ for pH > 12.3,Tc= -2.5, i_a	
			-1.42F			3.0F		0.7F		50F		-	C,r	
			-1.72F			2.0F		0.5F		56F			C	
C0028 1985	438 a		-	-	i_ℓ/i_d	0.75F	-	-	-	-	-	-	C,p	FB00
			-	-		0.90F	-	-	-	-	-	-	C,p	
			-	-		1.0F	-	-	-	-	-	-	C,p	
			-	-		0.60F	sttd	4	-	-	-	-	C,p	
			-	-		0.8F	-	-	-	-	-	-	C,p	
			-	-		1.00F	-	-	-	-	-	-	C,p	
		$E_{\frac{1}{2}}$	-1.77F	SCE		0.65F	sttd	4	-	-	-	-	C,p	
			-1.77F			0.74F		4	-	-	-	-	C,p	
			-1.77F			0.96F		4	-	-	-	-	C,p	
			-1.77F			1.00F		4	-	-	-	-	C,p	

FB01 $C_{10}H_{12}F_3NO_2$

Code No.	Empirical Formula	Name and C.A. Number	Structural Formula	Solvent	Tech.	Medium		μ, M	pH	T, °C	Electrodes	App.	Experimental Parameters
FB01	$C_{10}H_{12}F_3NO_2$	β-methoxy-β-(3-trifluoromethylphenyl)-ethyl C.A. 64756-74-7	Table II	MeOH 10	PY	BR	-	-	7	-	DME//SCE	2AO	C=ns,m=1.73, t=2.03,h=50, v=2,Pt Aux
					DI	BR	-	-	7.0	-	DME//SCE	2AO	C=0.001-0.1, m=1.73,t=0.3, h=50,ΔE=100, v=2,Pt Aux
FB02	$C_{10}H_{12}O$	1-methoxy-1-phenyl-propene C.A. 4541-69-9	$C_6H_5C(OCH_3):CH-CH_3$	DMF	PY	Bu_4NI	0.1	-	-	-	DME// Ag/AgI, I⁻ 0.1	3-0	C=1,Pt Aux
						Bu_4NI PHEN	0.1 0.01						
FB03 ar85	$C_{10}H_{12}O$	3-methoxy-1-phenyl-propene C.A. 16277-67-1	$C_6H_5CH:CHCH_2-OCH_3$	DMF	PY	Bu_4NI	0.1	-	-	-	DME// Ag/AgI, I⁻ 0.1	3-0	C=1,Pt Aux
						Bu_4NI PHEN	0.1 0.01						
FB04 ar86	$C_{10}H_{12}O$	2-methyl-1-phenyl-propanone C.A. 611-70-1	$C_6H_5COCH(CH_3)_2$	MeOH	PY	Me_4NCl TX100	0.1 0.05	-	-	25	DME// Ag/AgCl, KCl satd	O15- -	C=1
						1R,2S-C_6H_5CH-$OHCH(CH_3)$-$NH_2CH_3^+$ Cl^- TX100	0.1 0.05						
						1R,2S-C_6H_5CH-$OHCH(CH_3)$-$N(CH_3)_3^+$ Cl^- TX100	0.1 0.05						
						S-$C_6H_5CH_2CH$-$(CH_3)N$-$(CH_3)_3^+$ Cl^- TX100	0.1 0.05						
FB05 as13 CONT	$C_{10}H_{14}NO_5PS$	O,O-diethyl O(4-nitrophenyl)phosphorothioate C.A. 56-38-2	4-$O_2NC_6H_4OP(:S)$-$(OC_2H_5)_2$	MeOH %?	DI	BR	-	0.12	2.2	-	DME//SCE	2AO	C=ns,m=1.32, t=4.6,h=76, ΔE=100,t(c)= 1,Pt Aux

TABLE I. Electrochemical Data

$C_{10}H_{14}NO_5PS$ (CONT.) FB05

Ref.	C/M	Charact.		Potential vs.	Response Const.		n Tech.		Electrokinetic Data			Products and Identification	Description and Remarks	Code No.
			Value			Value			Parameter	Value	From			
AA092 0353	-	$E_{\frac{1}{2}}$	-	SCE	-	-	-	-	$dE_{\frac{1}{2}}/dpH$	90	-	β-methoxy-β-(3-tri-fluoromethylphenyl)-acetaldoxime,GLC, TLC;α-hydroxy-α-(3-trifluoromethyl-phenyl)acetalde-hyde;3-trifluoro-methylphenylglyoxal	A,i_d,pK'_a=6,pK=4.01,p	FB01
			-0.071F						Elog αn_a p=	45 1.3 2.2			C,$dE_{\frac{1}{2}}/dpH$ for pH<6 ≠ $dE_{\frac{1}{2}}/dpH$ for pH>6	
AA092 0353	-	E_{su}	-0.07 -0.856	SCE	-	-	-	-	-	-	-	-	A,i_{su}=kC,p C,$i_{su}=k_1+k_2C$	
EA020 0853	258 b	$E_{\frac{1}{2}}$	-2.11	Ag/AgI, I⁻ 0.1	i_ℓ	2.9	QE (-2.2V)	2.25	-	-	-	1-methoxy-1-phenyl-propane	C,r	FB02
			-2.06			3.2		-	-	-	-	-	C,r	
EA020 0853	33k	$E_{\frac{1}{2}}$	-1.82	Ag/AgI, I⁻ 0.1	i_ℓ	3.6	QE (-1.8V)	2.05	-	-	-	1-phenylpropene 65%, 1-phenylpropane 18%	C,r	FB03 ar85
			-2.11			3.0		-	-	-	-	-	C	
			-1.82			4.3	QE (-1.8V)	1.8	-	-	-	1-phenylpropene 62%, 3-methoxy-1-phenyl-propane 24%	C,r	
			-2.04			2.9		-	-	-	-	-	r	
EA019 0611	65y	$E_{\frac{1}{2}}$	-1.62	Ag/AgCl, KCl satd	-	-	-	-	-	-	-	isopropylphenyl-carbinol 69%,2,5-dimethyl-3,4-di-phenyl-3,4-hexan-diol 13%(meso-84%, DL-16%)	C,r	FB04 ar86
			-1.54		-	-	-	-	-	-	-	isopropylphenyl-carbinol 44%(S-2.7%);2,5-dimethyl-3,4-diphenyl-3,4-hexandiol 38%(meso-5%,DL-95%)	C,r	
			-1.60		-	-	-	-	-	-	-	isopropylphenyl-carbinol 64%(S-5.8%);2,5-dimethyl-3,4-diphenyl-3,4-hexandiol 18%(meso-70%,DL-30%)	C,r	
			-1.61		-	-	-	-	-	-	-	isopropylphenyl-carbinol 64%(S-1.9%);2,5-dimethyl-3,4-diphenyl-3,4-hexandiol 13%(meso-67%,DL-33%)	C,r	
AA096 0335	124 j	E_{su}	-0.12F	SCE	-	-	-	-	dE_{su}/dpH	62F	-	-	C,i_{su}=kC for C=1E-5 to 1E-2,r	FB05 as13
			-0.36							128F			C	CONT

FB05 (CONT.) $C_{10}H_{14}NO_5PS$

Code No.	Empirical Formula	Name and C.A. Number	Structural Formula	Solvent	Tech.	Medium		μ, M	pH	T, °C	Electrodes	App.	Experimental Parameters
FB05 as13	$C_{10}H_{14}NO_5PS$	O,O-diethyl O(4-nitrophenyl)phosphorothioate	$4\text{-}O_2NC_6H_5OP(:S)\text{-}(OC_2H_5)_2$	MeOH %?	DI	BR	-	0.12	4.0	-	DME//SCE	2AO	C=ns,m=1.32, t=4.6,h=76, ΔE=100,t(c)=1,Pt Aux
									4.8				
									5.8				
									7.8				
									9.6				
									10.5				
						Bu_4NBr	0.1		-				
					PP	Bu_4NBr	0.001	-	-	-	DME//SCE	2AO	C=ns,m=1.32, t=4.6,h=76, t(c)=1, Pt Aux
FB06 as14	$C_{10}H_{14}NO_6P$	diethyl 4-nitrophenyl phosphate C.A. 311-45-5	$4\text{-}O_2NC_6H_4OP(:O)\text{-}(OC_2H_5)_2$	MeOH %?	PP	BR	-	0.12	1.9	-	DME//SCE	2AO	C=ns,m=1.32, t=4.6,h=76, t(c)=1, Pt Aux
									4				
									6				
									8				
									10				
FB07 as58 cf66 dd32	$C_{10}H_{16}N_2$	Wurster's Blue C.A. 100-22-1	$4\text{-}(CH_3)_2NC_6H_4\text{-}N(CH_3)_2$	MeCN	XT	Et_4NClO_4	0.1	-	-	-30 ±1	HMDE Ag/ AgClO$_4$ 0.01	2-2	C=ns,v= (1.5-5.5)E2
										0±1			
										20± 0.5			
FB08 CONT	$C_{11}H_8N_2$	4-methylbenzylidene-malonitrile C.A. 2826-25-7	$4\text{-}CH_3C_6H_4CH{:}C(CN)_2$	MeCN 95	PY	Et_4NClO_4	0.1	-	-	25	DME// Ag/ AgClO$_4$ 0.01	12AO	C=1.1,Pt Aux

TABLE I. Electrochemical Data

$C_{11}H_8N_2$ (CONT.) FA08

Ref.	C/M	Charact. Potential		Response Const.		n	Tech.	Electrokinetic Data			Products and Identification	Description and Remarks	Code No.
		Value	vs.		Value			Parameter	Value	From			
AA096 0335	124 j	E_{su}	-0.22F	SCE	-	-	-	-	dE_{su}/dpH	62F	-	-	C,i_{su}=kC for C=1E-5 to 1E-2,r	FB05 as13
			-0.59F							128F			C	
			-0.28F		-	-	-	-		62F		-	C,pK=5.2,r	
	-		-0.36F		-	-	-	-		100F			C,r	
			-0.56F		-	-	-	-		39F			C,pK'=7.8,r	
			-0.63F		-	-	-	-		39F		-	C,i_{su}=kC for C=1E-5 to 1E-2,r	
	124 h		-0.63F		-	-	-	-	ΔE_{su}	0F 53	sttd	-	C,W,i_a,i_{su}=kC for C=1E-5 to 1E-2,r	
			-0.63F		5F	-	-	-	-	-	-	-	C,r	
			-0.83F		4F								C	
AA096 0335	-	E_{max}	-0.67F	SCE	i_{max}	13F	-	-	-	-	-	-	C,r	
AA096 0335	124 j	E_{max}	-0.1	SCE	-	-	-	-	-	-	-	-	C,Mx,i_ℓ=kC for C=1E-5 to 1E-2,r	FB06 as14
			-0.4										C,Mx	
			-0.3F		-	-	-	-	-	-	-	-	C,Mx,i_ℓ=kC for C=1E-5 to 1E-2,r	
			-0.8F										C,Mx	
	-		-0.5F		-	-	-	-	-	-	-	-	C,Mx,i_ℓ=kC for C=1E-5 to 1E-2,r	
			-										C,0 for pH > 6	
			-0.67F		-	-	-	-	-	-	-	-	C,Mx,i_ℓ=kC for C=1E-5 to 1E-3,r	
	124 h		-0.72F		-	-	-	-	-	-	-	-	C,Mx,i_ℓ=kC for C=1E-5 to 1E-3,r	
JE047 0115	-	XE	0.546 ±0.002	Ag/ AgClO$_4$ 0.01	-	-	-	-	-	-	-	-	C,XT=semiintegral chronoamperometry with linear potential sweep,XE= difference between potentials for the two steps,p	FB07 as58 cf66 dd32
			0.562 ±0.002		-	-	-	-	-	-	-	-	C,p	
			-										C	
			0.572 ±0.002		-	-	-	-	-	-	-	-	C,p	
			-										C	
JE042 0189	50s	$E_{\frac{1}{2}}$	-1.48F	Ag/ AgClO$_4$ 0.01	i_ℓ	3.4F	-	-	-	-	-	-	C,r	FB08
			-2.16F			1.8F							C	

CONT

FB08 (CONT.) $C_{11}H_8N_2$

Code No.	Empirical Formula	Name and C.A. Number	Structural Formula	Solvent	Tech.	Medium		μ, M	pH	T, °C	Electrodes	App.	Experimental Parameters
FB08	$C_{11}H_8N_2$	4-methylbenzylidene-malononitrile	4-$CH_3C_6H_4$CH:C(CN)$_2$	MeCN 97	PY	Et_4NClO_4	0.1	-	-	25	DME// Ag/ AgClO$_4$ 0.01	12A0	C=1.1, Pt Aux
				MeCN 97-99	PW	Et_4NClO_4	0.1	-	-	25	DME// Ag/ AgClO$_4$ 0.01	12A2	C=1, Pt Aux
				MeCN 99.95	PY	Et_4NClO_4	0.1	-	-	25	DME// Ag/ AgClO$_4$ 0.01	12A0	C=1.1, Pt Aux
					PW	Et_4NClO_4	0.1	-	-	25	DME// Ag/ AgClO$_4$ 0.01	12A2	C=0.23-2.25, Pt Aux
FB09 at14	$C_{11}H_{10}$	2-methylnaphthalene C.A. 91-57-6	Table II	DMF	PY	Bu_4NI	0.25	-	-	-	DME//SCE (o)	2-0	C=2
						Bu_4NI PHEN	0.25 0.01						
FB10 at50 cf93 ed72	$C_{11}H_{12}O_2$	ethyl cinnamate C.A. 103-36-6	C_6H_5CH:CHCOO-C_2H_5	MeCN 90	PY	Et_4NClO_4	0.1	-	-	25	DME// Ag/ AgClO$_4$ 0.01	12A0	C=3.4, Pt Aux
				MeCN 98	PY	Et_4NClO_4	0.1	-	-	25	DME// Ag/ AgClO$_4$ 0.01	12A0	C=3.4, Pt Aux
				MeCN 99	PW	Et_4NClO_4	0.1	-	-	25	DME// Ag/ AgClO$_4$ 0.01	12A2	C=0.38, Pt Aux
													C=3.78, Pt Aux
FB08 CONT													

TABLE I. Electrochemical Data

$C_{11}H_{12}O_2$ (CONT.) FB10

Ref.	C/M	Charact. Potential		Response Const.		n Tech.		Electrokinetic Data			Products and Identification	Description and Remarks	Code No.
		Value	vs.		Value			Parameter	Value	From			
JE042 0189	50s	$E_{\frac{1}{2}}$	-1.48F	Ag/AgClO$_4$ 0.01	i_ℓ	3.4F	-	-	-	-	-	-	C,r	FB08
			-2.12F			1.2F							C	
JE042 0189	50s	-	-	-	-	-	-	-	$dE_p/d\log v$	20	for v < 1E6	-	C, data for first wave only, r	
										0	for v > 1E6			
			-							-			C(?)	
JE042 0189	50s	$E_{\frac{1}{2}}$	-1.48F	Ag/AgClO$_4$ 0.01	i_ℓ	3.6F	i:i, QE	1 0.94	-	-	-	-	C,r	
			-					-					C(?)	
JE042 0189	50s	-	-	-	-	-	-	-	$dE_p/d\log v$	18	for v < 1E6	-	C,r	
										0	for v > 1E6			
									$dE_p/d\log C$	21	sttd			
			-							-			C(?)	
EA019 0629	-	$E_{\frac{1}{2}}$	-2.63	SCE(o)	i_ℓ	5.6	-	-	-	-	-	-	C,r	FB09 at14
			-2.53			12.4	QE (-2.7V)	4.2	-	-	-	2-methyl-1,4-di-hydronaphthalene 60%, 2-methyl-3,4-dihydronaphthalene 30%	C,r	
JE042 0189	50s	$E_{\frac{1}{2}}$	-2.0F	Ag/AgClO$_4$ 0.01	i_ℓ	40F	-	-	-	-	-	-	C,r	FB10 at50 cf93 ed72
			-2.2F			16.7F							C	
JE042 0189	50s	$E_{\frac{1}{2}}$	-2.04F	Ag/AgClO$_4$ 0.01	i_ℓ	30F	-	-	-	-	-	-	C,r	
			-2.2F			37F							C	
JE042 0189	50s	-	-	-	-	-	-	-	$dE_p/d\log v$	19	for v < 5E2	-	C, $dE_p/d\log[H_2O]=5$ for $[H_2O]=0.05-1\%$ (V/V) data for first wave only, r	
										0	for v > 5E2			
									$dE_p/d\log C$	20	sttd			
									$dE_p/d\log[H_2O]$	5				
			-							-			C(?)	
		-	-	-	-	-	-	-	$dE_p/d\log v$	18	for v < 5E3	-	C,r	
										0	for v > 5E3			
			-							-			C(?)	
														CONT

FB10 (CONT.) $C_{11}H_{12}O_2$

Code No.	Empirical Formula	Name and C.A. Number	Structural Formula	Solvent	Tech.	Medium	μ, M	pH	T, °C	Electrodes	App.	Experimental Parameters
FB10 at50 cf93 ed72	$C_{11}H_{12}O_2$	ethyl cinnamate	$C_6H_5CH{:}CHCOO\text{-}C_2H_5$	MeCN 99	PW	Et_4NClO_4 0.1	–	–	25	DME// Ag/ AgClO$_4$ 0.01	12A2	C=4.6, Pt Aux
				MeCN 99.95	PY	Et_4NClO_4 0.1	–	–	25	DME// Ag/ AgClO$_4$ 0.01	12A0	C=3.4, Pt Aux
					PW	Et_4NClO_4 0.1	–	–	25	DME// Ag/ AgClO$_4$ 0.01	12A2	C=4.61, Pt Aux
				DMF 90	PW	Et_4NClO_4 0.1	–	–	25	DME// Ag/ AgClO$_4$ 0.01	12A2	C=5.7, Pt Aux
				DMF 95	PW	Et_4NClO_4 0.1	–	–	25	DME// Ag/ AgClO$_4$ 0.01	12A2	C=5.7, Pt Aux
FB11 at70	$C_{11}H_{14}O$	2,2-dimethyl-1-phenyl-1-propanone C.A. 938-16-9	$C_6H_5COC(CH_3)_3$	MeOH	PY	Me_4NCl 0.1 TX100 0.05	–	–	25	DME// Ag/AgCl, KCl satd	O15- –	C=1
						1R,2S-$C_6H_5CH\text{-}OHCH(CH_3)\text{-}NH_2CH_3^+$ Cl^- 0.1 TX100 0.05						
						1R,2S-$C_6H_5CH\text{-}OHCH(CH_3)\text{-}N(CH_3)_3^+$ Cl^- 0.1 TX100 0.05						
						S-$C_6H_5CH_2CH\text{-}(CH_3)N\text{-}(CH_3)_3^+$ Cl^- 0.1 TX100 0.05						
FB12 at71 CONT	$C_{11}H_{14}O_2$	1,2-dimethoxy-4-propenylbenzene C.A. 54349-79-0	Table II	MeCN	VY	Et_4NClO_4 0.1	–	–	0	Pt: rodi// SCE	5-0	C=0.4, r_d=0.25, v=25, ω=9

TABLE I. Electrochemical Data

$C_{11}H_{14}O_2$ (CONT.) FB12

Ref.	C/M	Charact. Potential Value	vs.	Response Const. Value		Tech.	n	Electrokinetic Data Parameter	Value	From	Products and Identification	Description and Remarks	Code No.
JE042 0189	50s	-	-	-	-	-	-	$dE_p/dlogv$	21	for v < 6.8E3	-	C,r	FB10 at50 cf93 ed72
									0	for v > 6.8E3			
		-							-			C(?)	
JE042 0189	50r	$E_{\frac{1}{2}}$	-2.1F	Ag/ AgClO$_4$ 0.01	i_ℓ	30F	i:i	1	-	-	-	-	C,r
		-2.38F			37F		1					C	
JE042 0189	50r	-	-	-	-	-	-	$dE_p/logv$	20	for v < 3.5E3	-	C,data for first wave only,r	
									0	for v > 3.5E3			
		-							-			C(?)	
JE042 0189	50s	-	-	-	-	-	-	$dE_p/dlogv$	21	for v < 1.2E3	-	C,$dE_p/dlog[H_2O]$=20 for $[H_2O]$=5-10% (V/V),data for first wave only,r	
									0	for v > 1.2E3			
								$dE_p/dlog[H_2O]$	20	sttd			
		-							-			C(?)	
JE042 0189	50s	-	-	-	-	-	-	$dE_p/dlogv$	21	for v < 6E2	-	C,r	
									0	for v > 6E2			
		-							-			C(?)	
EA019 0611	65y	$E_{\frac{1}{2}}$	-1.66	Ag/AgCl, KCl satd	-	-	-	-	-	-	t-butylphenylcarbinol 89%,2,2,5,5-tetramethyl-3,4-diphenyl-3,4-hexandiol 2%	C,r	FB11 at70
		-1.60		-	-	-	-	-	-	-	-	C,M,r	
		-1.66		-	-	-	-	-	-	-	t-butylphenylcarbinol 73%(S-2.5%);2,2,5,5-tetramethyl-3,4-diphenyl-3,4-hexandiol 4%	C,r	
		-1.66		-	-	-	-	-	-	-	t-butylphenylcarbinol 71%(S-0.8%);2,2,5,5-tetramethyl-3,4-diphenyl-3,4-hexandiol 4%	C,r	
EA019 0681	-	$E_{\frac{1}{2}}$	1.14F	SCE	i_ℓ	9F	-	-	-	-	-	A,r	FB12 at71
		1.36F			8F							A	

CONT

FB12 (CONT.) $C_{11}H_{14}O_2$

Code No.	Empirical Formula	Name and C.A. Number	Structural Formula	Solvent	Tech.	Medium	μ, M	pH	T, °C	Electrodes	App.	Experimental Parameters
FB12 at71	$C_{11}H_{14}O_2$	1,2-dimethoxy-4-propenylbenzene	Table II	MeCN	VY	Et_4NClO_4 0.1 PYR 3E-4	-	-	0	Pt: rodi// SCE	5-O	C=0.4, r_d=0.25, v=25, ω=9
						Et_4NClO_4 0.1 PYR 4E-4						
						Et_4NClO_4 0.1 PYR 6E-4						
						Et_4NClO_4 0.1 PYR 2E-3						
						Et_4NClO_4 0.1 PYR 2E-2						
						Et_4NClO_4 0.1				Pt: rord// SCE		C=0.4, E_{disc}=1.2, $r_{r,i}$=0.275, $r_{r,o}$=0.335, v=25, ω=9, N_o=0.33±0.1
						Et_4NClO_4 0.1 PYR 5E-5						
						Et_4NClO_4 0.1 PYR 1E-4						
						Et_4NClO_4 0.1 PYR 1.25E-4						
						Et_4NClO_4 0.1 PYR 3E-4						
						Et_4NClO_4 0.1 PYR 4E-4						
						Et_4NClO_4 0.1 PYR 6E-4						
						Et_4NClO_4 0.1 PYR 2E-2						
FB13 cg05	$C_{11}H_{15}N_2O_8P$	nicotinamide mononucleotide C.A. 1094-61-7	Table II-3	DMSO	PY	Et_4NClO_4 0.1	-	-	-	DME//SCE //Rb^+	35AF0	C=ns, m=1.02, t=5.0, h=70
					VR	Et_4NClO_4 0.1	-	-	-	HMDE// SCE//Rb^+	35AF0	0→-2.2→ 0→-2.2 V, C=ns
FB12	$C_{11}H_{14}O_2$	1,2-dimethoxy-4-propenylbenzene		MeCN	VY							

TABLE I. Electrochemical Data

$C_{11}H_{15}N_2O_8P$ FB13

Ref.	C/M	Charact. Potential		Response Const.		n Tech.	n	Electrokinetic Data			Products and Identification	Description and Remarks	Code No.
		Value	vs.		Value			Parameter	Value	From			
EA019 0681	-	$E_{\frac{1}{2}}$	1.14F	SCE	i_ℓ	9F	-	-	-	-	-	-	A,r	FB12 at71
			1.38F			9F							A	
			1.15F			10	-	-	-	-	-	-	A,r	
			1.53F			10Ap							A	
			1.16F			10.4F	-	-	-	-	-	-	A,r	
			-			-							A,0	
			1.16F			14F	-	-	-	-	-	-	A,r	
			1.17F			15F	-	-	-	-	-	-	A,r	
		$E_{\frac{1}{2}}$, ring	0.04F			17.1F	-	-	-	-	-	-	C,r	
			0.08F			10.0F	-	-	-	-	-	-	C,r	
			-0.47F			6.4F							C	
			0.08F			2.9F	-	-	-	-	-	-	C,r	
			-0.47F			14.3							C	
			-0.49F			18.6F	-	-	-	-	-	-	C,r	
			-0.49F			15F	-	-	-	-	-	-	C,r	
			-1.29F			3.2F							C	
			-0.49F			11.4F	-	-	-	-	-	-	C,r	
			-1.23F			7.9F							C	
			-0.49F			7.9F	-	-	-	-	-	-	C,r	
			-0.98F			3.6F							C	
			-1.26F			10.0F							C	
			-0.95F			12.1F	-	-	-	-	-	-	C,r	
			-1.22F			14.3F							C	
			-			-							C	
JA095 5482	42e f	$E_{\frac{1}{2}}$	-0.38	SCE	I	1.25	-	-	-	-	-	-	C, i_ℓ=kh, i_a,r	FB13 cg05
			-0.99			-			Tomeš	45	sttd		C,R,i_d	
JA095 5482	42e f	E_p	-0.48	SCE	-	-	-	-	-	-	-	-	C,Pr,r	
			-1.02			2.1							C	
			-1.96			-							C,redn of Na(I),R	
			-1.90										A,R	
			-0.23										A,W	
			-0.13			0.5							A	

FB14 $C_{12}F_{10}$

Code No.	Empirical Formula	Name and C.A. Number	Structural Formula	Solvent	Tech.	Medium		μ, M	pH	T, °C	Electrodes	App.	Experimental Parameters
FB14 cg27	$C_{12}F_{10}$	decafluorobiphenyl C.A. 434-90-2	$(C_6F_5)_2$	HFSO$_3$	VA	HOAc	0.1	-	-	20	Pt: xxwi// Pd/H$_2$	25-0	C=ns,A=0.6, v=100,Pt Aux
FB15 ba06	$C_{12}H_4N_4$	2,2'-(2,5-cyclo-hexadiene-1,4-diyl-idene)bispropane-dinitrile C.A. 1518-16-7	Table II	MeCN	VA	LiClO$_4$	0.1	-	-	25	gw: xxdi// SCE		0.4→-0.5→ 0.4→-0.5V, C=1.24, d=0.62,v=67, Pt Aux
FB16 ba20	$C_{12}H_8N_2$	9,10-diazaphen-anthrene C.A. 230-17-1	Table II-2	MeCN	IL	Et$_4$NClO$_4$	ns	-	-	-	Pt: xxbu// SCE	2-0	C=ns,A=0.22
FB17 ba21 dd64	$C_{12}H_8N_2$	1,10-phenanthroline C.A. 66-71-7	Table II-2	MeCN	VA	Bu$_4$NBF$_4$	0.1	-	-	-	Pt: xxdi//Ag	2F-	0→-2.6→0V, C=ns,Pt Aux 0→2.1→0V
FB18 ba22 cg69	$C_{12}H_8N_2$	phenazine C.A. 92-82-0	Table II-2	MeCN	VA	Et$_4$NClO$_4$	ns	-	-	-	Pt: xxbu// SCE	2-0	0→-1.8→ 1.7→0V, C=ns,A=0.22
FB19 ba30	$C_{12}H_8N_2O_4$	2,2'-dinitrobiphenyl C.A. 2436-96-6	$(2-O_2NC_6H_4)_2$	MeCN	XT	Et$_4$NClO$_4$	0.1	-	-	-30 ±1	HMDE// Ag/ AgClO$_4$ 0.01	2-2	C=ns,v= (1.5-5.5)E3
										0			
										20± 0.05			
				DMF	XT	Et$_4$NClO$_4$	0.1	-	-	0±1	HMDE// SCE	2-2	C=ns,v= (1.5-5.5)E3
										30± 0.5			
										60± 0.5			
FB20 ba30 CONT	$C_{12}H_8N_2O_4$	3,3'-dinitrobiphenyl C.A. 958-96-3	$(3-O_2NC_6H_4)_2$	MeCN	XT	Et$_4$NClO$_4$	0.1	-	-	-30 ±1	HMDE// Ag/ AgClO$_4$ 0.01	2-2	C=ns,v= (1.5-5.5)E3

$C_{12}H_8N_2O_4$ (CONT.) FB20

Ref.	C/M	Charact. Potential		Response Const.		n Tech.		Electrokinetic Data			Products and Identification	Description and Remarks	Code No.
		Value	vs.		Value			Parameter	Value	From			
EA018 0331	-	$E_{p/2}$	1.14	-	i_p/C	43	sttd	-	ΔE_p	70	sttd	$R^{\pm}$ (stable), ESR	A,R,r	FB14 cg27
			1.39			143				-			A	
			-			-							C,R	
EA021 0973	50t	E_p	0.05F	SCE	i_p	106F	-	-	ΔE_p	130	sttd	radical anion	C,R,D=14.2,r	FB15 ba06
			-0.40F			94F				-		dianion	C,R	
			-0.23F			100F							A,R	
			0.22			100F							A,R	
JE049 0111	-	$E_{p/2}$	1.72	SCE	-	-	-	-	-	-	-	-	A,p	FB16 ba20
JA095 6582	-	E_p	-2.1	SCE	-	-	-	-	-	-	-	-	C,≠,r	FB17 ba21 dd64
	-	-	-	-	-	-	-	-	-	-	-	-	A,O,r	
JE049 0111	-	$E_{p/2}$	-	SCE	$j_p/Cv^{\frac{1}{2}}$	10.7	-	-	-	-	-	-	C,p	FB18 ba22 cg69
			1.91			8.99							A	
JE047 0115	-	XE	0.190 ±0.002	Ag/ AgClO$_4$ 0.01	-	-	-	-	-	-	-	-	C,XT=semiintegral chronoamperometry with linear potential sweep,XE=difference between potentials for the two steps,p	FB19 ba30
			-										C	
			0.224 ±0.002		-	-	-	-	-	-	-	-	C,p	
			-										C	
			0.252 ±0.002		-	-	-	-	-	-	-	-	C,p	
			-										C	
JE047 0115	-	XE	0.246 ±0.002	SCE	-	-	-	-	-	-	-	-	C,XT=semiintegral chronoamperometry with linear potential sweep,XE=difference between potentials for the two steps,p	
			0.269 ±0.002		-	-	-	-	-	-	-	-	C,p	
			-										C	
			0.302 ±0.002		-	-	-	-	-	-	-	-	C,p	
			-										C	
JE047 0115	-	XE	0.072 ±0.002	Ag/ AgClO$_4$ 0.01	-	-	-	-	-	-	-	-	C,XT=semiintegral chronoamperometry with linear potential sweep,XE=difference between potentials for the two steps,p	FB20
			-										C	CONT

FB20 (CONT.) $C_{12}H_8N_2O_4$

Code No.	Empirical Formula	Name and C.A. Number	Structural Formula	Solvent	Tech.	Medium	μ, M	pH	T, °C	Electrodes	App.	Experimental Parameters
FB20	$C_{12}H_8N_2O_4$	3,3'-dinitrobiphenyl	$(3-O_2NC_6H_4)_2$	MeCN	XT	Et_4NClO_4 0.1	-	-	20±0.5	HMDE// Ag/AgClO$_4$ 0.01	2-2	C=ns, v=(1.5-5.5)E3
				DMF	XT	Et_4NClO_4 0.1	-	-	0±1	HMDE// SCE	2-2	C=ns, v=(1.5-5.5)E3
									60±0.5			
FB21 ba32	$C_{12}H_8N_2O_4$	4,4'-dinitrobiphenyl C.A. 1528-74-1	$(4-O_2NC_6H_4)_2$	MeCN	XT	Et_4NClO_4 0.1	-	-	-30±1	HMDE// Ag/AgClO$_4$ 0.01	2-2	C=ns, v=(1.5-5.5)E
									20±0.5			
				DMF 98	XT	Et_4NClO_4 0.1	-	-	0	HMDE// Ag/AgI, Bu$_4$NI 0.1, Et$_4$NClO$_4$ 0.1	2-2	C=ns, v=(1.5-5.5)E3
				DMF	IL	Et_4NClO_4 0.1	-	-	40±0.5	HMDE// SCE	2-2	C=ns, v=3E3
					XT	Et_4NClO_4 0.1	-	-	0±1	HMDE// SCE	2-2	v=(1.5-5.5)E3
									50±0.5			
					XT	Et_4NClO_4 0.1 Li$^+$ 0.015	-	-	0	HMDE// Ag/AgI, Bu$_4$NI 0.1, Et$_4$NClO$_4$ 0.1	2-2	C=ns, v=(1.5-5.5)E3

TABLE I. Electrochemical Data

$C_{12}H_8N_2O_4$ FB21

Ref.	C/M	Charact. Value	Potential vs.	Response Const. Value		n Tech.	Electrokinetic Data Parameter	Value	From	Products and Identification	Description and Remarks	Code No.
JE047 0115	-	XE	0.091 ±0.002	Ag/ AgClO$_4$ 0.01	-	-	-	-	-	-	-	C,p	FB20
			-									C	
JE047 0115	-	XE	0.090 ±0.002	SCE	-	-	-	-	-	-	-	C,XT=semiintegral chronoamperometry with linear potential sweep,XE=difference between potentials for the two steps,p	
			-									C	
			0.110 ±0.002		-	-	-	-	-	-	-	C,p	
			-									C	
JE047 0115	-	XE	0.01 ±0.003	Ag/ AgClO$_4$ 0.01	-	-	-	-	-	-	-	C,XT=semiintegral chronoamperometry with linear potential sweep,XE=difference between potentials for the two steps,p	FB21 ba32
			-									C	
			0.021 ±0.002		-	-	-	-	-	-	-	C,p	
			-									C	
JE057 0027	-	XE	0.016	Ag/AgI Bu$_4$NI 0.1, Et$_4$NClO$_4$ 0.1	-	-	-	-	-	-	-	C,XT=semiintegral linear potential sweep chronoamperometry,XE=difference between potentials for the two steps, see reference for dependence of XE on [Li$^+$],[H$_2$O],and T,p	
			-									C	
JE047 0215	-	E$_p$	-0.98F	SCE	i$_p$	22.9F	-	-	-	-	-	C,p	
JE047 0115	-	XE	0.056 ±0.002	SCE	-	-	-	-	-	-	-	C,XT=semiintegral chronoamperometry with linear potential sweep,XE=difference between p potentials for the two steps,p	
			-									C	
			0.068 ±0.002		-	-	-	-	-	-	-	C,p	
			-									C	
JE057 0027	-	XE	0.015	Ag/AgI, Bu$_4$NI 0.1, Et$_4$NClO$_4$ 0.1	-	-	-	-	-	-	-	C,XT=semiintegral linear potential sweep chronoamperometry,XE=difference between potentials for the two steps, see reference for dependence of XE on [Li$^+$],[H$_2$O],and T,p	
			-									C	

FB22 $C_{12}H_8OS_2$

Code No.	Empirical Formula	Name and C.A. Number	Structural Formula	Solvent	Tech.	Medium		μ, M	pH	T, °C	Electrodes	App.	Experimental Parameters
FB22	$C_{12}H_8OS_2$	thianthrene 5-oxide C.A. 2362-50-7	Table II	HOAc 80	CP	$HClO_4$	0.2	-	-	-	Pt: nsgz// Ag/AgCl, KCl	2--	C=ns, A=100, E_{app}=1.6 V, Pt Aux
					IL	$HClO_4$	0.2	-	-	-	Pt: xxwi// Ag/AgCl, KCl	25-0	C=1, A=0.25, v=25, Pt Aux
FB23	$C_{12}H_8O_2S_2$	thianthrene 5,5-dioxide	Table II	HOAc 80	IL	$HClO_4$	0.2	-	-	-	Pt: xxwi// Ag/AgCl, KCl	25-0	C=ns, A=0.25, v=25, Pt Aux
FB24	$C_{12}H_8O_2S_2$	thianthrene 5,10-dioxide C.A. 951-02-0	Table II	HOAc 80	IL	$HClO_4$	0.2	-	-	-	Pt: xxwi// Ag/AgCl, KCl	25-0	C=ns, A=0.25, v=25, Pt Aux
FB25 ba42 cg75 dd67 ed90	$C_{12}H_8S_2$	thianthrene C.A. 92-85-3	Table II-2	HOAc 80	CP	$HClO_4$	0.2	-	-	-	Pt: nsgz// Ag/AgCl, KCl	29--	C=ns, A=100, E_{app}=1.15 V, Pt Aux
													E_{app}=1.6
					IL	$HClO_4$	0.2	-	-	-	Pt: xxwi// Ag/AgCl, KCl	25-0	C=0.4, A=0.25, v=25, Pt Aux
				MeCN	VA	Bu_4NBF_4 Al_2O_3	0.2 0.25 gcm^{-3}	-	-	-	Pt: xxbu// Ag/ $AgNO_3$ 0.1, Bu_4NBF_4 0.1	25AF0	1.0→2.0→ 1.0 V, C=1, v=86
				C_6H_5CN	VA	Bu_4NBF_4 Al_2O_3	0.2 0.25 gcm^{-3}	-	-	-	Pt: xxbu// Ag/ $AgNO_3$ 0.1, Bu_4NBF_4 0.1	25AF0	1.0→2.0→ 1.0 V, C=1, v=86
				2-PrCN	VA	Bu_4NBF_4 Al_2O_3	0.2 0.25 gcm^{-3}	-	-	-	Pt: xxbu// Ag/ $AgNO_3$ 0.1, Bu_4NBF_4 0.1	25AF0	1.0→2.0→ 1.0 V, C=1.0, v=86

CONT

TABLE I. Electrochemical Data

$C_{12}H_8S_2$ (CONT.) FB25

Ref.	C/M	Charact. Potential		Response Const.		n Tech.		Electrokinetic Data			Products and Identification	Description and Remarks	Code No.
		Value	vs.		Value			Parameter	Value	From			
EA018 0373	-	-	-	-	-	-	-	-	-	-	thianthrene cis-dioxide 44%, thianthrene trans-dioxide 28%; thianthrene sulfone 13%, thianthrene trioxide 10%, thianthrene tetroxide 5%	A,p	FB22
EA018 0373	-	E_p	1.45	Ag/AgCl, KCl	i_p	260F	-	-	-	-	-	A,p	
EA018 0373	-	-	-	-	-	-	-	-	-	-	-	A,O,p	FB23
EA018 0373	-	-	-	-	-	-	-	-	-	-	-	A,O,p	FB24
EA018 0373	371 d	-	-	-	-	-	-	-	-	-	thianthrene monoxide 98%, IR	A,p	FB25 ba42 cg75 dd67 ed90
		-	-	-	-	-	-	-	-	-	thianthrene cis-dioxide 44%, thianthrene trans-dioxide 28%, thianthrene sulfone 13%, thianthrene trioxide 10%, thianthrene tetroxide 5%	A,p	
EA018 0373	371 d	E_p	1.10 1.45	Ag/AgCl, KCl	i_p	100F 110ApF	-	-	-	-	-	A,W,p A	
EA018 0537	371 e	E_p	1.28 1.55 1.48 1.21	SCE	-	-	sttd	1 1 1 1	-	-	-	thianthrene radical cation thianthrene dication	A,R,Al$_2$O$_3$ added to remove H$_2$O,r A,R C,R C,R	
EA018 0537	371 e	E_p	1.32 1.81 1.71 1.24	SCE	-	-	sttd	1 1 1 1	-	-	-	thianthrene radical cation thianthrene dication	A,R A,R C,R C,R	
EA018 0537	371 e	E_p	1.33 1.75 1.68 1.27	SCE	-	-	sttd	1 1 1 1	-	-	-	thianthrene radical cation thianthrene dication	A,R A,R C,R C,R	

CONT

FB25 (CONT.) $C_{12}H_8S_2$

Code No.	Empirical Formula	Name and C.A. Number	Structural Formula	Solvent	Tech.	Medium		μ, M	pH	T, °C	Electrodes	App.	Experimental Parameters
FB25 ba42 cg75 dd67 ed90	$C_{12}H_8S_2$	thianthrene	Table II-2	CH_2Cl_2 88 F_3CCOOH 2 $(F_3CCO)_2O$ 10	VA	Bu_4NBF_4	0.2	-	-	-	Pt: xxbu// Ag/ $AgNO_3$ 0.1, Bu_4NBF_4 0.1	25A0	$1.0 \to 2.0 \to 1.0$ V, C=1, v=43
													v=86
													v=210
				CH_2Cl_2	VA	Bu_4NBF_4 Al_2O_3	0.2 0.25 gcm^{-3}	-	-	-	Pt: xxbu// Ag/ $AgNO_3$ 0.1, Bu_4NBF_4 0.1	25AF0	$1.0 \to 2.0 \to 1.0$ V, C=1, v=86
				$C_6H_5NO_2$	VA	Bu_4NBF_4 Al_2O_3	0.2 0.25 gcm^{-3}	-	-	-	Pt: xxbu// Ag/ $AgNO_3$ 0.1, Bu_4NBF_4 0.1	25AF0	C=1, v=86
				$MeNO_2$	VA	Bu_4NBF_4 Al_2O_3	0.2 0.25 gcm^{-3}	-	-	-	Pt: xxbu// Ag/ $AgNO_3$ 0.1, Bu_4NBF_4 0.1	25AF0	C=1, v=86
				EtCN	VA	Bu_4NBF_4 Al_2O_3	0.2 0.25 gcm^{-3}	-	-	-	Pt: xxbu// Ag/ $AgNO_3$ 0.1, Bu_4NBF_4 0.1	25AF0	C=1.0, v=86
				F_3CCOOH 90 $(F_3CCO)_2O$ 10	VA	Bu_4NBF_4	0.2	-	-	-	Pt: xxbu// Ag/ $AgNO_3$ 0.1, Bu_4NBF_4 0.1	25A0	C=1.0, v=86
FB26	$C_{12}H_9B_2N_2O$	4-[(4-bromophenyl)-azo]phenol C.A. 3035-94-7	Table II	H_2O	PY	BR		-	11.00 ±0.05	-	DME//SCE	05A01	C=ns, m=2.11, t=4.09, h=45

TABLE I. Electrochemical Data

$C_{12}H_9B_2N_2O$ FB26

Ref.	C/M	Charact. Potential		Response Const.		n Tech.		Electrokinetic Data			Products and Identification	Description and Remarks	Code No.
			Value	vs.		Value			Parameter	Value	From			
EA018 0537	371 e	-	-	-	$i_p/Cv^{\frac{1}{2}}$	7.0	sttd	1	-	-	-	thianthrene radical cation	A,R,r	FB25 ba42 cg75 dd67 ed90
						6.3		1				thianthrene dication	A,R	
						6.3		1					C,R	
						7.0		1					C,R	
		E_p	1.36	SCE		6.9	QE	1.0	-	-	-	thianthrene radical cation,ESR,IRS	A,R,r	
			1.96			6.3		1.0				thianthrene di-cation	A,R	
			1.88			6.3	sttd	1					C,R	
			1.28			6.9		1					C,R	
			-			6.6		1	-	-	-	thianthrene radical cation	A,R,r	
						6.0		1				thianthrene dication	A,R	
						6.0		1					C,R	
						6.6		1					C,R	
EA018 0537	371 e	E_p	1.33	SCE	-	-	sttd	1	-	-	-	thianthrene radical cation	A,R,r	
			1.83					1				thianthrene dication	A,R	
			-					1					C,dication strongly adsorbed,R	
			1.27					1					C,R	
EA018 0537	371 e	E_p	1.32	SCE	-	-	sttd	1	-	-	-	thianthrene radical cation	A,R,r	
			1.83					1				thianthrene dication	A,R	
			1.75					1					C,R	
			1.25					1					C,R	
EA018 0537	371 e	E_p	1.22	SCE	-	-	sttd	1	-	-	-	thianthrene radical cation	A,R,r	
			1.75					1				thianthrene dication	A,R	
			1.65					1					C,R	
			1.16					1					C,R	
EA018 0537	371 e	E_p	1.31	SCE	-	-	sttd	1	-	-	-	thianthrene radical cation	A,R,r	
			1.77					1				thianthrene dication	A,R	
			1.70					1					C,R	
			1.24					1					C,R	
EA018 0537	371 e	E_p	1.26	SCE	-	-	sttd	1	-	-	-	thianthrene radical cation	A,R,r	
			1.95					1				thianthrene dication	A,R	
			1.88					1					C,R,$i_{p,C}/i_{p,A}=1.0$	
			1.20					1					C,R,$i_{p,C}/i_{p,A}=1.0$	
EA018 0139	-	$E_{\frac{1}{2}}$	-0.72	SCE	-	-	-	-	$dE_{\frac{1}{2}}/dpH$	50	sttd	-	C,$\ast$,i_ℓ=kh$^{0.45-0.55}$, i_d,$dE_{\frac{1}{2}}/dpH$ for pH=9.5-12,$\Delta E_{\frac{1}{2}}$= 0.17σ,p	FB26
									k_h^o	5.90 E-13	sttd			
			-							-			C,X,H	

FB27 $C_{12}H_9ClN_2O$

Code No.	Empirical Formula	Name and C.A. Number	Structural Formula	Solvent	Tech.	Medium	μ, M	pH	T, °C	Electrodes	App.	Experimental Parameters
FB27	$C_{12}H_9ClN_2O$	4-[(4-chlorophenyl)-azo]phenyl C.A. 2703-27-7	Table II	H_2O	PY	BR	-	11.00 ±0.05	-	DME//SCE	05AO1	C=ns, m=2.11, t=4.09, h=45
FB28	$C_{12}H_9ClN_2O_2$	4-chloro-4'-nitrodiphenylamine C.A. 20983-67-9	Table II	MeCN	VY	Et_4NClO_4 0.1	-	-	-	Pt: rodi// Ag/Ag⁺ 0.01, Et_4NClO_4 0.1	2-O	C=ns, d=0.1, ω=10, Pt Aux
FB29	$C_{12}H_9ClN_2O_2$	N-(4-chlorophenyl)-N-(4-nitrosophenyl)-hydroxylamine C.A. 56795-80-3	Table II	MeCN	VA	Et_4NClO_4 0.1	-	-	-	Pt: xxdi// Ag/ $AgClO_4$ 0.01, Et_4NClO_4 0.1//Fc	2AO	C=0.9, Pt Aux
						Et_4NClO_4 0.1 2,6-lutidine 1.8E-3						
						Et_4NClO_4 0.1 Bu_4NOAc HOAc 9E-4						
					VY	Et_4NClO_4 0.1	-	-	-	Pt: rodi// Ag/ $AgClO_4$ 0.01, Et_4NClO_4 0.1//Fc	2AO	C=0.9, ω=10, Pt Aux
						Et_4NClO_4 0.1 2,6-lutidine 1.8E-3						
						Et_4NClO_4 0.1 Bu_4NOAc HOAc 9E-4						
FB30	$C_{12}H_9Cl_2N$	4,4'-dichlorodiphenylamine C.A. 6962-04-5	$(4-ClC_6H_4)_2NH$	MeCN	VY	Et_4NClO_4 0.1	-	-	-	Pt: rodi// Ag/ $AgClO_4$ 0.01, Et_4NClO_4 0.1//Fc	2-O	C=ns, d=0.1, ω=10, Pt Aux
FB31	$C_{12}H_9IN_2O$	4-[(4-iodophenyl)-azo]phenol C.A. 2703-28-8	Table II	H_2O	PY	BR	-	11.00 ±0.05	-	DME//SCE	05AO1	C=ns, m=2.11, t=4.09, h=45

TABLE I. Electrochemical Data

$C_{12}H_9IN_2O$ FB31

Ref.	C/M	Charact. Potential		Response Const.		n Tech.	Electrokinetic Data			Products and Identification	Description and Remarks	Code No.
		Value	vs.		Value		Parameter	Value	From			
EA018 0139	-	$E_{\frac{1}{2}}$	-0.71	SCE	-	-	-	$dE_{\frac{1}{2}}/dpH$	57	sttd	-	$C, \neq, i_\ell = kh^{0.45-0.55}$, $i_d, dE_{\frac{1}{2}}/dpH$ for pH=8.5-12, $\Delta E_{\frac{1}{2}} = 0.17\sigma, p$	FB27
							k_h^0	3.18 E-20				
			-					-			C,H	
EA021 0557	147 n	$E_{\frac{1}{2}}$	0.94	Ag/Ag+ 0.01, Et_4NClO_4 0.1	i_ℓ/C	7.7	sttd 1	-	-	-	radical cation, VIS	A, r	FB28
			1.48		-		1					A	
EA020 0469	424 c		-	-	-	-	-	ΔE_p	160	sttd	-	A, p	FB29
			-						75			A	
			-						-			C	
			-						-			C	
			-	-	-	-	-		210	-		A, p	
									70			A	
			-						-			C	
			-						-			C	
			-	-	-	-	-		160	-		A, p	
			-						80			A	
			-						-			C	
			-						-			C	
EA020 0469	424 c	$E_{\frac{1}{2}}$	0.51	Ag/AgClO$_4$ 0.01, Et_4NClO_4 0.1	-	-	-	-	-	-	-	A, p	
			0.81									A	
			0.13		-	-	-	-	-	-	-	A, p	
			0.81									A	
			-0.31		-	-	-	-	-	-	-	A, p	
			-0.81									A	
EA021 0557	147 h	$E_{\frac{1}{2}}$	0.74	Ag/AgClO$_4$ 0.01, Et_4NClO_4 0.1	i_ℓ/C	6.5	sttd 1	-	-	-	radical cation, VIS	A, $\Delta E_{\frac{1}{2}} = 0.46\sigma, r$	FB30
			1.46		-		1					A	
EA018 0139	-	$E_{\frac{1}{2}}$	-0.72	SCE	-	-	-	$dE_{\frac{1}{2}}/dpH$	58	sttd		$C, \neq, i_\ell = kh^{\frac{1}{2}}, i_d$, $dE_{\frac{1}{2}}/dpH$ for pH=9.5-12, $\Delta E_{\frac{1}{2}} = 0.17\sigma, p$	FB31
								k_h^0	1.10 E-10				
			-						-			C,X,H	

FB32 $C_{12}H_9N_3O_3$

Code No.	Empirical Formula	Name and C.A. Number	Structural Formula	Solvent	Tech.	Medium		μ, M	pH	T, °C	Electrodes	App.	Experimental Parameters
FB32	$C_{12}H_9N_3O_3$	4-[(2-nitrophenyl)-azo]phenol C.A. 2724-85-8	Table II	H_2O	PY	BR	—	—	9.10 ±0.05	—	DME//SCE	05A01	C=ns, m=2.11, t=4.09, h=45
									10.98 ±0.05				
FB33	$C_{12}H_9N_3O_3$	4-[(3-nitrophenyl)-azo]phenol C.A. 2011-53-2	Table II	H_2O	PY	BR	—	—	7.12	—	DME//SCE	05A01	C=0.03, m=2.11, t=4.09, h=45
									9.00 ±0.05				
									11.00 ±0.05				
FB34 ed97	$C_{12}H_9N_3O_3$	4-[(4-nitrophenyl-azo]phenol C.A. 1435-60-5		H_2O	PY	BR	—	—	11.05 ±0.05	—	DME//SCE	05A01	C=ns, m=2.11, t=4.09, h=45
FB35	$C_{12}H_9N_3O_4$	4,4'-dinitrodi-phenylamine C.A. 1821-27-8	$(4-O_2NC_6H_4)_2NH$	MeCN	VY	Et_4NClO_4 0.1		—	—	—	Pt: rodi// Ag/ AgClO$_4$ 0.01, Et$_4$NClO$_4$ 0.1//Fc	2-0	C=ns, d=0.1, ω=10, Pt Aux
FB36 ba64 cg90	$C_{12}H_{10}$	biphenyl C.A. 92-52-4	$C_6H_5C_6H_5$	DMF	PY	Bu_4NI 0.25		—	—	—	DME//SCE (o)	2-0	C=2
						Bu_4NI PHEN	0.25 0.01						
FB37	$C_{12}H_{10}BrN$	4-bromodiphenyl-amine C.A. 54446-36-5	$4-BrC_6H_4NHC_6H_5$	MeCN	VY	Et_4NClO_4 0.1		—	—	—	Pt: rodi// Ag/ AgClO$_4$ 0.01, Et$_4$NClO$_4$ 0.1	2-0	C=0.87, d=0.1, ω=10, Pt Aux
						Et_4NClO_4 H_2O	0.1 5E-3						

TABLE I. Electrochemical Data

$C_{12}H_{10}BrN$ FB37

Ref.	C/M	Charact. Potential Value	vs.	Response Const. Value	Tech.	n	Electrokinetic Data Parameter	Value	From	Products and Identification	Description and Remarks	Code No.
EA018 0139	-	-	-	D(∥k) 3.3	sttd	4(?)	$dE_{\frac{1}{2}}/dpH$ k_h^o	50 8.00 E-13	sttd	-	$C, \not\models, i_\ell = kh^{\frac{1}{2}}, i_d$, $dE_{\frac{1}{2}}/dpH$ for pH=8.0-11.8, $\Delta E_{\frac{1}{2}}=0.17\sigma,p$	FB32
		-0.54	SCE	6.8		4(?)	$dE_{\frac{1}{2}}/dpH$ k_h^o	50 1.50 E-20		-	$C,\not\models,i_d,\Delta E_{\frac{1}{2}}=0.17\sigma,p$	
EA018 0139	-	$E_{\frac{1}{2}}$	-0.34F -0.50F -1.2F	SCE	-	-	-	-	-	-	-	$C,\not\models,i_d,p$ C C	FB33
		-0.53F	-	-	sttd	4	$dE_{\frac{1}{2}}/dpH$ k_h^o	58 4.50 E-13	sttd	-	$C,\not\models,i_\ell=kh^{\frac{1}{2}},i_d$, $dE_{\frac{1}{2}}/dpH$ for pH=8.5-11.8, $\Delta E_{\frac{1}{2}}=0.17\sigma,p$	
		-1.2F		-		-		-		-	C,H	
		-0.63		-		-	$dE_{\frac{1}{2}}/dpH$	58	sttd	-	$C,\not\models,i_d,\Delta E_{\frac{1}{2}}=0.17\sigma,p$	
		-0.74						-			C,H	
EA018 0139	-	$E_{\frac{1}{2}}$	-0.56	SCE	-	-	-	$dE_{\frac{1}{2}}/dpH$ k_h^o	50 3.60 E-16	sttd	-	$C,\not\models,i_\ell=kh^{\frac{1}{2}},i_d$, $dE_{\frac{1}{2}}/dpH$ for pH=8.0-11.8, $\Delta E_{\frac{1}{2}}=0.17\sigma,p$	FB34 ed97
EA021 0557	147 p	$E_{\frac{1}{2}}$	1.22	Ag/AgClO$_4$ 0.01, Et$_4$NClO$_4$ 0.1	i_ℓ/C	9Ap	sttd	1	-	-	-	radical cation, unstable	A,$\Delta E_{\frac{1}{2}}=0.46\sigma,r$	FB35
EA019 0629	-	$E_{\frac{1}{2}}$	-2.76	SCE(o)	i_ℓ	5.0	-	-	-	-	-	-	C,r	FB36 ba64 cg90
		-2.71		14.0	QE (-2.8v)	4.5	-	-	-	cyclohexylbenzene 60%, cyclohexen-4-ylbenzene 40%	C,r	
EA021 1171	147 jkl	$E_{\frac{1}{2}}$	0.68F	Ag/AgClO$_4$ 0.01, Et$_4$NClO$_4$ 0.1	i_ℓ	7.6F	-	-	-	-	-	A,X,r	FB37
		0.67		6.2	-	-	-	-	-	radical cation of dimer	A,r	
		0.79		-						dication of dimer	A	

FB38 $C_{12}H_{10}ClN_3$

Code No.	Empirical Formula	Name and C.A. Number	Structural Formula	Solvent	Tech.	Medium		μ, M	pH	T, °C	Electrodes	App.	Experimental Parameters
FB38	$C_{12}H_{10}ClN_3$	1-(4-chlorophenyl)-3-phenyltriazene C.A. 19838-82-5	4-ClC$_6$H$_4$N:NNH-C$_6$H$_5$	MeOH 50	PY	UB GEL	ns 1E-2	-	7	21±1	DME//SCE	O-O	C=ns, m(oc)=4.9, t=1.9
									11				
FB39	$C_{12}H_{10}ClN_3S$	3,7-diaminophenothiazin-5-ium chloride C.A. 581-64-6	Table II	MeOH 50	PY	Me$_4$NOH	0.1	-	-	-	DME//SCE	O--	C=0.05, m=2.61, t=3.0, h=50
FB40 ba88 cg92 dd72 ed99	$C_{12}H_{10}N_2$	azobenzene C.A. 103-33-3	C$_6$H$_5$N:NC$_6$H$_5$	DMF	VA	LiCl	satd	-	-	-	Hg: xxsd// SCE	OA-	C=ns, v=300
						Me$_4$NBF$_4$	satd						
						Me$_4$NBF$_4$ (EtOCO)$_2$CH$_2$	satd ns						
						Me$_4$NBF$_4$ fluorene	satd ns						
				HMP	VA	LiCl	ns	-	-	-	Hg: xxsd// SCE	OA-	C=ns, v=300
FB41 ba93 dd73 ee02	$C_{12}H_{10}N_2O$	4-phenylazophenol C.A. 1689-82-3	4-HOC$_6$H$_4$N:N-C$_6$H$_5$	H$_2$O	PY	BR	-	-	8.45 ±0.05	-	DME//SCE	O5AO1	C=0.5, m=2.11, t=4.09, h=45
									9.85				
									11 ±0.05				
CONT													

TABLE I. Electrochemical Data

$C_{12}H_{10}N_2O$ (CONT.) FB41

Ref.	C/M	Charact. Potential		Response Const.		n Tech.		Electrokinetic Data			Products and Identification	Description and Remarks	Code No.
		Value	vs.		Value			Parameter	Value	From			
JE035 0369	386 a	$E_{\frac{1}{2}}$	-1.0c	SCE	-	-	-	-	$dE_{\frac{1}{2}}/dpH$	61	sttd	-	$C, E_{\frac{1}{2}} = -0.56-0.061$pH for pH=7-12,p	FB38
			-1.25c		-	-	-	-		61		-	C,p	
			-1.51c							0			$C, E_{\frac{1}{2}} \neq f(pH)$ for pH=8.5-14	
EA019 0215	-	$E_{\frac{1}{2}}$	-0.39	SCE	I	2.90	sttd	2	Tomeš	29	-	-	$C,R,Mc(E_{\frac{1}{2}}),p$	FB39
EA020 0033	-	E_p	-1.50	SCE	-	-	-	-	-	-	-	CP(-1.35 V, [MeI]= 1.07 M) → N-methyl-hydrazobenzene 20% (% ↑ on adding H_2O); N,N'-dimethyl-hydrazobenzene 80%	$C,\neq,r$	FB40
		"$E_{\frac{1}{2}}$"	-1.34		$i_{p,A}/i_{p,C}$	1	-	-	-	-	-	CP(-1.40 V), [MeI] = 5.3 M,[H_2O]=3.7 M → N-methylhydrazo-benzene 25%, hydrazobenzene 35%	C,R,r	
		E_p	-1.95			-							$C,\neq$	
			-										A,R	
			-1.3			-							$C,\neq, pK_a$=12-14 for diethyl malonate,r	
			-0.39										A,oxidn of malonate ion	
			-1.34			0.8							C,R, pK_a=20.5 for fluorene,r	
			-1.9(?)			-							$C,\neq$	
			-										A	
			-0.74										A,oxidn of fluorene anion	
EA020 0033	-	$E_{\frac{1}{2}}$	-1.30	SCE	-	-	-	-	ΔE_p	60	sttd	CP(-1.35 V) → N,N'-dimethyl-hydrazobenzene 100%, NMR	$C,R, i_{p,A}/i_{p,C} \neq f(v)$	
			-							-			A	
EA018 0139	-	$E_{\frac{1}{2}}$	-0.53F	SCE	i_ℓ	3.5F(?)	-	-	$dE_{\frac{1}{2}}/dpH$	57	sttd	-	$C,\neq,i_d,dE_{\frac{1}{2}}/dpH$ for pH=8.5-12, ΔE_p=0.17σ,p	FB41 ba93 dd73 ee02
		E_{max}	-1.27F			-							C,X,Mx, i_{max} ↓ on adding TX100	
		$E_{\frac{1}{2}}$	-0.60F			3.2F(?)	-	-		57			$C,\neq, i_\ell$ ↓ for pH > 9, $i_d, \Delta E_{\frac{1}{2}}$=0.17σ,p	
		E_{max}	-1.32F			-				-			C,X,Mx,0 on adding TX100	
			-0.71			-				57			$C,\neq, i_\ell = kh^{\frac{1}{2}}, i_\ell$ ↓ for pH > 9, $i_d, \Delta E_{\frac{1}{2}}$=0.17σ,p	
			-							-			C,H	

CONT

FB41 (CONT.) $C_{12}H_{10}N_2O$

Code No.	Empirical Formula	Name and C.A. Number	Structural Formula	Solvent	Tech.	Medium		μ, M	pH	T, °C	Electrodes	App.	Experimental Parameters
FB41 ba93 dd73 ee02	$C_{12}H_{10}N_2O$	4-phenylazophenol	4-HOC_6H_4N:N-C_6H_5	H_2O	PY	BR	-	-	11.98	-	DME//SCE	05A01	C=0.5, m=2.11, t=4.09, h=45
FB42 ee06	$C_{12}H_{10}N_2O_2$	4,4'-**azobisphenol** C.A. 2050-16-0	(4-HOC_6H_4N$)_2$	H_2O	PY	BR	-	-	12± 0.05	-	DME//SCE	05A01	m=2.11, t=4.09, h=45
FB43 ba99	$C_{12}H_{10}N_2O_2$	4-nitrodiphenylamine C.A. 836-30-6	4-$O_2NC_6H_4NHC_6H_5$	MeCN	VA	Et_4NClO_4 H_2O	0.1 5E-3	-	-	-	Pt: xxns// Ag/ $AgClO_4$ 0.01, Et_4NClO_4 0.1	2-0	C=0.75, d=0.5, v=41, Pt Aux
					VY	Et_4NClO_4 H_2O	0.1 5E-3	-	-	-	Pt: rodi// Ag/ $AgClO_4$ 0.01, Et_4NClO_4 0.1	2-0	C=1, d=0.1, ω=10, Pt Aux
FB44	$C_{12}H_{10}N_2O_2$	4-nitrosodiphenyl-hydroxylamine C.A. 28548-57-4	4-ONC_6H_4N(OH)-C_6H_5	MeCN	VA	Et_4NClO_4	0.1	-	-	-	Pt: xxdi// Ag/ $AgClO_4$ 0.01, Et_4NClO_4 0.1//Fc	2AO	0→1.4→0V, C=0.7, v=16, Pt Aux
						Et_4NClO_4 $HClO_4$	0.1 7E-4						
						Et_4NClO_4 $HClO_4$ H_2O	0.1 7E-4 7E-2						
						Et_4NClO_4 2,6-lutidine	0.1 1.8E-3						C=0.9, v=16
					VY	Et_4NClO_4	0.1	-	-	-	Pt: rodi// Ag/ $AgClO_4$ 0.01, Et_4NClO_4 0.1//Fc	2AO	C=0.9, ω=10, Pt Aux
						Et_4NClO_4 $HClO_4$	0.1 9E-4						
CONT													

TABLE I. Electrochemical Data

$C_{12}H_{10}N_2O_2$ (CONT.) FB44

Ref.	C/M	Charact. Potential		Response Const.		n Tech.		Electrokinetic Data			Products and Identification	Description and Remarks	Code No.
		Value	vs.		Value			Parameter	Value	From			
EA018 0139	-	$E_{\frac{1}{2}}$	-0.75F	SCE	i_ℓ	3.1F	-	-	$dE_{\frac{1}{2}}/dpH$	57	sttd	-	$C, \not\models, i_\ell \downarrow$ for pH > 9, i_d, i_ℓ do, $\Delta E_{\frac{1}{2}}=$ 0.17σ,p	FB41 ba93 dd73 ee02
EA018 0139	-	$E_{\frac{1}{2}}$	-0.81	SCE	-	-	-	-	$dE_{\frac{1}{2}}/dpH$ k_h^o	57 4.97 E-22	sttd	-	$C, \not\models, i_\ell = kh^{\frac{1}{2}}, i_d$, $dE_{\frac{1}{2}}/dpH$ for pH=6- 12, $\Delta E_{\frac{1}{2}}$=0.17σ,p	FB42 ee06
EA021 1171	147 jkl	E_p	0.69F	Ag/ AgClO$_4$ 0.01, Et$_4$NClO$_4$ 0.1	i_p	5F	-	-	-	-	-	-	A,0 on first scan,r	FB43 ba99
			0.91F			63F							A	
			0.64F			12F							C	
EA021 1171	147 jkl	$E_{\frac{1}{2}}$	0.88	Ag/ AgClO$_4$ 0.01, Et$_4$NClO$_4$ 0.1	i_ℓ/C	12	-	-	-	-	-	-	A,r	
EA020 0469	424 c	E_p	0.55F	Ag/ AgClO$_4$ 0.01, Et$_4$NClO$_4$ 0.1	i_p	15F	-	-	-	-	-	-	A,p	FB44
			0.83F			20F							A	
			1.06F			7F							A	
			1.0F			4F							C	
			0.77F			16F							C	
			0.42F			8F							C	
			0.89F			29F	-	-	-	-	-	-	A,p	
			1.02F			10 ApF							A	
			1.00F			30F							C	
			0.79F			57F	-	-	-	-	-	-	A,p	
			1.0F			2F							A	
			0.68F			12F							C	
			0.27F			16F	-	-	-	-	-	-	A,p	
			0.83F			30F							A	
			0.76F			25F							C	
			-0.1F			15F							C	
EA020 0469	424 c	$E_{\frac{1}{2}}$	0.50	Ag/ AgClO$_4$ 0.01, Et$_4$NClO$_4$ 0.1	i_ℓ	2.8F	-	-	-	-	-	-	A,p	
			0.79			3.8F							A	
			0.0FAp			5F	-	-	-	-	-	-	C,p	
			0.8FAp			-							A,X	
			1.1FAp			-							A,X	
		$E_{\frac{1}{2}}$												CONT

FB44 (CONT.) $C_{12}H_{10}N_2O_2$

Code No.	Empirical Formula	Name and C.A. Number	Structural Formula	Solvent	Tech.	Medium	μ, M	pH	T, °C	Electrodes	App.	Experimental Parameters
FB44	$C_{12}H_{10}N_2O_2$	4-nitrosodiphenyl-hydroxylamine	4-$ONC_6H_4N(OH)C_6H_5$	MeCN	VY	Et_4NClO_4 0.1 $HClO_4$ 9E-4 H_2O 0.018	-	-	-	Pt: rodi// Ag/ $AgClO_4$ 0.01, Et_4NClO_4 0.1//Fc	2AO	C=0.9, ω=10, Pt Aux
						Et_4NClO_4 0.1 Bu_4NOAc ns HOAc 9E-4						
						Et_4NClO_4 0.1 2,6-lutidine 1.8E-3						
FB45 bb03	$C_{12}H_{10}N_2O_4S$	4-[(4-hydroxyphenyl)azo]benzenesulfonic acid C.A. 2918-83-4	Table II	H_2O	PY	BR	-	2.90 ±0.05	-	DME//SCE	05AO1	C=ns, m=2.11, t=4.09, h=45
								4.0				
								4.7				
								8.10 ±0.05				
								11± 0.05				
FB46 bb16	$C_{12}H_{10}Se_2$	diphenyl diselenide C.A. 1666-13-3	$(C_6H_5Se)_2$	MeOH 50	PY	BOR ns	-	9.2	-	DME//SCE	OCO	C=0.06
												C=0.2
						BOR ns erythrosine 9E-5						C=0.06
												C=0.2
FB47 bb19 ch09	$C_{12}H_{11}N$	diphenylamine C.A. 122-39-4	$C_6H_5NHC_6H_5$	MeCN	VY	Et_4NClO_4 0.1 H_2O 5E-3	-	-	-	Pt: rodi// Ag/ $AgClO_4$ 0.01, Et_4NClO_4 0.1//Fc	2-0	C=1, d=0.1, ω=10, Pt Aux
FB48 bb20 ch10 ee18	$C_{12}H_{11}NO$	4-(phenylamino)-phenol C.A. 122-37-2	4-$HOC_6H_4NHC_6H_5$	MeCN	VY	Et_4NClO_4 0.1 H_2O 5E-3	-	-	-	Pt: rodi// Ag/ $AgClO_4$ 0.01, Et_4NClO_4 0.1//Fc	2-0	C=1, d=0.1, ω=10, Pt Aux

TABLE I. Electrochemical Data 117

$C_{12}H_{11}NO$ FB48

Ref.	C/M	Charact. Potential		Response Const.		n Tech.	n	Electrokinetic Data			Products and Identification	Description and Remarks	Code No.
		Value	vs.		Value			Parameter	Value	From			
EA020 0469	424 c	$E_{\frac{1}{2}}$	0.82F	Ag/AgClO$_4$ 0.01, Et$_4$NClO$_4$ 0.1	i_ℓ	5.9F	-	-	-	-	-	-	A,p	FB44
			1.10F			1.5F							A	
			-0.25			-	-	-	-	-	-	-	A,p	
			0.78			-							A	
			0.18			-	-	-	-	-	-	-	A,p	
			0.78			-							A	
EA018 0139	-	-	-	-	i_ℓ	4.9F	sttd	2	k_h^o	7.60 E-5	sttd	-	$C,\rightleftharpoons,i_\ell=kh^{\frac{1}{2}},i_d, \Delta E_{\frac{1}{2}}=0.17\sigma,p$	FB45 bb03
			-			-				-			C,X,H	
			-			3.9			$dE_{\frac{1}{2}}/dpH$	52	sttd	-	$C,\rightleftharpoons,i_d,dE_{\frac{1}{2}}/dpH$ for pH=4-9,r	
			-			-				-			C,X,H	
			-	-		3.3F	-	-		52		-	C,p	
						-				-			C,X,H	
			-	-		7F	sttd	4		52		-	$C,\rightleftharpoons,i_\ell\neq f(pH),i_d, \Delta E_{\frac{1}{2}}=0.17\sigma,p$	
									k_h^o	2.50 E-12				
			-			-				-			C,X,H	
			-0.66			-	-	-	-	-	-	-	$C,\rightleftharpoons,i_\ell,\Delta E_{\frac{1}{2}}=0.17\sigma,p$	
			-										C,X,H	
JE046 0391	-	$E_{\frac{1}{2}}$	-0.75F	SCE	i_ℓ	1.7F	-	-	-	-	-	-	C,sur,p	FB46 bb16
			-0.6F			3.7F							C,p	
			-0.79F			2.2F							C,Po	
			-0.65F			1.75F	-	-	-	-	-	-	C,p	
			-0.6F			3.7F	-	-	-	-	-	-	C,p	
			-0.7F			2.2F							C,Po	
EA021 1171	147 jkl	$E_{\frac{1}{2}}$	0.60	Ag/AgClO$_4$ 0.01, Et$_4$NClO$_4$ 0.1	i_ℓ/C	6.4	-	-	-	-	-	radical cation of dimer	A,r	FB47 bb19 ch09
			0.74			-						dication of dimer	$A,E_{\frac{1}{2}}$ and i_ℓ vary strongly with [H$_2$O]	
EA021 1171	147 jkl	$E_{\frac{1}{2}}$	0.32	Ag/AgClO$_4$ 0.01, Et$_4$NClO$_4$ 0.1	i_ℓ/C	6.5	-	-	-	-	-	radical cation of dimer	A,i_d for $\omega>20$,r	FB48 bb20 ch10 ee18
			0.57			-						dication of dimer	$A,E_{\frac{1}{2}}$ and i_ℓ vary strongly with [H$_2$O]	

FB49 $C_{12}H_{11}NO_2$

Code No.	Empirical Formula	Name and C.A. Number	Structural Formula	Solvent	Tech.	Medium	μ, M	pH	T, °C	Electrodes	App.	Experimental Parameters
FB49 bb23	$C_{12}H_{11}NO_2$	ethyl α-cyanocinnamate C.A. 14533-87-0	$C_6H_5CH{:}C(CN)\text{-}COOC_2H_5$	MeCN 90	PY	Et_4NClO_4 0.1	-	-	25	DME// Ag/ AgClO$_4$ 0.01	12A0	C=2.5, Pt Aux
				MeCN 95	PW	Et_4NClO_4 0.1	-	-	25	DME// Ag/ AgClO$_4$ 0.01	12A2	C=1.2, v=4.4E4, Pt Aux
				MeCN 96	PY	Et_4NClO_4 0.1	-	-	25	DME// Ag/ AgClO$_4$ 0.01	12A0	C=2.5, Pt Aux
				MeCN 98	PY	Et_4NClO_4 0.1	-	-	25	DME// Ag/ AgClO$_4$ 0.01	12A0	C=2.5, Pt Aux
				MeCN 99.95	PY	Et_4NClO_4 0.1	-	-	25	DME// Ag/ AgClO$_4$ 0.01	12A0	C=2.5, Pt Aux
					PW	Et_4NClO_4 0.1	-	-	25	DME// Ag/ AgClO$_4$ 0.01	12A2	C=0.34, Pt Aux
												C=1.20
												C=3.4
				MeCN	PY	Et_4NClO_4 0.1 PHEN 2E-4	-	-	25	DME// Ag/ AgClO$_4$ 0.01	12A0	C=1.9, Pt Aux
					PW	Et_4NClO_4 0.1 PHEN 0.05	-	-	25	DME// Ag/ AgClO$_4$ 0.01	12A2	C=1.89, Pt Aux
FB49	$C_{12}H_{11}NO_2$	ethyl α-cyanocinnamate C.A. 14533-87-0	$C_6H_5CH{:}C(CN)\text{-}COOC_2H_5$	MeCN 90	PY	Et_4NClO_4 0.1	-	-	25		12A0	

TABLE I. Electrochemical Data

$C_{12}H_{11}NO_2$ FB49

Ref.	C/M	Charact. Potential Value	vs.	Response Const. Value		Tech.	n	Electrokinetic Data Parameter	Value	From	Products and Identification	Description and Remarks	Code No.
JE042 0189	50s	$E_{\frac{1}{2}}$ -0.96F	Ag/ AgClO$_4$ 0.01	i_ℓ	8F	-	-	-	-	-	-	C,r	FB49 bb23
		-1.28F			8F							C	
JE042 0189	50r	-	-	-	-	-	-	$dE_p/d\log v$	20	for v < 4.4E4	-	C,data for first wave only,r	
									0	for v > 4.4E4			
		-							-			C(?)	
JE042 0189	50r	$E_{\frac{1}{2}}$ -0.96F	Ag/ AgClO$_4$ 0.01	i_ℓ	8F	-	-	-	-	-	-	C,r	
		-1.28F			4F							C	
JE042 0189	50r	$E_{\frac{1}{2}}$ -0.96F	Ag/ AgClO$_4$ 0.01	i_ℓ	8.8F	-	-	-	-	-	-	C,r	
		-1.28F			6.0F							C	
		-1.60F			1.2F							C	
JE042 0189	50r	$E_{\frac{1}{2}}$ -1.00F	Ag/ AgClO$_4$ 0.01	i_ℓ	9.6F	i:i QE	1 0.93	-	-	-	-	C,r	
		-1.48F			small							C	
JE042 0189	50r	-	-	-	-	-	-	$dE_p/d\log v$	19	for v < 1.1E4	-	C,data for first wave only,r	
									0	for v > 1.1E4			
								$dE_p/d\log C$	21				
		-							-			C(?)	
		-	-	-	-	-	-	$dE_p/d\log v$	19	for v < 4E4	-	C,r	
									0	for v > 4E4			
		-							-			C(?)	
		-	-	-	-	-	-		21	for v < 9E4	-	C,r	
									0	for v > 9E4			
		-							-			C(?)	
JE042 0189	50r	$E_{\frac{1}{2}}$ -1.1F	Ag/ AgClO$_4$ 0.01	i_ℓ	7.7F	-	-	-	-	-	-	C,r	
		-1.66F			8F							C	
JE042 0189	50r	-	-	-	-	-	-	$dE_p/d\log v$	18	for v < 5.7E4	-	C,data for first wave only,r	
									0	for v > 5.7E4			
		-							-			C(?)	

FB50 $C_{12}H_{11}N_3$

Code No.	Empirical Formula	Name and C.A. Number	Structural Formula	Solvent	Tech.	Medium		μ, M	pH	T, °C	Electrodes	App.	Experimental Parameters
FB50 bb27	$C_{12}H_{11}N_3$	4-aminoazobenzene C.A. 60-09-3	4-$H_2NC_6H_4$N: NC_6H_5	dioxane 50	VD	HCl KCl	ns ns	–	1.0	–	Pt: rodi // SCE	OAO	C=4.0
									3.1				
						ACET	ns		4.85				
									7.0				
						NH_3 NH_4Cl	ns ns		9.0				
					VY	HCl KCl	ns ns	–	1.0	–	Pt: rodi // SCE	OAO	C=4.0
									3.1				
						ACET	ns		4.85				
									7.0				
						NH_3 NH_4Cl	ns ns		9.0				
FB51	$C_{12}H_{11}N_3$	diazoaminobenzene C.A. 136-35-6	C_6H_5NHN:NC_6H_5	MeOH 40	PY	UB GEL	ns 4E-3	–	8.6	21±1	DME // SCE	0-0	C=0.5, m(oc)=4.9, t=1.9
						UB triphenyl-phosphine oxide	ns 0.1						
						UB GEL	ns 4E-2		9.22				
						UB triphenyl-phosphine oxide	ns 0.1						
				MeOH 50	PY	UB GEL	ns 1E-2	–	7.45	21±1	DME // SCE	0-0	C=0.5, m=4.9, t=1.9
						UB CsCl	ns 6E-2		9.3				
						UB Et_4NI	ns 2E-2						
						UB GEL	ns 1E-2		9.6				
									14				

TABLE I. Electrochemical Data

$C_{12}H_{11}N_3$ FB51

Ref.	C/M	Charact.	Potential	Response Const.		n Tech.	Electrokinetic Data			Products and Identification	Description and Remarks	Code No.
		Value	vs.		Value		Parameter	Value	From			
C0034 1615	430 a	E_{su}	0.95	NCE	-	-	-	-	-	-	-	A,p	FB50 bb27
			1.00		-	-	-	-	-	-	-	A,p	
			0.88		-	-	-	-	-	-	-	A,p	
			0.78		-	-	-	-	-	-	-	A,p	
			0.76		-	-	-	-	-	-	-	A,p	
			1.27									A	
C0034 1615	430 a	$E_{\frac{1}{2}}$	0.91	NCE	-	-	-	-	-	-	-	A,p	
			0.98		-	-	-	-	-	-	-	A,p	
			0.85		i(u)	21	-	-	-	-	-	A,p	
			0.73		-	13	-	-	-	-	soluble products, unidentified	A,p	
			0.73		-	13	-	-	-	-	-	A,p	
JE035 0369	386 a	$E_{\frac{1}{2}}$	-1.20F	SCE	i_ℓ	4.4F	-	-	-	-	-	-	C,r	FB51
			-1.55F			3.4F							C	
			-1.19F			1.4F	-	-	-	-	-	-	C,r	
			-1.58F			6.1F							C	
			-1.23F			2.5F	-	-	-	-	-	-	C,r	
			-1.50F			5.7F							C	
			-1.23F			0.8F	-	-	-	-	-	-	C,r	
			-1.58F			6.6F							C	
JE035 0369	386 a	$E_{\frac{1}{2}}$	-1.05F	SCE	i_ℓ	7.5F	-	-	$dE_{\frac{1}{2}}/dpH$	62	sttd	-	C, $E_{\frac{1}{2}}$ = -0.57-0.062pH for pH=7-12,r	
			-1.16F			3.2F	-	-		62		-	C,r	
			-1.50F			4.4F							C	
			-1.25F			7.6F	-	-		62		-	C,W,r	
			-1.18F			4.8F	-	-		62	-	-	C,r	
			-1.56F			3.0F					0		C,0 at pH < 8.5, $E_{\frac{1}{2}} \neq f(pH)$ for pH=8.5-14	
			-1.53F			5.5F	-	-			0	-	C, $E_{\frac{1}{2}} \neq f(pH)$ for pH=8.5-14,r	

FB52 $C_{12}H_{12}Cr$

Code No.	Empirical Formula	Name and C.A. Number	Structural Formula	Solvent	Tech.	Medium		μ, M	pH	T, °C	Electrodes	App.	Experimental Parameters
FB52	$C_{12}H_{12}Cr$	bis(η-benzene)chromium C.A. 1271-54-1	$(\eta-C_6H_6)_2Cr$	DMF	VY	Et_4NClO_4 GEL	1 0.04(?)	-	-	-	Pt: pdhb// SCE	3A-	C=1?, Pt Aux
FB53	$C_{12}H_{12}N_2$	4-aminodiphenylamine C.A. 101-54-2	$4-H_2NC_6H_4NHC_6H_5$	MeCN	VY	Et_4NClO_4 H_2O	0.1 5E-3	-	-	-	Pt: rodi// Ag/ $AgClO_4$ 0.01, Et_4NClO_4 0.1	2-0	C=1, d=0.1, ω=10, Pt Aux
FB54	$C_{12}H_{12}N_2O_2$	1,4-diacetyl-1,4-dihydroquinoxaline C.A. 60389-44-8	Table II	DMF	VA	$LiClO_4$	0.1	-	-	-	Pt: xxns// Ag/AgI, I- 0.1	2-0	C=5, A=0.015, v=100
FB55	$C_{12}H_{12}O$	1-(1-hydroxyethyl)-naphthalene C.A. 1517-72-2	Table II	DMF	PY	Et_4NBr	0.1	-	-	-	DME// SCE(o)	2-0	C=2
						Et_4NBr PHEN	0.1 0.01						
FB56	$C_{12}H_{12}O$	2-(1-hydroxyethyl)-naphthalene C.A. 7228-47-9	Table II	DMF	PY	Et_4NBr	0.1	-	-	-	DME// SCE(o)	2-0	C=2
						Et_4NBr PHEN	0.1 0.01						
FB57	$C_{12}H_{13}N_3$	2,4-diaminodiphenylamine C.A. 136-17-4	Table II	EtOH 10	VA	H_2SO_4	2.5	-	-	25	Cp: xxns// SCE	2-0	0→0.7→ -0.2→0V, C=0.1, A=0.18, v=13, Pt Aux
						H_2SO_4	0.5						0→0.6→ -0.2→0V
						H_2SO_4	0.05		1.29				0→0.5→ -0.2→0V
						H_2SO_4	5E-3		2.18				0→0.4→ -0.2→0V
						BR	-		2.97				0→0.3→ -0.2→0V
						ACET	ns		4.12				
									5.12				0→0.2→ -0.2→0V

TABLE I. Electrochemical Data

$C_{12}H_{13}N_3$ FB57

Ref.	C/M	Charact. Potential		Response Const.		n Tech.	n	Electrokinetic Data			Products and Identification	Description and Remarks	Code No.
		Value	vs.		Value			Parameter	Value	From			
JE051 0226	443 a	$E_{\frac{1}{2}}$	-0.82	SCE	-	-	-	-	Elog	69	-	-	A,QI,p	FB52
			1.06							143			A,≠	
									βn_b	0.41	Elog			
EA021 1171	-	$E_{\frac{1}{2}}$	0.06	Ag/AgClO$_4$ 0.01, Et$_4$NClO$_4$ 0.1	i_ℓ/C	6.0	sttd	1	-	-	-	radical cation	A,r	FB53
			0.64			-		-					A	
EA021 0345	434 a	E_p	1.38	Ag/AgI, I$^-$ 0.1	i_p	17.6	QE (1.5V)	1.75	-	-	-	quinoxaline 66%, NMR	A,≠,r	FB54
			0.63F			1.3F		-					C	
			0.09F			8F							C	
EA019 0629	-	$E_{\frac{1}{2}}$	-2.64	SCE(o)	i_ℓ	10.6	QE (-2.75V)	2	-	-	-	CP(-2.75 V; [Bu$_4$NI]= 0.25) → 1-ethyl-naphthalene 90%	C,r	FB55
			-2.54			14.4		-	-	-	-	-	C,r	
EA019 0629	-	$E_{\frac{1}{2}}$	-2.64	SCE(o)	i_ℓ	10.6	QE (-2.75V)	2Ap	-	-	-	CP(-2.75 V; [Bu$_4$NI]= 0.25) → 2-ethyl-naphthalene 90%	C,r	FB56
			-2.54			14.4		-	-	-	-	-	C,r	
JE043 0397	429 a	E_p	0.535	SCE	$i_p(u)$	1	QE	0.93	-	-	-	2-amino-4-benzo-quinone-4-mono-imine,VIS,VA,TLC; aniline,VA	A,r	FB57
			0.375			0.62		-					C	
			0.225			-							C	
			0.470			1		0.86	-	-	-	-	A,r	
			0.300			1.06		-					C	
			0.206			-							C	
	429 b		0.330			1		2.03	-	-	-	-	A,r	
			0.230			1.13		-					C	
			0.245			1		1.99	-	-	-	-	A,r	
			0.125			0.94		-					C	
			0.195			1		2.08	-	-	-	-	A,r	
			0.030			1.04		-					C	
			0.175			1		1.77	-	-	-	-	A,r	
			-0.027			0.99		-					C	
			0.110			1		1.78	-	-	-	-	A,r	
			-0.065			1.05		-					C	

FB58 $C_{12}H_{13}N_3$

Code No.	Empirical Formula	Name and C.A. Number	Structural Formula	Solvent	Tech.	Medium		μ, M	pH	T, °C	Electrodes	App.	Experimental Parameters
FB58	$C_{12}H_{13}N_3$	4,4'-diaminodiphenylamine C.A. 537-65-5	$(4\text{-}NH_2C_6H_4)_2NH$	MeCN	VY	Et_4NClO_4	0.1	–	–	–	Pt: rodi// Ag/ $AgClO_4$ 0.01, Et_4NClO_4 0.1	2-O	C=1, d=0.1, ω=10, Pt Aux
FB59	$C_{12}H_{13}N_5S$	3,5-dimethyl-4-(phenylazo)-1-thiocarbamoylpyrazole C.A. 3696-00-2	Table II	H_2O	PY	BR GEL	– 5E-3	–	8	–	DME//SCE	OAO	C=0.01, EC=3.75
FB60 bb96	$C_{12}H_{14}O_4$	diethyl phthalate C.A. 84-66-2	$2\text{-}C_2H_5OCOC_6H_4\text{-}COOC_2H_5$	EtOH 40	PY	BR	–	–	7.4 9.5 10.0	20	DME//SCE	O4AO	C=ns, m=1.77, t=5, h=60
FB61	$C_{12}H_{16}F_3NO_2$	N-ethyl-N-hydroxy-β-methoxy-β-(3-trifluoromethylphenyl)ethylamine C.A. 64756-75-8	$3\text{-}F_3CC_6H_4CH\text{-}(OCH_3)CH_2N(OH)C_2H_5$	MeOH 10	PY	BR	–	–	8	–	DME//SCE	2AO	C=ns, m=1.73, t=2.03, h=50, v=2, Pt Aux
					DI	BR	–	–	8	–	DME//SCE	2AO	C=0.01-0.1, ΔE=100, m=1.73, t=0.3, h=50, v=2, Pt Aux
FB62	$C_{12}H_{20}Cl_2N_2O$	trans-azoxy-2-chlorocyclohexane C.A. 59190-81-7	Table II	EtOH 30	PY	KNO_3 HCl	0.1 –	–	0.6	–	DME//SCE	O-O	C=0.36
						KNO_3 BR	0.1 –		3.9 4.4 5.8 7.3F 8.4 9.9				
						KNO_3 NaOH	0.1 –		13				
FB63	$C_{12}H_{22}N_2$	trans-azocyclohexane C.A. 2159-74-2	Table II	EtOH 30	PY	KNO_3	0.1	–	1.3	–	DME//SCE	O-O	C=0.36
						KNO_3 BR	0.1 –		7.4 8.4 9.5				

TABLE I. Electrochemical Data

$C_{12}H_{22}N_2$ FB63

Ref.	C/M	Charact. Potential		Response Const.		n Tech.	Electrokinetic Data			Products and Identification	Description and Remarks	Code No.
		Value	vs.		Value		Parameter	Value	From			
EA021 0557	147 n	$E_{\frac{1}{2}}$	-0.135	Ag/AgClO$_4$ 0.01, Et$_4$NClO$_4$ 0.1	i_ℓ/C	5.5	sttd 1	-	-	-	radical cation,VIS	A,$\Delta E_{\frac{1}{2}}$=0.46σ,r	FB58
			0.24			-	1					A	
JE054 0411	198 1	$E_{\frac{1}{2}}$	-0.54	SCE	-	-	-	-	-	-	-	C,$\overline{\tau}$,i_d,Tc=1.6,$E_{\frac{1}{2}}\rightarrow$ more neg as pH $\uparrow$ and C $\uparrow$,p	FB59
C0028 1985	438 a	-	-	-	i_ℓ/i_d	0.60F	sttd 4	-	-	-	-	C,p	FB60 bb96
		-	-			0.85F	4	-	-	-	-	C,p	
		$E_{\frac{1}{2}}$	-1.82	SCE		1.00F	4	-	-	-	-	C,$\Delta E_{\frac{1}{2}}$=0.39σ*,p	
AA092 0353	-	$E_{\frac{1}{2}}$	-0.03	SCE	-	-	-	αn_a p=	0.92 and 0.62 0.86	sttd	-	A,0 for pH < 7,X at pH=7,i_d,p	FB61
			-						-			C	
AA092 0353	-	E_{su}	-0.03 -0.80 -0.85	SCE	-	-	-	-	-	-	-	A,p C C	
EA021 0473	432 d	$E_{\frac{1}{2}}$	-0.45F	SCE	i_ℓ	2.6F	sttd 4	$dE_{\frac{1}{2}}/dpH$	82F	-	-	C,r	FB62
			-0.70F			2.6F	4		82F	-	-	C,r	
			-0.76F			2.5F	4		82F	-	-	C,r	
			-0.93F			2.3F	-	-	-	-	-	C,r	
			-1.13F			1.0F	-	-	-	-	-	C,r	
			-1.54F			1.7F			0F			C	
			-1.17F			0.2F	-	-	-	-	-	C,r	
			-1.51			1.6F			0F			C	
			-1.51F			1.7F	-	-	0F			C,r	
			-1.51F			1.7F	-	-	0F			C,r	
EA021 0473	432 g	$E_{\frac{1}{2}}$	-0.32F	SCE	i_ℓ	1.3F	sttd 2	$dE_{\frac{1}{2}}/dpH$	75F	-	-	C,$dE_{\frac{1}{2}}/dpH$ for pH 1.3-8.4,r	FB63
			-0.78F			1.4F	2		75F		-	C,r	
			-0.85F			-	2		136F			C,$dE_{\frac{1}{2}}/dpH$ for pH 8.4-9.5,r	
			-1.00F			0.6F	-		136F		-	C,r	

FB64 $C_{12}H_{22}N_2O$

Code No.	Empirical Formula	Name and C.A. Number	Structural Formula	Solvent	Tech.	Medium		μ, M	pH	T, °C	Electrodes	App.	Experimental Parameters
FB64	$C_{12}H_{22}N_2O$	cis-azoxycyclohexane C.A. 57497-40-2	Table II	EtOH 30	PY	KNO_3 HCl	0.1	-	1.2	-	DME//SCE	O-O	C=0.36
						KNO_3 BR	0.1		4.0				
									6.0				
									7.5				
									8.4				
									10.8				
						KNO_3 NaOH	0.1 -		13				
FB65	$C_{12}H_{22}N_2O$	trans-azoxycyclohexane C.A. 58379-14-9	Table II	EtOH 30	PY	KNO_3 HCl	0.1 -	-	1.2	-	DME//SCE	O-O	C=0.36
						KNO_3 BR	0.1 -		3.8				
									6.0				
									7.5				
									8.5				
FB66	$C_{12}H_{22}N_2O_2$	trans-azodioxycyclohexane C.A. 26049-06-9	Table II	EtOH 30	PY	KNO_3 BR	0.1 -	-	2.46	-	DME//SCE	---	C=0.36
									10.41				
						NaOH	0.1		-				C=0.5
													C=1.0
													C=2
CONT		cis-azoxycyclohexane C.A. 57497-40-2		EtOH 30	PY								

$C_{12}H_{22}N_2O_2$ (CONT.) FB66

Ref.	C/M	Charact. Potential Value	vs.	Response Const. Value		n Tech.		Electrokinetic Data Parameter	Value	From	Products and Identification	Description and Remarks	Code No.
EA021 0473	432 d	$E_{\frac{1}{2}}$	-0.41F	SCE	i_ℓ	2.8F	sttd	4	$dE_{\frac{1}{2}}/dpH$	54F	-	-	C, $dE_{\frac{1}{2}}/dpH$ for pH 1.2-6	FB64
			-0.56F			2.9F		4		54F		-	C, r	
			-0.67F			2.8F		4		-		-	C, $dE_{\frac{1}{2}}/dpH=f(pH)$ for pH=6-10.5, r	
			-0.88F			2.6F		4		-		-	C, $dE_{\frac{1}{2}}/dpH=f(pH)$, r	
			-1.05F			2.5F		-		-		-	C, $dE_{\frac{1}{2}}/dpH=f(pH)$, r	
	432 f		-1.20F			2.1	-	-		0F		-	C, $E_{\frac{1}{2}} \neq f(pH)$ for pH > 10.5, r	
			-1.20F			2.1F	QE (-1.5V)	3.2		0F		trans-azocyclo-hexane 17.5%, hy-drazocyclohexane 63%	C, r	
EA021 0473	432 d	$E_{\frac{1}{2}}$	-0.40F	SCE	i_ℓ	2.8F	sttd	4	-	-	-	-	C, r	FB65
			-0.60F			2.8F	sttd QE (-0.77V)	4 3.9	-	-	-	hydrazocyclohexane 98%	C, r	
			-0.75F			2.8F		4	-	-	-	-	C, r	
			-0.94F			1.4F	-	-	-	-	-	-	C, r	
			-1.26F			0.6F	-	-	-	-	-	-	C, r	
EA021 0479	432 a	$E_{\frac{1}{2}}$	-0.33F	SCE	i_ℓ	1.5F	i:i	2	$dE_{\frac{1}{2}}/dpH$	126	sttd	-	C, i_d, i_ℓ ↓ as pH ↑, p	FB66
			-0.5			3.2F		4		86		-	C, Mx, i_d, i_ℓ ↑ as pH ↑	
			-0.75F			4.6F		6		-	-	-	C, p	
	432 c		-0.98F			1.6F		2		-	-	-	C, p	
			-1.21F			0.8F		2					C	
			-1.45F			1.0F							C	
			-1.01F			2.1F	-	-		-	-	-	C, p	
			-1.28F			1.1F							C	
			-1.56F			1.1F							C	
			-1.06F			4.8F	-	-		-	-	-	C, p	
			-1.20F			2.0F							C	
			-1.55F			2.0F							C	
			-			16F	-	-		-	-	-	C, X, wave begins at -1.0 V and has Mx at -1.7 V, p	
														CONT

FB66 (CONT.) $C_{12}H_{22}N_2O_2$

Code No.	Empirical Formula	Name and C.A. Number	Structural Formula	Solvent	Tech.	Medium		μ, M	pH	T, °C	Electrodes	App.	Experimental Parameters
FB66	$C_{12}H_{22}N_2O_2$	trans-azodixycyclohexane	Table II	EtOH 30	QE	NaOH		0.1	-	-	-	Hg:ns po//SCE	---	C=0.30, E_{app}= -1.7 V
														C=0.44, E_{app}= -1.05V
														C=1.0, E_{app}= -1.7 V
				EtOH 75	QE	HCl	ns	-	-	-	Hg:ns po//SCE	---	C=4.6, E_{app}= -0.30V
FB67	$C_{13}H_8Cl_2OS$	3-(2,6-dichlorophenyl)-1-(2-thienyl)-2-propen-1-one C.A. 32348-13-3	Table II	EtOH 50	PY	NaOAc HCl	ns ns	0.2	0.77	25±0.1	DME//SCE	O3AO	C=0.2, m=1.86, t=3.8, h=70
FB68	$C_{13}H_8N_2Na_2O_6S$	3-carboxy-4-hydroxy-4'-sulfo-azobenzene disodium salt C.A. 6054-99-5	Table II	H_2O	PY	buffer	ns	-	1.8	35±0.1	DME//SCE	OAO	C=0.5, EC=3.75, h=50
									4.8				
									5.8				
									6.8				
									9.15				
									10.40				
FB69	$C_{13}H_8N_4O_7$	2-hydroxy-5-(2,4-dinitrophenylazo)-benzoic acid C.A. 40650-95-1	Table II	H_2O	PY	BR	-	-	11.00 ±0.05	-	DME//SCE	O5AO1	C=ns, m=2.11, t=4.09, h=45
FB70 CONT	$C_{13}H_8N_4O_8$	bis(2,4-dinitrophenyl)methane C.A. 1817-76-1	Table II	MeCN	XT	Et_4NClO_4	0.1	-	-	-20±1	HMDE Ag/AgClO_4 0.01	2-2	C=ns, v=(1.5-5.5)E3
										0±1			
										20±0.5			

$C_{13}H_8N_4O_8$ (CONT.) FB70

Ref.	C/M	Charact. Potential Value	vs.	Response Const. Value	Tech.	n	Electrokinetic Data Parameter	Value	From	Products and Identification	Description and Remarks	Code No.
EA021 0479	432 c	-	-	-	QE	3.6	-	-	-	hydrazocyclohexane 35%, trans-azoxy-cyclohexane 15%, trans-azocyclohexane 25%	C,r	FB66
		-	-	-		2.3	-	-	-	trans-azoxycyclohexane 82%, cis-azoxycyclohexane 9%	C,r	
		-	-	-		4.1	-	-	-	hydrazocyclohexane 45%, trans-azoxy-cyclohexane 55%	C,r	
EA021 0479	432 a	-	-	-	QE	2.10	-	-	-	trans-azoxycyclohexane 97%, hydrazocyclohexane 3%	C,r	
JE053 0439	156 k	$E_{1/2}$	-0.455 -0.687	SCE	-	-	-	-	-	-	-	C, $\Delta E_{1/2}=0.22\sigma$, p C	FB67
JE054 0417	-	$E_{1/2}$	-0.050	SCE	-	-	PQ	2	αn_a $k_{s,h}$	0.74 4.7E-6	-	-	C, $\frac{1}{t}$, i_d, r	FB68
			-0.140		-	-	-	-		0.74 3.1E-6	-	-	C, $\frac{1}{t}$, i_d, r	
			-0.160		-	-	-	-		1.03 5.3E-6	-	-	C, $\frac{1}{t}$, i_d, r	
			-0.180		-	-	-	-		1.03 4.2E-6	-	-	C, $\frac{1}{t}$, i_d, r	
			-0.250		-	-	-	-		1.03 1.9E-6	-	-	C, $\frac{1}{t}$, i_d, r	
			-0.270		-	-	-	-		1.03 1.0E-6	-	-	C, $\frac{1}{t}$, i_d, r	
EA018 0139	-	$E_{1/2}$	-0.43	SCE	-	-	-	-	$dE_{1/2}/dpH$	42	sttd	-	C, $\frac{1}{t}$, $i_\ell=kh^{1/2}$, i_d, $dE_{1/2}/dpH$ for pH=8.2-11.9, $\Delta E_{1/2}=0.17\sigma$, p	FB69
JE047 0115	-	XE	0.110 ±0.002	Ag/AgClO$_4$ 0.01	-	-	-	-	-	-	-	C, XT=semiintegral chronoamperometry with linear potential sweep, XE=difference between potentials for the two steps, p C	FB70
			0.120 ±0.002		-	-	-	-	-	-	-	C, p C	
			0.129 ±0.002		-	-	-	-	-	-	-	C, p C	

FB70 (CONT.) $C_{13}H_8N_4O_8$

Code No.	Empirical Formula	Name and C.A. Number	Structural Formula	Solvent	Tech.	Medium		μ, M	pH	T, °C	Electrodes	App.	Experimental Parameters
FB70	$C_{13}H_8N_4O_8$	bis(2,4-dinitrophenyl)methane	Table II	DMF	XT	Et_4NClO_4	0.1	-	-	10±1	HMDE// SCE	2-2	C=ns, v= (1.5-5.5)E3
										30± 0.05			
										60± 0.5			
FB71 bc67 dd99	$C_{13}H_8O$	9-fluorenone C.A. 486-25-9	Table II-2	MeCN	XT	Et_4NClO_4	0.1	-	-	22	HMDE Ag/Ag$^+$ 0.01	2-2	C=1.8-4, v=100-3.2E5
				DMF	IL	Bu_4NI PHEN	0.1 0.01	-	-	-	HMDE// Ag/AgI, I$^-$ 0.1	2A0	C=ns, v=400
					VA	Bu_4NI	0.1	-	-	-	HMDE// Ag/AgI, I$^-$ 0.1	5A0	0→-2.0→ 0 V, C=1, v=400
FB72	$C_{13}H_9ClN_2O_3$	5-[(2-chlorophenyl)-azo]-2-hydroxybenzoic acid C.A. 21461-10-9	Table II	H_2O	PY	BR		-	11.00 ±0.05	-	DME//SCE	05A0 1	C=ns, m=2.11, t=4.09, h=45
FB73	$C_{13}H_9ClN_2O_3$	5-[(4-chlorophenyl)-azo]-2-hydroxybenzoic acid C.A. 21461-12-1	Table II	H_2O	PY	BR		-	11.00 ±0.05	-	DME//SCE	05A0 1	C=ns, m=2.11, t=4.09, h=45
FB74	$C_{13}H_9FOS$	3-(4-fluorophenyl)-1-(2-thienyl)-2-propen-1-one C.A. 367-00-0	Table II	EtOH 50	PY	NaOAc HCl	ns ns	0.2	0.77	25± 0.1	DME//SCE	03A0	C=0.2, m=1.86, t=3.8, h=70
FB75	$C_{13}H_9N_3O_5$	2-hydroxy-5-(2-nitrophenylazo)-benzoic acid C.A. 21461-09-6	Table II	H_2O	PY	BR		-	11.00 ±0.05	-	DME//SCE	05A0 1	C=ns, m=2.11, t=4.09, h=45
FB76	$C_{13}H_9N_3O_5$	2-hydroxy-5-(4-nitrophenylazo)-benzoic acid C.A. 2243-76-7	Table II-2	H_2O	PY	BR		-	11.0	-	DME//SCE	05A0 1	C=ns, m=2.11, t=4.09, h=45

TABLE I. Electrochemical Data

$C_{13}H_9N_3O_5$ FB76

Ref.	C/M	Charact. Potential		Response Const.		n Tech.	Electrokinetic Data			Products and Identification	Description and Remarks	Code No.
		Value	vs.		Value		Parameter	Value	From			
JE047 0115	-	XE	0.128 ±0.002	SCE	-	-	-	-	k^o_h	-	-	-	C,XT=semiintegral chronoamperometry with linear potential sweep,XE=difference between potentials for the two steps,p	FB70
			-									C	
			0.138 ±0.002	-	-	-	-	-	-	-	-	C,p	
			-									C	
			0.154 ±0.002	-	-	-	-	-	-	-	-	C,p	
			-									C	
JE052 0403	-	$E_{\frac{1}{2}}$	-1.610 ±0.003	Ag/Ag$^+$ 0.01	-	-	-	-	Elog	58.5 ±1.5	-	-	C,R,XT=semiintegral chronoamperometry with linear potential sweep,p	FB71 bc67 dd99
EA020 0143	65z	E_p	-0.75F	Ag/AgI, I= 0.1	$i_p(u)$	15F	-	-	-	-	-	-	C,r	
			-2.00F			21F							C	
			-2.2F			20F							C	
EA020 0143	65z	E_p	-0.82F	Ag/AgI, I= 0.1	$i_p(u)$	11F	-	-	-	-	-	-	C,r	
			-1.53F			9F							C	
			-1.53F			8F							A	
			-0.63F			21F							A	
FA018 0139	-	$E_{\frac{1}{2}}$	-0.69	SCE	-	-	-	-	$dE_{\frac{1}{2}}/dpH$	58	sttd	-	C,$\neq$,$i_\ell=kh^{\frac{1}{2}}$,i_d, $dE_{\frac{1}{2}}/dpH$ for pH=5-12, $\Delta E_{\frac{1}{2}}=0.17\sigma$,p	FB72
									k^o_h	1.33 E-12				
			-							-			C,H	
EA018 0139	-	$E_{\frac{1}{2}}$	-0.67	SCE	-	-	-	-	$dE_{\frac{1}{2}}/dpH$	68	sttd	-	C,$\neq$,$i_\ell=kh^{\frac{1}{2}}$,i_d, $dE_{\frac{1}{2}}/dpH$ for pH=5-12, $\Delta E_{\frac{1}{2}}=0.17\sigma$,p	FB73
									k^o_h	1.35 E-14				
			-							-			C,H	
JE053 0439	156 k	$E_{\frac{1}{2}}$	-0.537	SCE	-	-	-	-	-	-	-	-	C,$\Delta E_{\frac{1}{2}}=0.22\sigma$,p	FB74
			-0.950										C	
EA018 0139	-	$E_{\frac{1}{2}}$	-0.56	SCE	-	-	-	-	$dE_{\frac{1}{2}}/dpH$	71	sttd	-	C,$\neq$,$i_\ell=kh^{\frac{1}{2}}$,i_d, $dE_{\frac{1}{2}}/dpH$ for pH=8.1-10.7, $\Delta E_{\frac{1}{2}}=0.17\sigma$,p	FB75
EA018 0139	-	-	-0.56	-	D	3.91	-	-	$dE_{\frac{1}{2}}/dpH$	50	sttd	-	C,$\neq$,$i_\ell=kh^{\frac{1}{2}}$,i_d, $dE_{\frac{1}{2}}/dpH$ for pH=6-10.7, $\Delta E_{\frac{1}{2}}=0.17\sigma$,p	FB76
									k^o_h	4.70 E-14				

FB77 $C_{13}H_{10}$

Code No.	Empirical Formula	Name and C.A. Number	Structural Formula	Solvent	Tech.	Medium		μ, M	pH	T, °C	Electrodes	App.	Experimental Parameters
FB77 bd00 ch68	$C_{13}H_{10}$	fluorene C.A. 86-73-7	Table II-2	DMF	PY	Bu_4NI	0.25	-	-	-	DME// SCE(o)	2-0	C=2
						Bu_4NI PHEN	0.25 0.01						
					IL	Bu_4NI PHEN	0.1 0.01	-	-	-	HMDE// Ag/AgI, I^- 0.1	2AO	C=ns, v=400
					VA	Bu_4NI	0.1	-	-	-	HMDE// Ag/AgI, I^- 0.1	2AO	-1.5 → -2.5 → -1.5 V, C=1, v=400
FB78	$C_{13}H_{10}N_2O$	1-(2-pyridyl)-3-(4-pyridyl)-2-propen-1-one C.A. 13309-09-6	Table II	EtOH 50	PY	BR GEL	- 0.02	-	2.6	22±1	DME//SCE	OAO	C=0.5, m=1.15, t=5.2, h=64
									6.9				
									8.0				
									11.0				
									12.6				
				dioxane 75	PY	Bu_4NI GEL	0.175 0.02	-	-	22±1	DME//SCE	OAO	C=0.5, m=1.15, t=5.2, h=64
FB79	$C_{13}H_{10}N_2O_3$	2-hydroxy-5-phenyl-azobenzoic acid C.A. 3147-53-3	Table II	H_2O	PY	BR		-	5.00 ±0.05	-	DME//SCE	O5AO 1	C=ns, m=2.11, t=4.09, h=45
									7				
									8.20 ±0.05				
CONT													

TABLE I. Electrochemical Data

$C_{13}H_{10}N_2O_3$ (CONT.) FB79

Ref.	C/M	Charact. Potential		Response	Const. Value	Tech.	n	Electrokinetic Data			Products and Identification	Description and Remarks	Code No.
		Value	vs.					Parameter	Value	From			
EA019 0629	-	$E_{\frac{1}{2}}$	-2.78	SCE(o)	i_ℓ	3.8	-	-	-	-	-	-	C,r	FB77 bd00 ch68
			-2.71			12.4	QE (-2.7V)	6?	-	-	-	1,4-dihydrofluorene 30%, 1,2,3,4-tetra-hydrofluorene 30%, 1,1a,2,3,4,4a-hexa-hydrofluorene 30%	C,r	
EA020 0143	33f	E_p	-2.3F	Ag/AgI, I⁻ 0.1	-	-	-	-	-	-	-	-	C,r	
EA020 0143	33f	E_p	-2.25F	Ag/AgI, I⁻ 0.1	$i_p(u)$	20F	-	-	-	-	-	-	C,r	
			-2.25F			10F							A	
JE039 0419	-	$E_{\frac{1}{2}}$	-0.20	SCE	-	-	-	-	-	-	-	-	C,i_d,r	FB78
			-0.55										C,i_d	
			-1.23										C	
			-0.54		-	-	-	-	-	-	-	-	C,r	
			-0.86										C	
			-1.42										C	
			-0.6		-	-	-	-	-	-	-	-	C,r	
			-0.7										C	
			-0.93										C	
			-1.75										C,i_{cat},H	
			-0.82		-	-	-	-	-	-	-	-	C,r	
			-0.88										C	
			-1.15										C	
			-1.83										C,i_{cat},H	
			-0.88		-	-	-	-	-	-	-	-	C,r	
			-0.98										C	
			-1.15										C	
			-1.28										C	
JE039 0419	-	$E_{\frac{1}{2}}$	-0.80	SCE	-	-	-	-	-	-	-	-	C,$\Delta E_{\frac{1}{2}}=0.72\Sigma\sigma^*$,r	
			-1.35										C	
			-1.92										C	
EA018 0139	-	-	-	-	i_ℓ D	5.3F(?) 3.90	-	-	k_h^o	5.40 E-7	sttd	-	C,≒,$i_\ell=kh^{\frac{1}{2}}$,i_d, $\Delta E_{\frac{1}{2}}=0.17\sigma$,p	FB79
			-		-	-				-			C,X,H	
			-		i_ℓ	3.5F(?)	-	-		-			C,p	
			-			-							C,X,H	
			-		i_ℓ D	3.1F(?) 9.05	-	-	$dE_{\frac{1}{2}}/dpH$ k_h^o	59 1.10 E-18	sttd	-	C,≒,$i_\ell=kh^{\frac{1}{2}}$,i_d, $dE_{\frac{1}{2}}/dpH$ for pH=8.2-11.9,$\Delta E_{\frac{1}{2}}=0.17\sigma$,p	

CONT

FB79 (CONT.) $C_{13}H_{10}N_2O_3$

Code No.	Empirical Formula	Name and C.A. Number	Structural Formula	Solvent	Tech.	Medium		μ, M	pH	T, °C	Electrodes	App.	Experimental Parameters
FB79	$C_{13}H_{10}N_2O_3$	2-hydroxy-5-phenyl-azobenzoic acid	Table II	H_2O	PY	BR	-	-	11.00 ±0.05	-	DME//SCE	05A01	C=ns, m=2.11, t=4.09, h=45
FB80	$C_{13}H_{10}N_2O_3$	4(4-hydroxyphenyl-azo)benzoic acid C.A. 2497-38-3	Table II	H_2O	PY	BR	-	-	8.00 ±0.05	-	DME//SCE	05A01	C=ns, m=2.11, t=4.09, h=45
									11.00 ±0.05				
FB81 bd11	$C_{13}H_{10}N_2O_4$	bis(4-nitrophenyl)-methane C.A. 1817-74-9	$(4-O_2NC_6H_4)_2CH_2$	MeCN	XT	Et_4NClO_4 0.1		-	-	-30 ±1	HMDE// Ag/AgClO$_4$ 0.01	2-2	C=ns, v=(1.5-5.5)E3
										0±1			
										20± 0.5			
				DMF	XT	Et_4NClO_4 0.1		-	-	0±1	HMDE// SCE	2-2	C=ns, v=(1.5-5.5)E3
										30± 0.5			
										60± 0.5			
FB82	$C_{13}H_{10}N_4$	4-cyanodiazoamino-benzene C.A. 34529-60-7	$4-NCC_6H_4NHN:N-C_6H_5$	MeOH 50	PY	UB GEL	ns 1E-2	-	4.0	21±1	DME//SCE	0-0	C=0.5, m(oc)=4.9, t=1.9
									5.1				
									8.05				
									9.55				
									12				
									14				

TABLE I. Electrochemical Data

$C_{13}H_{10}N_4$ FB82

Ref.	C/M	Charact. Potential		Response	Const. Value	n	Tech.	Electrokinetic Data			Products and Identification	Description and Remarks	Code No.
		Value	vs.					Parameter	Value	From			
EA018 0139	-	$E_{\frac{1}{2}}$	-0.70	SCE	i_ℓ	2.9F(?)	-	-	$dE_{\frac{1}{2}}/dpH$	59	sttd	-	$C, \doteqdot, i_d, \Delta E_{\frac{1}{2}}=0.17\sigma, p$	FB79
		-			-				-				C,X,H	
EA018 0139	-	-	-	-	-	-	-	-	$dE_{\frac{1}{2}}/dpH$	54	sttd	-	$C,\doteqdot,i_\ell=kh^{\frac{1}{2}},i_d,$ $dE_{\frac{1}{2}}/dpH$ for pH=5.0- 9.0, $\Delta E_{\frac{1}{2}}=0.17\sigma, p$	FB80
									k_h^o	2.08 E-13				
		-							-				C,X,H	
		-0.67			-	-	-	-		3.18 E-21		-	$C,\doteqdot,i_d,\Delta E_{\frac{1}{2}}=0.17\sigma, p$	
		-								-			C,X,H	
JE047 0115	-	XE	0.060 ±0.002	Ag/ AgClO₄ 0.01	-	-	-	-	-	-	-	-	C,XT=semiintegral chronoamperometry with linear potential sweep, XE=difference between potentials for the two steps, p	FB81 bd11
		-											C	
		0.065 ±0.002			-	-	-	-	-	-	-	-	C,p	
		-											C	
		0.070 ±0.002			-	-	-	-	-	-	-	-	C,p	
		-											C	
JE047 0115	-	XE	0.068 ±0.002	SCE	-	-	-	-	-	-	-	-	C,XT=semiintegral chronoamperometry with linear potential sweep, XE=difference between potentials for the two steps, p	
		-											C	
		0.075 ±0.002			-	-	-	-	-	-	-	-	C,p	
		-											C	
		0.083 ±0.002			-	-	-	-	-	-	-	-	C,p	
		-											C	
JE035 0369	386 a	$E_{\frac{1}{2}}$	-0.76F	SCE	i_ℓ	6.8F	-	-	$dE_{\frac{1}{2}}/dpH$	55	-	-	$C, E_{\frac{1}{2}}= -0.55-0.055pH$ for pH=4-10, r	FB82
		-0.83F				7.7F	-	-		55		-	C,r	
		-1.00F				7.3F	-	-		55		-	C,r	
		-1.08F				3.0F	-	-		55		-	C,r	
		-1.29F				6.0F	-	-		76		-	$C, E_{\frac{1}{2}}= -0.37-0.076pH$ for pH=10-14, r	
		-1.45F				5.4F	-	-		76		-	C,r	

FB83 $C_{13}H_{10}O$

Code No.	Empirical Formula	Name and C.A. Number	Structural Formula	Solvent	Tech.	Medium		μ, M	pH	T, °C	Electrodes	App.	Experimental Parameters
FB83 bd15 ch75 de02 ee44	$C_{13}H_{10}O$	benzophenone C.A. 119-61-9	$(C_6H_5)_2C:O$	Me_2CO	PY	NaCl PHOS	0.5 4E-3 or 8E-3	–	8.0 or 7.2	–	DME//SCE	2AO	C=0.16, t(c)=1,h=66, Pt Aux
FB84	$C_{13}H_{10}O$	9-fluorenol C.A. 1689-64-1	Table II	DMF	PY	Et_4NBr	0.1	–	–	–	DME//SCE(o)	2-O	C=2
						Et_4NBr PHEN	0.1 0.01						
						Bu_4NI PHEN	0.1 1E-3	–	–	–	DME// Ag/AgI, I⁻ ns	5AO	C=1
					IL	Bu_4NI PHEN	0.1 0.01	–	–	–	HMDE// Ag/AgI, I⁻ 0.1	2AO	v=400
					VA	Bu_4NI	0.1	–	–	–	HMDE// Ag/AgI, I⁻ 0.1	2AO	-1.5→-2.5 →-1.5V,C=1, v=400
FB85	$C_{13}H_{10}OS$	3-phenyl-1-(2-thienyl)-2-propen-1-one C.A. 3988-77-0	Table II	EtOH 50	PY	NaOAc HCl	ns ns	0.2	0.77	25± 0.1	DME//SCE	O3AO	C=0.2, m=1.86, t=3.8,h=70
						PHTH	ns	–	2.25				C=ns
						ACET?	ns		5.3				
						PHOS	ns		6.6				
									10.0				
						NaOH	ns		11.4				
FB86	$C_{13}H_{11}NO_2$	N-benzoyl-N-phenyl-hydroxylamine C.A. 304-88-1	$C_6H_5CON(OH)C_6H_5$	MeCN	VA	Et_4NClO_4	0.1	–	–	–	glC: xxdi// Ag/ AgClO₄ 0.01, Et₄NClO₄ 0.1	2AO	-0.1→1.2→ -0.1V, C=1.25,v=33, Pt Aux

CONT

TABLE I. Electrochemical Data

$C_{13}H_{11}NO_2$ (CONT.) FB86

Ref.	C/M	Charact. Potential		Response	Const.	n Tech.		Electrokinetic Data			Products and Identification	Description and Remarks	Code No.
		Value	vs.		Value			Parameter	Value	From			
AA098 0093	–	$E_{\frac{1}{2}}$	-1.2	SCE	i_ℓ	0.7F	–	–	–	–	–	–	$C, i_\ell = kC, r$	FB83 bd15 ch75 de02 ee44
EA019 0629	–	$E_{\frac{1}{2}}$	-2.445	SCE(o)	i_ℓ	8.2	QE (-2.6V)	2Ap	–	–	–	CP(-2.6 V, [Bu_4NI]= 0.25) → fluorene 50%	C,r	FB84
			-2.445			10.7	–	–	–	–	–	–	C,r	
			-2.73			7.7							C	
EA020 0143	33f	$E_{\frac{1}{2}}$	-1.96	Ag/AgI, I⁻ ns	–	–	–	–	–	–	–	–	C,r	
			-2.05										C	
EA020 0143	33f	E_p	-2.0F	Ag/AgI, I⁻ 0.1	$i_p(u)$	15F	–	–	–	–	–	–	C,r	
			-2.2F			13F							C	
EA020 0143	33f	E_p	-1.95F	Ag/AgI, I⁻ 0.1	$i_p(u)$	7F	–	–	–	–	–	–	C,⊬,r	
			–			–							A,O	
JE053 0439	156 k	$E_{\frac{1}{2}}$	-0.537	SCE	–	–	–	–	$dE_{\frac{1}{2}}/dpH$	60	–	–	$C, \Delta E_{\frac{1}{2}} = 0.22\sigma, p$	FB85
			-0.900							66			C	
			-0.63F		i_ℓ	0.5F	sttd	1		60	–	–	C,R,p	
			-1.08F			0.55F		1		66			C,R,M	
			-0.81F			–	–	–		60		–	C,R,p	
			-1.18F							0			C,R	
			–							–			C,M	
			-0.90F			0.45F	sttd	1		31	–	–	$C,R, i_\ell=kh^{\frac{1}{2}}, i_d$, Tc=1.75±0.25, p	
			-1.15F			0.45F		1		0			$C, i_\ell=kh^{\frac{1}{2}}, i_d$, Tc=1.75±0.25, $E_{\frac{1}{2}} \neq$ f(pH) for pH >5.5	
			-1.43F			0.5F		1		41			$C,⊬, i_\ell \neq f(h), Tc=5, i_k$, M	
			-1.00F			–	–	–		31	–	–	C,R, i_d, p	
			-1.18F							0			C,O for pH > 10	
			-1.57F							0			$C,M,⊬, i_k$	
			-1.05F			0.55F		1		31	–	–	C,R,p	
			-1.58F			0.6F		1		0			C,M	
EA020 0469	425 b	E_p	0.93F	Ag/ AgClO₄ 0.01, Et₄NClO₄ 0.1	i_p	4.8F	–	–	–	–	–	–	A,p	FB86
			–			–							C,P	

CONT

FB86 (CONT.) $C_{13}H_{11}NO_2$

Code No.	Empirical Formula	Name and C.A. Number	Structural Formula	Solvent	Tech.	Medium	μ, M	pH	T, °C	Electrodes	App.	Experimental Parameters
FB86	$C_{13}H_{11}NO_2$	N-benzoyl-N-phenyl-hydroxylamine	$C_6H_5CON(OH)C_6H_5$	MeCN	VA	Et_4NClO_4 0.1 2,6-lutidine 3.8E-3	–	–	–	glC: xxdi // Ag/AgClO$_4$ 0.01, Et$_4$NClO$_4$ 0.1	2AO	$-0.1 \to 1.0 \to -0.1V, C=1.25, v=33,$ Pt Aux
					VY	Et_4NClO_4 0.1	–	–	–	glC: rodi // Ag/AgClO$_4$ 0.01, Et$_4$NClO$_4$ 0.1	2AO	$C=1.25, \omega=10,$ Pt Aux
						Et_4NClO_4 0.1 ACET 3.8E-3						
						Et_4NClO_4 0.1 4-cyano-pyridine 8.8E-3						
						Et_4NClO_4 0.1 2,6-lutidine 3.8E-3						
						Et_4NClO_4 0.1				Pt: rodi // Ag/AgClO$_4$ 0.01, Et$_4$NClO$_4$ 0.1		
						Et_4NClO_4 0.1 ACET 3.8E-3						
						Et_4NClO_4 0.1 4-cyano-pyridine 8.8E-3						
						Et_4NClO_4 0.1 2,6-lutidine 3.8E-3						
FB87 ch91 de04	$C_{13}H_{11}NS$	10-methylphenothiazine C.A. 1207-72-3	Table II-3	MeCN	IL	Bu_4NBF_4 0.1	–	–	–	Pt: xxdi // SCE	2F-	Pt Aux
FB88 CONT	$C_{13}H_{11}N_3O$	1,3-bis(3-pyridyl)-2-propen-1-one oxime C.A. 51210-88-9	Table II	DMF 20	PY	BR	–	2.3	–	DME // SCE	OAO	$C=0.25, m=3.45, t=2.72, h=40$
								3.2				
								4.3				
							5.15					

$C_{13}H_{11}N_3O$ (CONT.) FB88

Ref.	C/M	Charact. Potential		Response Const.		n		Electrokinetic Data			Products and Identification	Description and Remarks	Code No.
		Value	vs.		Value	Tech.		Parameter	Value	From			
EA020 0469	425 a	E_p	0.26F	Ag/ AgClO$_4$ 0.01, Et$_4$NClO$_4$ 0.1	i_p	3.1F	-	-	ΔE_p	130	sttd	-	A,R,p	FB86
		0.78F			2.6F				-			A, ⇌	
		0.1F			2.3F							C,R	
EA020 0469	425 b	$E_{\frac{1}{2}}$	0.80	Ag/ AgClO$_4$ 0.01, Et$_4$NClO$_4$ 0.1	i_ℓ		-	-	-	-	-	-	A,p	
					30Ap									
		0.95											A	
		1.50			8								A	
	425 a		-0.20		-		-	-	-	-	-	-	A,p	
			0.63										A	
			0.47		-		-	-	-	-	-	-	A,p	
			0.79										A	
			0.19		14.5F	QE	1	-	-	-	radical 60%	A,p	
	425 b		0.785		14.5F		-					A	
			1.09 ±0.03		2.1F		2	-	-	-	orange solution($E_{\frac{1}{2}}$= 1.49 V,A; and 0.32 V and -0.50 V, C)	A,p	
			1.50 ±0.03		1.3F		-					A	
	425 a		0.04		-	-	-	-	-	-	-	A,p	
			0.89									A	
			0.53		-	-	-	-	-	-	-	A,p	
			0.80									A	
			0.36		-	-	-	-	-	-	-	A,p	
			0.80									A	
JA095 6582	-	E_p	0.83	SCE	-	-	-	-	-	-	-	-	A,R,r	FB87 ch91 de04
			1.45										A	
JE048 0297	-	$E_{\frac{1}{2}}$	-0.39	SCE	-	-	i:i	4	-	-	-	-	C, redn of =NOH, i_d, r	FB88
			-0.54		Σi_ℓ	4.66		2					C, redn of C=C, i_d	
			-0.47			-		-					C, i_d, r	
			-0.61			4.66							C, i_d	
			-0.58			-		-					C, r	
			-0.71			4.72							C	
			-1.22			-							C	
			-1.34										C	
			-0.65			-		-					C, r	
			-0.75			4.72							C	
			-1.23			-							C	
			-1.36										C	CONT

FB88 (CONT.) $C_{13}H_{11}N_3O$

Code No.	Empirical Formula	Name and C.A. Number	Structural Formula	Solvent	Tech.	Medium		μ, M	pH	T, °C	Electrodes	App.	Experimental Parameters
FB88	$C_{13}H_{11}N_3O$	1,3-bis(3-pyridyl)-2-propen-1-one oxime	Table II	DMF 20	PY	BR		-	6.1	-	DME//SCE	OAO	C=0.25, m=3.45, t=2.72, h=40
									7.15				
									9.9				
									11.5				
FB89	$C_{14}H_{13}ClN_6O_6$	1,5-di(5-nitro-2-furyl)-1,4-pentadien-3-one amidinohydrazone hydrochloride C.A. 2315-20-0	Table II	MeOH 45 N(CH$_2$CH$_2$OH)$_3$ 5	PY	H_2SO_4	ns	-	1.40	-	DME// Ag/AgCl, KCl satd	-A-	C=0.0086, t=0.5
						CITR	ns		3.85				
						ACET	0.1		5.65				
						CITR	ns		6.15				
						PHOS	ns		8.35				
						NH_3	ns		11.10				
					DI	H_2SO_4	ns	-	1.40	-	DME// Ag/AgCl, KCl satd	-A-	C=0.0086, t=0.5, v=5, ΔE=50
						CITR	ns		3.85				
						ACET	0.1		5.65				
						CITR	ns		6.15				
						PHOS	ns		8.35				
						NH_3	ns		11.10				
						BR		-	6.1				C=0.25, m=3.45,

TABLE I. Electrochemical Data   141

$C_{13}H_{12}ClN_5O_6$ FB89

Ref.	C/M	Charact. Potential		Response Const.		n		Electrokinetic Data			Products and Identification	Description and Remarks	Code No.
		Value	vs.		Value	Tech.		Parameter	Value	From			
JE048 0297	-	$E_{\frac{1}{2}}$	-0.78	SCE	i_ℓ	4.72	i:i	6	-	-	-	-	c,r	FB88
			-1.26			-		-					c	
			-1.41										c	
			-0.87		-	4.72	-	-	-	-	-	-	c,r	
			-1.31			-							c	
			-1.46										c	
			-1.12			3.84	-	-	-	-	-	-	c,r	
			-1.70			-							c	
			-1.26			3.6	-	-	-	-	-	-	c,r	
AA094 0461	-	$E_{\frac{1}{2}}$	0.07	Ag/AgCl, KCl satd	i_ℓ	0.258	-	-	-	-	-	-	c, i_d, r	FB89
			-0.10			0.299	-	-	-	-	-	-	c, i_d, r	
			-1.2			-							c	
			-0.20			0.237	-	-	p= αn_a	1.86 1.76	sttd	-	$c, i_\ell = kh^{\frac{1}{2}}, i_d, r$	
			-0.78F			0.04F				-			c	
			-1.0F			0.04F							c	
			-1.18F			0.16F							c	
			-0.22			0.254	-	-	-	-	-	-	c, i_d, r	
			-1.2			-							c	
			-0.37			0.267	-	-	-	-	-	-	c, i_d, r	
			-0.48			0.211	-	-	-	-	-	-	c, i_d, r	
AA094 0461	-	E_{su}	0.07	Ag/AgCl, KCl satd	i_{su}	3.95	-	6	-	-	-	-	c, i_d, r	
			-0.10			3.67	-	-	-	-	-	-	c, i_d, r	
			-1.2			-							c	
			-0.21F			3.94	-	-	-	-	-	-	$c, i_d, i_{su}/C = 404 \pm 14$ Av(10) for C=2E-4 to 5E-2, $i_{su}/C = 290$ for C=7E-5, r	
			-0.81F			0.2F							c	
			-1.00F			0.1F							c	
			-1.18F			1.0F							c	
			-0.22			3.49	-	-	-	-	-	-	c, i_d, r	
			-1.2			-							c	
			-0.37			3.15	-	-	-	-	-	-	c, i_d, r	
			-0.48			2.72	-	-	-	-	-	-	c, i_d, r	

FB90 $C_{13}H_{12}N_2O$

Code No.	Empirical Formula	Name and C.A. Number	Structural Formula	Solvent		Tech.	Medium		μ, M	pH	T, °C	Electrodes	App.	Experimental Parameters
FB90	$C_{13}H_{12}N_2O$	4-[(3-methylphenyl)-azo]phenol C.A. 6676-96-6	Table II	H_2O		PY	BR	-	-	11.00 ±0.05	-	DME//SCE	05AO1	C=ns,m=2.11, t=4.09,h=45
FB91	$C_{13}H_{12}N_2O$	4-[(4-methylphenyl)-azo]phenol C.A. 2497-33-8	Table II	H_2O		PY	BR	-	-	11.00 ±0.05	-	DME//SCE	05AO1	C=ns,m=2.11, t=4.09,h=45
FB92	$C_{13}H_{12}N_4O_2$	1-(2-pyridyl)-3-(4-pyridyl)-1,3-propanedione dioxime C.A. 36475-32-8	Table II	DMF	20	PY	BR	-	-	2.3	-	DME//SCE	OAO	C=0.25, m=3.45, t=2.72,h=40
										3.2				
										4.3				
										5.15				
										6.1				
										7.15				
										9.9				
										11.5				
FB93	$C_{13}H_{12}N_4O_2$	1-(3-pyridyl)-3-(2-pyridyl)-1,3-propanedione dioxime C.A. 36257-01-9	Table II	DMF	20	PY	BR	-	-	2.3	-	DME//SCE	OAO	C=0.25, m=3.45, t=2.72,h=40
										3.20				
CONT										4.3				

TABLE I. Electrochemical Data

$C_{13}H_{12}N_4O_2$ (CONT.) FB93

Ref.	C/M	Charact. Potential		Response Const.		n Tech.	n	Electrokinetic Data			Products and Identification	Description and Remarks	Code No.
		Value	vs.		Value			Parameter	Value	From			
EA018 0139	-	$E_{\frac{1}{2}}$	-0.71	SCE	-	-	-	-	$dE_{\frac{1}{2}}/dpH$	65	sttd	-	$C,\neq,i=kh^{\frac{1}{2}},i_d,E_{\frac{1}{2}}=$ 0.010-0.065pH for pH=4-12,$\Delta E_{\frac{1}{2}}$=0.17σ,p	FB90
								k^o_h	4.34 E-23				
			-						-			C,H	
EA018 0139	-	$E_{\frac{1}{2}}$	-0.69	SCE	-	-	-	-	$dE_{\frac{1}{2}}/dpH$	50	sttd	-	$C,\neq,i=kh^{\frac{1}{2}},i_d,E_{\frac{1}{2}}=$ 0.015-0.050pH for pH=9-12,$\Delta E_{\frac{1}{2}}$=0.17σ,p	FB91
								k^o_h	1.80 E-20				
			-						-			C,H	
JE048 0297	-	$E_{\frac{1}{2}}$	-0.45	SCE	i_ℓ	3.12	i:i	4	-	-	-	-	C,i_d,r	FB92
			-0.63			3.12		4					C,i_d	
			-1.03			4.48		-					C	
			-0.49			3.12	-						C,i_d,r	
			-0.69			3.12							C,i_d	
			-1.08			3.82							C	
			-0.56			3.12	i:i	4	-	-	-	-	C,i_d,r	
			-0.76			2.88		4					C,i_d	
			-1.16			2.56		-					C	
			-1.32			-							C	
			-0.66		Σi_ℓ	-	-	-	-	-	-	-	C,r	
			-0.85			6.04							C	
			-1.19			7.82							C	
			-1.35			-							C	
			-			-	-	-	-	-	-	-	C,S,r	
			-0.85			5.98							C,P	
			-1.23			-							C	
			-1.39			-							C	
			-			-	-	-	-	-	-	-	C,S,r	
			-1.00			5.94							C,P	
			-1.31			-							C	
			-1.44			-							C	
			-1.29			5.94	-	-	-	-	-	-	C,r	
			-			-							C	
			-1.44			5.6	-	-	-	-	-	-	C,r	
JE048 0297	-	$E_{\frac{1}{2}}$	-0.71	SCE	i_ℓ	6.16	i:i	8	-	-	-	-	C,i_d,r	FB93
			-1.10			4.34		-					C	
			-0.77			6.16	-						C,i_d,r	
			-1.13			3.70							C	
			-0.85			6.16	-						C,i_d,r	
			-1.21			2.24							C	CONT

FB93 (CONT.) $C_{13}H_{12}N_4O_2$

Code No.	Empirical Formula	Name and C.A. Number	Structural Formula	Solvent	Tech.	Medium	μ, M	pH	T, °C	Electrodes	App.	Experimental Parameters
FB93	$C_{13}H_{12}N_4O_2$	1-(3-pyridyl)-3-(2-pyridyl)-1,3-propanedione dioxime	Table II	DMF 20	PY	BR	–	5.15	–	DME//SCE	OAO	C=0.25, m=3.45, t=2.72, h=40
								6.1				
								7.15				
								9.9				
								11.5				
FB94	$C_{13}H_{12}N_4O_2$	1-(3-pyridyl)-3-(4-pyridyl)-1,3-propanedione dioxime C.A. 36257-03-1	Table II	DMF 20	PY	BR	–	2.3	–	DME//SCE	OAO	C=0.25, m=3.45, t=2.72, h=40
								3.2				
								4.3				
								5.15				
								6.1				
								7.15				
								9.9				
								11.5				
FB95 bd58	$C_{13}H_{12}O$	benzhydrol C.A. 91-01-0	$(C_6H_5)_2CHOH$	DMF	PY	Et_4NBr 0.1	–	–	–	DME//SCE(o)	2-0	C=2
						Et_4NBr 0.1 PHEN 0.01						

TABLE I. Electrochemical Data

$C_{13}H_{12}O$ FB95

| Ref. | C/M | Charact. Potential | | Response | Const. | n | Electrokinetic Data | | | Products and | Description and | Code |
		Value	vs.		Value	Tech.	Parameter	Value	From	Identification	Remarks	No.
JE048 0297	–	$E_{\frac{1}{2}}$	-0.93	SCE	i_ℓ	6.16	–	–	–	–	–	–	C, i_d, r	FB93
			-1.30			–						C	
			-1.04			6.16	–	–	–	–	–	–	C, i_d, r	
			-1.36			–						C	
			-1.18			6.16	–	–	–	–	–	–	C, i_d, r	
			-1.47			–						C	
			-1.40			4.64	–	–	–	–	–	–	C, i_d, r	
			–			–						C	
			-1.49			–	–	–	–	–	–	–	C, r	
JE048 0297	–	$E_{\frac{1}{2}}$	-0.50	SCE	i_ℓ	3.14	i:i	4	–	–	–	–	C, i_d, r	FB94
			-0.67			3.14		4					C, i_d	
			-1.10			4.10		–					C	
			-0.56			3.14	–	–	–	–	–	–	C, i_d, r	
			-0.77			3.14							C, i_d	
			-1.12			3.36							C	
			-0.62			3.14	–	–	–	–	–	–	C, i_d, r	
			-0.89			3.14							C, i_d	
			-1.12			2.60							C	
			-0.68			3.14	–	–	–	–	–	–	C, i_d, r	
			-0.97			3.04							C	
			-1.12			2.24							C	
			-0.77			3.04	–	–	–	–	–	–	C, i_d, r	
			-1.07		?	6.24							C	
			-1.25										C	
			-0.90			3.04	–	–	–	–	–	–	C, i_d, r	
			–		?	6.30							C, s	
			-1.28										C, p	
			-1.24			3.04	–	–	–	–	–	–	C, i_d, r	
			-1.60			3.04							C	
			-1.38			2.74	–	–	–	–	–	–	C, i_d, r	
			-1.70			–							C	
EA019 0629	–	$E_{\frac{1}{2}}$	-2.90	SCE(o)	i_ℓ	9	QE (-2.90V)	2.4	–	–	–	QE(-2.90 V, [Bu$_4$NI]= 0.25) → diphenyl- methane 80%	C, r	FB95 bd58
			-2.865			9	–	–	–	–	–	–	C, r	

FB96 $C_{13}H_{13}N$

Code No.	Empirical Formula	Name and C.A. Number	Structural Formula	Solvent	Tech.	Medium	μ, M	pH	T, °C	Electrodes	App.	Experimental Parameters
FB96	$C_{13}H_{13}N$	4-methyldiphenyl-amine C.A. 620-84-8	4-$CH_3C_6H_4NHC_6H_5$	MeCN	VR	Et_4NClO_4 0.1	-	-	-	Pt: xxns// Ag/ $AgClO_4$ 0.01, Et_4NClO_4 0.1	2-0	$0 \to 0.9 \to 0 \to 0.9$ V, C=0.87, d=0.5, v=25, Pt Aux
						Et_4NClO_4 0.1 $HClO_4$ 1.9E-3						
					VY	Et_4NClO_4 0.1	-	-	-	Pt: rodi// Ag/ $AgClO_4$ 0.01, Et_4NClO_4 0.1	2-0	C=1.02, d=0.1, ω=10, Pt Aux
						Et_4NClO_4 0.1 H_2O 5E-3						C=1
FB97	$C_{13}H_{13}NO$	4-methoxydiphenyl-amine C.A. 1208-86-2	4-$CH_3OC_6H_4NH$-C_6H_5	MeCN	VA	Et_4NClO_4 0.1 H_2O 5E-3	-	-	-	Pt: xxns// Ag/ $AgClO_4$ 0.01, Et_4NClO_4 0.1	2-0	$-0.5 \to 0.7 \to -0.5$ V, C=0.90, d=0.5, v=41, Pt Aux
						Et_4NClO_4 0.1 H_2O 0.52						
					VY	Et_4NClO_4 0.1 H_2O 5E-3	-	-	-	Pt: rodi// Ag/ $AgClO_4$ 0.01, Et_4NClO_4 0.1	2-0	C=1, d=0.1, ω=10, Pt Aux
						Et_4NClO_4 0.1 $HClO_4$ 1.6E-3						C=0.8
						Et_4NClO_4 0.1 H_2O 1.6E-2 $HClO_4$ 2E-3						
						Et_4NClO_4 0.1 H_2O 6.4E-3 $HClO_4$ 2E-3						
						Et_4NClO_4 0.1 H_2O 6.4E-2 $HClO_4$ 2E-3						
FB97 CONT												

TABLE I. Electrochemical Data

$C_{13}H_{13}NO$ (CONT.) FB97

| Ref. | C/M | Charact. Potential | | Response Const. | | n Tech. | Electrokinetic Data | | | Products and Identification | Description and Remarks | Code No. |
		Value	vs.		Value		Parameter	Value	From			
EA021 1171	147 jkl	E_p	0.35F	Ag/ AgClO$_4$ 0.01, Et$_4$NClO$_4$ 0.1	i_p	2F	-	-	-	-	-	-	A,0 on first scan,r	FB96
			0.55F			34F							A	
			0.76F			12F							A	
			0.43F			4F							C	
			0.33F			2F							C	
			0.57F			4F	-	-	-	-	-	-	A,r	
			0.78F			31F							A	
			0.53F			8F							C	
EA021 1171	147 jkl	$E_{\frac{1}{2}}$	0.54F	Ag/ AgClO$_4$ 0.01, Et$_4$NClO$_4$ 0.1	i_ℓ	7.6F	-	-	-	-	-	-	A,r	
			0.54			6.4	-	-	-	-	-	radical cation of dimer	A,r	
			0.7Ap			-						dication of dimer	A,$E_{\frac{1}{2}}$ and i_ℓ vary strongly with [H$_2$O]	
EA021 1171	147 jkl	E_p	0.47F	Ag/ AgClO$_4$ 0.01, Et$_4$NClO$_4$ 0.1	i_p	53F	-	-	-	-	-	-	A,r	FB97
			0.39F			42F							C	
			0.47F			62F	-	-	-	-	-	-	A,r	
			0.49F			35F							C	
			0.20F			15F							C	
EA021 1171	147 jkl	$E_{\frac{1}{2}}$	0.435	Ag/ AgClO$_4$ 0.01, Et$_4$NClO$_4$ 0.1	i_ℓ/C	6.1	-	-	-	-	-	radical cation of dimer	A,r	
			0.98			-						dication of dimer	A	
			0.60			-	-	-	-	-	-	-	A,i_k,r	
			0.63F			1.7F	-	-	-	-	-	-	A,i_k,r	
			0.92F			1.7F							A,i_k	
			0.63F			0.3F	-	-	-	-	-	-	A,i_k,r	
			0.98F			0.5F							A,i_k	
			0.56F			3.4F	-	-	-	-	-	-	A,r	
			0.88F			3.4F							A	

FB97 (CONT.) $C_{13}H_{13}NO$

Code No.	Empirical Formula	Name and C.A. Number	Structural Formula	Solvent	Tech.	Medium		μ, M	pH	T, °C	Electrodes	App.	Experimental Parameters
FB97	$C_{13}H_{13}NO$	4-methoxydiphenyl-amine	4-$CH_3OC_6H_4NH$-C_6H_5	MeCN	QE	Et_4NClO_4	0.1	–	–	–	Pt?: nsns// Ag/ AgClO$_4$ 0.01, Et$_4$NClO$_4$ 0.1	2-0	C=1.0, E_{app}=0.70, Pt Aux
FB98	$C_{13}H_{13}N_3$	4-methylaminoazo-benzene C.A. 621-90-9	4-$CH_3NHC_6H_4N$: NC_6H_5	dioxane 50	VD	HCl KCl	ns ns	–	1.0	–	Pt: rodi// SCE	OAO	C=4.0
						ACET	ns		4.85				
									7.0				
						NH_3 NH_4Cl	ns ns		9.0				
					VY	HCl KCl	ns ns	–	1.0	–	Pt: rodi// SCE	OAO	C=4.0
						ACET	ns		4.85				
									7.0				
						NH_3 NH_4Cl	ns ns		9.0				
FB99	$C_{13}H_{13}N_3O$	4-methoxydiazoamino-benzene C.A. 4625-55-2	4-$CH_3OC_6H_4N$:N-NHC_6H_5	MeOH 50	PY	UB GEL	ns 0.01	–	7.47	21±1	DME//SCE	0-0	C=0.5, m(oc)=4.9, t=1.9
									9.52				
									14				
FC00 bd83 ch98 ee57	$C_{13}H_{14}N_2O$	4-amino-4'-methoxy-diphenylamine C.A. 101-64-4	Table II	MeCN	VY	Et_4NClO_4	0.1	–	–	–	Pt: rodi// Ag/ AgClO$_4$ 0.01, Et$_4$NClO$_4$ 0.1	2-0	C=ns, d=0.1, ω=10, Pt Aux

TABLE I. Electrochemical Data

$C_{13}H_{14}N_2O$ FC00

Ref.	C/M	Charact. Potential		Response Const.		n Tech.	Electrokinetic Data			Products and Identification	Description and Remarks	Code No.
		Value	vs.		Value		Parameter	Value	From			
EA021 1171	147 jkl	-	-	-	-	-	1.4 ± 0.1	-	-	-	for n < 0.5, 4-methoxydiphenyl-amine radical cation, blue-green; 5,10-di(4-methoxy-phenyl)-5,10-di-hydro-2,7-dimeth-oxyphenazine radical cation; for n > 0.5, 4,4'-di(4-methoxyphenylamino)-biphenyldication and 5,10-di(4-methoxyphenyl)-5,-10-dihydro-2,7-di-methoxyphenazine dication 13-48%	A,r	FB97
C0034 1615	430 a	$E_{\frac{1}{2}}$	0.90	NCE	-	-	-	-	-	-	-	A,p	FB98
			0.825		-		-	-	-	-	-	A,p	
			1.25									A	
			0.75		-		-	-	-	-	-	A,p	
			1.37									A	
			0.76		-		-	-	-	-	-	A,p	
C0034 1615	430 a	$E_{\frac{1}{2}}$	0.82	NCE	-	-	-	-	-	-	-	A,p	
			0.825			49	-	-	-	-	-	A,$i_\ell \neq f(pH)$ for pH=3-4.8,r	
			0.75			19	-	-	-	-	-	CP→4-aminoazoben-zene	A,p	
			0.75			18	-	-	-	-	-	-	A,p	
JE035 0369	386 a	$E_{\frac{1}{2}}$	-1.08F	SCE	i_ℓ	7.2F	-	-	$dE_{\frac{1}{2}}/dpH$	62	sttd	-	C, $E_{\frac{1}{2}}$ = -0.595-0.062pH for pH=7-12,r	FB99
			-1.63F			small				0			C, $E_{\frac{1}{2}} \neq f(pH)$	
			-1.2F			5.4F	-	-		62		-	C,r	
			-1.6F			2.1F				0			C	
			-1.60F			5.8F	-	-		0		-	C,r	
EA021 0557	147 n	$E_{\frac{1}{2}}$	-0.015	Ag/AgClO₄ 0.01, Et₄NClO₄ 0.1	i_ℓ/C	6.0	sttd	1	-	-	-	radical cation, VIS	A, $\Delta E_{\frac{1}{2}}$=0.30σ,r	FC00 bd83 ch98 ee57
			0.49			-		1					A	

FC01 $C_{13}H_{14}N_2O_2$

Code No.	Empirical Formula	Name and C.A. Number	Structural Formula	Solvent	Tech.	Medium		μ, M	pH	T, °C	Electrodes	App.	Experimental Parameters
FC01	$C_{13}H_{14}N_2O_2$	1,2-diacetyl-1,2-dihydro-4-methyl-cinnoline C.A. 60389-46-0	Table II	DMF	VA	$LiClO_4$	0.1	—	—	—	Pt: xxns// Ag/AgI, I^- 0.1	2-0	C=ns, A=0.015, v=100
FC02	$C_{13}H_{14}N_4O_2$	1-(2-pyridyl)-3-hydroxyamino-3-(3-pyridyl)-1-propanone oxime C.A. 36256-97-0		DMF 20	PY	BR	—	—	2.3	—	DME//SCE	OAO	C=0.25, m=3.45, t=2.72, h=40
									2.90				
									3.95				
									5.20				
									6.8				
									7.8				
									10.0				
									13.2				
FC03	$C_{13}H_{15}N_5OS$	3,5-dimethyl-4-(2-methoxyphenylazo)-1-thiocarbamoyl-pyrazole C.A. 29147-34-0		H_2O	PY	BR GEL	— 5E-3	—	8	—	DME//SCE	OAO	C=0.01, EC=3.75
FC04	$C_{13}H_{15}N_5OS$	3,5-dimethyl-4-(4-methoxyphenylazo)-1-thiocarbamoyl-pyrazole C.A. 35872-22-1		H_2O	PY	BR GEL	— 5E-3	—	8	—	DME//SCE	OAO	C=0.01, EC=3.75
FC05	$C_{13}H_{22}BrN$	triethyl(benzyl)-ammonium bromide C.A. 5197-95-5	$(CH_3CH_2)_3(C_6H_5-CH_2)NBr$	MeCN	QE	—		—	—	-40 to 25	Al: nsns// Ag/Ag$^+$ 0.01, Et$_4$NBr 0.2	2--	E_{app} = -3.25 to -3.50 V, C=100, j=7.5E2-3E4, t=0, Pt Aux
										-37	Hg: srpo// Ag/Ag$^+$ 0.01, Et$_4$NBr 0.2		E_{app} = -2.7V, C=100, j=5.0E4-0, Pt Aux

TABLE I. Electrochemical Data

$C_{13}H_{22}BrN$ FC05

Ref.	C/M	Charact. Potential		Response Const.		n Tech.	Electrokinetic Data			Products and Identification	Description and Remarks	Code No.
		Value	vs.		Value		Parameter	Value	From			
EA021 0345	434 a	E_p	1.80	Ag/AgI, I^- 0.1	i_p	17.2	sttd 2	-	-	-	-	A,≠,r	FC01
JE048 0297	109 d	$E_{\frac{1}{2}}$	-0.45	SCE	i_ℓ	-	i:i -	-	-	-	-	C,i_a,Pr,r	FC02
			-0.65			2.76	4	-	-	-	-	C,W,i_d	
			-1.11			3.80	-	-	-	-	-	C,M,0 for pH ≥ 3.95	
			-0.48			-	-	-	-	-	-	C,i_a,Pr,r	
			-0.70			2.80	-	-	-	-	-	C,W,i_d	
			-1.16			3.12	-	-	-	-	-	C	
			-0.52			-	-	-	-	-	-	C,i_a,Pr,r	
			-0.78			2.76	-	-	-	-	-	C	
			-1.08			0.8	-	-	-	-	-	C	
			-0.60			-	-	-	-	-	-	C,i_a,Pr,r	
			-0.88			-	-	-	-	-	-	C	
			-1.15			-	-	-	-	-	-	C	
			-0.70			-	-	-	-	-	-	C,i_a,Pr,r	
			-1.04			2.76	-	-	-	-	-	C	
			-1.28			0.5	-	-	-	-	-	C	
			-1.12			2.76	-	-	-	-	-	C,slight splitting of wave,r	
			-1.31			2.76	-	-	-	-	-	C,r	
			-1.56			0.4	-	-	-	-	-	C,reversed S-shaped wave,r	
JE054 0411	198 l	$E_{\frac{1}{2}}$	-0.53	SCE	-	-	-	-	-	-	-	C,≠,i_d,Tc=1.6, $E_{\frac{1}{2}}$ → more neg as pH and C ↑,p	FC03
JE054 0411	198 l	$E_{\frac{1}{2}}$	-0.64	SCE	-	-	-	-	-	-	-	C,=,i_d,Tc=1.6, $E_{\frac{1}{2}}$ → more neg as pH and C ↑,p	FC04
JE035 0013	180 c	-	-	-	-	-	2	-	-	-	toluene(80% at -40°, 40% at 25°),GLC;bi- benzyl(19% at -40°, 35% at 25°),GLC;no bibenzyl at E_{app}= -2.3±0.1V and 25°	C,p	FC05
		-	-	-	-	-	QE 2	-	-	-	toluene(70% at -2.7V,98% at -2.4V and 25°),no bibenzyl or di- benzylmercury,GLC	C,p	

FC06 $C_{14}H_7N_2O$

Code No.	Empirical Formula	Name and C.A. Number	Structural Formula	Solvent	Tech.	Medium		μ, M	pH	T, °C	Electrodes	App.	Experimental Parameters
FC06	$C_{14}H_7N_2O$	pyrazolo[a,b,c-m,n]-9(10H)anthrone anion	Table 11	EtOH 50	PY	NaOH GEL	0.65 0.005	-	-	-	DME//SCE	04A0	C=1.0, t=0.87, h=44
FC07 be54 ci35 ee62	$C_{14}H_8O_2$	9,10-anthraquinone C.A. 84-65-1	Table 11-2	DMF	PY	Et_4NClO_4	0.1	-	-	-	DME//SCE	26A0	C=1, m=3.38, t=3.05
						$Ba(ClO_4)_2$	0.05						
						$LiClO_4$	0.1						
						$Sr(ClO_4)_2$	0.05						
FC08 be55	$C_{14}H_8O_2$	9,10-phenanthraquinone C.A. 84-11-7	Table 11-2	DMF	PY	Et_4NClO_4	0.1	-	-	-	DME//SCE	26A0	C=1, m=3.38, t=3.05
						$Ba(ClO_4)_2$	0.05						
						$LiClO_4$	0.1						
						$KClO_4$	0.1						
						$NaClO_4$	0.1						
						$Sr(ClO_4)_2$	0.05						
	$C_{14}H_7N_2O$	pyrazolo[a,b,c-m,n]-9(10H)anthrone anion	Table 11	EtOH 50	PY								

TABLE I. Electrochemical Data

$C_{14}H_8O_2$ FC08

Ref.	C/M	Charact. Potential		Response Const.		n		Electrokinetic Data			Products and Identification	Description and Remarks	Code No.
		Value	vs.	Response	Value		Tech.	Parameter	Value	From			
C0026 1763	433 a	$E_{\frac{1}{2}}$	-1.24	SCE	-	-	i:i	2	-	-	-	-	C, i_d, P	FC06
JE055 0277	-	$E_{\frac{1}{2}}$	-0.88	SCE	i_ℓ	3.8	-	-	Elog	60	-	-	$C, R, E_{\frac{1}{2}} \neq f(t$ or $C)$ for $t=1.02-5.5$ and $C=0.02-1, r$	FC07 be54 ci35 ee62
					-	-				-			$C, \neq$	
			-0.85			4	-	-	Elog $dE_{\frac{1}{2}}/$ $d\log[Ba^{2+}]$	65 +52	-	-	$C, R, E_{\frac{1}{2}} \neq f(t$ or $C)$ for $t=1.02-5.5$ and $C=0.02-1, r$	
					-	-				-			$C, \neq$	
			-0.83			4.1	-	-	Elog $dE_{\frac{1}{2}}/$ $d\log[Li^+]$	60 +48	-	-	$C, R, E_{\frac{1}{2}} \neq f(t$ or $C)$ for $t=1.02-5.5$ and $C=0.02-1, r$	
					-	-				-			$C, \neq$	
			-0.83			3.9	-	-	Elog	65	-	-	$C, R, E_{\frac{1}{2}} \neq f(t$ or $C)$ for $t=1.02-5.5$ and $C=0.02-1, r$	
					-	-				-			$C, \neq$	
JE055 0277	-	$E_{\frac{1}{2}}$	-0.66	SCE	i_ℓ	4.2	-	-	Elog	64	-	-	$C, R, E_{\frac{1}{2}} \neq f(t$ or $C)$ for $t=1.02-5.5$ and $C=0.02-1, r$	FC08 be55
					-	-				-			$C, \neq$	
			-0.38			4	-	-	Elog $dE_{\frac{1}{2}}/$ $d\log[Ba^{2+}]$	60 +64	-	-	$C, R, E_{\frac{1}{2}} \neq f(t$ or $C)$ for $t=1.02-5.5$ and $C=0.02-1, r$	
					-	-				-			$C, \neq$	
			-0.43			4.2	-	-	Elog $dE_{\frac{1}{2}}/$ $d\log[Li^+]$	63 +60	-	-	$C, R, E_{\frac{1}{2}} \neq f(t$ or $C)$ for $t=1.02-5.5$ and $C=0.02-1, r$	
					-	-				-			$C, \neq$	
			-0.53			4.2	-	-	Elog	61	-	-	$C, R, E_{\frac{1}{2}} \neq f(t$ or $C)$ for $t=1.02-5.5$ and $C=0.02-1, r$	
					-	-				-			$C, \neq$	
			-0.51			4.1	-	-		66	-	-	$C, R, E_{\frac{1}{2}} \neq f(t$ or $C)$ for $t=1.02-5.5$ and $C=0.02-1, r$	
					-	-				-			$C, \neq$	
			-0.37			4.3	-	-		65	-	-	$C, R, E_{\frac{1}{2}} \neq f(t$ or $C)$ for $t=1.02-5.5$ and $C=0.02-1, r$	
					-	-				-			$C, \neq$	

FC09 $C_{14}H_9N_3S_2$

Code No.	Empirical Formula	Name and C.A. Number	Structural Formula	Solvent	Tech.	Medium		μ, M	pH	T, °C	Electrodes	App.	Experimental Parameters
FC09	$C_{14}H_9N_3S_2$	4,4'-dithiocyanato-diphenylamine C.A. 5339-39-9	$(4-NCSC_6H_4)_2NH$	MeCN	VY	Et_4NClO_4	0.1	-	-	-	Pt: rodi// Ag/ $AgClO_4$ 0.01, Et_4NClO_4 0.1	2-0	C=ns,d=0.1, ω=10,Pt Aux
FC10 be82 ci50 de32	$C_{14}H_{10}$	anthracene C.A. 120-12-7	Table II-2	MeCN	VY	$AlCl_3$	1	-	-	-	Pt: rodi// Ag/AgCl, Me_4NCl satd	2AO	ns
				DMF	PY	Bu_4NI	0.25	-	-	-	DME// SCE(o)	2-0	C=2
						Bu_4NI PHEN	0.25 0.01						
				DMSO	VA	Bu_4NClO_4	0.1	-	-	-	HMDE// Ag/AgCl	12AO 2	-1.5 → -2.0 → -1.5 V, C=0.1,v=50, Pt Aux
						Bu_4NClO_4 $C_6H_5CH_2Cl$	0.1 1E-4						
						Bu_4NClO_4 $C_6H_5CH_2Cl$	0.1 5E-4						
						Bu_4NClO_4 PHEN	0.1 1E-3						C=0.5, v=200, Pt Aux
						Bu_4NClO_4 PHEN	0.1 1E-3						v=2E4
				HMP	VA	$LiClO_4$	0.1	-	-	-	glC: xxdi// Li/LiCl, satd	25AO	2.5 → 0 → 2.5 V,C=2, d=0.3,v=300
						Et_4NClO_4	0.1						
						Bu_4NClO_4	0.1						
				HF	VY	KF	0.1	-	-	0	glC: rodi// Cu/CuF_2, KF 0.2	2-0	C=satd, A=0.07, v=100,ω=1, Pt Aux

CONT

TABLE I. Electrochemical Data

$C_{14}H_{10}$ (CONT.) FC10

Ref.	C/M	Charact. Value	Potential vs.	Response Value	Const.	n Tech.	Electrokinetic Data Parameter	Value	From	Products and Identification	Description and Remarks	Code No.
EA021 0557	147 n	$E_{\frac{1}{2}}$ 0.86	Ag/ AgClO$_4$ 0.01, Et$_4$NClO$_4$ 0.1	i_ℓ/C	5.5	sttd 1	-	-	-	radical cation,VIS	A,$\Delta E_{\frac{1}{2}}=0.46\sigma$,r	FC09
		1.33			-	1					A	
JE039 0385	-	$E_{\frac{1}{2}}$ 0.90	Fc	-	-	-	-	-	-	-	A,Σ,p	FC10 be82 ci50 de32
EA019 0629	-	$E_{\frac{1}{2}}$ -2.06	SCE(o)	i_ℓ	3.4	-	-	-	-	-	C,r	
		-2.53			3.0						C	
		-1.99			8.4	QE (-2.2V) 4.2	-	-	-	9,10-dihydroanthracene 90%	C,r	
		2.71			3.4	-				-	C	
JE056 0259	418 a	E_p -1.87F	Ag/AgCl	i_p	1.4F	-	-	-	-	-	C,p	
		-1.81F			0.86F						A	
		-1.87F			2.1F	-	-	-	-	-	C,p	
		-1.81F			0.5F						A	
		-1.87F			4F	-	-	-	-	-	C,p	
		-			-						A,0	
		-1.87F			4.05F	-	-	-	-	-	C,p	
		-			-						A,0	
		-1.85F			30F	-	-	-	-	-	C,p	
		-1.79F			18.9F						A	
JE041 0405	215 j	E_p 1.23	Li/LiCl satd	-	-	-	-	-	-	-	C,p	
		0.33									C	
		1.29									A	
		1.23		-	-	-	-	-	-	-	C,p	
		0.58									C	
		1.28									A,related to peak at 1.23 V	
		1.95									A	
		-									C,0 on first scan, related to peak at 1.95 V	
		1.22		-	-	-	-	-	-	-	C,p	
		0.47									C	
		1.28									A,related to peak at 1.22 V	
		1.89									A	
		1.95 ApF									C,0 on first scan, related to peak at 1.89 V	
JE054 0232	-	$E_{\frac{1}{2}}$ 1.045	NHE	j_ℓ	43.1	-	-	Elog	71	-	-	A,p	

CONT

FC10 (CONT.) $C_{14}H_{10}$

Code No.	Empirical Formula	Name and C.A. Number	Structural Formula	Solvent	Tech.	Medium		μ, M	pH	T, °C	Electrodes	App.	Experimental Parameters
FC10 be82 ci50 de32	$C_{14}H_{10}$	anthracene	Table II-2	$MeNO_2$	VY	$AlCl_3$	1	-	-	-	Pt: rodi // Ag/AgCl, Me_4NCl satd	2AO	C=ns, A=0.01
						Et_4NClO_4	0.1						
FC11 be84 ci52	$C_{14}H_{10}$	phenanthrene C.A. 85-01-8	Table II-2	HF	VY	KF	0.1	+	-	0	Pt: rodi // Cu/CuF_2, KF 0.2	2-0	C=satd, A=5E-3, v=100, ω=1, Pt Aux
FC12 bf19	$C_{14}H_{10}N_2$	3-imino-2-phenyl-3H-indole C.A. 6339-33-9	Table II	DMF	PY	$KClO_4$	0.02	-	-	20±0.1	DME // SCE (NaCl, aq)	23AO	C=0.1-1.0, m=4.84, t=2.9, h=80, Hg Aux
FC13 be91 ee66	$C_{14}H_{10}N_2O_4$	1,2-bis(4-nitrophenyl)ethene C.A. 2501-02-2	$[4-O_2NC_6H_4CH:]_2$	DMF	XT	Et_4NClO_4	0.1	-	-	10±1	HMDE // SCE	2-2	C=ns, v=(1.5-5.5)E3
										60±0.5			
FC14 bf19	$C_{14}H_{11}NO_2$	1-(2-hydroxyphenyl)-3-(2-pyridyl)-2-propen-1-one C.A. 2875-24-3	Table II-2	EtOH 50	PY	BR GEL	0.02	-	2.6	22±1	DME // SCE	OAO	C=0.5, m=1.15, t=5.2, h=64
									5.6				
									6.9				
									8.0				
									9.2				
									10.0				

CONT

TABLE I. Electrochemical Data

$C_{14}H_{11}NO_2$ (CONT.) FC14

Ref.	C/M	Charact. Potential		Response Const.		n Tech.	n	Electrokinetic Data			Products and Identification	Description and Remarks	Code No.
		Value	vs.		Value			Parameter	Value	From			
JE039 0385	-	$E_{\frac{1}{2}}$	0.83	Fc	-	-	-	-	-	-	-	-	A,r	FC10 be82 ci50 de32
		0.83		-	-	-	-	-	-	-	-	A,r	
JE054 0232	-	$E_{\frac{1}{2}}$	1.295	NHE	i_ℓ	2.2F	-	-	Elog	77	-	-	A,p	FC11 be84 ci52
JE036 0147	427 ce	$E_{\frac{1}{2}}$	-1.114	SCE (NaCl,aq)	I	0.97	QE	-	Tomeš	63	-	2-phenyl-3-aminoindole,λ_{max}=308	C,R,i_ℓ=kC for C=0.1-1,i_d,Tc=1 for T=15-30,r	FC12
		-1.712			0.96		2± 0.1		85			C,$\neq$,i_d,i_ℓ=kC for C=0.1-1	
JE047 0115	-	XE	0.007 ±0.003	SCE	-	-	-	-	-	-	-	-	C,XT=semiintegral chronoamperometry with linear potential sweep,XE=difference between potentials for the two steps,p	FC13 be91 ee66
		-										C	
		0.023 ±0.002		-	-	-	-	-	-	-	-	C,p	
		-										C	
JE039 0407	-	$E_{\frac{1}{2}}$	-0.55	SCE	i_ℓ	1.1F	-	-	$dE_{\frac{1}{2}}$/dpH	67F	-	-	C,i_d,$dE_{\frac{1}{2}}$/dpH for pH=2.6-4.7,r	FC14 bf19
		-1.14			1.3F				53F			C,i_d,$dE_{\frac{1}{2}}$/dpH for pH=2.6-10	
		-0.56			1.1F	-	-		82F		-	C,i_d,$dE_{\frac{1}{2}}$/dpH for pH=4.7-8,r	
		-1.32			1.6F				53F			C,i_d	
		-0.67			1.1F	-	-		82F		-	C,i_d,r	
		-1.42			1.4F				53F			C,i_d	
		-0.76			0.53F	-	-		53F		-	C,i_d,i_ℓ=kC for C=0-0.8,$dE_{\frac{1}{2}}$/dpH for pH=8-11.2,r	
		-1.48			0.63F				53F			C,i_d,i_ℓ=kC for C=0-0.8	
		-1.79			2.2F				-			C,i_{cat},i_ℓ=f(C),i_ℓ ↓ as h ↑	
		-0.83			0.94F	-	-		53F		-	C,r	
		-1.53			0.67F				53F			C,Mx	
		-0.86			0.5F	-	-		53F		-	C,r	
		-1.53			0.78F				0F			C	

CONT

FC14 (CONT.) $C_{14}H_{11}NO_2$

Code No.	Empirical Formula	Name and C.A. Number	Structural Formula	Solvent	Tech.	Medium		μ, M	pH	T, °C	Electrodes	App.	Experimental Parameters
FC14 bf19	$C_{14}H_{11}NO_2$	1-(2-hydroxyphenyl)-3-(2-pyridyl)-2-propen-1-one	Table II-2	EtOH 50	PY	BR GEL	- 0.02	-	11.2	22±1	DME//SCE	OAO	C=0.5,m=1.15, t=5.2,h=64
									12.5				
				dioxane 75	PY	Bu$_4$NI GEL	0.175 0.02	-	-	22±1	DME//SCE	OAO	C=0.5,m=1.15, t=5.2,h=64
FC15 bf20	$C_{14}H_{11}NO_2$	1-(2-hydroxyphenyl)-3-(3-pyridyl)-2-propen-1-one C.A. 2875-25-4	Table II-2	EtOH 50	PY	BR GEL	- 0.02	-	2.6	22±1	DME//SCE	OAO	C=0.5,m=1.15, t=5.2,h=64
									5.6				
									8.0				
									9.2				
									11.2				
									12.15				
				dioxane 75	PY	Bu$_4$NI GEL	0.175 0.02	-	-	22±1	DME//SCE	OAO	C=0.5,m=1.15, t=5.2,h=64
FC16 bf21	$C_{14}H_{11}NO_2$	1-(2-hydroxyphenyl)-3-(4-pyridyl)-2-propen-1-one C.A. 2875-27-6	Table II-2	EtOH 50	PY	BR GEL	- 0.02	-	2.6	22±1	DME//SCE	OAO	C=0.5,m=1.15, t=5.2,h=64
									5.6				
									8.0				

TABLE I. Electrochemical Data

$C_{14}H_{11}NO_2$ (CONT.) FC16

Ref.	C/M	Charact. Potential		Response Const.		n	Tech.	Electrokinetic Data			Products and Identification	Description and Remarks	Code No.
		Value	vs.		Value			Parameter	Value	From			
JE039 0407	-	$E_{1/2}$	-0.93	SCE	i_ℓ	0.17F	-	-	$dE_{1/2}/dpH$	53F	-	-	C,r	FC14 bf19
			-1.53			0.72F				0F			C	
			-1.05			0.94F	-	-	-	-	-	-	C,r	
JE039 0407	-	$E_{1/2}$	-0.90	SCE	I	1.36	-	-	-	-	-	-	C,$\Delta E_{1/2}=0.72\Sigma\sigma^*$,r	
			-1.63			1.38							C	
			-2.03			-							C	
JE039 0407	-	$E_{1/2}$	-0.57	SCE	-	-	-	-	$dE_{1/2}/dpH$	48F	-	-	C,$dE_{1/2}/dpH$ for pH=2.6-4.7,r	FC15 bf20
			-1.13							50F			C,$dE_{1/2}/dpH$ for pH=2.6-5.6	
			-0.75		-	-	-	-		95F	-	-	C,$dE_{1/2}/dpH$ for pH=4.7-6.9	
			-1.28							79F			C,$dE_{1/2}/dpH$ for pH=5.6-8	
			-0.93		-	-	-	-	-	-	-	-	C,i_d,r	
			-1.47										C,i_d	
			-1.75										C,i_{cat},i_ℓ ↓ as h ↑, $i_\ell=f(C)$	
			-0.96		-	-	-	-	-	-	-	-	C,r	
			-1.50										C	
			-1.03		-	-	-	-	-	-	-	-	C,r	
			-1.53										C	
			-1.15		-	-	-	-	-	-	-	-	C,r	
			-1.32										C	
JE039 0407	-	$E_{1/2}$	-0.97	SCE	I	1.37	-	-	-	-	-	-	C,p	
			-1.33			1.37							C	
			-1.66			-							C	
JE039 0407	-	$E_{1/2}$	-0.33	SCE	-	-	-	-	$dE_{1/2}/dpH$	60F	-	-	C,$dE_{1/2}/dpH$ for pH=2.6-5.6	FC16 bf21
			-1.18							50F			C,$dE_{1/2}/dpH$ for pH=2.6-5.6	
			-0.51		-	-	-	-		83F	-		C,$dE_{1/2}/dpH$ for pH=5.6-8,r	
			-1.33							71F			C,$dE_{1/2}/dpH$ for pH=5.6-8	
			-0.71		-	-	-	-		60F			C,i_d,$dE_{1/2}/dpH$ for pH=8-10	
			-1.50							0F			C,i_d,$dE_{1/2}/dpH$ for pH > 8	
			-1.88							-			C,i_{cat}	

CONT

FC16 (CONT.) $C_{14}H_{11}NO_2$

Code No.	Empirical Formula	Name and C.A. Number	Structural Formula	Solvent	Tech.	Medium		μ, M	pH	T, °C	Electrodes	App.	Experimental Parameters
FC16 bf21	$C_{14}H_{11}NO_2$	1-(2-hydroxyphenyl)-3-(4-pyridyl)-2-propen-1-one	Table II-2	EtOH 50	PY	BR GEL	- 0.02	-	10.0	22±1	DME//SCE	OAO	C=0.5, m=1.15, t=5.2, h=64
									11.2				
									12.15				
				dioxane 75	PY	Bu_4NI GEL	0.175 0.02	-	-	22±1	DME//SCE	OAO	C=0.5, m=1.15, t=5.2, h=64
FC17	$C_{14}H_{11}NO_2S$	3-(4-carbamoylphenyl)-1-(2-thienyl)-2-propen-1-one C.A. 53094-49-8	Table II	EtOH 50	PY	NaOAc HCl	ns ns	0.2	0.77	25±0.1	DME//SCE	O3AO	C=0.2, m=1.86, t=3.8, h=70
FC18	$C_{14}H_{11}NO_3$	1-(2-hydroxyphenyl)-3-hydroxy-3-(2-pyridyl)-2-propen-1-one	Table II	EtOH 50	PY	BR GEL	- 0.02	-	3.6	22±1	DME//SCE	OAO	C=0.5, m=1.15, t=5.2, h=64
									5.6				
									8.0				
									9.2				
									11.2				
									12.15				
FC19 bf32	$C_{14}H_{12}$	1,1-diphenylethene C.A. 530-48-3	$(C_6H_5)_2C{:}CH_2$	DMF	VA	Et_4NBF_4	satd	-	-	-	Hg: xxsd//SCE	5AO	C=ns, v=300
				HMP	VA	Bu_4NBF_4	satd	-	-	-	Hg: xxsd//SCE	5AO	C=ns, v=300
FC20	$C_{14}H_{12}$	phenylcyclooctatetraene C.A. 4603-00-3	Table II	MeCN	PY	Et_4NClO_4	ns	-	-	0	DME//SCE (NaCl, aq)	2AO	C=ns, m=2.19, h=65, EC=2.05, Pt Aux
										22			
				PrCN	PY	Et_4NClO_4	ns	-	-	22	DME//SCE (NaCl, aq)	2AO	C=ns, m=2.19, h=65, EC=2.05, Pt Aux

CONT

$C_{14}H_{12}$ (CONT.) FC20

Ref.	C/M	Charact. Potential		Response Const.		n Tech.	Electrokinetic Data			Products and Identification	Description and Remarks	Code No.
		Value	vs.		Value		Parameter	Value	From			
JE039 0407	-	$E_{\frac{1}{2}}$	-0.83	SCE	-	-	-	-	-	-	-	-	C,r	FC16 bf21
			-1.50									C	
			-0.9		-	-	-	-	-	-	-	C,i_d,r	
			-1.50									C,i_d	
			-									C,i_{cat}	
			-1.01		-	-	-	-	-	-	-	C,r	
JE039 0407	-	$E_{\frac{1}{2}}$	-0.80	SCE	I	1.39	-	-	-	-	-	C,$\Delta E_{\frac{1}{2}}=0.72\Sigma\sigma^*$,r	
			-1.55			1.43						C	
			-2.00			-						C	
JE053 0439	-	$E_{\frac{1}{2}}$	-0.512	SCE	-	-	-	-	-	-	-	C,p	FC17
			-0.887									C	
JE039 0407	-	$E_{\frac{1}{2}}$	-1.18F	SCE	i_ℓ	1.7F	sttd	2	-	-	-	-	C,r	FC18
			1.26F			1.4F	-	-	-	-	-	-	C,r	
			1.41F			1.2F	-	-	-	-	-	-	C,r	
			-1.45F			0.63F	-	-	-	-	-	-	C,r	
			-1.05F			-	-	-	-	-	-	-	C,Pr,i_ℓ ↑ on standing,r	
			-1.46F			0.50F							C	
			-1.05F			0.4F	-	-	-	-	-	-	C,Pr,i_ℓ ↑ on standing,r	
			-			-							C,M	
EA019 0951	-	XE	-2.43	SCE	-	-	-	-	-	-	-	C,A,XE=($E_{p,A}+E_{p,C}$)/2, ⇌,ece,p	FC19 bf32
EA019 0951	-	XE	-2.42	SCE	-	-	-	-	-	-	-	C,A,XE=($E_{p,A}+E_{p,C}$)/2, ⇌,0 in satd LiCl, ece,p	
JE056 0409	-	$E_{\frac{1}{2}}$	-1.730	SCE (NaCl,aq)	-	-	sttd	2	-	-	-	-	C,p	FC20
			-1.758		I	3.5	1:1	2	st	2.66	sttd	dianion	C,$i_\ell=kh^{\frac{1}{2}}$,i_d,p	
JE056 0409	-	$E_{\frac{1}{2}}$	-1.78	SCE (NaCl,aq)	-	-	sttd	2	st	1.45	sttd	dianion	C,$i_\ell=kh^{\frac{1}{2}}$,i_d,i_ℓ=kC for C=0.9-5.5,p	

CONT

FC20 (CONT.) $C_{14}H_{12}$

Code No.	Empirical Formula	Name and C.A. Number	Structural Formula	Solvent	Tech.	Medium	μ, M	pH	T, °C	Electrodes	App.	Experimental Parameters
FC20	$C_{14}H_{12}$	phenylcyclooctatetraene	Table II	DMF	PY	Et_4NClO_4 ns	-	-	0	DME//SCE (NaCl,aq)	2AO	C=ns, m=2.19, h=65, EC=2.05, Pt Aux
									22			
FC21 bf34 (ci75)	$C_{14}H_{12}$	trans-stilbene C.A. 103-30-0	$(C_6H_5CH:)_2$	DMF	VA	Et_4NBF_4 satd	-	-	-	Hg: xxsd// SCE	5AO	C=ns, v=300
				HMP	VA	Bu_4NBF_4 satd	-	-	-	Hg: xxsd// SCE	5AO	C=ns, v=300
FC22 (de38)	$C_{14}H_{12}Br_2$	dl-1,2-dibromo-1,2-diphenylethane C.A. 13027-48-0	dl-$(C_6H_5CHBr)_2$	DMF	PY	Et_4NClO_4 0.1	-	-	20± 0.1	DME//SCE	23AO	C=0.5-1, m=0.7, t(oc)=19.6
					VA	Et_4NClO_4 0.1	-	-	-	HMDE// SCE	23AO	0.16 → -0.64 → 0.16V, C=3, Hg Aux
FC23	$C_{14}H_{12}Br_2$	meso-1,2-dibromo-1,2-diphenylethane C.A. 13440-24-9	meso-$(C_6H_5CHBr)_2$	DMF	PY	Et_4NClO_4 0.1	-	-	20± 0.1	DME//SCE	23AO	C=0.5-1, m=0.7, t(oc)=19.6
					VA	Et_4NClO_4 0.1	-	-	20± 0.1	HMDE// SCE	23AO	0.16 → -0.64 → 0.16 V, C=3, Hg Aux
FC24 bf46 de39	$C_{14}H_{12}N_2O_4$	1,2-bis(4-nitrophenyl)ethane C.A. 736-30-1	$(4-O_2NC_6H_4CH_2)_2$	MeCN	XT	Et_4NClO_4 0.1	-	-	-30 ±1	HMDE// Ag/ $AgClO_4$ 0.01	2-2	C=ns, v= (1.5-5.5)E3
									0±1			
									20± 0.5			
				DMF	XT	Et_4NClO_4 0.1	-	-	0±1	HMDE// SCE	2-2	C=ns, v= (1.5-5.5)E3

CONT

$C_{14}H_{12}N_2O_4$ (CONT.) FC24

Ref.	C/M	Charact. Potential		Response Const.		n	Electrokinetic Data			Products and Identification	Description and Remarks	Code No.
		Value	vs.		Value	Tech.	Parameter	Value	From			
JE056 0409	–	$E_{\frac{1}{2}}$ -1.640	SCE (NaCl,aq)	–	–	–	–	–	–	–	C,p	FC20
		-1.652		I	3.44	1:1	2 St	1.16	sttd	dianion	$C, i_\ell = kh^{\frac{1}{2}}, i_d, p$	
		-2.35			small		–				C	
EA019 0951	–	XE -2.17	SCE	–	–	–	–	–	–	–	$C,A,XE=(E_{p,A}+E_{p,C})/2$, R,eec,p	FC21 bf34 (ci75)
		E_p -2.60									C,≠	
EA019 0951	–	XE -2.11	SCE	–	–	–	–	–	–	–	$C,A,XE=(E_{p,A}+E_{p,C})/2$, R,O in satd LiCl, eec,p	
		E_p -2.65									C,≠	
JE054 0289	94h	$E_{\frac{1}{2}}$ -0.23	SCE	I	3.6	QE 1:1	2 αn_a	0.52	Elog	CP→trans-stilbene 100%Ap, PY, UVS	C,W,i_d,r	FC22 (de38)
		-2.18			–		–	–			C,W,i_d, redn of trans-stilbene	
		-2.54									C,W,i_d	
JE054 0289	94h	E_p -0.294	SCE	–	–	QE	2	–	–	trans-stilbene	C,r	
		-0.105					–				A,oxidn of Br⁻	
JE054 0289	94h	$E_{\frac{1}{2}}$ -0.23	SCE	I	3.6	QE 1:1	2 αn_a	0.52	Elog	CP→trans-stilbene 100%Ap, PY, UVS	C,W,i_d,r	FC23
		-2.18			–		–	–			C,W,i_d, redn of trans-stilbene	
		-2.54									C,W,i_d	
JE054 0289	94h	E_p -0.294	SCE	–	–	–	–	–	–	trans-stilbene	C,r	
		-0.105									A,oxidn of Br⁻	
JE047 0115	–	XE 0.035 ±0.002	Ag/ AgClO₄ 0.01	–	–	–	–	–	–	–	C,XT=semiintegral chronoamperometry with linear potential sweep, XE=difference between potentials for the two steps,p	FC24 bf46 de39
		–									C	
		0.042 ±0.002		–	–	–	–	–	–	–	C,p	
		–									C	
		0.046 ±0.002		–	–	–	–	–	–	–	C,p	
		–									C	
JE047 0115	–	XE 0.061 ±0.002	SCE	–	–	–	–	–	–	–	C,XT=semiintegral chronoamperometry with linear potential sweep, XE=difference between potentials for the two steps,p	
		–									C	

CONT

FC24 (CONT.) $C_{14}H_{12}N_2O_4$

Code No.	Empirical Formula	Name and C.A. Number	Structural Formula	Solvent	Tech.	Medium		μ, M	pH	T, °C	Electrodes	App.	Experimental Parameters
FC24 bf46 de39	$C_{14}H_{12}N_2O_2$	1,2-bis(4-nitrophenyl)ethane	$(4-O_2NC_6H_4-CH_2)_2$	DMF	XT	Et_4NClO_4	0.1	-	-	30±0.5	HMDE// SCE	2-2	C=ns, v= (1.5-5.5)E3
										60±0.5			
FC25	$C_{14}H_{12}O$	9-hydroxy-9,10-dihydroanthracene C.A. 611-63-2	Table II	DMF	PY	Et_4NBr	0.1	-	-	-	DME//SCE (o)	2-0	C=2
						Et_4NBr PHEN	0.1 0.01						
FC26	$C_{14}H_{12}O$	9-hydroxy-9-methyl-fluorene C.A. 6311-22-4	Table II	DMF	PY	Bu_4NI	0.1	-	-	-	DME//SCE (o)	2-0	C=2
						Bu_4NI PHEN	0.1 ns						
FC27	$C_{14}H_{12}OS$	3-(2-methylphenyl)-1-(2-thienyl)-2-propen-1-one C.A. 53094-46-5	Table II	EtOH 50	PY	NaOAc HCl	ns ns	0.2	0.77	25±0.1	DME//SCE	03A0	C=0.2, m=1.86, t=3.8, h=70
FC28	$C_{14}H_{12}OS$	3-(4-methylphenyl)-1-(2-thienyl)-2-propen-1-one C.A. 6028-94-0	Table II	EtOH 50	PY	NaOAc HCl	ns ns	0.2	0.77	25±0.1	DME//SCE	03A0	C=0.2, m=1.86, t=3.8, h=70
FC29	$C_{14}H_{13}ClN_2O$	4-chloro-4'-acetyl-aminodiphenylamine C.A. 61236-16-6	Table II	MeCN	VY	Et_4NClO_4	0.1	-	-	-	Pt: rodi// Ag/ AgClO_4 0.01, Et_4NClO_4 0.1	2-0	C=ns, d=0.1, ω=10, Pt Aux
						Et_4NClO_4 2,6-lutidine ACET	0.1 ns ns						
	$C_{14}H_{13}ClN_6O_6$ see FB89												
FC30 bf65	$C_{14}H_{13}NO$	4-acetyldiphenyl-amine C.A. 23600-83-1	$4-CH_3COC_6H_4NH C_6H_5$	MeCN	VY	Et_4NClO_4	0.1	-	-	-	Pt: rodi// Ag/ AgClO_4 0.01, Et_4NClO_4 0.1	2-0	C=0.98, d=0.1, ω=10, Pt Aux
						Et_4NClO_4 H_2O	0.1 5E-3						C=1

TABLE I. Electrochemical Data

$C_{14}H_{13}NO$ FC30

Ref.	C/M	Charact.	Potential Value	vs.	Response	Const. Value	Tech.	n	Electrokinetic Data Parameter	Value	From	Products and Identification	Description and Remarks	Code No.
JE047 0115	-	XE	0.065 ±0.002	SCE	-	-	-	-	-	-	-	-	C,p	FC24 bf46 de39
			-										C	
			0.068 ±0.002	-	-	-	-	-	-	-	-	-	C,p	
			-										C	
EA019 0629	-	$E_{\frac{1}{2}}$	-2.05	SCE(o)	i_ℓ	4.4	QE (-2.3V)	1.75	-	-	-	QE(-2.3 V, [Bu$_4$NI]= 0.25) → 9,10-di-hydroanthracene 95%	C,r	FC25
			-2.53			6.8	-	-				-	C	
			-2.00			8.1	QE (-2.5V)	2.0				-	C,r	
			-2.58			0.8	-	-					C	
			-2.805			3.4							C	
EA019 0629	-	$E_{\frac{1}{2}}$	-2.47	SCE(o)	-	-	QE (-2.4V)	2.1	-	-	-	-	C,r	FC26
			-2.45				QE (-2.3V)	2.0				9-methylfluorene 90%	C,r	
			-2.69					-				-	C	
JE053 0439	-	$E_{\frac{1}{2}}$	-0.552	SCE	-	-	-	-	-	-	-	-	C,p	FC27
			-0.937										C	
JE053 0439	156 k	$E_{\frac{1}{2}}$	-0.587	SCE	-	-	-	-	-	-	-	-	C,$\Delta E_{\frac{1}{2}}$=0.22σ,p	FC28
			-0.950										C	
EA021 0557	147 n	$E_{\frac{1}{2}}$	0.505	Ag/ AgClO$_4$ 0.01, Et$_4$NClO$_4$ 0.1	i_ℓ/C	5.9	sttd	1	-	-	-	radical cation,VIS	A,r	FC29
			0.91			-		1					A	
			0.32			-	-	-				-	A,r	
			0.09			-	-	-				-	A,r	
EA021 1171	147 jkl	$E_{\frac{1}{2}}$	0.79F	Ag/ AgClO$_4$ 0.01, Et$_4$NClO$_4$ 0.1	i_ℓ	11.2F	-	-	-	-	-	-	A,r	FC30 bf65
			0.75			11Ap	-	-				radical cation of dimer	A,r	
			0.85			-						dication of dimer	A	

FC31 $C_{14}H_{13}N_3O_3$

Code No.	Empirical Formula	Name and C.A. Number	Structural Formula	Solvent	Tech.	Medium		μ, M	pH	T, °C	Electrodes	App.	Experimental Parameters
FC31	$C_{14}H_{13}N_3O_3$	1-(2-hydroxyphenyl)-3-(4-pyridyl)-1,3-propanedione dioxime C.A. 36256-98-1	Table II	DMF 20	PY	BR	-	-	2.3	-	DME//SCE	OAO	c=0.25, m=3.45, t=2.72, h=40
									3.20				
									4.3				
									5.15				
									6.10				
									7.15				
									9.90				
									11.5				
FC32	$C_{14}H_{14}$	4-ethylbiphenyl C.A. 5707-44-8	4-$CH_3CH_2C_6H_4$-C_6H_5	DMF	PY	Bu_4NI	0.25	-	-	-	DME//SCE(o)	2-0	c=2
						Bu_4NI PHEN	0.25 0.01						
FC33	$C_{14}H_{14}BrN_3$	3-bromo-4'-dimethylaminoazobenzene C.A. 17576-88-4	Table II	Me_2CO 50	VY	ACET	1	-	7.0	-	Pt: rodi//SCE	OAO	c=4.0
FC34	$C_{14}H_{14}BrN_3$	4-bromo-4'-dimethylaminoazobenzene C.A. 3805-65-0	Table II	dioxane 50	VD	ACET	1	-	7.0	-	Pt: rodi//SCE	OAO	c=4.0
					VY	ACET	1	-	7.0	-	Pt: rodi//SCE	OAO	c=4.0
FC35	$C_{14}H_{14}ClN_3$	3-chloro-4'-dimethylaminoazobenzene C.A. 3789-77-3	Table II	Me_2CO 50	VY	ACET	1.0	-	7.0	-	Pt: rodi//SCE	OAO	c=4.0
FC36 CONT	$C_{14}H_{14}ClN_3$	4-chloro-4'-dimethylaminoazobenzene C.A. 2491-76-1	Table II	dioxane 50	VD	ACET	1.0	-	7.0	-	Pt: rodi//SCE	OAO	c=4.0

TABLE I. Electrochemical Data

$C_{14}H_{14}ClN_3$ (CONT.) FC36

Ref.	C/M	Charact. Potential		Response	Const. Value	n Tech.		Electrokinetic Data			Products and Identification	Description and Remarks	Code No.
		Value	vs.					Parameter	Value	From			
JE048 0297	–	$E_{\frac{1}{2}}$	–0.43	SCE	i_ℓ	3.04	i:i	4	–	–	–	–	c,i_d,r	FC31
			–0.83			3.04		4					c,i_d	
			–1.12			1.12		–					c	
			–0.51			3.04		4	–	–	–	–	c,i_d,r	
			–0.91			3.04		4					c,i_d	
			–1.14			1.12		–					c	
			–0.60			3.04	–	–	–	–	–	–	c,r	
			–1.00			4.16							c	
			–1.22										c	
			–0.67			2.88	–	–	–	–	–	–	c,r	
			–1.07			4.10							c	
			–1.25										c	
			–0.76			2.88	–	–	–	–	–	–	c,r	
			–1.26			4.10							c	
			–										c	
			–0.87			2.80	–	–	–	–	–	–	c,r	
			–1.31			3.80							c	
			–										c	
			–1.26			2.64	–	–	–	–	–	–	c,r	
			–1.60			1.42							c	
			–										c	
			–1.48			1.8	–	–	–	–	–	–	c,r	
EA019 0629	–	$E_{\frac{1}{2}}$	–2.77	SCE(o)	i_ℓ	5.0	–	–	–	–	–	–	c,r	FC32
			–2.725			14.6	QE (–2.8V)	4.5	–	–	–	4-ethyl-1-(3-hex-enyl)benzene 95%	c,r	
C0035 2944	–	$E_{\frac{1}{2}}$	0.82	SCE	–	–	–	–	–	–	–	–	A,p	FC33
C0034 3952	–	E_{su}	0.88	SCE	–	–	–	–	–	–	–	–	A,p	FC34
C0034 3952	–	$E_{\frac{1}{2}}$	0.87	SCE	–	–	–	–	–	–	–	–	A,p	
C0035 2944	–	$E_{\frac{1}{2}}$	0.80	SCE	–	–	–	–	–	–	–	–	A,p	FC35
C0034 3952	–	E_{su}	0.88	SCE	–	–	–	–	–	–	–	–	A,p	FC36

CONT

FC36 (CONT.) $C_{14}H_{14}ClN_3$

Code No.	Empirical Formula	Name and C.A. Number	Structural Formula	Solvent	Tech.	Medium		μ, M	pH	T, °C	Electrodes	App.	Experimental Parameters
FC36	$C_{14}H_{14}ClN_3$	4-chloro-4'-dimethylaminoazobenzene	Table II	dioxane 50	VY	ACET	1.0	-	7.0	-	Pt: rodi // SCE	OAO	C=4.0
FC37	$C_{14}H_{14}ClN_3$	3,6-diamino-10-methylacridinium chloride C.A. 86-40-8	Table II	MeOH 50	PY	Me_4NOH	0.1	-	-	-	DME // SCE	O--	C=0.05, m=2.61, t=3.0, h=50
FC38	$C_{14}H_{14}FN_3$	4-dimethylamino-3'-fluoroazobenzene C.A. 332-54-7	Table II	Me_2CO 50	VY	ACET	1.0	-	7.0	-	Pt: rodi // SCE	OAO	C=4.0
FC39	$C_{14}H_{14}FN_3$	4-dimethylamino-4'-fluoroazobenzene C.A. 150-74-3	Table II	dioxane 50	VD	ACET	1.0	-	7.0	-	Pt: rodi // SCE	OAO	C=4.0
					VY	ACET	1.0	-	7.0	-	Pt: rodi // SCE	OAO	C=4.0
FC40	$C_{14}H_{14}IN_3$	4-dimethylamino-3'-iodoazobenzene C.A. 3808-71-7	Table II	Me_2CO 50	VY	ACET	1.0	-	7.0	-	Pt: rodi // SCE	OAO	C=4.0
FC41	$C_{14}H_{14}IN_3$	4-dimethylamino-4'-iodoazobenzene C.A. 3805-67-2	Table II	dioxane 50	VD	ACET	1.0	-	7.0	-	Pt: rodi // SCE	OAO	C=4.0
					VY	ACET	1.0	-	7.0	-	Pt: rodi // SCE	OAO	C=4.0
FC42	$C_{14}H_{14}NO$	bis(4-methoxyphenyl)-nitroxide C.A. 2643-00-7	$(4-CH_3OC_6H_4)_2-NO\cdot$	MeCN	PY	Et_4NClO_4	0.1	-	-	-	DME // Ag/ $AgClO_4$ 0.01, Et_4NClO_4 0.1	2AO	C=1, Pt Aux
					VR	Et_4NClO_4	0.1	-	-	-	Pt: xxdi // Ag/ $AgClO_4$ 0.01, Et_4NClO_4 0.1	2AO	0.5 → -1.5 → 0.5 → -1.5 V, C=0.8, d=5, v=25, Pt Aux, PE
						Et_4NClO_4 $MeNO_2$	0.1 0.1						
						Et_4NClO_4 HOAc	0.1 6E-3						
CONT													

TABLE I. Electrochemical Data

$C_{14}H_{14}NO$ (CONT.) FC42

Ref.	C/M	Charact. Potential		Response Const.		n Tech.		Electrokinetic Data			Products and Identification	Description and Remarks	Code No.
		Value	vs.		Value			Parameter	Value	From			
C0034 3952	-	$E_{\frac{1}{2}}$	0.88	SCE	-	-	-	-	-	-	-	-	A,p	FC36
EA019 0215	-	$E_{\frac{1}{2}}$	1.05	SCE	-	-	-	-	Tomeš	64	sttd	-	C,R,Mc(formula and $E_{\frac{1}{2}}$),p	FC37
			-		I	2.15	sttd	2	-				C	
C0035 2944	-	$E_{\frac{1}{2}}$	0.81	SCE	-	-	-	-	-	-	-	-	A,p	FC38
C0034 3952	-	E_{su}	0.89	SCE	-	-	-	-	-	-	-	-	A,p	FC39
C0034 3952	-	$E_{\frac{1}{2}}$	0.88	SCE	-	-	-	-	-	-	-	-	A,p	
C0035 2944	-	$E_{\frac{1}{2}}$	0.79	SCE	-	-	-	-	-	-	-	-	A,p	FC40
C0034 3952	-	E_{su}	0.89	SCE	-	-	-	-	-	-	-	-	A,p	FC41
C0034 3952	-	$E_{\frac{1}{2}}$	0.88	SCE	-	-	-	-	-	-	-	-	A,p	
EA020 0469	424 b	$E_{\frac{1}{2}}$	0.20	Ag/ AgClO$_4$ 0.01, Et$_4$NClO$_4$ 0.1	i_ℓ/C	5.8	-	-	Elog	59	sttd	-	A,$E_{\frac{1}{2}}$ and $i_\ell/C \neq f(C)$ for C=0.1-4,p	FC42
			-1.31			-				-			C	
EA020 0469	424 b	E_p	-1.38F	Ag/ AgClO$_4$ 0.01, Et$_4$NClO$_4$ 0.1	i_p	26F	-	-	ΔE_p	75	sttd	-	C,p	
			-1.30F			26F				-			A	
			0.22F			30F				62			A,R	
			0.16F			27F				-			C,R	
			-1.16F			20	-	-		380(?)			C,p	
			-0.82F			17				-			A	
			0.22F			30				62			A	
			0.16F			27				-			C,p	
			-0.74F			26F	QE (-0.8V)	1.0		-		di(4-methoxyphenyl)- nitroxide mono- cation	C	
			-0.47F			17F	-			-			A	
			0.22F			30F				62		di(4-methoxyphenyl)- hydroxylamine	A	
			0.16F			27F				-			C	

CONT

FC42 (CONT.) $C_{14}H_{14}NO$

Code No.	Empirical Formula	Name and C.A. Number	Structural Formula	Solvent	Tech.	Medium		μ, M	pH	T, °C	Electrodes	App.	Experimental Parameters
FC42 CONT	$C_{14}H_{14}NO$	bis(4-methoxy-phenyl)nitroxide	$(4-CH_3OC_6H_4)_2$-NO·	MeCN	VY	Et_4NClO_4	0.1	-	-	-	glC: rodi // Ag/ AgClO$_4$ 0.01, Et_4NClO_4 0.1	2AO	C=ns, d=1.3, ω=10, Pt Aux
											Pt: rodi // Ag/ AgClO$_4$ 0.01, Et_4NClO_4 0.1		
													C=1, ω=10, Pt Aux
						Et_4NClO_4 HOAc	0.1 0.01						
						Et_4NClO_4 $MeNO_2$	0.1 0.1						
				$MeNO_2$	VY	Et_4NClO_4	0.1	-	-	-	Pt: rodi // Ag/ AgClO$_4$ 0.01, Et_4NClO_4 0.1	2AO	C=1, Pt Aux
FC43	$C_{14}H_{14}N_2$	trans-α,α'-azotoluene C.A. 3395-76-4	trans-$[C_6H_5CH_2N:]_2$	EtOH 25	PY	KNO_3 BR	0.1	-	1.5	-	DME // SCE	0-0	C=0.25
									2.5				
									3.6				
									7.1F				
									8.4F				
									9.9				
									10.9				C=0.1
									12				C=0.25
FC44	$C_{14}H_{14}N_2O$	4-acetylaminodi-phenylamine C.A. 38674-90-7	4-$CH_3CONHC_6H_4$-NHC_6H_5	MeCN	VY	Et_4NClO_4	0.1	-	-	-	Pt: rodi // Ag/ AgClO$_4$ 0.01, Et_4NClO_4 0.1	2-0	C=0.96, d=0.1, ω=10, Pt Aux
						Et_4NClO_4 H_2O	0.1 5E-3							C=1

$C_{14}H_{14}N_2O$ (CONT.) FC44

Ref.	C/M	Charact. Potential		Response Const.		n Tech.	n	Electrokinetic Data			Products and Identification	Description and Remarks	Code No.
		Value	vs.		Value			Parameter	Value	From			
EA020 0469	424 b	$E_{\frac{1}{2}}$	0.20	Ag/ AgClO$_4$ 0.01, Et$_4$NClO$_4$ 0.1	i_ℓ/C	5.8	–	–	Elog	59	sttd	–	A,R,$E_{\frac{1}{2}}$ and $i_\ell/C \neq f(C)$ for C=0.1-4,p	FC42
			-1.31			–				–			C	
			0.20			5.8	QE	1.0		59		QE(0.45 V) → di(4-methoxyphenyl)oxo-ammonium cation, [λ=256(logε=3.86), 390nm(logε=3.90), and 594nm(logε=4.65)]	A,R,$E_{\frac{1}{2}}$ and $i_\ell/C \neq f(C)$ for C=0.1-4,p	
			-1.33			–				–			C,R,$E_{\frac{1}{2}} \neq f(C)$	
			0.115	Fc	–	–	–	–	–	–	–	–	A,p	
			-1.415										C	
			0.115		–	–	–	–	–	–	–	–	A,p	
			-0.75										C	
			0.115		–	–	–	–	–	–	–	–	A,p	
			-1.235										C	
EA020 0469	–	$E_{\frac{1}{2}}$	0.130	Fc	–	–	–	–	–	–	–	–	A,p	
			-1.13										C	
EA021 0473	432 g	$E_{\frac{1}{2}}$	-0.30F	SCE	i_ℓ	0.8F	–	–	$dE_{\frac{1}{2}}/dpH$	78F	–	–	C,r	FC43
			-0.38F			1.0F	–	–		78F		–	C,r	
			-0.47F			1.2F	–	–		78F		–	C,r	
			-0.73F			1.2F	–	–		78F		–	C,r	
			–			1.0F	–	–		–		–	C,r	
						0.3F				0			C	
			–			0.3F	–	–		–		–	C,r	
			-1.45F			0.75F				0			C	
			-1.45F			0.95F	–	–		0		–	C,r	
			-1.45F			0.9F	–	–		0		–	C,r	
EA021 1171	147 jlm	$E_{\frac{1}{2}}$	0.48F	Ag/ AgClO$_4$ 0.01, Et$_4$NClO$_4$ 0.1	i_ℓ	5.3F	–	–	–	–	–	–	A,r	FC44
			0.89F			4.3F							A	
			0.46			5.8	–	–	–	–	–	dimer of radical cation	A,r	
			0.90			–						dimer of dication	A	

CONT

FC44 (CONT.) $C_{14}H_{14}N_2O$

Code No.	Empirical Formula	Name and C.A. Number	Structural Formula	Solvent	Tech.	Medium		μ, M	pH	T, °C	Electrodes	App.	Experimental Parameters
FC44	$C_{14}H_{14}N_2O$	4-acetylaminodi-phenylamine	4-$CH_3COC_6H_4$NH-C_6H_5	MeCN	VY	Et_4NClO_4 ACET	0.1 2E-3?	-	-	-	Pt: rodi// Ag/ $AgClO_4$ 0.01, Et_4NClO_4 0.1	2-0	C=1?, d=0.1, ω=10, Pt Aux
						Et_4NClO_4 2,6-lutidine	0.1 2E-3						
FC45	$C_{14}H_{14}N_2O$	trans-α,α'-azoxy-toluene	Table II	EtOH 10	VR	BR	-	-	6.9	-	DME//SCE	0-0	0.15 → -1.40 → 0.15 → -1.40 V, C=0.1, v=2E3
				EtOH 25	PY	BR KNO_3	- 0.1	-	2.1	-	DME//SCE	0-0	C=0.25
									4.3				
									5.5				
									7.0				
									8.1				
									10				
									12				
FC46 bf84	$C_{14}H_{14}N_2O$	2-methyl-1,2-di(3-pyridyl)-1-propanone C.A. 54-36-4	Table II-2	H_2O	PY	BR	-	-	1-5.5	-	DME//SCE	-	-
FC47	$C_{14}H_{14}N_2O_2$	trans-α,α'-azodioxy-toluene	Table II	EtOH 25	PY	KNO_3 HCl	0.1 ns	-	0.5	-	DME//SCE	---	C=0.25
									1				
									1.4				
									2.1				
									5.6				
									8.25				
CONT													

TABLE I. Electrochemical Data   173

$C_{14}H_{14}N_2O_2$ (CONT.) FC47

Ref.	C/M	Charact. Potential		Response	Const. Value	n Tech.	n	Electrokinetic Data			Products and Identification	Description and Remarks	Code No.
		Value	vs.					Parameter	Value	From			
EA021 0557	147 n	$E_{\frac{1}{2}}$	0.03	Ag/AgClO$_4$ 0.01, Et$_4$NClO$_4$ 0.1	-	-	-	-	-	-	-	-	A,r	FC44
			0.33		-	-	-	-	-	-	-	-	A,r	
EA021 0473	432 d	E_p	-0.62F	SCE	i_p	0.5F	-	-	-	-	-	-	C,0 on first scan,r	FC45
			-0.85F			0.8F							C,0 on first scan	
			-1.23F			4F							C	
			0.06F			4F							A	
EA021 0473 EA021 0479	432 d	$E_{\frac{1}{2}}$	-0.46F	SCE	i_ℓ	2.3F	sttd	4	dE$_{\frac{1}{2}}$/dpH	122F	-	-	C,r	
			-0.71F			2.3F		4		122F		-	C,r	
			0.88F			2.0F	-	-		122F		-	C,r	
			-1.06F			1.6F	-	-		122F		-	C,r	
			-			0.5F				0F			C	
			-			0.3F	-	-		122F		-	C,r	
	432 e		-1.38F			1.6F				0F			C	
			-1.38F			1.9F	QE (-1.6V)	3.9		0F		α,α'-hydrazotoluene 100%	C,r	
			-1.38F			1.7F	-	-		0F		-	C,r	
C0027 0483	-	-	-	-	-	-	-	-	-	-	-	-	see BF84	FC46 bf84
EA021 0479	432 a	$E_{\frac{1}{2}}$	0.00F	SCE	i_ℓ	1.2F	i:i	2	dE$_{\frac{1}{2}}$/dpH	100F	-	-	C,r	FC47
			-0.30F			2.4F	QE (-0.65V)	4 5.9		86F		α,α'-hydrazotoluene 100%	C	
			-0.07F			1.04F	-			100F			C,r	
			-0.36F			2.4F				86F				
			-0.09F			0.8F	-			-		-	C,r	
			-0.38F			2.6F				86F			C,0 for pH > 1.8	
			-0.43F			3.4F	-	-		86F		-	C,r	
			0.74F			3.4F	-	-		86F		-	C,r	
			-0.79F			1.33F	QE (-0.94V)	1.95		0F		benzaldoximes 20%, trans-α,α'-azoxytoluene 40%	C,r	
			-1.00F			1.17F	-			0F		-	C	
			-1.38F			0.75F	QE (-1.5V)	6.0		0F		α,α'-hydrazotoluene 100%	C	

CONT

FC47 (CONT.) $C_{14}H_{14}N_2O_2$

Code No.	Empirical Formula	Name and C.A. Number	Structural Formula	Solvent	Tech.	Medium		μ, M	pH	T, °C	Electrodes	App.	Experimental Parameters
FC47	$C_{14}H_{14}N_2O_2$	trans-α,α'-azodioxy-toluene	Table II	EtOH 25	PY	KNO_3 BR	0.1 -	-	10	-	DME//SCE	---	C=0.25
									10.8				
					QE	NaCl BR	0.1 -	-	4.25	-	Hg: nspo//SCE	9?--	C=0.25, E_{app}=-1.00 V
FC48	$C_{14}H_{14}N_4O_2$	4-dimethylamino-3'-nitroazobenzene C.A. 3837-55-6	Table II	Me_2CO 50	VY	ACET	1.0	-	7.0	-	Pt: rodi//SCE	OAO	C=4.0
FC49	$C_{14}H_{14}N_4O_2$	4-dimethylamino-4'-nitroazobenzene C.A. 2491-74-9	Table II	dioxane 50	VY	ACET	1.0	-	7.0	-	Pt: rodi//SCE	OAO	C=4.0
FC50 de43	$C_{14}H_{14}O$	benzhydryl methyl-ether C.A. 1016-09-7	$(C_6H_5)_2CHOCH_3$	DMF	PY	Bu_4NI	0.1	-	-	-	DME//Ag/AgI, I^- 0.1	3-O	C=1, Pt Aux
					QE	Bu_4NI PHEN	0.1 0.01	-	-	-	Hg?: nsns//Ag/AgI I^- 0.1	3-O	C=1, E_{app}=-2.4 V, Pt Aux
FC51 de44	$C_{14}H_{14}O$	benzyl ether C.A. 103-50-4	$(C_6H_5CH_2)_2O$	DMF	PY	Bu_4NI	0.1	-	-	-	DME//Ag/AgI, I^- 0.1	3-O	C=1, Pt Aux
FC52	$C_{14}H_{14}O$	1-(4-biphenylyl)-ethanol C.A. 3562-73-0	$4-C_6H_5C_6H_4CHOH-CH_3$	DMF	PY	Et_4NBr	0.1	-	-	-	DME//SCE(o)	2-O	C=2
						Et_4NBr PHEN	0.1 0.01						
						Et_4NBr PHEN	0.1 0.02	-	-	-	DME//SCE(o)	2AO	C=2.10
					QE	Bu_4NI PHEN	0.25 0.01	-	-	-	-//SCE(o)	2-O	C=2
FC53	$C_{14}H_{14}O_2$	1,2-dihydro-1,2-dimethylcyclobuta-[b]naphthalene-3,8-diol C.A. 50982-47-3	Table II	EtOH 50	PY	ACET	0.1	-	5.6	-	DME//SCE	---	ns
FC54 bf91	$C_{14}H_{14}O_2$	4,4'-dimethoxybi-phenyl C.A. 2132-80-1	$(4-CH_3OC_6H_4)_2$	MeCN	VA	Bu_4NBF_4 Al_2O_3	0.2 0.25 gcm^{-3}	-	-	-	Pt: xxbu//Ag/$AgNO_3$ 0.01, Bu_4NBF_4 0.1	5AFO	0.9→1.8→0.9 V, C=1.0, v=86

CONT

$C_{14}H_{14}O_2$ (CONT.) FC54

Ref.	C/M	Charact. Potential		Response Const.		n Tech.		Electrokinetic Data			Products and Identification	Description and Remarks	Code No.
		Value	vs.		Value			Parameter	Value	From			
EA021 0479	432 b	$E_{\frac{1}{2}}$	-0.80F	SCE	i_ℓ	1.25F	-	-	$dE_{\frac{1}{2}}/dpH$	0F	-	-	C,r	FC47
			-1.03F			1.08F				-			C	
			-1.40F			0.75F							C	
			-0.80			1.11F	-	-		0F		-	C,r	
			-1.03			0.88F				-			C	
			-1.40			0.63F							C	
EA021 0479	432 ab	-	-	-	-	-	QE	5.9	-	-	-	α,α'-hydrazotoluene 100%	C,r	
C0035 2944	-	$E_{\frac{1}{2}}$	0.81	SCE	-	-	-	-	-	-	-	-	A,p	FC48
C0034 3952	-	$E_{\frac{1}{2}}$	-	SCE	-	-	-	-	-	-	-	-	A,0,p	FC49
EA020 0853	33k	$E_{\frac{1}{2}}$	-2.40	Ag/AgI, I⁻ 0.1	i_ℓ	4.6	QE (-2.40V)	2.0	-	-	-	diphenylmethane 78%, benzhydrol 30%	C,r	FC50 de43
EA020 0853	33k	-	-	-	-	-	QE	2.2	-	-	-	diphenylmethane 90%	C,r	
EA020 0853	33j	$E_{\frac{1}{2}}$	-2.41	Ag/AgI, I⁻ 0.1	i_ℓ	6.9	QE (-2.4V)	2.0	-	-	-	benzyl alcohol 50%, toluene presumed lost in workup	C,r	FC51 de44
EA019 0629	33e	$E_{\frac{1}{2}}$	-2.725	SCE(o)	i_ℓ	10.7	QE (-2.75V)	≈2	-	-	-	QE(-2.75 V, [Bu₄NI] = 0.25) → 4-ethyl-biphenyl 95%	C,r	FC52
			-2.70			18.1	QE (-2.8V)	6	-	-	-	4-ethyl-1-(3-hexenyl)benzene 95%	C,r	
EA020 0143	-	$E_{\frac{1}{2}}$	-2.69	SCE(o)	i_ℓ	17.0	-	-	-	-	-	-	C,r	
EA019 0629	33e	-	-	-	-	-	QE (-2.8V)	6	-	-	-	4-ethyl-1-(3-hexenyl)benzene 95%	C,r	
JA092 4139	-	$E_{\frac{1}{2}}$	-0.153	SCE	-	-	-	-	-	-	-	-	C,Σ,p	FC53
EA018 0537	419 e	E_p	1.28	SCE	-	-	sttd	1	-	-	-	-	A,R,Al₂O₃ added to remove H₂O,r	FC54 bf91
			1.55					1					A,R	
			1.48					1					C,R	
			1.21					1					C,R	

CONT

FC54 (CONT.) $C_{14}H_{14}O_2$

Code No.	Empirical Formula	Name and C.A. Number	Structural Formula	Solvent	Tech.	Medium	μ, M	pH	T, °C	Electrodes	App.	Experimental Parameters
FC54 bf91	$C_{14}H_{14}O_2$	4,4'-dimethoxybiphenyl	$(4-CH_3OC_6H_4)_2$	EtCN	VA	Bu_4NBF_4 0.2 Al_2O_3 0.25 gcm^{-3}	-	-	-	Pt: xxbu// Ag/ $AgNO_3$ 0.1, Bu_4NBF_4 0.1	25AFO	$0.9 \to 1.8 \to 0.9$ V, C=1.0, v=86
				2-PrCN	VA	Bu_4NBF_4 0.2 Al_2O_3 0.25 gcm^{-3}	-	-	-	Pt: xxbu// Ag/ $AgNO_3$ 0.1, Bu_4NBF_4 0.1	25AFO	$0.9 \to 1.8 \to 0.9$ V, C=1.0, v=86
				C_6H_5CN	VA	Bu_4NBF_4 0.2 Al_2O_3 0.25 gcm^{-3}	-	-	-	Pt: xxbu// Ag/ $AgNO_3$ 0.1, Bu_4NBF_4 0.1	25AFO	$0.9 \to 1.8 \to 0.9$ V, C=1.0, v=86
				$MeNO_2$	VA	Bu_4NBF_4 0.2 Al_2O_3 0.25 gcm^{-3}	-	-	-	Pt: xxbu// Ag/ $AgNO_3$ 0.1, Bu_4NBF_4 0.1	25AFO	$0.9 \to 1.8 \to 0.9$ V, C=1.0, v=86
				$C_6H_5NO_2$	VA	Bu_4NBF_4 0.2 Al_2O_3 0.25 gcm^{-3}	-	-	-	Pt: xxbu// Ag/ $AgNO_3$ 0.1, Bu_4NBF_4 0.1	25AFO	$0.9 \to 1.8 \to 0.9$ V, C=1.0, v=86
				CH_2Cl_2	VA	Bu_4NBF_4 0.2 Al_2O_3 0.25 gcm^{-3}	-	-	-	Pt: xxbu// Ag/ $AgNO_3$ 0.1, Bu_4NBF_4 0.1	25AFO	$0.9 \to 1.8 \to 0.9$ V, C=1.0, v=86
				CH_2Cl_2 88 F_3CCOOH 2 $(F_3CCO)_2O$ 10	VA	Bu_4NBF_4 0.2	-	-	-	Pt: xxbu// Ag/ $AgNO_3$ 0.1, Bu_4NBF_4 0.1	25AO	C=1.0, v=43
												v=86
												v=161
												v=210

CONT

TABLE I. Electrochemical Data

$C_{14}H_{14}O_2$ (CONT.) FC54

Ref.	C/M	Charact. Potential		Response Const.		n Tech.		Electrokinetic Data			Products and Identification	Description and Remarks	Code No.
		Value	vs.		Value			Parameter	Value	From			
EA018 0537	419 e	E_p	1.31	SCE	-	-	sttd	1	-	-	-	-	A,R,r	FC54 bf91
			1.58					1					A,R	
			1.52					1					C,R	
			1.25					1					C,R	
EA018 0537	419 e	E_p	1.32	SCE	-	-	sttd	1	-	-	-	radical cation	A,R,r	
			1.60					1				dication	A,R	
			1.53					1					C,R	
			1.26					1					C,R	
EA018 0537	419 e	E_p	1.32	SCE	-	-	sttd	1	-	-	-	radical cation	A,R,r	
			1.63					1				dication	A,R	
			1.56					1					C,R	
			1.25					1					C,R	
EA018 0537	419 e	E_p	1.24	SCE	-	-	sttd	1	-	-	-	radical cation	A,R,r	
			1.53					1				dication	A,R	
			1.47					1					C,R	
			1.18					1					C,R	
EA018 0537	419 e	E_p	1.31	SCE	-	-	sttd	1	-	-	-	radical cation	A,R,r	
			1.65					1				dication	A,R	
			1.59					1					C,R	
			1.25					1					C,R	
EA018 0537	419 e	E_p	1.33	SCE			sttd	1	-	-	-	radical cation	A,R,r	
			1.70					1				dication	A,R	
			1.63					1					C,R	
			1.26					1					C,R	
EA018 0537	419 e	-	-	-	$i_p/v^{\frac{1}{2}}C$	6.9	QE	1.0	-	-	-	radical cation, ESR, IRS	A,R,r	
			-			7.5		1.0				dication, VIS(λ=520)	A,R	
			-			7.5	sttd	1					C,R	
			-			6.9		1					C,R	
		E_p	1.37	SCE		6.7		1	-	-	-	radical cation	A,R,r	
			1.78			7.2		1				dication	A,R	
			1.70			7.2		1					C,R	
			1.29			6.7		1					C,R	
			-			6.5		1	-	-	-	radical cation	A,R,r	
			-			7.0		1				dication	A,R	
			-			7.0		1					C,R	
			-			6.5		1					C,R	
			-			6.4		1	-	-	-	radical cation	A,R,r	
			-			6.9		1				dication	A,R	
			-			6.9		1					C,R	
			-			6.4		1					C,R	
		E_p	1.37	SCE										CONT

FC54 (CONT.) $C_{14}H_{14}O_2$ (1)

Code No.	Empirical Formula	Name and C.A. Number	Structural Formula	Solvent	Tech.	Medium		μ, M	pH	T, °C	Electrodes	App.	Experimental Parameters
FC54 bf91	$C_{14}H_{14}O_2$	4,4'-dimethoxybi-phenyl	$(4-CH_3OC_6H_4)_2$	F_3CCOOH 90 $(F_3CCO)_2O$ 10	VA	Bu_4NBF_4	0.2	-	-	-	Pt: xxbu// Ag/ $AgNO_3$ 0.01, Bu_4NBF_4 0.1	25AO	C=1.0,v=86
FC55	$C_{14}H_{15}N$	4,4'-dimethyldi-phenylamine C.A. 620-93-9	$(4-CH_3C_6H_4)_2NH$	MeCN	VY	Et_4NClO_4	0.1	-	-	-	Pt: rodi// Ag/ $AgClO_4$ 0.01, Et_4NClO_4 0.1	2-O	C=ns,d=0.1, ω=10,Pt Aux
FC56	$C_{14}H_{15}NO_2$	4,4'-dimethoxydi-phenylamine C.A. 101-70-2	$(4-CH_3OC_6H_4)_2NH$	MeCN	VA	Et_4NClO_4	0.1	-	-	-	Pt: xxdi// Ag/ $AgClO_4$ 0.01, Et_4NClO_4 0.1	2-O	0→0.6→ -0.5→0 V, C=1.3,d=0.5, v=30,Pt Aux
						Et_4NClO_4 2,6-lutidine	0.1 1.3						
						Et_4NClO_4 2,6-lutidine	0.1 2.0						
					VY	Et_4NClO_4	0.1	-	-	-	Pt: rodi// Ag/ $AgClO_4$ 0.01, Et_4NClO_4 0.1	2-O	C=ns,d=0.1, ω=10,Pt Aux
FC57 bg06 de46	$C_{14}H_{15}N_3$	4-(dimethylamino)-azobenzene C.A. 60-11-7	$4-(CH_3)_2NC_6H_4N$: NC_6H_5	Me_2CO 50	VY	ACET	1.0	-	7.0	-	Pt: rodi// SCE	OAO	C=4.0
				dioxane 50	VD	HCl KCl	ns ns	-	1.0	-	Pt: rodi// SCE	OAO	C=4.0
						ACET	ns		4.85				
									7.0				
						NH_3 NH_4Cl	ns ns		9.0				
					VY	HCl KCl	ns ns	-	1.0	-	Pt: rodi// SCE	OAO	C=4.0
FC54 CONT						ACET	ns		4.85				

TABLE I. Electrochemical Data

$C_{14}H_{15}N_3$ (CONT.) FC57

Ref.	C/M	Charact.	Potential	Response Const.		n Tech.		Electrokinetic Data			Products and Identification	Description and Remarks	Code No.
		Value	vs.		Value			Parameter	Value	From			
EA018 0537	419 e	E_p	1.33	SCE	–	–	sttd	1	–	–	–	radical cation	A,R,r	FC54 bf91
			1.74					1				dication	A,R	
			1.66					1					$C,R,i_{p,C}/i_{p,A}=1.0$	
			1.26					1					$C,R,i_{p,C}/i_{p,A}=1.0$	
EA021 0557	147 n	$E_{\frac{1}{2}}$	0.49	Ag/ AgClO$_4$ 0.01, Et$_4$NClO$_4$ 0.1	i_ℓ/C	6.4	sttd	1	–	–	–	radical cation, VIS	$A,\Delta E_{\frac{1}{2}}=0.46\sigma,r$	FC55
			1.30		–			1					A	
EA021 0557	147 n	E_p	0.33F	Ag/ AgClO$_4$ 0.01, Et$_4$NClO$_4$ 0.1	i_p	63F	–	–	–	–	–	–	A,r	FC56
			0.27F			63F							C	
			0.27F			102F	–	–	–	–	–	–	A,r	
			0.38F			5F							A	
			0.38F			–							C	
			0.19F			–							C	
			-0.29F			8F							C	
			0.28F			138F	–	–	–	–	–	–	A,r	
			0.39F			5F							C	
			-0.29F			10F							C	
			-0.24F			10F							A	
EA021 0557	147 n	$E_{\frac{1}{2}}$	0.295	Ag/ AgClO$_4$ 0.01, Et$_4$NClO$_4$ 0.1	i_ℓ/C	5.7	sttd	1	–	–	–	radical cation, VIS	$A,\Delta E_{\frac{1}{2}}=0.30\sigma,r$	
			0.85		–			1					A	
C0035 2944	–	$E_{\frac{1}{2}}$	0.79	SCE	–	–	–	–	–	–	–	–	A,p	FC57 bg06 de46
C0034 1615	430 a	E_{su}	0.94	NCE	–	–	–	–	–	–	–	–	A,p	
			0.93		–	–	–	–	–	–	–	–	A,p	
C0034 1615 (C0034 3952)	430 a		0.90		–	–	–	–	–	–	–	–	A,p	
			0.95		–	–	–	–	–	–	–	–	A,p	
C0034 1615	430 a	$E_{\frac{1}{2}}$	0.88	NCE	–	–	–	–	–	–	–	–	A,p	
		E_p	0.93		–	70	sttd	–	–	–	–	–	A,p	CONT

FC57 (CONT.) $C_{14}H_{15}N_3$

Code No.	Empirical Formula	Name and C.A. Number	Structural Formula	Solvent	Tech.	Medium		μ, M	pH	T, °C	Electrodes	App.	Experimental Parameters
FC57 bg06 de46	$C_{14}H_{15}N_3$	4-(dimethylamino)-azobenzene	4-$(CH_3)_2NC_6H_4N$: NC_6H_5	dioxane 50	VY	ACET	ns	-	7.0	-	Pt: rodi// SCE	OAO	C=4.0
						NH_3 NH_4Cl	ns ns		9.0				
FC58	$C_{14}H_{15}N_3O$	4-dimethylamino-2'-hydroxyazobenzene C.A. 6396-84-5	Table II	dioxane 50	VD	ACET	ns	-	4.85	-	Pt: rodi// SCE	OAO	C=4.0
						NH_3 NH_4Cl	ns ns		9.0				
					VY	ACET	ns	-	4.85	-	Pt: rodi// SCE	OAO	C=4.0
									7.0				
						NH_3 NH_4Cl	ns ns		9.0				
FC59	$C_{14}H_{15}N_3O$	4-dimethylamino-4'-hydroxyazobenzene C.A. 2496-15-3	Table II	dioxane 50	VD	ACET	ns	-	4.85	-	Pt: rodi// SCE	OAO	C=4.0
									7.0				
					VY	HCl KCl	ns ns	-	3.1	-	Pt: rodi// SCE	OAO	C=4.0
						ACET	ns		4.85				
									7.0				
						NH_3 NH_4Cl	ns ns		9.0				
FC60	$C_{14}H_{15}N_3O_3$	3-hydroxyamino-1-(2-hydroxyphenyl)-3-(2-pyridyl)-1-propanone oxime C.A. 36256-94-7	Table II	DMF 20	PY	BR	-	-	2.3	-	DME//SCE	OAO	C=0.25, m=3.45, t=2.72, h=40
									2.90				
									3.95				
									5.20				
									6.80				
									7.8				
FC61	$C_{14}H_{15}N_3O_3$	3-hydroxyamino-1-(2-hydroxyphenyl)-3-(3-pyridyl)-1-propanone oxime C.A. 36257-04-2	Table II	DMF 20	PY	BR	-	-	2.3	-	DME//SCE	OAO	C=0.25, m=3.45, t=2.72, h=40

CONT

TABLE I. Electrochemical Data

$C_{14}H_{15}N_3O_3$ (CONT.) FC61

Ref.	C/M	Charact. Potential		Response Const.		n Tech.	n	Electrokinetic Data			Products and Identification	Description and Remarks	Code No.
		Value	vs.		Value			Parameter	Value	From			
C0034 1615 (C0034 3952)	432 a	$E_{\frac{1}{2}}$	0.90	NCE	-	69	-	-	-	-	-	4-methylaminoazobenzene,CP,LSC;4-aminoazobenzene, trace,CP,LSC	A,p	FC57 bg06 de46
			0.88		-	55	-	-	-	-	-	-	A,p	
C0034 1615	430 bc	E_{su}	0.80	NCE	-	-	-	-	-	-	-	-	A,p	FC58
			0.77		-	-	-	-	-	-	-	-	A,p	
C0034 1615	430 bc	$E_{\frac{1}{2}}$	0.76	NCE	$i_\ell(u)$	54	-	-	-	-	-	-	A,p	
			0.46		-	44	-	-	-	-	-	-	A,p	
			0.74		-	15	-	-	-	-	-	-	A,p	
C0034 1615	430 bc	E_{su}	0.55	NCE	-	-	-	-	-	-	-	-	A,p	FC59
			0.75		-	-	-	-	-	-	-	-	A,p	
C0034 1615		$E_{\frac{1}{2}}$	0.64	NCE	-	46	-	-	-	-	-	-	A,p	
			0.55		-	42	-	-	-	-	-	-	A,p	
			0.70		-	48	-	-	-	-	-	-	A,p	
			0.38		-	19	-	-	-	-	-	-	A,p	
JE048 0297	109 d	$E_{\frac{1}{2}}$	-0.85	SCE	i_ℓ	2.8	i:i	4	-	-	-	-	C,W,i_d,$i_\ell=kC$,$i_\ell=kh^{\frac{1}{2}}$,r	FC60
			-0.90			2.80	-	-	-	-	-	-	C,i_d,r	
			-0.96			2.68	-	-	-	-	-	-	C,r	
			-1.07			2.58	i:i	4	-	-	-	-	C,r	
			-1.39			1.40	-	-	-	-	-	-	C,r	
			-1.46			-	-	-	-	-	-	-	C,r	
JE048 0297	109 d	$E_{\frac{1}{2}}$	-0.75	SCE	i_ℓ	-	i:i	-	-	-	-	-	C,i_a,Pr,r	FC61
			-0.88			2.88		4					C,W,i_d,$i_\ell=kC$,$i_\ell=kh^{\frac{1}{2}}$	

CONT

FC61 (CONT.) $C_{14}H_{15}N_3O_3$

Code No.	Empirical Formula	Name and C.A. Number	Structural Formula	Solvent	Tech.	Medium	μ, M	pH	T, °C	Electrodes	App.	Experimental Parameters
FC61	$C_{14}H_{15}N_3O_3$	3-hydroxyamino-1-(2-hydroxyphenyl)-3-(3-pyridyl)-1-propanone oxime	Table II	DMF 20	PY	BR	-	2.90	-	DME//SCE	OAO	C=0.25, m=3.45, t=2.72, h=40
								3.95				
								5.20				
								6.80				
								7.8				
FC62	$C_{14}H_{15}N_3O_3S$	4-dimethylaminoazobenzene-3-sulfonic acid C.A. 18625-21-3	Table II	Me_2CO 50	VY	ACET 1.0	-	7.0	-	Pt: rodi// SCE	OAO	C=4.0
FC63 bg08	$C_{14}H_{15}N_3O_3S$	4-dimethylaminoazobenzene-4-sulfonic acid C.A. 502-02-3	Table II	dioxane 50	VY	ACET 1.0	-	7.0	-	Pt: rodi// SCE	OAO	C=4.0
FC64	$C_{14}H_{16}Cr$	bis(η^6-toluene)-chromium(0) C.A. 12087-58-0	$(C_6H_5CH_3)_2Cr$	DMF	VY	Et_4NClO_4 1 GEL 0.04?	-	-	-	Pt: pdhb// SCE	3A-	C=1(?), Pt Aux
FC65	$C_{14}H_{16}N_2$	trans-α,α'-hydrazotoluene C.A. 7626-68-8	trans-$(C_6H_5CH_2-NHNHCH_2C_6H_5)$	EtOH 25	PY	KNO_3 0.1 BR -	-	5.5	-	DME//SCE	O-O	C=0.25
								12				
FC66	$C_{14}H_{17}N_5OS$	3,5-dimethyl-4-(2-ethoxyphenylazo)-1-thiocarbamoyl-pyrazole C.A. 29147-32-8	Table II	H_2O	PY	BR - GEL 5E-3	-	3.0	-	DME//SCE	OAO	C=0.1, EC=3.75
								5.0				
								6.0				C=0.01, EC=3.75
								7				C=0.1
												C=0.3

CONT

$C_{14}H_{17}N_5OS$ (CONT.) FC66

Ref.	C/M	Charact. Potential		Response Const.		n Tech.		Electrokinetic Data			Products and Identification	Description and Remarks	Code No.
		Value	vs.		Value			Parameter	Value	From			
JE048 0297	109 d	$E_{\frac{1}{2}}$ -0.78	SCE	i_ℓ	-	-	-	-	-	-	-	C,i_a,Pr,r	FC61
		-0.93			2.88							C,W,i_d	
		-0.85			-							$C,i_a,Pr,0$ for pH ≥ 4, r	
		-1.02			2.8							C	
		-1.12			2.80	-	-	-	-	-	-	C,r	
		-1.25			1.40	-	-	-	-	-	-	C,r	
		-1.40			-	-	-	-	-	-	-	C,r	
C0035 2944	-	$E_{\frac{1}{2}}$ 0.80	SCE	-	-	-	-	-	-	-	-	A,p	FC62
C0034 3952	-	$E_{\frac{1}{2}}$ -	SCE	-	-	-	-	-	-	-	-	A,0,p	FC63 bg08
JE051 0226	443 a	$E_{\frac{1}{2}}$ -0.89	SCE	-	-	-	-	Elog	81	-	-	A,QI,p	FC64
		0.97							211			A,≠	
								βn_b	0.28	Elog			
EA021 0473	-	$E_{\frac{1}{2}}$ 0.05F	SCE	i_ℓ	1.1F	-	-	$dE_{\frac{1}{2}}/dpH$	62	plot	-	A,r	FC65
		-0.35F			1.1F	-	-		62		-	A,r	
JE054 0411	198 l	$E_{\frac{1}{2}}$ -0.24F	SCE	-	-	-	-	-	-	-	-	$C,\neq,W,i_\ell=kh^{\frac{1}{2}},i_d,$ $i_\ell \neq f(pH),Tc=1.6,r$	FC66
		-1.6										C,H,i_{cat}	
		-0.40F		-	-	-	-	-	-	-	-	$C,\neq,W,i_d,r$	
		-1.6										C,H,i_{cat}	
		-0.49F										$C,\neq,W,i_d,r$	
		-1.6										C,H,i_{cat}	
		-0.52		-	-	-	-	-	-	-	-	$C,\neq,W,i_d,r$	
		-1.6										C,H,i_{cat}	
		-0.56		-	-	-	-	-	-	-	-	$C,\neq,W,i_d,r$	
		-1.6										C,H,i_{cat}	
		-0.63		-	-	-	-	-	-	-	-	$C,\neq,W,i_d,r$	
		-1.6										C,H,i_{cat}	

CONT

FC66 (CONT.) $C_{14}H_{17}N_5OS$

Code No.	Empirical Formula	Name and C.A. Number	Structural Formula	Solvent	Tech.	Medium		μ, M	pH	T, °C	Electrodes	App.	Experimental Parameters
FC66	$C_{14}H_{17}N_5OS$	3,5-dimethyl-4-(2-ethoxyphenylazo)-1-thiocarbamoyl-pyrazole	Table II	H_2O	PY	BR GEL	– 5E-3	–	9.0	–	DME//SCE	OAO	C=0.1
									10.0				
FC67	$C_{14}H_{17}N_5OS$	3,5-dimethyl-4-(4-ethoxyphenylazo)-1-thiocarbamoyl-pyrazole C.A. 29147-33-9	Table II	H_2O	PY	BR GEL	– 5E-3	–	3.0	–	DME//SCE	OAO	C=0.1, EC=3.75
									5.0				
									7.0				
									8.0				C=0.01, EC=3.75
													C=0.1
													C=0.3
									9.0				C=0.1?
									10.0				
FC68	$C_{15}H_9Br_3O_2$	1-(2,4,6-tribromo-3-hydroxyphenyl)-3-phenyl-2-propen-1-one C.A. 28073-96-3	Table II	EtOH 50	PY	PHTH	ns	0.2	0.84	25±0.1	DME//SCE	O3AO	C=ns, m=1.80, t=3.8, h=65
									2.65				
									3.32				
FC69	$C_{15}H_9Br_3O_3$	1-(2,4,6-tribromo-3-hydroxyphenyl)-3-(4-hydroxyphenyl)-2-propen-1-one C.A. 28073-99-6	Table II	EtOH 50	PY	PHTH	ns	0.2	0.84	25±0.1	DME//SCE	O3AO	C=ns, m=1.80, t=3.8, h=65

CONT

TABLE I. Electrochemical Data

$C_{15}H_9Br_3O_3$ (CONT.) FC69

| Ref. | C/M | Charact. | Potential | | Response Const. | | n | Electrokinetic Data | | | Products and | Description and | Code |
		Value	vs.		Value	Tech.	Parameter	Value	From	Identification	Remarks	No.
JE054 0411	198 1	$E_{\frac{1}{2}}$	-0.68F	SCE	-	-	-	-	-	-	-	$C, \not\models, W, i_d, r$	FC66
			-1.6									C, H, i_{cat}	
			-0.70F		-	-	-	-	-	-	-	$C, \not\models, W, i_d, r$	
			-1.6									C, H, i_{cat}	
JE054 0411	198 1	$E_{\frac{1}{2}}$	-0.34F	SCE	-	-	-	-	-	-	-	$C, \not\models, W, i_\ell=kh^{\frac{1}{2}}, i_d,$ $i_\ell \neq f(pH), Tc=1.6, r$	FC67
			-1.6									C, H, i_{cat}	
			-0.49F		-	-	-	-	-	-	-	$C, \not\models, W, i_d, r$	
			-1.6									C, H, i_{cat}	
			-0.63F		-	-	-	-	-	-	-	$C, \not\models, W, i_d, r$	
			-1.6									C, H, i_{cat}	
			-0.64		-	-	-	-	-	-	-	$C, \not\models, W, i_d, r$	
			-1.6									C, H, i_{cat}	
			-0.68		-	-	-	-	-	-	-	$C, \not\models, W, i_d, r$	
			-1.6									C, H, i_{cat}	
			-0.76		-	-	-	-	-	-	-	$C, \not\models, W, i_d, r$	
			-1.6									C, H, i_{cat}	
			-0.74F		-	-	-	-	-	-	-	$C, \not\models, W, i_d, r$	
			-1.6									C, H, i_{cat}	
			-0.77F		-	-	-	-	-	-	-	$C, \not\models, W, i_d, r$	
			-1.6										
JE053 0449	156 g	$E_{\frac{1}{2}}$	-0.600	SCE	-	-	-	-	$dE_{\frac{1}{2}}/dpH$	88	sttd	-	$C, E_{\frac{1}{2}}=-0.530-0.088pH$ for pH=0-4, i_d, $i_\ell=kC, \Delta E_{\frac{1}{2}}=0.24\sigma, p$	FC68
			-0.855							-			$C, i_d, i_\ell=kC, \Delta E_{\frac{1}{2}}=0.31\sigma$	
			-0.735		-	-	-	-		88		-	$C, i_\ell=kC, i_d, p$	
			-1.010							-			$C, i_d, i_\ell=kC$	
			-0.810		-	-	-	-		88		-	$C, i_d, i_\ell=kC, \Delta E_{\frac{1}{2}}=0.29\sigma, p$	
			-1.065							-			$C, i_d, i_\ell=kC, \Delta E_{\frac{1}{2}}=0.25\sigma$	
JE053 0449	156 g	$E_{\frac{1}{2}}$	-0.697	SCE	-	-	-	-	$dE_{\frac{1}{2}}/dpH$	88	sttd	-	$C, E_{\frac{1}{2}}=-0.570-0.088pH$ for pH=0-4, $i_d, i_\ell= kC, \Delta E_{\frac{1}{2}}=0.24\sigma, p$	FC69
			-0.937							-			$C, i_d, i_\ell=kC, \Delta E_{\frac{1}{2}}=0.31\sigma$	

CONT

FC69 (CONT.) $C_{15}H_9Br_3O_3$

Code No.	Empirical Formula	Name and C.A. Number	Structural Formula	Solvent	Tech.	Medium		μ, M	pH	T, °C	Electrodes	App.	Experimental Parameters
FC69	$C_{15}H_9Br_3O_3$	1-(2,4,6-tribromo-3-hydroxyphenyl)-3-(4-hydroxyphenyl)-2-propen-1-one	Table II	EtOH 50	PY	PHTH	ns	0.2	2.65	25±0.1	DME//SCE	03A0	C=ns, m=1.80, t=3.8, h=65
									3.32				
FC70	$C_{15}H_{10}O_2$	2-benzoylbenzofuran C.A. 6272-40-8	Table II	EtOH 48	PY	BR	–	–	2.0	–	DME//SCE	04A0	C=0.2
									4.1				
									6.1				
									7.1				
									11.8				
FC71	$C_{15}H_{11}N_3$	2,2',2''-terpyridine C.A. 1148-79-4	Table II	MeCN	VA	Bu_4NBF_4	0.1	–	–	–	Pt: xxdi//Ag	2F-	C=ns, Pt Aux
FC72	$C_{15}H_{12}BrNO_2$	3-(3-bromophenyl)-3-oxo-N-phenylpropanamide C.A. 25559-51-7	3-$BrC_6H_4COCH_2$-$CONHC_6H_5$	EtOH 40.0	PY	BR GEL	– 0.01	–	5.0	–	DME//SCE	5A-	C=ns, m=1.8, t=3.7, h=59
									8.0				
FC73	$C_{15}H_{12}BrNO_2$	3-(4-bromophenyl)-3-oxo-N-phenylpropanamide C.A. 25559-52-8	4-$BrC_6H_4COCH_2$-$CONHC_6H_5$	EtOH 40.0	PY	BR GEL	– 0.01	–	5.0	–	DME//SCE	5A-	C=ns, m=1.8, t=3.7, h=59
									8.0				
FC74	$C_{15}H_{12}ClNO_2$	3-(3-chlorophenyl)-3-oxo-N-phenylpropanimide C.A. 38505-24-7	3-$ClC_6H_4COCH_2$-$CONHC_6H_5$	EtOH 40.0	PY	BR GEL	– 0.01	–	5.0	–	DME//SCE	5A-	C=ns, m=1.8, t=3.7, h=59
									6.0				
CONT									8.0				

TABLE I. Electrochemical Data

$C_{15}H_{12}ClNO_2$ (CONT.) FC74

Ref.	C/M	Charact. Potential		Response Const.		n Tech.	Electrokinetic Data			Products and Identification	Description and Remarks	Code No.
		Value	vs.		Value		Parameter	Value	From			
JE053 0449	156 g	$E_{\frac{1}{2}}$	-0.840	SCE	-	-	-	-	$dE_{\frac{1}{2}}/dpH$	88	sttd	-	$C,i_d,i_\ell=kC,\Delta E_{\frac{1}{2}}=0.28\sigma,p$	FC69
			-1.090								-		$C,i_d,i_\ell=kC,\Delta E_{\frac{1}{2}}=0.27\sigma$	
			-0.922		-	-	-	-		88		-	$C,i_d,i_\ell=kC,\Delta E_{\frac{1}{2}}=0.29\sigma,p$	
			-1.170							-			$C,i_d,i_\ell=kC,\Delta E_{\frac{1}{2}}=0.25\sigma$	
C0027 1861	-	$E_{\frac{1}{2}}$	-0.7F -0.97F	SCE	-	-	-	-	-	-	-	-	C,p C	FC70
			-0.86F -1.13F		-	-	-	-	-	-	-	-	C,p C	
			-1.00F -1.13F		-	-	-	-	-	-	-	-	C,p C	
			-1.10F		-	-	-	-	-	-	-	-	C,dr,p	
			-1.10F		-	-	-	-	-	-	-	-	C,p	
JA095 6582	-	E_p	-2.15	SCE	-	-	-	-	-	-	-	-	$C,\neq,Mc(structure),r$ A, 0 to 2.1V	FC71
EA021 0831	65c	$E_{\frac{1}{2}}$	-1.141	SCE	-	-	sttd	2	αn_a	0.790	sttd	-	$C,E_{\frac{1}{2}} Av(20)$ from $Elog, i_d, i_\ell=kh^{\frac{1}{2}}$ for pH=3-10, r	FC72
			-1.292		-	-		2		1.121		-	$C,E_{\frac{1}{2}} Av(20)$ from Elog, pK'=8.75, $E_{\frac{1}{2}}=-1.379+0.28\sigma,r$	
EA021 0831	65c	$E_{\frac{1}{2}}$	-1.163	SCE	-	-	sttd	2	αn_a	0.784	sttd	-	$C,E_{\frac{1}{2}} Av(20)$ from Elog, r	FC73
			-1.313		-	-		2		1.904		-	$C,E_{\frac{1}{2}} Av(20)$ from Elog, pK'=9.30, $E_{\frac{1}{2}}=-1.379+0.28\sigma$, $i_\ell=kh^{\frac{1}{2}}$ for pH=3-10, i_d, r	
EA021 0831	65c	$E_{\frac{1}{2}}$	-1.157	SCE	i_ℓ	0.43F	sttd	2	αn_a $dE_{\frac{1}{2}}/dpH$	0.920 54F	sttd -	-	$C,E_{\frac{1}{2}} Av(20)$ from $Elog, i_\ell=kh^{\frac{1}{2}}$ for pH=3-10, i_d, r	FC74
			-1.26F			0.45F		2		54F		-	$C,E_{\frac{1}{2}} Av(20)$ from Elog, r	
			-1.286			0.44F		2	αn_a $dE_{\frac{1}{2}}/dpH$	1.040 54F	sttd -	-	$C,E_{\frac{1}{2}} Av(20)$ from Elog, pK'=8.60, $E_{\frac{1}{2}}=-1.379+0.28\sigma, i_\ell=kC, r$	CONT

FC74 (CONT.) $C_{15}H_{12}ClNO_2$

Code No.	Empirical Formula	Name and C.A. Number	Structural Formula	Solvent	Tech.	Medium		μ, M	pH	T, °C	Electrodes	App.	Experimental Parameters
FC74	$C_{15}H_{12}ClNO_2$	3-(3-chlorophenyl)-3-oxo-N-phenylpropanamide	$3\text{-}ClC_6H_4COCH_2CO\text{-}NHC_6H_5$	EtOH 40.0	PY	BR GEL	0.01	-	9.3	-	DME//SCE	5A-	C=ns, m=1.8, t=3.7, h=59
									11.0				
FC75	$C_{15}H_{12}ClNO_2$	3-(4-chlorophenyl)-3-oxo-N-phenylpropanamide C.A. 962-05-0	$4\text{-}ClC_6H_4COCH_2CO\text{-}NHC_6H_5$	EtOH 40.0	PY	BR GEL	0.01	-	5.0	-	DME//SCE	5A-	C=ns, m=1.8, t=3.7, h=59
									7.0				
									8.0				
									8.6				
									9.8				
FC76 bh22 cj65 de88	$C_{15}H_{12}O$	<u>trans</u>-1,3-diphenyl-2-propen-1-one C.A. 614-47-1	$C_6H_5CH:CHCOC_6H_5$	EtOH	PY	Bu_4NOH	0.1	-	-	25	DME//SCE	12A0	C=1, Pt Aux
						$Bu_4N^+C_6H_5O^-$ PHEN	0.1						
				MeCN 90	PY	Et_4NClO_4	0.1	-	-	25	DME// Ag/ AgClO_4 0.01	12A0	C=4.7, Pt Aux
				MeCN 95	PW	Et_4NClO_4	0.1	-	-	25	DME// Ag/ AgClO_4 0.01	12A2	C=1.15, v=1.26E5, Pt Aux
				MeCN 99	PY	Et_4NClO_4	0.1	-	-	25	DME// Ag/ AgClO_4 0.01	12A0	C=4.7, Pt Aux
				MeCN 99.95	PY	Et_4NClO_4	0.1	-	-	25	DME// Ag/ AgClO_4 0.01	12A0	C=4.7, Pt Aux

CONT

$C_{15}H_{12}O$ (CONT.) FC76

Ref.	C/M	Charact. Potential		Response	Const. Value	n Tech.		Electrokinetic Data			Products and Identification	Description and Remarks	Code No.
		Value	vs.					Parameter	Value	From			
EA021 0831	65c	$E_{\frac{1}{2}}$	-1.39F	SCE	i_ℓ	0.36F	sttd	-	$dE_{\frac{1}{2}}/dpH$	54F	-	-	$C, E_{\frac{1}{2}}$ Av(20) from Elog,r	FC74
		-	-	-	0.11F	-		-	-	-	-	$C, E_{\frac{1}{2}}$ Av(20) from Elog,r	
EA021 0831	65c	$E_{\frac{1}{2}}$	-1.198	SCE	i_ℓ	0.40	sttd	2	αn_a $dE_{\frac{1}{2}}/dpH$	0.870 55F	sttd -	-	$C, i_\ell = kC, E_{\frac{1}{2}}$ Av(20) from Elog, $i_\ell = kh^{\frac{1}{2}}$ for pH=3-10, i_d, r	FC75
			-1.29F			0.39F		2		55F	-	-	$C, E_{\frac{1}{2}}$ Av(20) from Elog,r	
			-1.345			0.37F		2	αn_a $dE_{\frac{1}{2}}/dpH$	0.900 55F	sttd -	-	$C, E_{\frac{1}{2}}$ Av(20) from Elog, pK'=9.30, $E_{\frac{1}{2}}$ = -1.379+0.28σ, $i_\ell \propto C$, r	
			-1.34F			0.24F		2		55F	-	-	$C, E_{\frac{1}{2}}$ Av(20) from Elog,r	
			-1.46F			0.09F		2		55F	-	-	$C, E_{\frac{1}{2}}$ Av(20) from Elog,r	
JE042 0189	50s	$E_{\frac{1}{2}}$	-1.03F	SCE	i_ℓ	1.2F	-	-	-	-	-	-	C,X,p	FC76 bh22 cj65 de88
			-1.19F			0.8F							C,X	
			-1.46F			-							C,X	
			-1.64F			-							C,X,M	
			-1.03F			2.6F	i:i	1	-	-	-	-	C,r	
			-1.44F			3.4F		1					C	
JE042 0189	50s	$E_{\frac{1}{2}}$	-1.6F	Ag/ AgClO$_4$ 0.01	i_ℓ	17F	-	-	-	-	-	-	C,r	
			-1.8F			7F							C,Mx	
			-2.4F			37F							C	
JE042 0189	50s	-	-	-	-	-	-	-	$dE_p/d\log v$ $dE_p/d\log[H_2O]$	21 23	-	-	C, $dE_p/d\log[H_2O]$ for $[H_2O]$=0.6-2.8, data for first wave only, r	
			-										C(?)	
JE042 0189	50s	$E_{\frac{1}{2}}$	-1.7F	Ag/ AgClO$_4$ 0.01	i_ℓ	17F	-	-	-	-	-	-	C,r	
			-2.1F			7F							C	
			-2.4F			7F							C	
			-2.8F			43F							C	
JE042 0189	50s	$E_{\frac{1}{2}}$	-1.8F	Ag/ AgClO$_4$ 0.01	i_ℓ	17F	i:i	1	-	-	-	-	C,r	
			-2.3F			13F		1Ap					C	
			-2.7F			15F		-					C	CONT

FC76 (CONT.) $C_{15}H_{12}O$

Code No.	Empirical Formula	Name and C.A. Number	Structural Formula	Solvent	Tech.	Medium		μ, M	pH	T, °C	Electrodes	App.	Experimental Parameters
FC76 bh22 cj65 de88	$C_{15}H_{12}O$	trans-1,3-diphenyl-2-propen-1-one	$C_6H_5CH:CHCOC_6H_5$	DMF 99.9	PW	Et_4NClO_4	0.1	-	-	25	DME// Ag/ AgClO$_4$ 0.01	12A2	C=2.8, v=(5-5.7)E3, Pt Aux
FC77	$C_{15}H_{12}O$	1,1-diphenylprop-2-yn-1-ol C.A. 3923-52-2	$(C_6H_5)_2C(OH)C:CH$	DMF	PY	Bu_4NI	0.1	-	-	-	DME// SCE(o)	2-O	C=2
						Bu_4NI PHEN	0.1 ns						
FC78	$C_{15}H_{13}NO_2$	β-oxo-N-phenylbenzenepropanamide C.A. 959-66-0	$C_6H_5COCH_2CONH-C_6H_5$	EtOH 40.0	PY	BR GEL	- 0.01	-	5.0	-	DME//SCE	5A-	C=ns, m=1.8, t=3.7, h=59
									8.0				
FC79	$C_{15}H_{13}NO_3$	1-(2-hydroxy-4-methoxyphenyl)-3-(2-pyridyl)-2-propen-1-one C.A. 6344-90-7	Table II	EtOH 50	PY	BR GEL	- 0.02	-	2.6	22±1	DME//SCE	OAO	C=0.5, m=1.149, t=5.2, h=64
									4.7				
									6.9				
									8.0				
									9.2				
									10.0				
									12.15				
				dioxane 75	PY	Bu_4NI GEL	0.175 0.02	-	-	22±1	DME//SCE	OAO	C=0.5, m=1.149, t=5.2, h=64

TABLE I. Electrochemical Data

$C_{15}H_{13}NO_3$ FC79

Ref.	C/M	Charact. Potential		Response Const.		n Tech.		Electrokinetic Data			Products and Identification	Description and Remarks	Code No.
		Value	vs.		Value			Parameter	Value	From			
JE042 0189	50r	-	-	-	-	-	-	$dE_p/d\log v$	20	-	-	C, data for first wave only, r	FC76 bh22
		-							-			C(?)	cj65 de88
EA019 0629	33c	$E_{\frac{1}{2}}$	-2.71	SCE(o)	-	-	QE (-2.70v)	0.6	-	-	-	-	C, r	FC77
		-2.70		-	-		5.6	-	-	-	1,1-diphenylpropane 95%	C, r	
EA021 0831	65c	$E_{\frac{1}{2}}$	-1.208	SCE	-	-	sttd	2	αn_a	1.201	sttd	-	C, $E_{\frac{1}{2}}$ Av(20) from Elog, i_ℓ=kC, r	FC78
		-1.361		-	-	QE	1.98		0.985		3-phenyl-3-hydroxy-propionanilide, IRS, MPT	C, $E_{\frac{1}{2}}$ Av(20) from Elog, pK'=9.40, $E_{\frac{1}{2}}$= -1.379+0.28σ, i_ℓ ↓ as pH ↑, i_ℓ=kh$^{\frac{1}{2}}$ for pH=3-10, i_d, r	
JE039 0407	-	$E_{\frac{1}{2}}$	-0.39	SCE	-	-	-	-	-	-	-	-	C, r	FC79
		-1.24											C	
		-0.54		-	-	-	-	$dE_{\frac{1}{2}}/dpH$	79F	-	-	C, $dE_{\frac{1}{2}}/dpH$ for pH=4.7-8.0, r	
		-1.37							69F			C, $dE_{\frac{1}{2}}/dpH$ for pH=4.7-9.2	
		-0.72		-	-	-	-		79F	-	-	C, r	
		-1.53							69F			C	
		-0.80		-	-	-	-		-	-	-	C, i_d, r	
		-1.60							69F			C, i_d	
		-1.80							-			C, i_{cat}	
		-0.86		-	-	-	-		50F	-	-	C, $dE_{\frac{1}{2}}/dpH$ for pH=9.2-11.2	
		-1.68							-			C	
		-0.90		-	-	-	-		50F	-	-	C, r	
		-1.68							-			C	
		-1.08		-	-	-	-		-	-	-	C, r	
JE039 0407	-	$E_{\frac{1}{2}}$	-0.95	SCE	l	1.37	-	-	-	-	-	-	C, $\Delta E_{\frac{1}{2}}$=0.72Σσ*, r	
		-0.85			1.32								C	

FC80 $C_{15}H_{13}NO_3$

Code No.	Empirical Formula	Name and C.A. Number	Structural Formula	Solvent	Tech.	Medium		μ, M	pH	T, °C	Electrodes	App.	Experimental Parameters
FC80	$C_{15}H_{13}NO_3$	1-(2-hydroxy-4-methoxyphenyl)-3-(3-pyridyl)-2-propen-1-one C.A. 6622-61-3	Table II	EtOH 50	PY	BR GEL	– 0.02	–	2.6	22±1	DME//SCE	OAO	C=0.5, m=1.6, t=5.2, h=64
									5.6				
									6.9				
									8.0				
									9.2				
									12.15				
				dioxane 75	PY	Bu_4NI	0.175	–	–	22±1	DME//SCE	OAO	C=0.5, m=1.15, t=5.2, h=64
FC81	$C_{15}H_{13}NO_3$	1-(2-hydroxy-4-methoxyphenyl)-3-(4-pyridyl)-2-propen-1-one C.A. 7401-39-0	Table II	EtOH 50	PY	BR GEL	– 0.02	–	2.6	22±1	DME//SCE	OAO	C=0.5, m=1.15, t=5.2, h=64
									5.6				
									6.9				
									8.0				
									9.2				
									10.0				
									12.5				
				dioxane	PY	Bu_4NI GEL	0.175 0.02	–	–	22±1	DME//SCE	OAO	C=0.5, m=1.15, t=5.2, h=64

TABLE I. Electrochemical Data

$C_{15}H_{13}NO_3$ FC81

Ref.	C/M	Charact.	Potential	Response Const.		n		Electrokinetic Data			Products and	Description and	Code
		Value	vs.	Value		Tech.		Parameter	Value	From	Identification	Remarks	No.
JE039 0407	-	$E_{\frac{1}{2}}$	-0.62	SCE	-	-	-	-	$dE_{\frac{1}{2}}/dpH$	67F	-	-	$C, dE_{\frac{1}{2}}/dpH$ for pH=2.6-5.6	FC80
			-1.20							50F			$C, dE_{\frac{1}{2}}/dpH$ for pH=2.6-5.6	
			-0.82		-	-	-	-		75F		-	C,r	
			-1.35							-			C	
			-0.92		-	-	-	-		75F		-	C,r	
			-1.45							-			C	
			-1.0		-	-	-	-		75F		-	C, i_d, r	
			-1.55							-			C, i_d	
			-1.74										C, i_{cat}	
			-1.01		-	-	-	-	-	-	-	-	C,r	
			-1.62										C	
			-1.25		-	-	-	-	-	-	-	-	C,r	
JE039 0407	-	$E_{\frac{1}{2}}$	-1.07	SCE	I	1.38	-	-	-	-	-	-	$C, \Delta E_{\frac{1}{2}}=0.72\sigma^*, r$	
			-1.35			1.41							C	
			-1.90			-							C	
JE039 0407	-	$E_{\frac{1}{2}}$	-0.37	SCE	-	-	-	-	$dE_{\frac{1}{2}}/dpH$	63F	-	-	C,r	FC81
			-1.28							-			C	
			-0.56		-	-	-	-		65F		-	C,r	
			-1.42							70F			C	
			-0.64		-	-	-	-		65F		-	C,r	
			-1.52							70F			C	
			-0.72		-	-	-	-		65F		-	C, i_d, r	
			-1.59							-			C, i_d	
			-1.78										C, i_{cat}	
			-0.80		-	-	-	-		65F		-	C,r	
			-1.65							-			C	
			-0.85		-	-	-	-		65F		-	C, i_d, r	
			-1.60							-			C, i_d	
			-1.68										C, i_{cat}	
			-1.05		-	-	-	-		65F		-	C,r	
JE039 0407	-	$E_{\frac{1}{2}}$	-0.88	SCE	I	1.32	-	-	-	-	-	-	$C, \Delta E_{\frac{1}{2}}=0.72\sigma^*, r$	
			-1.72			1.31							C	
			-2.05			-							C	

FC82 $C_{15}H_{14}$

Code No.	Empirical Formula	Name and C.A. Number	Structural Formula	Solvent	Tech.	Medium	μ, M	pH	T, °C	Electrodes	App.	Experimental Parameters
FC82	$C_{15}H_{14}$	1,2-diphenylpropene C.A. 833-81-8 or 1017-22-7	$C_6H_5C(CH_3):CH-C_6H_5$	DMF	VA	Et_4NBF_4 satd	-	-	-	Hg: xxsd// SCE	5AO	C=ns, v=300
				HMP	VA	Bu_4NBF_4 satd	-	-	-	Hg: xxsd// SCE	5AO	C=ns, v=300
FC83	$C_{15}H_{14}N_2O_4$	4,4'-dinitrodiphenyl-propane C.A. 10368-11-3	$(4-O_2NC_6H_4CH_2)_2-CH_2$	MeCN	XT	Et_4NClO_4 0.1	-	-	-30±1	HMDE// Ag/ AgClO$_4$ 0.01	2-2	C=ns, v=(1.5-5.5)E3
									0±1			
									20±0.5			
				DMF	XT	Et_4NClO_4 0.1	-	-	0±1	HMDE// SCE	2-2	C=ns, v=(1.5-5.5)E3
									60±0.5			
FC84	$C_{15}H_{14}O$	1,1-diphenylprop-2-en-1-ol C.A. 3923-51-1	$(C_6H_5)_2C(OH)CH:CH_2$	DMF	PY	Bu_4NI 0.1	-	-	-	DME// SCE (o)	2-0	C=2
FC85	$C_{15}H_{14}O$	3,3-diphenylprop-2-en-1-ol C.A. 4801-14-3	$(C_6H_5)_2C:CH\ CH_2OH$	DMF	PY	Bu_4NI 0.1	-	-	-	DME// SCE (o)	2-0	C=2
FC86	$C_{15}H_{14}O$	9-methoxy-9-methyl-fluorene C.A. 39194-34-8	Table II	DMF	PY	Et_4NClO_4 0.1	-	-	-	DME// Ag/AgI, I$^-$ 0.1	3-0	C=1, Pt Aux
						Et_4NClO_4 0.1 PHEN 0.01						

$C_{15}H_{14}O$ FC86

Ref.	C/M	Charact. Potential		Response Const.		n		Electrokinetic Data			Products and Identification	Description and Remarks	Code No.
		Value	vs.	Value	Tech.			Parameter	Value	From			
EA019 0951	-	XE -2.32	SCE	-	-	-	-	-	-	-	-	$C;XE=(E_{p,A}+E_{p,C})/2;$ $R;ece;i_{p,C,1}/$ $i_{p,C,2}=45$ for v= 100,1.1 for v=3000; p	FC82
		E_p -2.52										$C,\neq$	
		-										A,R	
		-0.94										A, oxidn of monoprotonated dianion	
EA019 0951	-	XE -2.18	SCE	-	-	-	-	-	-	-	-	$C,XE=(E_{p,A}+E_{p,C})/2,$ R,eec,0 in LiCl satd,p	
		E_p -2.58										$C,\neq$	
		-										A,R	
JE047 0115	-	XE 0.033 ±0.002	Ag/ AgClO$_4$ 0.01	-	-	-	-	-	-	-	-	C,XT=semiintegral chronoamperometry with linear potential sweep,XE=difference between potentials for the two steps,p	FC83
		-										C	
		0.036 ±0.002		-	-	-	-	-	-	-	-	C,p	
		-										C	
		0.038 ±0.002		-	-	-	-	-	-	-	-	C,p	
		-										C	
JE047 0115	-	XE 0.037 ±0.002	SCE	-	-	-	-	-	-	-	-	C,XT=semiintegral chronoamperometry with linear potential sweep,XE=difference between potentials for the two steps,p	
		-										C	
		0.042 ±0.002		-	-	-	-	-	-	-	-	C,p	
		-										C	
EA019 0629	33d	$E_{\frac{1}{2}}$ -2.82	SCE(o)	-	-	QE (-2.80V)	3.5	-	-	-	1,1-diphenylpropane 90%	$C,Mc(E_{\frac{1}{2}}),r$	FC84
EA019 0629	33b	$E_{\frac{1}{2}}$ -2.45	SCE(o)	-	-	QE (-2.60V)	1.92	-	-	-	1,1-diphenyl-2-propene 95%	C,r	FC85
		-2.68										C	
EA020 0853	331	$E_{\frac{1}{2}}$ -1.71	Ag/AgI, I$^-$ 0.1	i_ℓ	3.5	QE (-1.90V)	2.0	-	-	-	methyl fluorenol, dimeric product, methyl fluorene < 20%	C,r	FC86
		-1.71			3.4		2.9	-	-	-	methyl fluorene 90%	C,r	
		-2.10			3.4		-					C	

FC87 $C_{15}H_{14}O_2S$

Code No.	Empirical Formula	Name and C.A. Number	Structural Formula	Solvent	Tech.	Medium		μ, M	pH	T, °C	Electrodes	App.	Experimental Parameters
FC87	$C_{15}H_{14}O_2S$	3-(4-ethoxyphenyl)-1-(2-thienyl)-2-propen-1-one C.A. 46910-91-2	Table II	EtOH 50	PY	NaOAc HCl	ns ns	0.2	0.77	25±0.1	DME//SCE	O3AO	C=0.2, m=1.86, t=3.8, h=70
FC88	$C_{15}H_{15}NOS$	3-(4-dimethylaminophenyl)-1-(2-thienyl)-2-propen-1-one C.A. 14385-59-2	Table II	EtOH 50	PY	NaOAc HCl	ns ns	0.2	0.77	25±0.1	DME//SCE	O3AO	C=0.2, m=1.86, t=3.8, h=70
FC89 bh71	$C_{15}H_{15}NO_2$	4-acetyl-4'-methoxydiphenylamine C.A. 23689-01-2	Table II	MeCN	VY	Et_4NClO_4	0.1	–	–	–	Pt: rodi// Ag/ $AgClO_4$ 0.01, Et_4NClO_4 0.1	2-O	C=ns, d=0.1, ω=10, Pt Aux
FC90	$C_{15}H_{15}N_3O_2$	3-(4-dimethylaminophenylazo)benzoic acid C.A. 20691-84-3	Table II	Me_2CO 50	VY	ACET	1.0	–	7.0	–	Pt: rodi// SCE	OAO	C=4.0
FC91	$C_{15}H_{15}N_3O_2$	4-(4-dimethylaminophenylazo)benzoic acid C.A. 6268-49-1	Table II	dioxane 50	VD	ACET	1.0	–	7.0	–	Pt: rodi// SCE	OAO	C=4.0
					VY	ACET	1.0	–	7.0	–	Pt: rodi// SCE	OAO	C=4.0
FC92	$C_{15}H_{15}N_3O_2$	3-(4-ethoxycarbonylphenyl)-1-phenyltriazene C.A. 34529-59-4	$4-C_2H_5OCOC_6H_4$ $NHN:NC_6H_5$	MeOH 50	PY	UB GEL	ns 0.01	–	4 7.0 9 14	21±1	DME//SCE	O-O	C=ns, m(oc)=4.9, t=1.9
FC93	$C_{15}H_{15}N_3O_4$	1-(2-hydroxy-4-methoxyphenyl)-3-(4-pyridyl)-1,3-propanedione dioxime C.A. 36256-99-2	Table II	DMF 20	PY	BR	–	–	2.3 3.20 4.3 5.15	–	DME//SCE	OAO	C=0.25, m=3.45, t=2.72, h=40
CONT			Table II										

$C_{15}H_{15}N_3O_4$ (CONT.) FC93

Ref.	C/M	Charact. Potential		Response Const.		n Tech.	n	Electrokinetic Data			Products and Identification	Description and Remarks	Code No.
		Value	vs.		Value			Parameter	Value	From			
JE053 0439	156 k	$E_{\frac{1}{2}}$	-0.537	SCE	-	-	-	-	-	-	-	-	$C, \Delta E_{\frac{1}{2}} \neq 0.22\sigma, p$	FC87
			-0.600										C	
JE053 0439	-	$E_{\frac{1}{2}}$	-0.520	SCE	-	-	-	-	-	-	-	-	C,p	FC88
			-0.712										C	
EA021 0557	147 n	$E_{\frac{1}{2}}$	0.575	Ag/ AgClO$_4$ 0.01, Et$_4$NClO$_4$ 0.1	i_ℓ/C	5.9	sttd	1	-	-	-	radical cation,VIS	$A, \Delta E_{\frac{1}{2}} = 0.30\sigma, r$	FC89 bh71
			1.025			-		1					A	
C0035 2944	-	$E_{\frac{1}{2}}$	0.78	SCE	-	-	-	-	-	-	-	-	A,p	FC90
C0034 3952	-	E_{su}	0.82	SCE	-	-	-	-	-	-	-	-	A,p	FC91
C0034 3952	-	$E_{\frac{1}{2}}$	0.80	SCE	-	-	-	-	-	-	-	-	A,p	
JE035 0369	386 a	$E_{\frac{1}{2}}$	-0.78F	SCE	-	-	-	-	$dE_{\frac{1}{2}}/dpH$	55	sttd	-	$C, E_{\frac{1}{2}} = -0.560-0.055pH$ for pH=4-8, p	FC92
			-0.93F		-	-	-	-		55		-	C,p	
			-1.07F		-	-	-	-		76		-	$C, E_{\frac{1}{2}} = -0.380-0.076pH$ for pH=8-14, p	
			-1.45F		-	-	-	-		76		-	C,p	
JE048 0297	-	$E_{\frac{1}{2}}$	-0.45	SCE	i_ℓ	3.04	i:i	4	-	-	-	-	C, i_d, r	FC93
			-0.85			2.88		4					C, i_d	
			-1.12			1.12		-					$C, i_\ell = kC, i_\ell = kh^{\frac{1}{2}}$	
			-0.53			3.04	-	-	-	-	-	-	C, i_d, r	
			-0.93			2.88							C, i_d	
			-1.14			1.12							C	
			-0.62			3.04	-	-	-	-	-	-	C, i_d, r	
			-1.02		}	4.16							C, i_d, S	
			-1.22										C,P	
			-0.69			3.04	-	-	-	-	-	-	C,r	
			-1.08										C,S	
			-1.25		}	3.04							C,P	

CONT

FC93 (CONT.) $C_{15}H_{15}N_3O_4$

Code No.	Empirical Formula	Name and C.A. Number	Structural Formula	Solvent		Tech.	Medium		μ, M	pH	T, °C	Electrodes	App.	Experimental Parameters
FC93	$C_{15}H_{15}N_3O_4$	1-(2-hydroxy-4-methoxyphenyl)-3-(4-pyridyl)-1,3-propanedione dioxime	Table II	DMF	20	PY	BR		-	-	6.10	-	DME//SCE	OAO	C=0.25, m=3.45, t=2.72,h=40
										7.15				
										9.90				
										11.5				
FC94	$C_{15}H_{16}N_2O_2$	4-acetylamino-4'-methoxydiphenylamine C.A. 17785-89-6	Table II	MeCN		VY	Et_4NClO_4	0.1	-	-	-	Pt: rodi// Ag/ $AgClO_4$ 0.01, Et_4NClO_4 0.1	2-0	C=ns,d=0.1, ω=10,Pt Aux
							Et_4NClO_4 2,6-lutidine	0.1						
							ACET	ns						
FC95	$C_{15}H_{16}N_2O_4$	3-hydroxy-1-(2-hydroxy-4-methoxyphenyl)-3-(2-pyridyl)-1-propanone oxime C.A. 35590-50-2	Table II	DMF	20	PY	BR		-	-	2.30	-	DME//SCE	OAO	C=0.25, m=3.45, t=2.72,h=40
										2.90				
										3.95				
										5.20				
										6.80				
FC96	$C_{15}H_{16}N_2O_4$	3-hydroxy-1-(2-hydroxy-4-methoxyphenyl)-3-(3-pyridyl)-1-propanone oxime C.A. 35590-57-9	Table II	DMF	20	PY	BR		-	-	2.30	-	DME//SCE	OAO	C=0.25, m=3.45, t=2.72,h=40
										2.90				
										3.95				
										5.20				
										6.80				
										7.80				
FC97 CONT	$C_{15}H_{16}N_2O_4$	3-hydroxy-1-(2-hydroxy-4-methoxyphenyl)-3-(4-pyridyl)-1-propanone oxime C.A. 35590-59-1	Table II	DMF	20	PY	BR		-	-	2.30	-	DME//SCE	OAO	C=0.25, m=3.45, t=2.72,h=40
										2.90				
										3.95				
										5.20				

$C_{15}H_{16}N_2O_4$ (CONT.) FC97

Ref.	C/M	Charact. Potential Value	vs.	Response Const. Value		n Tech.	Electrokinetic Data Parameter	Value	From	Products and Identification	Description and Remarks	Code No.
JE048 0297	-	$E_{\frac{1}{2}}$	-0.77	SCE	i_ℓ	3.04	-	-	-	-	-	-	C,r	FC93
			-1.25			4.16							C,S	
			-			-							C,P	
			-0.88			2.92	-	-	-	-	-	-	C,r	
			-1.32			3.82							C	
			-1.15			2.08	-	-	-	-	-	-	C,r	
			-1.55			-							C	
			-1.47			2.08	-	-	-	-	-	-	C,r	
EA021 0557	147 n	$E_{\frac{1}{2}}$	0.32	Ag/ AgClO$_4$ 0.01, Et$_4$NClO$_4$ 0.1	i_ℓ/C	5.2	sttd	1	-	-	-	radical cation, VIS	A, $\Delta E_{\frac{1}{2}}=0.30\sigma$	FC94
			0.88			-		1					A	
			0.22			-	-	-	-	-	-	-	A, $\Delta E_{\frac{1}{2}}=0.30\sigma$, r	
			-0.04			-	-	-	-	-	-	-	A,r	
JE048 0297	109 d	$E_{\frac{1}{2}}$	-0.97	SCE	i_ℓ	2.72	i:i	4	-	-	-	-	C,W,i_d,$i_\ell=kC$,$i_\ell=kh^{\frac{1}{2}}$, r	FC95
			-1.02			2.72	-	-	-	-	-	-	C,r	
			-1.09			2.28	-	-	-	-	-	-	C,r	
			-1.18			1.98	-	-	-	-	-	-	C,r	
			-1.4			-	-	-	-	-	-	-	C,r	
JE048 0297	-	$E_{\frac{1}{2}}$	-0.99	SCE	i_ℓ	2.88	-	-	-	-	-	-	C,W,i_d,$i_\ell=kC$,$i_\ell=kh^{\frac{1}{2}}$, r	FC96
			-1.03			2.88	-	-	-	-	-	-	C,r	
			-1.08			2.80	-	-	-	-	-	-	C,r	
			-1.17			2.32	-	-	-	-	-	-	C,r	
			-1.32			1.0	-	-	-	-	-	-	C, wave has reverse S shape, r	
			-1.40			-	-	-	-	-	-	-	C,r	
JE048 0297	-	$E_{\frac{1}{2}}$	-0.96	SCE	i_ℓ	2.76	i:i	4	-	-	-	-	C,W,i_d,$i_\ell=kC$,$i_\ell=kh^{\frac{1}{2}}$, r	FC97
			-1.01			2.76	-	-	-	-	-	-	C,r	
			-1.06			2.36	-	-	-	-	-	-	C,r	
			-1.15			2.00	-	-	-	-	-	-	C,r	

CONT

FC97 (CONT.) $C_{15}H_{16}N_2O_4$

Code No.	Empirical Formula	Name and C.A. Number	Structural Formula	Solvent	Tech.	Medium		μ, M	pH	T, °C	Electrodes	App.	Experimental Parameters
FC97	$C_{15}H_{16}N_2O_4$	3-hydroxy-1-(2-hydroxy-4-methoxyphenyl)-3-(4-pyridyl)-1-propanone oxime	Table II	DMF 20	PY	BR	-	-	6.80	-	DME//SCE	OAO	C=0.25, m=3.45, t=2.72, h=40
FC98	$C_{15}H_{16}O_2$	bis(3-methoxyphenyl)-methane C.A. 51095-48-8	$(2-CH_3OC_6H_4)_2-CH_2$	MeCN	VR	$NaClO_4$	0.1	-	-	-	Pt: xxbe//SCE	2A2	C=1, v=86
				CH_2Cl_2 80 CF_3COOH 20	VR	Bu_4NBF_4	0.25	-	-	-	Pt: xxbe//SCE	2A2	C=1, v=86
FC99	$C_{15}H_{16}O_2$	(4-methoxyphenyl)-phenylmethyl methyl ether C.A. 7364-21-8	$4-CH_3OC_6H_4CH(OCH_3)C_6H_5$	DMF	PY	Bu_4NI	0.1	-	-	-	DME// Ag/AgI, I⁻ 0.1	3-O	C=1, Pt Aux
					QE	Bu_4NI PHEN	0.1 0.01	-	-	-	Hg?: nsns// Ag/AgCl, I⁻ 0.1	3-O	C=1, E_{app}=-2.45V, Pt Aux
FD00	$C_{15}H_{17}N_3$	4-dimethylamino-3'-methylazobenzene C.A. 55-80-1	Table II	Me_2CO 50	VY	ACET	1.0	-	7.0	-	Pt: rodi//SCE	OAO	C=4.0
FD01	$C_{15}H_{17}N_3$	4-dimethylamino-4'-methylazobenzene C.A. 3010-57-9	Table II	dioxane 50	VD	ACET	1.0	-	7.0	-	Pt: rodi//SCE	OAO	C=4.0
					VY	ACET	1.0	-	7.0	-	Pt: rodi//SCE	OAO	C=4.0
FD02	$C_{15}H_{17}N_3O$	4-dimethylamino-3'-methoxyazobenzene C.A. 20691-83-2	Table II	Me_2CO 50	VY	ACET	1.0	-	7.0	-	Pt: rodi//SCE	OAO	C=4.0
FD03	$C_{15}H_{17}N_3O$	4-dimethylamino-4'-methoxyazobenzene C.A. 3009-50-5	Table II	dioxane 50	VD	ACET	1.0	-	7.0	-	Pt: rodi//SCE	OAO	C=4.0
					VY	ACET	1.0	-	7.0	-	Pt: rodi//SCE	OAO	C=4.0

TABLE I. Electrochemical Data   201

$C_{15}H_{17}N_3O$ FD03

Ref.	C/M	Charact. Potential		Response Const.		n Tech.		Electrokinetic Data			Products and Identification	Description and Remarks	Code No.
		Value	vs.		Value			Parameter	Value	From			
JE048 0297	-	$E_{\frac{1}{2}}$	-1.28	SCE	i_ℓ	1.10	-	-	-	-	-	-	C,r	FC97
JA095 7132	437 a	E_p	1.68	SCE	-	-	-	-	-	-	-	QP(i_{app}=200mA, [LiClO$_4$]=0.1) → black ppt (blue in MeCN, green in MeNO$_2$), ESR; redn. with Zn → starting material 26%; 3,7-dimethoxyfluorene 39%, polymers 44%	A,≠,r	FC98
			1.01									-	A,R,O on first scan	
			0.95									-	C,R,O on first scan	
JA095 7132	437 a	E_p	1.81	SCE	-	-	QP	3	-	-	-	QP(i_{app}=200mA) followed by current reversal → colorless solution, n=0.7 for cathodic part, starting material 20%, and 3,7-dimethsxyfluorene 70%	A,≠,r	
			1.34									-	A,R,O on first scan	
			1.27									-	C,R,O on first scan	
EA020 0853	33k	$E_{\frac{1}{2}}$	-2.40	Ag/AgI, I$^-$ 0.1	i_ℓ	4.0	QE (-2.45V)	2.0	-	-	-	(4-methoxyphenyl)-phenylmethane 70%, 4-methoxybenzhydrol 30%	C,r	FC99
EA020 0853	33k	-					QE	1.05	-	-	-	(4-methoxyphenyl)-phenylmethane	C,r	
C0035 2944	-	$E_{\frac{1}{2}}$	0.77	SCE	-	-	-	-	-	-	-	-	A,p	FD00
C0034 3952	-	E_{su}	0.86	SCE	-	-	-	-	-	-	-	-	A,p	FD01
C0034 3952	-	$E_{\frac{1}{2}}$	0.84	SCE	-	-	-	-	-	-	-	-	A,p	
C0035 2944	-	$E_{\frac{1}{2}}$	0.79	SCE	-	-	-	-	-	-	-	-	A,p	FD02
C0034 3952	-	E_{su}	0.82	SCE	-	-	-	-	-	-	-	-	A,p	FD03
C0034 3952	-	$E_{\frac{1}{2}}$	0.80	SCE	-	-	-	-	-	-	-	-	A,p	

FD04 $C_{15}H_{17}N_3O_4$

Code No.	Empirical Formula	Name and C.A. Number	Structural Formula	Solvent	Tech.	Medium	μ, M	pH	T, °C	Electrodes	App.	Experimental Parameters
FD04	$C_{15}H_{17}N_3O_4$	3-hydroxyamino-1-(2-hydroxy-4-methoxyphenyl)-3-(3-pyridyl)-1-propanone oxime C.A. 36256-96-9	Table II	DMF 20	PY	BR	–	2.30	–	DME∥SCE	OAO	C=0.25, m=3.45, t=2.72,h=40
								2.90				
								3.95				
								5.20				
								6.80				
								7.80				
FD05	$C_{15}H_{24}O_3$	2-tert-butyl-5-(2-methoxy-2-methyl-1-propyl)hydroquinone C.A. 17208-03-6	Table II	EtOH 50	PY	ACET	0.2	–	0	25± 0.1	DME∥SCE	O4AO	ns
FD06	$C_{16}H_4N_6O_8$	(2,4,5,7-tetranitro-9H-fluoren-9-ylidene)propanedinitrile C.A. 15517-55-2	Table II	MeCN	VA	LiClO$_4$	0.1	–	–	25	gw: xxdi∥SCE	25AO	0.5→-0.5→0.5 V,C=ns, d=0.62,v=1, Pt Aux
													v=6
													v=12
													v=66
FD07	$C_{16}H_5N_5O_6$	(2,4,7-trinitro-9H-fluoren-9-ylidene)-propanedinitrile C.A. 1172-02-7	Table II	MeCN	VA	LiClO$_4$	0.1	–	–	–	gw: xxdi∥SCE	25AO	0.5→-0.5→0.5 V,C=ns, d=0.62,v=3, Pt Aux
													v=6
													v=12
													v=67
													0.5→1.2→0.5 V,C=0.94, d=0.62,v=67, Pt Aux

TABLE I. Electrochemical Data   203

$C_{16}H_5N_5O_6$ FD07

Ref.	C/M	Charact. Potential		Response Const.		n Tech.	n	Electrokinetic Data			Products and Identification	Description and Remarks	Code No.
		Value	vs.		Value			Parameter	Value	From			
JE048 0297	109 d	$E_{\frac{1}{2}}$	−0.78	SCE	i_ℓ	−	−	−	−	−	−	−	C,Pr,i_a,r	FD04
			−0.92			2.88							$C,W,i_d, i_\ell=kC, i_\ell=kh^{\frac{1}{2}}, i_\ell \downarrow$ as pH $\uparrow$	
			−0.82			−	−	−	−	−	−	−	C,Pr,i_a,r	
			−0.97			2.88							C,W,i_d	
			−0.88			−	−	−	−	−	−	−	C,Pr,i_a,r	
			−1.06			2.76							C	
			−0.95			−	−	−	−	−	−	−	C,Pr,i_a,r	
			−1.17			2.64							$C,W,i_d, i_\ell=kC, i_\ell=kh^{\frac{1}{2}}$	
			−1.26			1.36	−	−	−	−	−	−	C,r	
			−1.35			−	−	−	−	−	−	−	C,r	
C0032 2140	−	$E_{\frac{1}{2}}^o$	0.610	NHE	−	−	−	−	−	−	−	−	$A,E_{\frac{1}{2}} \rightarrow \rightarrow$ from data for pH 3.8-5.6	FD05
EA021 0973	50t	−	−	−	D $i_{p,A}/i_{p,C}$	14.0 0.87F	−	−	ΔE_p	75	sttd	−	C,R,r	FD06
													A,R	
		−	−	−		−	−	−	−	−	−	−	C,R,r	
						1.1F							A,R	
		−	−	−		−	−	−	−	−	−	−	C,R,r	
						1.16F							A,R	
		−	−	−		−	−	−	ΔE_p	180	sttd	−	C,R,r	
						1.17							A,R	
EA021 0973	50t	−	−	−	D $i_{p,A}/i_{p,C}$	13.0 0.98F	−	−	ΔE_p	75	sttd	−	C,R,r A,R	FD07
		−	−	−		− 1.07	−	−		80		−	C,R,r A,R	
		−	−	−		− 1.12F	−	−		86		−	C,R,r A,R	
		−	−	−		− 1.14	−	−		118		−	C,R,r A,R	
		E_p	0.16F	SCE	i_p	150F	sttd	1		118		radical anion	C,R,r	
			1.02F			1100F	−			−			C,X	
			0.06F			150F							A,R	

FD08 $C_{16}H_9ClO_2$

Code No.	Empirical Formula	Name and C.A. Number	Structural Formula	Solvent	Tech.	Medium		μ, M	pH	T, °C	Electrodes	App.	Experimental Parameters
FD08	$C_{16}H_9ClO_2$	2-(3-chlorobenzyl-idene)-1,3-indan-dione C.A. 15875-55-5	Table II	EtOH 20	PY	GLYC	ns	0.1	9.5	-	DME//SCE	O4AO	C=0.1
FD09	$C_{16}H_9ClO_2$	2-(4-chlorobenzyl-idene)-1,3-indan-dione C.A. 15875-54-4	Table II	EtOH 20	PY	GLYC	ns	0.1	9.5	-	DME//SCE	O4AO	C=0.1
FD10	$C_{16}H_{10}O_2$	2-benzylidene-1,3-indandione C.A. 5381-33-9	Table II	EtOH 20	PY	GLYC	ns	0.1	9.5	-	DME//SCE	O4AO	C=0.1
FD11	$C_{16}H_{10}O_3$	2-(3-hydroxybenzyl-idene)-1,3-indan-dione C.A. 25299-18-7	Table II	EtOH 20	PY	GLYC	ns	0.1	9.5	-	DME//SCE	O4AO	C=0.1
FD12	$C_{16}H_{10}O_4$	2-benzoyloxy-1,3-indandione C.A. 59410-64-9	Table II	H_2O	PY	MB	0.3	-	2.5	22	DME//SCE	2-0	C=1,h=50
									3				
									4.2				
									5.3				
									6.0				
									7.0				
									7.6				
									8.3				
						CARB	0.3		11				
FD13	$C_{16}H_{11}Br_3O_2$	3-(4-methylphenyl)-1-(2,4,6-tribromo-3-hydroxyphenyl)-2-propen-1-one C.A. 28074-05-7	Table II	EtOH 50	PY	PHTH	ns	0.2	0.84	25±0.1	DME//SCE	O3AO	C=ns,m=1.80, t=3.8,h=65
									2.65				
FD08 CONT													

TABLE I. Electrochemical Data

$C_{16}H_{11}Br_3O_2$ (CONT.) FD 13

Ref.	C/M	Charact. Potential		Response Const.		n Tech.	n	Electrokinetic Data			Products and Identification	Description and Remarks	Code No.
		Value	vs.		Value			Parameter	Value	From			
C0036 1406	-	$E_{\frac{1}{2}}$	-1.220	SCE	-	-	i:i	4	-	-	-	-	$C, \Delta E_{\frac{1}{2}}=0.14\sigma, i_d, p$	FD08
			-					-				C	
			-									C?	
C0036 1406	-	$E_{\frac{1}{2}}$	-1.245	SCE	-	-	i:i	4	-	-	-	-	$C, \Delta E_{\frac{1}{2}}=0.14\sigma, i_d, p$	FD09
			-					-				C	
			-									C?	
C0036 1406	-	$E_{\frac{1}{2}}$	-1.265	SCE	-	-	i:i	4	-	-	-	-	$C, \Delta E_{\frac{1}{2}}=0.14\sigma, i_d, p$	FD10
			-					-				C	
			-									C?	
C0036 1406	-	$E_{\frac{1}{2}}$	-1.285	SCE	-	-	i:i	4	-	-	-	-	$C, \Delta E_{\frac{1}{2}}=0.14\sigma, i_d, p$	FD11
			-					-				C	
			-									C?	
EA020 0973	358 a	$E_{\frac{1}{2}}$	-0.77F	SCE	i_ℓ	4.5F	-	-	$dE_{\frac{1}{2}}/dpH$	51	sttd	-	C,r	FD12
			-0.80F			4.6F	-	-		51		-	C,r	
			-1.02F			4.6F				42			C	
	358 b		-0.84F			4.2F	-	-		51		-	C,r	
			-1.08F			4.7F				42			C	
			-0.92F			3.2F	-	-		51		-	C,r	
			-1.14F			5.9F				42			C	
			-0.95F			2.1F	-	-		51		-	C,r	
			-1.15F			6.5F				42			C	
			-1.00F			0.4F	-	-		51		-	C,O for pH > 7, r	
			-1.18F			5.4F				42			C	
			-1.36F			1.1F				0			C	
			-1.21F			3.2F	-	-		42		-	C,r	
			-1.36F			4.9F				0			C	
	358 c		-1.24F			0.4F	-	-		42		-	C,r	
			-1.36F			7.0F				33			C	
			-1.45			7.0F	-	-		33		-	C,r	
JE053 0449	156 g	$E_{\frac{1}{2}}$	-0.625	SCE	-	-	-	-	$dE_{\frac{1}{2}}/dpH$	92	sttd	-	$C, E_{\frac{1}{2}}= -0.520-0.092pH$ for pH=0-4, $i_\ell=kC$, $i_d, \Delta E_{\frac{1}{2}}=0.24\sigma, p$	FD13
			-0.907									-	$C, i_\ell=kC, i_d, \Delta E_{\frac{1}{2}}= 0.31\sigma$	
			-0.75			-	-	-		92		-	$C, i_\ell=kC, i_d, \Delta E_{\frac{1}{2}}= 0.28\sigma, p$	
			-1.01							-			$C, i_\ell=kC, i_d, \Delta E_{\frac{1}{2}}=0.27\sigma$	

CONT

FD13 (CONT.) $C_{16}H_{11}Br_3O_2$

Code No.	Empirical Formula	Name and C.A. Number	Structural Formula	Solvent	Tech.	Medium		μ, M	pH	T, °C	Electrodes	App.	Experimental Parameters
FD13	$C_{16}H_{11}Br_3O_2$	3-(4-methylphenyl)-1-(2,4,6-tribromo-3-hydroxyphenyl)-2-propen-1-one	Table II	EtOH 50	PY	PHTH	ns	0.2	3.32	25±0.1	DME//SCE	O3AO	C=ns,m=1.80, t=3.8,h=65
FD14	$C_{16}H_{11}Br_3O_3$	3-(3-methoxyphenyl)-1-(2,4,6-tribromo-3-hydroxyphenyl)-2-propen-1-one C.A. 28074-00-2	Table II	EtOH 50	PY	PHTH	ns	0.2	0.84	25±0.1	DME//SCE	O3AO	C=ns,m=1.80, t=3.8,h=65
									2.65				
									3.32				
FD15	$C_{16}H_{11}Br_3O_3$	3-(4-methoxyphenyl)-1-(2,4,6-tribromo-3-hydroxyphenyl)-2-propen-1-one C.A. 28074-01-3	Table II	EtOH 50	PY	PHTH	ns	0.2	0.84	25±0.1	DME//SCE	O3AO	C=ns,m=1.80, t=3.8,h=65
									2.65				
									3.32				
FD16	$C_{16}H_{11}Br_3O_4$	3-(4-hydroxy-3-methoxyphenyl)-1-(2,4,-6-tribromo-3-hydroxyphenyl)-2-propen-1-one C.A. 28074-02-4	Table II	EtOH 50	PY	PHTH	ns	0.2	0.84	25±0.1	DME//SCE	O3AO	C=ns,m=1.80, t=3.8,h=65
									2.65				
									3.32				

$C_{16}H_{11}Br_3O_4$ FD16

Ref.	C/M	Charact. Potential		Response Const.		n	Electrokinetic Data			Products and Identification	Description and Remarks	Code No.
		Value	vs.		Value	Tech.	Parameter	Value	From			
JE053 0449	156 g	$E_{\frac{1}{2}}$ -0.825	SCE	-	-	-	-	$dE_{\frac{1}{2}}/dpH$	92	sttd	-	$C, i_{\ell}=kC, i_d, \Delta E_{\frac{1}{2}}=0.29\sigma, p$	FD13
		-1.103							-			$C, i_{\ell}=kC, i_d, \Delta E_{\frac{1}{2}}=0.25\sigma$	
JE053 0449	156 g	$E_{\frac{1}{2}}$ -0.577	SCE	-	-	-	-	$dE_{\frac{1}{2}}/dpH$	85	sttd	-	$C, E_{\frac{1}{2}}=-0.520-0.085$pH for pH=0-4, $i_{\ell}=kC, i_d, \Delta E_{\frac{1}{2}}=0.24\sigma, p$	FD14
		-0.795							-			$C, i_{\ell}=kC, i_d, \Delta E_{\frac{1}{2}}=0.31\sigma$	
		-0.72							85			$C, i_{\ell}=kC, i_d, \Delta E_{\frac{1}{2}}=0.28\sigma, p$	
		-0.95							-			$C, i_{\ell}=kC, i_d, \Delta E_{\frac{1}{2}}=0.27\sigma$	
		-0.802							85			$C, i_{\ell}=kC, i_d, \Delta E_{\frac{1}{2}}=0.29\sigma, p$	
		-1.042							-			$C, i_{\ell}=kC, i_d, \Delta E_{\frac{1}{2}}=-0.25\sigma$	
JE053 0449	156 g	$E_{\frac{1}{2}}$ -0.615	SCE	-	-	-	-	$dE_{\frac{1}{2}}/dpH$	87	sttd	-	$C, E_{\frac{1}{2}}=-0.530-0.087$pH for pH=0-4, $i_{\ell}=kC, i_d, \Delta E_{\frac{1}{2}}=0.24\sigma, p$	FD15
		-0.930							-			$C, i_{\ell}=kC, i_d, \Delta E_{\frac{1}{2}}=0.31\sigma$	
		-0.750							87			$C, i_{\ell}=kC, i_d, \Delta E_{\frac{1}{2}}=0.28\sigma, p$	
		-1.070							-			$C, i_{\ell}=kC, i_d, \Delta E_{\frac{1}{2}}=0.27\sigma$	
		-0.840							87			$C, i_{\ell}=kC, i_d, \Delta E_{\frac{1}{2}}=0.29\sigma, p$	
		-1.125							-			$C, i_{\ell}=kC, i_d, \Delta E_{\frac{1}{2}}=0.25\sigma$	
JE053 0449	156 g	$E_{\frac{1}{2}}$ -0.660	SCE	-	-	-	-	$dE_{\frac{1}{2}}/dpH$	86	sttd	-	$C, E_{\frac{1}{2}}=-0.500-0.086$pH for pH=0-4, $i_{\ell}=kC, i_d, \Delta E_{\frac{1}{2}}=0.24\sigma, p$	FD16
		-0.930							-			$C, i_{\ell}=kC, i_d, \Delta E_{\frac{1}{2}}=0.31\sigma$	
		-0.820							86			$C, i_{\ell}=kC, i_d, \Delta E_{\frac{1}{2}}=0.28\sigma, p$	
		-1.090							-			$C, i_{\ell}=kC, i_d, \Delta E_{\frac{1}{2}}=0.27\sigma$	
		-0.885							86			$C, i_{\ell}=kC, i_d, \Delta E_{\frac{1}{2}}=0.29\sigma, p$	
		-1.125							-			$C, i_{\ell}=kC, i_d, \Delta E_{\frac{1}{2}}=0.25\sigma$	
		$E_{\frac{1}{2}}$ -0.825						$dE_{\frac{1}{2}}/dpH$	92	sttd			

FD17 $C_{16}H_{11}N_2NaO_6S$

Code No.	Empirical Formula	Name and C.A. Number	Structural Formula	Solvent	Tech.	Medium		μ, M	pH	T, °C	Electrodes	App.	Experimental Parameters
FD17	$C_{16}H_{11}N_2NaO_6S$	3-[(2,5-dihydroxy-naphth-1-yl)azo]-4-hydroxybenzenesulfonic acid, sodium salt C.A. 53151-14-7	Table II	H_2O	PY	buffer	ns	-	1.8	35±0.1	DME//SCE	OAO	C=0.5, EC=3.75, h=50
									3.3				
									5.8				
									8.0				
									11.5				
FD18	$C_{16}H_{12}FeN_3O_7$	ferricinium picrate C.A. 11077-21-7	$(\pi-C_5H_5)_2Fe^+$ $2,4,6-(O_2N)_3-C_6H_2O^-$	H_2O	PVQ	Et_4NClO_4	0.1	-	-	-	Pt: xxbu// SCE	013-0	C=2.0, A=0.20, f=80, PE
					VA	Et_4NClO_4	0.1	-	-	-	Pt: xxbu// SCE	013-0	C=2.0, A=0.30, v=10, PE
													v=1000
FD19	$C_{16}H_{12}N_2O_2$	3-(4-cyanophenyl)-3-oxo-N-phenylpropan-amide C.A. 38505-26-9	$4-NCC_6H_4COCH_2-CONHC_6H_5$	EtOH 40	PY	BR GEL	- 0.01	-	5.0	-	DME//SCE		C=ns, m=1.8, t=3.7, h=59
									8.0				
FD20 bi39 ck14 ee94	$C_{16}H_{12}O_2$	trans(?)-1,2-di-benzoylethylene C.A. 4070-75-1	$C_6H_5COCH:CHCO-C_6H_5$	DMF 96	PY	Et_4NClO_4	0.1	-	-	25	DME//SCE	12AO	C=2.5, Pt Aux
				DMF 99	PY	Et_4NClO_4	0.1	-	-	25	DME//SCE	12AO	C=2.5, Pt Aux
				DMF 99.9	PY	Et_4NClO_4	0.1	-	-	25	DME//SCE	12AO	C=2.5, Pt Aux
FD21	$C_{16}H_{12}O_4$	2-benzoyloxy-1-hydroxy-3-indanone	Table II	H_2O	PY	MB	0.3	-	3.2	22	DME//SCE	2-0	C=1, h=50
									6.0				
CONT									6.9				

$C_{16}H_{12}O_4$ (CONT.) FD21

Ref.	C/M	Charact. Potential		Response Const.		n Tech.	Electrokinetic Data			Products and Identification	Description and Remarks	Code No.
		Value	vs.		Value		Parameter	Value	From			
JE054 0417	-	$E_{\frac{1}{2}}$ -0.07	SCE	-	-	PQ 2	αn_a $k_{s,h}$ $dE_{\frac{1}{2}}/dpH$	0.74 4.5E-6 31	-	-	$C,\frac{1}{r},E_{\frac{1}{2}}=0.068-0.031pH$ (± 0.003 V), $i_\ell=kh^{\frac{1}{2}}$, $i_\ell=kC, i_d$, $Mc(dE_{\frac{1}{2}}/dpH),p$	FD17
		-0.130		-	-	- -		0.74 4.6E-6 31	-	-	$C,\frac{1}{r},p$	
		-0.200		-	-	- -		0.74 4.6E-6 31	-	-	$C,\frac{1}{r},p$	
		-0.270		-	-	- -		0.74 4.6E-6 31	-	-	$C,\frac{1}{r},p$	
		-0.400		-	-	- -		0.74 5.2E-6 31	-	-	$C,\frac{1}{r},p$	
EA018 0975	-	E_{su} 0.170 ± 0.005	SCE	-	-	- -	$\Delta E_{su/2}$	90	sttd	-	C,r	FD18
EA018 0975	-	E_p 0.139 ± 0.005	SCE	-	-	- -	ΔE_p k	60 0.01	sttd N-S	-	C,r	
		-	-	-	-	- -	ΔE_p	75	sttd	-	C,r	
EA021 0831	65c	$E_{\frac{1}{2}}$ -1.000	SCE	-	-	sttd 2	αn_a	0.922	sttd	-	$C,E_{\frac{1}{2}}$ Av(20) from Elog, $i_\ell=kh^{\frac{1}{2}}$ for pH=3-10, i_d,r	FD19
		-1.078		-	-	2		1.187		-	$C,E_{\frac{1}{2}}$ Av(20) from Elog, pK'=8.45, $E_{\frac{1}{2}}=-1.379+0.284\sigma,r$	
JE042 0189	50s	$E_{\frac{1}{2}}$ -0.76F -0.96F	SCE	i_ℓ	1.7F 1.4F	- -	-	-	-	-	C,r C	FD20
JE042 0189	50s	$E_{\frac{1}{2}}$ -0.76F -1.14F	SCE	i_ℓ	1.7F 0.42F	- -	-	-	-	-	C,r C	
JE042 0189	50s	$E_{\frac{1}{2}}$ -0.8F -1.12F	SCE	i_ℓ	1.9F small	i:i 1 -	-	-	-	-	C,r C,Mn	
EA020 0973	-	$E_{\frac{1}{2}}$ -1.03F	SCE	-	-	sttd 2	$dE_{\frac{1}{2}}/dpH$	41	sttd	-	C,r	FD21
		-1.14F		-	-	-		41		-	C,r	
		-1.18F		-	-	-		27		-	C,r	CONT

FD21 (CONT.) $C_{16}H_{12}O_4$

Code No.	Empirical Formula	Name and C.A. Number	Structural Formula	Solvent	Tech.	Medium		μ, M	pH	T, °C	Electrodes	App.	Experimental Parameters
FD21	$C_{16}H_{12}O_4$	2-benzoyloxy-1-hydroxy-3-indanone	Table II	H_2O	PY	MB	0.3	-	8.0	22	DME//SCE	2-0	C=1, h=50
FD22 df09	$C_{16}H_{13}ClN_2O$	Diazepam C.A. 439-14-5	Table II-4	DMF 10	PY	BR	-	-	2.1	-	DME// Ag/AgCl	O4AO	C=0.1
									4.9				
									6.7				
									8.3				
									10.3				
FD23	$C_{16}H_{14}$	1,2-diethyliden-acenaphthene C.A. 13152-79-9	Table II	DMF	PY	Et_4NClO_4 PHEN	0.1 4.5E-5	-	-	25± 0.1	DME//Hg	---	C=0.75, m=2.38, t=3.2, h=50
						Et_4NClO_4 PHEN	0.1 6.3E-4						
						Et_4NClO_4 PHEN	0.1 6.5E-3						
						Et_4NClO_4 PHEN	0.1 0.04						
FD24	$C_{16}H_{14}N_2O_2$	5,6-diacetyl-5,6-dihydro-5,6-phen-anthroline	Table II	DMF	VA	$LiClO_4$	0.1	-	-	-	Pt// Ag/AgI, I⁻ 0.1	2-0	C=ns, A=0.015, v=100
FD25	$C_{16}H_{14}N_2O_2$	5,10-diacetyl-5,10-dihydrophenazine C.A. 7478-17-3	Table II	DMF	VA	$LiClO_4$	0.1	-	-	-	Pt// Ag/AgI, I⁻ 0.1	2-0	C=ns, A=0.015, v=100
FD26	$C_{16}H_{14}O$	9-(1-hydroxyethyl)-anthracene C.A. 7512-20-1	Table II	DMF	PY	Et_4NBr	0.1	-	-	-	DME// SCE(o)	2-0	C=2
						Et_4NBr PHEN	0.1 0.01						
FD27 df29	$C_{16}H_{15}ClN_2$	7-chloro-2,3-dihydro-1-methyl-5-phenyl-1H-1,4-benzodiazepine C.A. 2898-12-6	Table II-4	DMF 10	PY	BR	-	-	2.0	-	DME// Ag/AgCl	O4AO	C=0.1(?)
CONT													

TABLE I. Electrochemical Data   211

$C_{16}H_{15}ClN_2$ (CONT.) FD27

Ref.	C/M	Charact. Potential		Response Const.		n Tech.	n	Electrokinetic Data			Products and Identification	Description and Remarks	Code No.
		Value	vs.		Value			Parameter	Value	From			
EA020 0973	–	$E_{\frac{1}{2}}$	-1.21F	SCE	–	–	–	–	$dE_{\frac{1}{2}}/dpH$	27	sttd	–	C,r	FD21
C0031 1264	–	$E_{\frac{1}{2}}$	-0.68F	Ag/AgCl	–	–	sttd	2	p=	2	sttd	C=N → CH-NH(CP,UVS, PY)	C,p	FD22 df09
			-0.87F		–	–		2		2		C=N → CH-NH(CP,UVS, PY)	C,p	
			-0.97F		–	–		2		2		C=N → CH-NH(CP,UVS, PY)	C,p	
			-1.08		–	–		2		2		C=N → CH-NH(CP,UVS, PY)	C,p	
			-1.22F		–	–		2		2		C=N → CH-NH(CP,UVS, PY)	C,p	
JE055 0407	–	$E_{\frac{1}{2}}$	-1.6F	Hg	i_ℓ	1.0F	–	–	Elog	74	–	–	C,⇌,p	FD23
			-2.23F			0.85F				74			C,⇌	
			-2.45F			0.78F				85			C,⇌	
			-1.6F			1.0F	–	–		–	–	–	C,p	
			-2.23F			1.9F							C	
			-2.53F			0.3F							C	
			-1.6F			1F	–	–		–	–	–	C,p	
			-2.23F			2.8F							C	
			-2.6F			0.65F							C	
			-1.68F			1.23F	–	–		–	–	–	C,p	
			-2.23F			2.6F							C	
EA021 0345	434 a	E_p	1.56	Ag/AgI, I⁻ 0.1	i_p	14.8	sttd	2	–	–	–	–	A,⇌,r	FD24
EA021 0345	434 a	E_p	2.03	Ag/AgI, I⁻ 0.1	i_p	13.3	QE (2.1V)	1.87	–	–	–	phenazine 84%, NMR	A,⇌,r	FD25
EA019 0629	441 a	$E_{\frac{1}{2}}$	-2.02	SCE(o)	i_ℓ	5.0	QE (-2.5V)	2.1	–	–	–	CP(-2.5 V,[Bu₄NI]= 0.25) → 9,10-dihydro- anthracene 60%,9- ethylanthracene 25%	C,r	FD26
			-2.475			7.3		–						
			-1.95			8.8		2.0	–	–	–	9-(1-hydroxyethyl)- 9,10-dihydroanthra- cene 90%,9-ethyl- anthracene 5%	C,r	
			-2.475			5.2		–				–	C	
C0031 1264	–	$E_{\frac{1}{2}}$	-0.69F	Ag/AgCl	–	–	sttd	2	p=	2	sttd	7-chloro-1-methyl- 5-phenylperhydro- 1,4-benzodiazepine	C,p	FD27 df29

CONT

FD27 (CONT.) $C_{16}H_{15}ClN_2$

Code No.	Empirical Formula	Name and C.A. Number	Structural Formula	Solvent		Tech.	Medium		μ, M	pH	T, °C	Electrodes	App.	Experimental Parameters
FD27 df29	$C_{16}H_{15}ClN_2$	7-chloro-2,3-dihydro-1-methyl-5-phenyl-1H-1,4-benzodiazepine	Table II-4	DMF	10	PY	BR		-	-	4.9	-	DME// Ag/AgCl	O4AO	C=0.1(?)
										6.7				
										8.3				
										10.3				
FD28	$C_{16}H_{15}ClN_2O$	7-chloro-1,3,4,5-tetrahydro-1-methyl-5-phenyl-3H-1,4-benzodiazepin-2-one C.A. 2619-57-0	Table II	DMF	10	PY	BR		-	-	2-10	-	DME// Ag/AgCl	O4AO	C=0.1
FD29	$C_{16}H_{15}NO_2$	3-(3-methylphenyl)-3-oxo-N-phenylpropanamide C.A. 962-01-6	$3-CH_3C_6H_4COCH_2-CONHC_6H_5$	EtOH	40	PY	BR GEL	0.01	-	-	5.0	-	DME//SCE	5A-	C=ns,m=1.8, t=3.7,h=59
										8.0				
FD30	$C_{16}H_{15}NO_2$	3-(4-methylphenyl)-3-oxo-N-phenylpropanamide C.A. 3422-75-1	$4-CH_3C_6H_4COCH_2-CONHC_6H_5$	EtOH	40	PY	BR GEL	0.01	-	-	5.0	-	DME//SCE	5A-	C=ns,m=1.8, t=3.7,h=59
										8.0				
FD31	$C_{16}H_{15}NO_3$	3-(3-methoxyphenyl)-3-oxo-N-phenylpropanamide C.A. 38505-21-4	$3-CH_3OC_6H_4COCH_2-CONHC_6H_5$	EtOH	40	PY	BR GEL	0.01	-	-	5.0	-	DME//SCE	5A-	C=ns,m=1.8, t=3.7,h=59
										8.0				
FD32	$C_{16}H_{15}NO_3$	3-(4-methoxyphenyl)-3-oxo-N-phenylpropanamide C.A. 965-01-5	$4-CH_3OC_6H_4COCH_2-CONHC_6H_5$	EtOH	40	PY	BR GEL	0.01	-	-	5.0	-	DME//SCE	5A-	C=ns,m=1.8, t=3.7,h=59
										8.0				

TABLE I. Electrochemical Data

$C_{16}H_{15}NO_3$ FD32

Ref.	C/M	Charact. Potential		Response Const.		n Tech.		Electrokinetic Data			Products and Identification	Description and Remarks	Code No.
		Value	vs.		Value			Parameter	Value	From			
C0031 1264	-	$E_{\frac{1}{2}}$	-0.86F	Ag/AgCl	-	-	sttd	2	p=	2	sttd	7-chloro-1-methyl-5-phenylperhydro-1,4-benzodiazepine	C,p	FD27 df29
			-0.94F	-	-	-		2		2		7-chloro-1-methyl-5-phenylperhydro-1,4-benzodiazepine	C,p	
			-1.11F	-	-	-		2		2		7-chloro-1-methyl-5-phenylperhydro-1,4-benzodiazepine	C,p	
			-1.24F	-	-	-		2		2		7-chloro-1-methyl-5-phenylperhydro-1,4-benzodiazepine	C,p	
C0031 1264	-	-	-	-	-	-	-	-	-	-	-	-	C,0,p	FD28
EA021 0831	65c	$E_{\frac{1}{2}}$	-1.214	SCE	-	-	sttd	2	αn_a	1.17	sttd	-	C,$E_{\frac{1}{2}}$ Av(20) from Elog,r	FD29
			-1.386	-	-	-				0.98	-	-	C,$E_{\frac{1}{2}}$ Av(20) from Elog,pK'=9.60, $E_{\frac{1}{2}}$=-1.379+0.28σ, i_ℓ=kC, i_ℓ=kh$^{\frac{1}{2}}$ for pH=3-10, i_d,r	
EA021 0831	65c	$E_{\frac{1}{2}}$	-1.315	SCE	-	-	sttd	2	αn_a	1.00	sttd	-	C,$E_{\frac{1}{2}}$ Av(20) from Elog,r	FD30
			-1.406	-	-	-		2		1.19		-	C,$E_{\frac{1}{2}}$ Av(20) from Elog,pK'=9.90, $E_{\frac{1}{2}}$=-1.379+0.28σ, i_ℓ=kC, i_ℓ=kh$^{\frac{1}{2}}$ for pH=3-10, i_d,r	
EA021 0831	65c	$E_{\frac{1}{2}}$	-1.214	SCE	-	-	sttd	2	αn_a	1.08	sttd	-	C,$E_{\frac{1}{2}}$ Av(20) from Elog,r	FD31
			-1.360	-	-	-		2		1.05		-	C,$E_{\frac{1}{2}}$ Av(20) from Elog,pK'=9.35, $E_{\frac{1}{2}}$=-1.379+0.28σ, i_ℓ=kh$^{\frac{1}{2}}$ for pH=3-10, i_d,r	
EA021 0831	65c	$E_{\frac{1}{2}}$	-1.293	SCE	-	-	sttd	2	αn_a	1.003	sttd	-	C,$E_{\frac{1}{2}}$ Av(20) from Elog, i_ℓ=kh$^{\frac{1}{2}}$ for pH=3-10, i_d,r	FD32
			-1.460	-	-	-		2		0.809		-	C,$E_{\frac{1}{2}}$ Av(20) from Elog,pK'=10.50, $E_{\frac{1}{2}}$=-1.379+0.28σ,r	

FD33 $C_{16}H_{16}N_2O_4$

Code No.	Empirical Formula	Name and C.A. Number	Structural Formula	Solvent	Tech.	Medium		μ, M	pH	T, °C	Electrodes	App.	Experimental Parameters
FD33	$C_{16}H_{16}N_2O_4$	4,4'-dinitro-1,4-diphenylbutane C.A. 41854-80-2	$(4-O_2NC_6H_4CH_2-CH_2)_2$	MeCN	XT	Et_4NClO_4	0.1	–	–	-30 ±1	HMDE// Ag/ AgClO₄ 0.01	2-2	C=ns, v=(1.5-5.5)E3
										20± 0.5			
				DMF	XT	Et_4NClO_4	0.1	–	–	10±1	HMDE// SCE	2-2	C=ns, v=(1.5-5.5)E3
										60± 0.5			
FD34	$C_{16}H_{16}O_4S$	1-(2-thienyl)-3-(2,3,4-trimethoxyphenyl)-2-propen-1-one C.A. 53094-47-6	Table II	EtOH 50	PY	NaOAc HCl	ns ns	0.2	0.77	25± 0.1	DME//SCE	03A0	C=0.2, m=1.86, t=3.8,h=70
FD35	$C_{16}H_{16}O_4S$	1-(2-thienyl)-3-(3,4,5-trimethoxyphenyl)-2-propen-1-one C.A. 53094-48-7	Table II	EtOH 50	PY	NaOAc HCl	ns ns	0.2	0.77	25± 0.1	DME//SCE	03A0	C=0.2, m=1.86, t=3.8,h=70
FD36	$C_{16}H_{17}N_3O_2$	4,4'-bis(acetylamino)diphenylamine C.A. 61236-15-5	$(4-CH_3CONH-C_6H_4)_2NH$	MeCN	VY	Et_4NClO_4	0.1	–	–	–	Pt: rodi// Ag/ AgClO₄ 0.01, Et_4NClO_4 0.1	2-0	C=ns,d=0.1, ω=10, Pt Aux
						Et_4NClO_4 2,4-lutidine	0.1 ns						
						ACET	ns						
FD37	$C_{16}H_{18}ClN_3O_4S$	Capri Blue C.A. 13018-79-6	Table II	MeOH 50	PY	MB	ns	–	7	–	DME//SCE	0--	C=0.05, m=2.61, t=3.0,h=50
FD38 bj40 df38	$C_{16}H_{18}ClN_3S$	Methylene Blue C.A. 61-73-4	Table II-2	MeOH 40	VA	BOR	ns	–	8.4	–	HMDE// SCE	OCO	0→-0.4→0V, C=0.04
						BOR erythrosine	ns 5E-5						
CONT													

$C_{16}H_{18}ClN_3S$ (CONT.) FD38

Ref.	C/M	Charact. Potential		Response Const.		n Tech.		Electrokinetic Data			Products and Identification	Description and Remarks	Code No.
		Value	vs.		Value			Parameter	Value	From			
JE047 0115	-	XE	0.041 ±0.002	Ag/ AgClO₄ 0.01	-	-	-	-	-	-	-	-	C,XT=semiintegral chronoamperometry with linear potential sweep,XE=difference between potentials for the two steps,p	FD33
			-										C	
			0.046 ±0.002		-	-	-	-	-	-	-	-	C,p	
			-										C	
JE047 0115	-	XE	0.040 ±0.002	SCE	-	-	-	-	-	-	-	-	C,XT=semiintegral chronoamperometry with linear potential sweep,XE=difference between potentials for the two steps,p	
			-										C	
			0.046 ±0.002		-	-	-	-	-	-	-	-	C,p	
			-										C	
JE053 0439	-	$E_{\frac{1}{2}}$	-0.520	SCE	-	-	-	-	-	-	-	-	$C,\Delta E_{\frac{1}{2}} \neq \rho\sigma,p$	FD34
			-0.850										C	
JE053 0439	-	$E_{\frac{1}{2}}$	-0.505	SCE	-	-	-	-	-	-	-	-	$C,\Delta E_{\frac{1}{2}} \neq \rho\sigma,p$	FD35
			-0.887										C	
EA021 0557	147 n	$E_{\frac{1}{2}}$	0.375	Ag/ AgClO₄ 0.01, Et₄NClO₄ 0.1	i_ℓ/C	5.1	sttd	1	-	-	-	radical cation,VIS	$A,\Delta E_{\frac{1}{2}}=0.46\sigma,r$	FD36
			0.88			-		1					A	
			0.25			-	-	-	-	-	-	-	A,r	
			-0.05			-	-	-	-	-	-	-	A,r	
EA019 0215	-	$E_{\frac{1}{2}}$	-0.36	SCE	I	2.45	sttd	2	Tomeš	30	sttd	-	$C,R,Mc(E_{\frac{1}{2}}),p$	FD37
JE046 0391	-	E_p	-0.18F	SCE	i_p	0.026F	-	-	-	-	-	-	C,i_a,r	FD38 bj40 df38
			-0.28F			0.047F							C	
			-0.23F			0.013F							A	
			-0.16F			0.017F							A,related to peak at -0.18 V	
			-0.3F			0.042F	-	-	-	-	-	-	C,p	
			-0.24F			0.016F							A	

CONT

FD38 (CONT.) $C_{16}H_{18}ClN_3S$

Code No.	Empirical Formula	Name and C.A. Number	Structural Formula	Solvent	Tech.	Medium	μ, M	pH	T, °C	Electrodes	App.	Experimental Parameters
FD38	$C_{16}H_{18}ClN_3S$	Methylene Blue	Table II-2	MeOH 45	PY	BOR ns	-	8.5	-	DME//SCE	OCO	C=0.2, t(c)=4.5
						BOR ns erythrosine 1.5E-4						
				MeOH 50	PY	Me_4NOH 0.1	-	-	-	DME//SCE	0--	C=0.05, m=2.61, t=3.0,h=50
FD39	$C_{16}H_{18}N_2O_4S$	benzylpenicillenic acid	Table II	EtOH 5	PY	CITR ns	-	2.3	22±1	DME//SCE	2AO	C=ns,m=2.57, t=3.0,h=78.6, Pt Aux
								4.0				
								6.6				
								8.5				
								11.6				
FD40	$C_{16}H_{18}O_2$	1,2-di(3-methoxyphenyl)ethane C.A. 36707-27-4	$(3-CH_3OC_6H_4CH_2)_2$	MeCN	VR	$NaClO_4$ 0.1	-	-	-	Pt: xxbe// SCE	2A2	0→1.8?→ 0.5→1.8 V, C=1,v=86
				CH_2Cl_2 80 CF_3COOH 20	VR	Bu_4NBF_4 0.25	-	-	-	Pt: xxbe// SCE	2A2	0→1.8→ 0.5→1.8 V, C=1,v=86
FD41	$C_{16}H_{19}NO_4$	2,2',4,4'-tetramethoxydiphenylamine C.A. 7093-78-9	Table II	MeCN	QE	Et_4NBF_4 0.1 2,6-lutidine ns	-	-	-	Pt: nsns// Ag/ $AgClO_4$ 0.01, Et_4NClO_4 0.1	25-0	C=ns,A=25, E_{app}=0.10→ 0.30 V
					VA	Et_4NClO_4 0.1 H_2O 0.005	-	-	-	Pt: xxdi// Ag/ $AgClO_4$ 0.01, Et_4NClO_4 0.1	2-0	0→1.0→ -0.5V,C=1.0, v=30,Pt Aux

CONT

TABLE I. Electrochemical Data   217

$C_{16}H_{19}NO_4$ (CONT.) FD41

Ref.	C/M	Charact. Potential		Response Const.	n Tech.	Electrokinetic Data			Products and Identification	Description and Remarks	Code No.
		Value	vs.	Value		Parameter	Value	From			
JE046 0391	-	$E_{\frac{1}{2}}$ -0.18F	SCE	i_ℓ 1.2F, Ap	-	-	-	-	-	C,i_a,S,p	FD38
		-0.26F		2.5F, Ap						C	
		-0.28F		3.1F	-	-	-	-	-	C,p	
EA019 0215	-	$E_{\frac{1}{2}}$ -0.35	SCE	I 2.57	sttd 2	Tomeš	31	sttd	-	$C,R,Mc(E_{\frac{1}{2}}),p$	
AA096 0143	-	$E_{\frac{1}{2}}$ -0.30F	SCE	-	-	$dE_{\frac{1}{2}}/dpH$	55	sttd	-	$A,Pr,i_a,Hg\ salt,p$	FD39
		-					-			$A,i_\ell=kC,Hg\ salt$	
		-0.40F		-	-		55		-	$A,Pr,i_a,Hg\ salt,p$	
		-					-			$A,i_\ell=kC,Hg\ salt$	
		-0.54F		-	-		-	-	-	$A,Pr,i_a,E_{\frac{1}{2}}=f(pH)$, Hg salt,p	
		-								A,Hg salt	
		-0.62F		-	-		-	-	-	$A,Pr,i_a,Hg\ salt,p$	
		-								A,Hg salt	
		-0.71F		-	-		-	-	-	$A,Pr,i_a,Hg\ salt,p$	
		-								A,Hg salt	
JA095 7132	437 a	E_p 1.62	SCE	-	-	-	-	-	-	$A,\rightleftarrows,r$	FD40
		1.07								A,R,0 on first scan	
		1.01								C,R,0 on first scan	
JA095 7132	437 a	E_p 1.73	SCE	-	-	-	-	-	green soln.,current reversal at 0.8 F/mol → 2,7-di-methoxy 9,10-dihy-drophenanthrene (93%,LSC) + starting material(brown soln.)	$A,\rightleftarrows,r$	
		1.22								A,R,0 on first scan	
		1.16								C,R,0 on first scan	
EA021 0557	147 q	-	-	-	-	-	>1.7	-	-	heptamethoxy-5,10-diphenyl-5,10-phenazine ($E_{\frac{1}{2}}$= -0.27 V)	A,r	FD41
EA021 0557	147 q	E_p 0.27F	Ag/Ag+ 0.01	i_p 30F	QE (0.4V) 1	ΔE_p	62	sttd	radical cation 100%, ESR	A,r	
		0.78F		23F	QE (1.0V) 1		65		bis(2,4-dimethoxy-phenyl)amine cation(Ar_2N^+)	A,R	
		0.71F		23F	-		-			C	
		0.16F		27F						C	

CONT

FD41 (CONT.) $C_{16}H_{19}NO_4$

Code No.	Empirical Formula	Name and C.A. Number	Structural Formula	Solvent	Tech.	Medium	μ, M	pH	T, °C	Electrodes	App.	Experimental Parameters
FD41	$C_{16}H_{19}NO_4$	2,2',4,4'-tetra-methoxydiphenylamine	Table II	MeCN	VA	Et_4NClO_4 0.1 H_2O 0.005 4-cyanopyridine 0.001	-	-	-	Pt: xxdi // Ag/AgClO$_4$ 0.01, Et$_4$NClO$_4$ 0.1	2-O	0→1.0→0.5V, C=1.0, v=30, Pt Aux
						Et_4NClO_4 0.1 H_2O 0.005 2,6-lutidine 0.002						0→0.7→-0.5V
					VY	ACET ns	-	-	-	Pt: rodi // Ag/AgClO$_4$ 0.01, Et$_4$NClO$_4$ 0.1	2-O	C=ns, Pt Aux
						Et_4NClO_4 0.1 H_2O 0.005						C=0.1
FD42	$C_{16}H_{20}Cr$	bis(η^6-xylene)-chromium(0) C.A. 12171-29-8	$(CH_3C_6H_4CH_3)_2Cr$	DMF	VY	Et_4NClO_4 1 GEL 0.04?	-	-	-	Pt: pdhb // SCE	3A-	C=1?, Pt Aux
FD43	$C_{16}H_{22}O_4$	dibutyl phthalate C.A. 84-74-2	Table II	EtOH 40	PY	BR	-	8.3 8.7 9.6 10.0	20	DME // SCE	O4AO	C=ns, m=1.77, t=5, h=60
FD44	$C_{16}H_{26}O_3$	2-(1,1-dimethyl-ethyl)-5-(2-ethoxy-2-methylpropyl)-1,4-hydroquinone C.A. 3490-43-5	Table II	EtOH 50	PY	ACET 0.2	-	-	25±0.6	DME // SCE	O4AO	ns
FD45	$C_{16}H_{26}O_3$	2-(1,1-dimethyl-ethyl)-6-(2-ethoxy-2-methylpropyl)1,4-hydroquinone C.A. 3490-39-9	Table II	EtOH 50	PY	ACET 0.2	-	-	25±0.6	DME // SCE	O4AO	ns
FD46	$C_{16}H_{26}O_3$	2-(1,1-dimethyl-propyl)-5-(2-methoxy-2-methylpropyl)-hydroquinone	Table II	EtOH 50	PY	ACET 0.2	-	-	25±0.6	DME // SCE	O4AO	ns
FD47	$C_{16}H_{26}O_4$	2,5-bis(2-methoxy-2-methylpropyl)-1,4-hydroquinone	Table II	EtOH 50	PY	ACET 0.2	-	-	25±0.6	DME // SCE	O4AO	ns

TABLE I. Electrochemical Data

$C_{16}H_{26}O_4$ FD47

Ref.	C/M	Charact. Potential		Response Const.		n Tech.		Electrokinetic Data			Products and Identification	Description and Remarks	Code No.
		Value	vs.		Value			Parameter	Value	From			
EA021 0557	147 q	E_p	0.23F	Ag/Ag$^+$ 0.01	i_p	30F	-	-	-	-	-	-	A,r	FD41
			0.60			15F			ΔE_p	220	sttd		A	
			0.37F			~6F							C	
			0.16F			29F							C	
			0.17F			60F	QE (-0.40V)	2		100		bis(2,4-dimethoxy-phenyl)amine cation(Ar_2N^+)	A,($E_{\frac{1}{2}}$=0.12 VY)	
							sttd	2						
			0.09F			60		-					C	
EA021 0557	147 q	$E_{\frac{1}{2}}$	-0.04	Ag/ AgClO$_4$ 0.01, Et$_4$NClO$_4$ 0.1	-	-	QE	2	-	-	-	bis(2,4-dimethoxy-phenyl)amine cation(Ar_2N^+)	A,r	
			0.18		-	-	i:i	1	-	-	-	-	A,r	
			0.73					-					A	
			1.22										A	
JE051 0226	443 a	$E_{\frac{1}{2}}$	-0.95	SCE	-	-	-	-	Elog	68	-	-	A,QI,p	FD42
			0.75							226			A,≠	
									βn_b	0.26	Elog			
C0028 1985	438 a	-	-	-	i/i_d	0.60F	sttd	4	-	-	-	-	C,p	FD43
		-	-	-		0.64F		4	-	-	-	-	C,p	
		-	-	-		0.80F		4	-	-	-	-	C,p	
		$E_{\frac{1}{2}}$	-1.84	SCE		0.90F		4	-	-	-	-	C,$\Delta E_{\frac{1}{2}}$=0.39σ^*,p	
C0032 2140	-	$E_{\frac{1}{2}}^O$	0.612	NHE	-	-	-	-	-	-	-	-	A,$E_{\frac{1}{2}}$→from data for pH=3.8-5.6,$\Delta E_{\frac{1}{2}}$= -0.07σ^*,p	FD44
C0032 2140	-	$E_{\frac{1}{2}}^O$	0.617	NHE	-	-	-	-	-	-	-	-	A,$E_{\frac{1}{2}}$→from data for pH=3.8-5.6,p	FD45
C0032 2140	-	$E_{\frac{1}{2}}^O$	0.604	NHE	-	-	-	-	-	-	-	-	A,$E_{\frac{1}{2}}$→from data for pH=3.8-5.6,p	FD46
C0032 2140	-	$E_{\frac{1}{2}}^O$	0.649	NHE	-	-	-	-	-	-	-	-	A,$E_{\frac{1}{2}}$→from data for pH=3.8-5.6,p	FD47

FD48 $C_{17}H_{10}O_4$

Code No.	Empirical Formula	Name and C.A. Number	Structural Formula	Solvent	Tech.	Medium		μ, M	pH	T, °C	Electrodes	App.	Experimental Parameters
FD48	$C_{17}H_{10}O_4$	2-(3,4-methylene-dioxybenzylidene)-1,3-indandione C.A. 29874-34-8	Table II	EtOH 20	PY	GLYC	ns	0.1	9.5	-	DME//SCE	04AO	C=0.1
FD49	$C_{17}H_{11}ClO_3$	3-[5-(4-chlorophenyl)-2-furyl]-1-(2-furyl)-2-propen-1-one C.A. 20005-35-0	Table II	DMF	PY	Bu_4NClO_4	0.1	-	-	-	DME//SCE	126-0	C=1
						$LiClO_4$	0.1						
					PK	Bu_4NClO_4	0.1	-	-	-	DME//SCE	126-0	C=ns, E_{app}=-1.5 V, t=3-f=12.5
FD50	$C_{17}H_{11}NO_5$	1-(2-furyl)-3-[5-(4-nitrophenyl)-2-furyl]-2-propen-1-one C.A. 20005-37-2	Table II	DMF	PY	Bu_4NClO_4	0.1	-	-	-	DME//SCE	126-0	C=1
						$LiClO_4$	0.1						
					PK	Bu_4NClO_4	0.1	-	-	-	DME//SCE	126-0	C=1, E_{app}= -1.1 V?, f=53.5/t
						$LiClO_4$	0.1						f=46.9/t
FD51	$C_{17}H_{12}OS$	3-(1-naphthyl)-1-(2-thienyl)-2-propen-1-one C.A. 20894-65-9	Table II	EtOH 50	PY	NaOAc HCl	ns ns	-	0.65	25± 0.1	DME//SCE	03AO	C=ns, m=1.86, t=3.8, h=70
						PHOS	ns		6.7				
FD52	$C_{17}H_{12}O_2$	2-(4-methylbenzylidene)-1,3-indandione C.A. 15875-51-1	Table II	EtOH 20	PY	GLYC	ns	0.1	9.5	-	DME//SCE	04AO	C=0.1
FD53	$C_{17}H_{12}O_3$	1-(2-furyl)-3-(5-phenyl-2-furyl)-2-propen-1-one C.A. 40940-98-5	Table II	DMF	PY	Bu_4NClO_4	0.1	-	-	-	DME//SCE	126-0	C=1
						$LiClO_4$	0.1						
CONT					PK	Bu_4NClO_4	0.1	-	-	-	DME//SCE	126-0	C=1, E_{app}= -1.5 V, t=3, f=12.5

TABLE I. Electrochemical Data

$C_{17}H_{12}O_3$ (CONT.) FD53

Ref.	C/M	Charact. Potential		Response Const.		n Tech.		Electrokinetic Data			Products and Identification	Description and Remarks	Code No.
		Value	vs.	Value				Parameter	Value	From			
C0036 1406	-	$E_{\frac{1}{2}}$	-1.300	SCE	-	-	i:i	4	-	-	-	-	$C,i_d, \Delta E_{\frac{1}{2}}=0.14\sigma$, redn of C=C, p	FD48
			-				-						C	
													C(?)	
JE050 0351	-	$E_{\frac{1}{2}}$	-1.250	SCE	-	-	i:i	1	Elog	55	-	CP→ radical anion, ESR	C,W,i_d,p	FD49
			-1.830					-		69			C,W,i_d	
			-2.7							-			C	
			-1.110		-	-		1		54	-	-	C,W,i_d,p	
			-1.585					-		60			C,W,i_d	
JE050 0351	-	$E_{\frac{1}{2}}$	-1.25	SCE	XR	2.12	-	-	-	-	-	-	$A,XR=i_\ell,A,PK/i_\ell,C,PY,p$	
JE050 0351	-	$E_{\frac{1}{2}}$	-0.900	SCE	-	-	-	-	Elog	43	-	CP→ radical anion, ESR	C,p	FD50
			-1.125							56			C	
			-0.905		-	-	-	-		-		CP→ radical anion	C,p	
			-										C	
JE050 0351	-	$E_{\frac{1}{2}}$	-0.9	SCE	XR	3.74	-	-	-	-	-	-	$A,XR=i_\ell,A,PK/i_\ell,C,PY,p$	
			-			2.33	-	-	-	-	-	-	A,p	
JE053 0439	156 k	$E_{\frac{1}{2}}$	-0.46F	SCE	i_ℓ	0.4F	sttd	1	-	-	-	-	C,p	FD51
			-0.86F			0.4F, Ap		1					C,M	
			-0.94F			0.4F		1					C,R,i_d,p	
			-1.21F			0.45F		1					C,R,i_d	
			-1.45F			0.55F		1					$C,\not{v},i_k$	
C0036 1406	-	$E_{\frac{1}{2}}$	-1.285	SCE	-	-	i:i	4	-	-	-	-	$C,i_d,\Delta E_{\frac{1}{2}}=0.14\sigma$, redn of C=C, p	FD52
			-					-					C	
			-										C(?)	
JE050 0351	-	$E_{\frac{1}{2}}$	-1.260	SCE	-	-	i:i	1	Elog	63	-	CP→ radical anion, ESR	C,W,i_d,p	FD53
			-1.845					-		85		-	C,W,i_d	
			-2.7							-			C	
			-1.120					1		55	-		C,W,i_d,p	
			-1.525					-		63			C,W,i_d	
JE050 0351	-	$E_{\frac{1}{2}}$	-1.26	SCE	XR	2.41	-	-	-	-	-	-	$A,XR=i_\ell,A,PK/i_\ell,C,PY,p$	CONT

FD53 (CONT.) $C_{17}H_{12}O_3$

Code No.	Empirical Formula	Name and C.A. Number	Structural Formula	Solvent	Tech.	Medium		μ, M	pH	T, °C	Electrodes	App.	Experimental Parameters
FD53	$C_{17}H_{12}O_3$	1-(2-furyl)-3-(5-phenyl-2-furyl)-2-propen-1-one	Table II	DMF	PK	Bu_4NClO_4 H_2O	0.1 5%	-	-	-	DME//SCE	126-0	C=1, E_{app}=-1.5 V, t=3, f=12.5
						$LiClO_4$	1E-3						
FD54	$C_{17}H_{12}O_3$	2-(4-methoxybenzylidene)-1,3-indandione C.A. 7421-76-3	Table II	EtOH 20	PY	GLYC	ns	0.1	9.5	-	DME//SCE	O4AO	C=0.1
FD55	$C_{17}H_{12}O_4$	2-(benzoyloxy)-3-methoxy-1H-inden-one C.A. 59410-65-0	Table II	H_2O	PY	MB	0.3	-	2.5	22	DME//SCE	2-0	C=1, h=50
									3.3				
									4.2				
									5.0				
									6.7				
									7.1				
									8.0				
FD56	$C_{17}H_{12}S_3$	2,5-diphenyl-3,3a,4-trithiapentalene C.A. 1033-90-5	Table II	MeCN 50 CH_2Cl_2 50	IL	$NaClO_4$	0.1	-	-	-	Pt: xxbu// SCE	---	C=ns
FD57	$C_{17}H_{16}O_2$	1,3-dibenzoylpropane C.A. 6263-83-8	$C_6H_5CO(CH_2)_3CO$-C_6H_5	MeCN 98	PW	Et_4NClO_4	0.1	-	-	-	DME// Ag/Ag^+ 0.01	12-02	C=0.3, v=170
													v=1E4
				MeCN 99.9	PY	Et_4NClO_4	0.1	-	-	-	DME// Ag/Ag^+ 0.01	12-0	C=2
					PW	Et_4NClO_4	0.1	-	-	-	DME// Ag/Ag^+ 0.01	12-02	C=0.3, v=170
CONT													v=1E4

TABLE I. Electrochemical Data

$C_{17}H_{16}O_2$ (CONT.) FD57

| Ref. | C/M | Charact. | Potential | Response | Const. | n | Electrokinetic Data | | | Products and | Description and | Code |
		Value	vs.		Value	Tech.	Parameter	Value	From	Identification	Remarks	No.
JE050 0351	-	$E_{\frac{1}{2}}$	-1.26	SCE	XR	1.41	-	-	-	-	-	A,XR=$i_{\ell,A}$,PK/ $i_{\ell,C}$,PY,p	FD53
			-1.26			0.83	-	-	-	-	-	-	A,p	
C0036 1406	-	$E_{\frac{1}{2}}$	-1.315	SCE	-	-	i:i	4	-	-	-	-	C,i_d,$\Delta E_{\frac{1}{2}}$=0.14σ,redn of C=C,p	FD54
			-					-					C	
													C(?)	
EA020 0973	156 g	$E_{\frac{1}{2}}$	-0.64F	SCE	-	-	-	-	$dE_{\frac{1}{2}}$/dpH	62F	-	-	C,r	FD55
			-0.68F		-	-	-	-		62F	-	-	C,r	
			-0.85F				QE	2.01		53F			C,$dE_{\frac{1}{2}}$/dpH for pH=3.3-5.0	
			-0.76F		-	-	-	-		62F	-	-	C,r	
			-0.92F							53F			C	
			-1.11F							37F			C	
			-0.80F		-	-	-	-		62F	-	-	C,r	
			-0.94							0F			C,$dE_{\frac{1}{2}}$/dpH for pH > 5	
			-1.13F							37F			C	
			-0.89F		-	-	-	-		62F	-	-	C,r	
			-0.95F							0F			C,0 for pH > 7	
			-1.18F							37F			C	
			-0.94F		-	-	-	-		62F	-	-	C,r	
			-1.21F							37F			C	
			-0.98F		-	-	-	-		62F	-	-	C,r	
			-1.25F							37F			C	
JE038 0479	-	E_p	0.87	SCE	-	-	QE	1	-	-	-	-	A,p	FD56
JE053 0407	440 a	E_p	-2.118	Ag/Ag$^+$ 0.01	-	-	-	-	-	-	-	-	C,E_p≠f(C) for C=0.3-2,p	FD57
			-2.15		-	-	-	-	-	-	-	-	C,E_p≠f(C) for C=0.3-2,p	
JE053 0407	440 a	$E_{\frac{1}{2}}$	-2.2	Ag/Ag$^+$ 0.01	-	-	QE	2± 0.1	-	-	-	CP→cis-1,2-diphenyl-1,2-cyclopentane-diol,MAS,NMR	C,≠,ec,p	
JE053 0407	440 a	E_p	-2.234	Ag/Ag$^+$ 0.01	-	-	-	-	-	-	-	-	C,E_p≠f(C) for C=0.3-2,p	
			-2.269		-	-	-	-	-	-	-	-	C,E_p≠f(C) for C=0.3-2,p	CONT

FD57 (CONT.) $C_{17}H_{16}O_2$

Code No.	Empirical Formula	Name and C.A. Number	Structural Formula	Solvent	Tech.	Medium		μ, M	pH	T, °C	Electrodes	App.	Experimental Parameters
FD57	$C_{17}H_{16}O_2$	1,3-dibenzoylpropane	$C_6H_5CO(CH_2)_3CO-C_6H_5$	MeCN 99.9	PW	Et_4NClO_4 $LiClO_4$	0.1 0.01	-	-	-	DME// Ag/Ag^+ 0.01	12-02	C=1, v=100
													v=1E4
				MeCN	PW	Bu_4NOH	0.01	-	-	-	DME// Ag/Ag^+ 0.01	12-02	C=0.475-5, v=600
													v=6000
FD58	$C_{17}H_{18}O_3$	2,3,7-trimethoxy-9,10-dihydrophenanthrene C.A. 51095-51-3	Table II	MeCN	VA	$NaClO_4$	0.1	-	-	-	Pt: xxbe// SCE	2A2	0→1.1→0 V, C=ns, v=86
													0→1.4→0 V
FD59	$C_{17}H_{19}N_3O$	1-benzhydryl-4-nitrosopiperazine C.A. 1698-25-5	Table II	MeOH 10	PT	BR	-	0.12	1.5	-	DME//SCE	2A0	C=ns, m=2.57, t=3.46, h=55
									4				
									6.0				
									7.0				
									11.0				
FD60	$C_{17}H_{20}O_3$	3,3',4-trimethoxybibenzyl	Table II	MeCN	IM	$NaClO_4$	0.1	-	-	-28	Pt: xxbe// SCE	2A2	1.22(30s)→ 0.6 V, C=ns, v=86
													1.42(30s)→ 0.6 V
													1.62(30s)→ 0.6 V
					VR	$NaClO_4$	0.1	-	-	-28	Pt: xxbe// SCE	2A2	0→1.8→ 0.6 → 1.1 V
					VY	$NaClO_4$	0.1	-	-	-28	Pt: rons// SCE	2A0	ns
CONT													

TABLE I. Electrochemical Data

$C_{17}H_{20}O_3$ (CONT.) FD60

Ref.	C/M	Charact.	Potential	Response	Const.	n	Electrokinetic Data			Products and	Description and	Code
		Value	vs.		Value	Tech.	Parameter	Value	From	Identification	Remarks	No.
JE053 0407	440 a	E_p	-1.88F	Ag/Ag$^+$ 0.01	-	-	-	-	$dE_p/dlogv$	37	-	-	C,p	FD57
			-1.953F		-	-	-	-		37	-	-	C,p	
JE053 0407	440 a	E_p	-2.23	Ag/Ag$^+$ 0.01	-	-	-	-	-	-	-	-	C,p	
			-2.25		-	-	-	-	-	-	-	-	C,p	
JA095 7132	437 b	E_p	0.99F	SCE	$i_p(u)$	14F	-	-	-	-	-	-	A,R,r	FD58
			0.93F			8F	-	-	-	-	-	-	C,R	
			0.99F			14F	-	-	-	-	-	-	A,R,r	
			1.26F			15F							A,⊬	
			0.93F			8F							C,R	
AA094 0119	82f	$E_{\frac{1}{2}}$	-0.60F	SCE	-	-	sttd	4	$dE_{\frac{1}{2}}/dpH$	80	sttd	-	C,W,r	FD59
			-0.92		-	-		4		80		-	C,r	
									Elog	60 and 27.5				
			-0.97		-	-		4	$dE_{\frac{1}{2}}/dpH$	80		-	C,pK'=6.4,W,r	
	82g		-1.02		-	-		2		37.5		-	C,r	
			-1.16		-	-		2		37.5		-	C,Mx,r	
JA095 7132	437 a	E_p	1.12F	SCE	$i_p(u)$	2.9F	-	-	-	-	-	-	C,r	FD60
			0.85F			0.1F							C	
			1.12F			0.9F	-	-	-	-	-	-	C,r	
			0.85F			0.6F							C	
			1.12F			0.8F	-	-	-	-	-	-	C,r	
			0.86F			4.5F							C	
JA095 7132	437 a	E_p	1.22F	SCE	$i_p(u)$	18F	-	-	-	-	-	-	A,r	
			1.30F			0.5F, Ap							A	
			1.63F			19F							A	
			1.50F			3F							C	
			1.13F			3F							C	
			0.83F			2F							C,R	
			0.90			2F							A,R,0 on first scan	
JA095 7132	437 a	$E_{\frac{1}{2}}$	1.20F			10F	-	-	-	-	-	-	A,r	
			1.59F			12F							A	

CONT

FD60 (CONT.) $C_{17}H_{20}O_3$

Code No.	Empirical Formula	Name and C.A. Number	Structural Formula	Solvent	Tech.	Medium		μ, M	pH	T, °C	Electrodes	App.	Experimental Parameters
FD60	$C_{17}H_{20}O_3$	3,3',4-trimethoxy-bibenzyl	Table II	CH_2Cl_2 80 CF_3COOH 20	QP	Bu_4NBF_4	0.1	-	-	-25	Pt: xfgz// SCE	5A-	C=20,A=75, j=3E-5, Pt Aux
													j=0.1
FD61	$C_{17}H_{28}O_3$	2-(1,1-dimethyl-ethyl)-5-[2-methyl-2-(1-methylethoxy)-propyl]-1,4-hydro-quinone C.A. 17208-10-5	Table II	EtOH 50	PY	ACET	0.2	-	-	25± 0.6	DME//SCE	O4AO	ns
FD62	$C_{17}H_{28}O_3$	2-(1,1-dimethyl-propyl)-5-(2-ethoxy-2-methylpropyl)-1,4-hydroquinone	Table II	EtOH 50	PY	ACET	0.2	-	-	25± 0.6	DME//SCE	O4AO	ns
FD63	$C_{18}H_{10}O_6$	bi(2-hydroxy-1,3-dioxoinden-2-yl) C.A. 5103-42-4		H_2O	PY	buffer	ns	-	3.05	-	DME//SCE	2-0	C=0.2
													C=2
					PA	buffer	ns	-	5.40	-	DME//SCE	2-0?	-0.1→-0.6 →0.1 V,C=1, v=1E3
				DMF	PY	buffer	ns	-	2.13	-	DME//SCE	2-0	C=1
									3.10				
									3.81				
									4.60				
									5.42				
									5.92				
FD64	$C_{18}H_{12}N_6$	2,4,6-tri(2-pyridyl)-s-triazine C.A. 3682-35-7	Table II	MeCN	VA	Bu_4NBF_4	0.1	-	-	-	Pt: xxdi//Ag	2F-	0→-2.6→0 V, C=ns,Pt Aux
													0→2.1→0 V

TABLE I. Electrochemical Data   227

$C_{18}H_{12}N_6$ FD64

Ref.	C/M	Charact. Potential Value	vs.	Response Const. Value		Tech.	n	Electrokinetic Data Parameter	Value	From	Products and Identification	Description and Remarks	Code No.
JA095 7132	437 a	-	-	-	-	sttd	2	-	-	-	blue soln.: adding Zn + H_2O → 6,6'-bis(3,3',4-trimethoxybibenzyl) 46% + starting material 25% (yellow soln., LSC, NMR, MPT)	A,r	FD60
		-	-	-	-	sttd	6	-	-	-	starting material 39%, 6,6'-bis(3,3',-4-trimethoxybibenzyl) 12%, 2,3,7-trimethoxyphenanthrene 1%	A,r	
C0032 2140	-	$E_{\frac{1}{2}}^{o}$ 0.623	NHE	-	-	-	-	-	-	-	-	A, $E_{\frac{1}{2}}$ →→ from data for pH=3.8-5.6, $\Delta E_{\frac{1}{2}}$ = -0.07 σ*,p	FD61
C0032 2140	-	$E_{\frac{1}{2}}^{o}$ 0.606	NHE	-	-	-	-	-	-	-	-	A, $E_{\frac{1}{2}}$ →→ from data for pH=3.8-5.6, p	FD62
EA020 0981	-	$E_{\frac{1}{2}}$ 0.05F -0.05F	SCE	i_ℓ	0.8F 0.4	-	-	-	-	-	-	A,r C	FD63
		0.10F -0.05F			4.4F 2.1F	-	-	-	-	-	-	A,r C	
EA020 0981	-	$E_{\frac{1}{2}}$ -0.2F -0.2F 0.0F	SCE	i_p	2F 1.7F 3.7F	-	-	-	-	-	-	C,r A A	
EA020 0981	-	$E_{\frac{1}{2}}$ 0.13F 0.00F	SCE	i_ℓ	2.9F 2.9F	sttd	1.3 -	-	-	-	-	A,r C	
		0.10F -0.05F			4F 3.5F	-	-	-	-	-	-	A,r C	
		0.06F -0.07F			3.9F 4.0F	-	-	-	-	-	-	A,r C	
		0.03F -0.11F			2.8F 3.1F	-	-	-	-	-	-	A,Mx,r C	
		-0.04F -0.13F			2.7F 2.7F	-	-	-	-	-	-	A,Mx,r C	
		0.06Ap -0.14F			3.4F 2.5F	-	-	-	-	-	-	A,Mx,r C	
JA095 6582	-	E_p -1.56 -2.05	SCE	-	-	-	-	-	-	-	-	C,r C,≠	FD64
		-	-	-	-	-	-	-	-	-	-	A,0,r	

FD65 $C_{18}H_{13}ClN_2$

Code No.	Empirical Formula	Name and C.A. Number	Structural Formula	Solvent	Tech.	Medium	μ, M	pH	T, °C	Electrodes	App.	Experimental Parameters
FD65	$C_{18}H_{13}ClN_2$	5-phenylphenazinium chloride	Table II	MeOH 50	PY	Me_4NOH	0.1	-	-	-	DME//SCE	0--	C=0.05, m=2.61, t=3.0, h=50
FD66	$C_{18}H_{14}O_3$	1-(2-furyl)-2-methyl-3-(5-phenyl-2-furyl)-2-propen-1-one C.A. 51930-46-2	Table II	DMF	PY	Bu_4NClO_4	0.1	-	-	-	DME//SCE	12-0	C=1
					PK	Bu_4NClO_4	0.1	-	-	-	DME//SCE	---	C=ns, E_{app}=-1.5 V, t=3, f=12.5
FD67	$C_{18}H_{14}S_3$	2-(4-methylphenyl)-5-phenyl-3,3a,4-trithiapentalene C.A. 38443-42-4	Table II	MeCN 50 CH_2Cl_2 50	IL	$NaClO_4$	0.1	-	-	-	Pt: xxbu//SCE	---	C=ns
FD68	$C_{18}H_{15}ClN_4$	3,7-diamino-5-phenylphenazinium chloride C.A. 81-93-6	Table II	MeOH 50	PY	Me_4NOH	0.1	-	-	-	DME//SCE	0--	C=0.05, m=2.61, t=3.0, h=50
FD69 b117	$C_{18}H_{15}OP$	triphenylphosphine oxide C.A. 791-28-6	$(C_6H_5)_3PO$	DMF	PY	Bu_4NI	0.1	-	-	20	DME//SCE	2AO	C=0.75, Hg Aux
					PR	Bu_4NI	0.1	-	-	20	DME//SCE	2AO	-1.2→-2.9→-1.2→-2.9 V, C=0.75, Hg Aux
FD70	$C_{18}H_{15}O_4P$	triphenyl phosphate C.A. 115-86-6	$PO(OC_6H_5)_3$	MeOH 50	PO	KCl	0.5	-	-	-	DME//ns	--2	C=0.3
FD71 b118	$C_{18}H_{15}P$	triphenylphosphine C.A. 603-35-0	$(C_6H_5)_3P$	MeCN	VA	$NaClO_4$	0.2	-	-	25±0.1	Pt: xxbe//Ag/AgNO₃ 0.1	25AO	C=4, v=100
				DMF	PY	Bu_4NI	0.1	-	-	20	DME//SCE	2AO	C=0.75, Hg Aux
					PR	Bu_4NI	0.1	-	-	20	DME//SCE	2AO	-1.2→-2.9→-1.2→-2.9 V, C=0.75, Hg Aux
FD72	$C_{18}H_{16}AsClO_5$	hydroxytriphenylarsonium perchlorate C.A. 13787-49-0	$(C_6H_5)_3AsOH^+$ ClO_4^-	MeCN	PY	Et_4NClO_4	0.1	-	-	25±0.1	DME//Ag/AgNO₃ 0.1	25-0	C=5

CONT

TABLE I. Electrochemical Data

$C_{18}H_{16}AsClO_5$ (CONT.) FD72

Ref.	C/M	Charact. Potential		Response	Const. Value	n Tech.	n	Electrokinetic Data			Products and Identification	Description and Remarks	Code No.
		Value	vs.					Parameter	Value	From			
EA019 0215	-	$E_{\frac{1}{2}}$	-0.63	SCE	I	3.02	sttd	2	Tomeš	40	-	-	$C,\bar{\tau},Mc(E_{\frac{1}{2}}),p$	FD65
JE050 0351	-	$E_{\frac{1}{2}}$	-1.425	SCE	-	-	i:i	1	Elog	62	-	CP→ radical anion, ESR	C,W,i_d,p	FD66
			-1.850					-		80		-	C,W,i_d	
			-2.7										C	
JE050 0351	-	$E_{\frac{1}{2}}$	-1.42	SCE	XR	2.98	-	-	-	-	-	-	$A,XR=i_\ell,A,PK'/i_\ell,C,PY',p$	
JE038 0479	-	E_p	0.73	SCE	-	-	QE	1.0	-	-	-	-	A,p	FD67
EA019 0215	-	$E_{\frac{1}{2}}$	-0.65	SCE	I	2.28	sttd	2	Tomeš	30	sttd	-	$C,Mc(E_{\frac{1}{2}}),p$	FD68
EA020 0021	-	$E_{\frac{1}{2}}$	-2.5F	SCE	i_ℓ/C	2.0F	-	-	-	-	-	-	C,r	FD69 b117
			-2.8F			2.2F							C,M	
EA020 0021	-	E_p	-2.5F	SCE	i_p	1.6F	-	-	-	-	-	-	C,r	
			-2.9F			-							C,X	
			-2.5F			1.7F							A	
CZ016 0316	-	Q	0.70F	-	-	-	-	-	-	-	-	-	C,i_{cat},p	FD70
JE048 0425	-	E_p	0.83	Ag/ AgNO$_3$ 0.1	i_p	7.8F	QE	0.66 ± 0.03	-	-	-	triphenylphosphonium cation, triphenylphosphate, PY, IRS, UVS; CP with $[C_6H_6]$= 0.12→ triphenylphosphonium cation, triphenylphosphine oxide, tetraphenylphosphonium cation, UVS, IRS	$A,i_p=kC,r$	FD71 b118
			-0.68			3F		-					C, redn of triphenylphosphonium cation	
			-0.37			0.6F							A, related to peak at -0.68 V	
EA020 0021	-	$E_{\frac{1}{2}}$	-2.7F	SCE	i_p	1.6F	-	-	-	-	-	-	C,r	
EA020 0021	-	E_p	-2.8F	SCE	i_p	1.0F	-	-	-	-	-	-	C,r	
			-2.7F			1.0F							A	
JE052 0459	-	$E_{\frac{1}{2}}$	-1.15	Ag/ AgNO$_3$ 0.1	i_p	11.7F	QE (-1.33V)	0.66	-	-	-	CP→ $[(C_6H_5)_3AsO]_2 \cdot HClO_4$, IRS, PY, MAS; $(C_6H_5)_3As$, IRS, PY, MAS	C,r	FD72
			-1.6			12F	(-1.7V)	1				CP→ $(C_6H_5)_3As$, IRS; $(C_6H_5)_3AsO$, IRS	C	
			-2.69F			30F		-				-	C, redn of $(C_6H_5)_3AsO$	CONT

FD72 (CONT.) $C_{18}H_{16}AsClO_5$

Code No.	Empirical Formula	Name and C.A. Number	Structural Formula	Solvent	Tech.	Medium		μ, M	pH	T, °C	Electrodes	App.	Experimental Parameters
FD72	$C_{18}H_{16}AsClO_5$	hydroxytriphenyl-arsonium perchlorate	$(C_6H_5)_3AsOH^+$ ClO_4^-	MeCN	VA	Et_4NClO_4	0.1	-	-	25±0.1	Pt(Pt): xxfl// Ag/ AgNO$_3$ 0.1	25-O	C=2.5, A=0.15Ap, v=100, PE
					VY	Et_4NClO_4	0.1	-	-	25±0.1	Pt: pdfl// Ag/ AgNO$_3$ 0.1	25-O	C=4.7, A=0.15Ap
											glC: pdns// Ag/ AgNO$_3$ 0.1		C=5, A=0.1Ap
FD73	$C_{18}H_{20}Cl_2N_2O$	trans-azoxydi(1-chloro-1-phenyl)-2-propyl) C.A. 59190-83-9	Table II	EtOH 50	PY	KNO$_3$ HCl	0.1	-	0.8	-	DME//SCE	O-O	C=0.20
									1.3				
									2.8				
									4.5				
									5.3				
									6.5				
									7.8				
FD74	$C_{18}H_{22}O_4$	3,3',4,4'-tetra-methoxybibenzyl C.A. 5963-51-9	Table II	MeCN	VR	NaClO$_4$	0.1	-	-	-	Pt: xxbe// SCE	2A2	0→1.4→0.6→1.4V, C=1, v=86
				CH_2Cl_2 80 DF_3COOH 30	VR	Bu_4NBF_4	0.25	-	-	-	Pt: xxbe// SCE	2A2	C=1, v=86
FD75	$C_{18}H_{24}Cr$	bis(mesitylene)-chromium(0)	Table II	DMF	VY	Et_4NClO_4 GEL	1 0.04?	-	-	-	Pt: pdhb// SCE	3A-	C=1(?), Pt Aux
FD76	$C_{18}H_{30}O_3$	2-(2-tert-butoxy-2-methylpropyl)-5-tert-butylhydroquinone C.A. 17208-13-8	Table II	EtOH 50	PY	ACET	0.2	-	→→0	25±0.6	DME//SCE	O4AO	ns

TABLE I. Electrochemical Data

$C_{18}H_{30}O_3$ FD76

Ref.	C/M	Charact. Potential Value	vs.	Response Const. Value		n Tech.		Electrokinetic Data Parameter	Value	From	Products and Identification	Description and Remarks	Code No.
JE052 0459	-	E_p -0.57F	Ag/AgNO$_3$ 0.1	i_p	730F	-	-	-	-	-	-	C,r	FD72
		-0.86F			130F							C	
		-0.56F			400F							A	
		-0.22F			150F							A	
JE052 0459	-	$E_{\frac{1}{2}}$ -0.85	Ag/AgNO$_3$ 0.1	i_ℓ	138F	QE	0.5	-	-	-	CP→H$_2$,VY; [(C$_6$H$_5$)$_3$AsO]$_2$HClO$_4$, IRS,MPT,CHA	C,r	
		-1.24			138F		1				CP→H$_2$,VY; (C$_6$H$_5$)$_3$AsO,MPT,IRS	C	
		-2.66F			276F		-				-	C,X,M,redn of (C$_6$H$_5$)$_3$AsO	
		-1.7F			100F		-				CP→[(C$_6$H$_5$)$_3$AsO]$_2$· HClO$_4$,IRS,PY,MAS; (C$_6$H$_5$)$_3$As,IRS,PY, MAS	C,r	
		-2.15F			77F						CP→(C$_6$H$_5$)$_3$As,IRS; (C$_6$H$_5$)$_3$AsO,IRS	C	
		-			-						-	C,M	
EA021 0473	432 dh	$E_{\frac{1}{2}}$ -0.40F	SCE	i_ℓ	1.6F	sttd	-	$dE_{\frac{1}{2}}/dpH$	105F	-	-	C,$dE_{\frac{1}{2}}/dpH$ for pH=0.8-6.5,r	FD73
		-0.45F			1.6F		-		105F		-	C,r	
		-0.59F			4.8F		-		105F		-	C,Pl distorted by Mx,r	
		-0.75			2.9F		-		105F		-	C,Mx,r	
		-0.88F			2.0F		-		105F		-	C,Mx,r	
		-0.98F			0.5F		-		8F		-	C,$dE_{\frac{1}{2}}/dpH$ for pH>6.5,r	
		-0.99F			0.1F	-	-		8F		-	C,r	
JA095 7132	437 a	E_p 1.23	SCE	$i_p(u)$	47F	-	-	-	-	-	-	A,⇌,r	FD74
		0.89			4F							A,R,O on first scan	
		0.84			5F							C,R,O on first scan	
JA095 7132	437 a	E_p 1.42	SCE	-	-	QP	2	-	-	-	QP→2,3,6,7-tetra- methoxyphenanthrene 97%,LSC	A,⇌,r	
		1.10					-					A,R,O on first scan	
		1.04										C,R,O on first scan	
JE051 0226	443 a	$E_{\frac{1}{2}}$ -0.97	SCE	-	-	-	-	Elog	59	-	-	A,R,p	FD75
		0.9						βn_b	200 0.29	Elog		A,⇌	
C0032 2140	-	$E^0_{\frac{1}{2}}$ 0.6265	NHE	-	-	-	-	-	-	-	-	A,$E_{\frac{1}{2}}$→ from data at pH=3.8-5.6, $\Delta E_{\frac{1}{2}}$= -0.07 σ*,p	FD76

FD77 $C_{18}H_{30}O_4$

Code No.	Empirical Formula	Name and C.A. Number	Structural Formula	Solvent	Tech.	Medium	μ, M	pH	T, °C	Electrodes	App.	Experimental Parameters
FD77	$C_{18}H_{30}O_4$	2,5-bis(2-methyl-2-ethoxypropyl)hydroquinone C.A. 3490-44-6	Table II	EtOH 50	PY	ACET 0.2	-	→→0	25±0.6	DME//SCE	04A0	ns
FD78	$C_{18}H_{30}O_4$	2,6-bis(2-methyl-2-ethoxypropyl)hydroquinone C.A. 17208-16-1	Table II	EtOH 50	PY	ACET 0.2	-	→→0	25±0.6	DME//SCE	04A0	ns
FD79	$C_{18}H_{39}NO_2$	tetrabutylammonium acetate C.A. 10534-59-5	$(CH_3CH_2CH_2CH_2)_4$-N^+ $CH_3CO_2^-$	MeCN	PY	Bu_4NClO_4 0.05	-	-	25±0.1	DME//SCE	2A0	C=1,m=1.47, t=5.48,h=60, Pt Aux
						$LiClO_4$ 0.05						
FD80	$C_{19}H_{13}N_3O$	2-phenyl-3-(2-pyridylimino)-3H-indole 1-oxide C.A. 18852-47-6	Table II	DMF	PY	Bu_4NClO_4 0.03	-	-	17.5±0.1	DME//SCE (NaCl, aq)	23A0	C=0.4, m=1.71, t=2.9,h=80, Hg Aux
						Et_4NClO_4 0.1 $CH_2ClCOOH$ 0.018						C=0.55
					VA	Bu_4NClO_4 0.03	-	-	17.5±0.1	HMDE//SCE(NaCl, aq)	23A0	-0.2→-1.1 →-0.2 V, C=0.197, v=150, Hg Aux
FD81	$C_{19}H_{13}N_3O_2$	2-phenyl-3-(4-pyridylimino)-3H-indole 1,1-dioxide C.A. 19808-49-2		DMF	PY	Bu_4NClO_4 0.03	-	-	17.5±0.1	DME//SCE (NaCl, aq)	23A0	C=0.4, m=1.71, t=2.9,h=80, Hg Aux
						Bu_4NClO_4 0.03 $CH_2ClCOOH$ 6.1E-5			20.0±0.1			C=0.18
						Bu_4NClO_4 0.03 $CH_2ClCOOH$ 4.9E-4						
CONT												

TABLE I. Electrochemical Data

$C_{19}H_{13}N_3O_2$ (CONT.) FD81

Ref.	C/M	Charact. Potential		Response Const.		n Tech.	Electrokinetic Data			Products and Identification	Description and Remarks	Code No.
		Value	vs.		Value		Parameter	Value	From			
C0032 2140	–	$E_{\frac{1}{2}}^o$	0.656	NHE	–	–	–	–	–	–	–	A,$E_{\frac{1}{2}}\to\to$ from data for pH=3.8-5.6,p	FD77
C0032 2140	–	$E_{\frac{1}{2}}^o$	0.656	NHE	–	–	–	–	–	–	–	A,$E_{\frac{1}{2}}\to\to$ from data for pH=3.8,5-6,p	FD78
JE036 0179	–	$E_{\frac{1}{2}}$	-0.07F	SCE	i_ℓ	3.4F	–	–	–	–	mercury(II) acetate	A,p	FD79
			0.15F			1.5F						A	
			0.33F			1.9F					mercury(II) acetate	A,i_d+i_k,p	
JE041 0429	427 d	$E_{\frac{1}{2}}$	-0.919	SCE(NaCl, aq)	I	1.67	QE }2	Tomeš	64	–	–	C,i_d,Tc=1 for T=15-30,i_ℓ=kC for C=0.1-1,r	FD80
			-1.276			1.31			98			C,i_d,Tc=1 for T=15-30,i_ℓ=kC for C=0.1-1	
			-0.16F		i_ℓ	3.4F	2		–	–	CP→ 1-hydroxy-2-phenyl-3-(2-pyridyl)aminoindole, λ_{max}=307	C,Mx,i_d,r	
			-0.76F			0.8F	2.7 ±0.1				CP→ 1-hydroxy-2-phenyl-3-(2-pyridyl)aminoindole, λ_{max}=307;2-phenyl-3-(2-pyridyl)aminoindole	C,Mx,i_d,$i_{\ell,2}/i_{\ell,1}\neq$ f(t,T, or C)	
JE041 0429	427 d	E_p	-0.86F	SCE(NaCl, aq)	i_p	1.4F	QE }2	–	–	–	–	C,$E_p\neq f(v)$,r	
			-1.17F			1.1F		$dE_p/d\log v$	43			C,$i_p/v^{\frac{1}{2}}\neq f(v)$	
			-1.04F			small						A	
			-0.79F			1.1F						A,related to peak at -0.86 V,$E_p\neq f(v)$	
			-0.6F			0.4F						A	
JE041 0429	427 a	$E_{\frac{1}{2}}$	-0.787	SCE(NaCl, aq)	I	1.57	QE }2	Tomeš	64	–	–	C,i_d,Tc=1 for T=15-30,i_ℓ=kC for C=0.1-1,r	FD81
			-1.417			1.4			68			C,i_d,Tc=1 for T=15-30,i_ℓ=kC for C=0.1-1	
			-0.34F		i_ℓ	0.3F	–		–			C,r	
			-0.83F			0.45F						C	
			-1.18F			0.25F						C	
			-0.26F			1.1F	2		–	–	CP→ 1-hydroxy-2-phenyl-3-(4-pyridylimino)-1H-indole 1'-oxide,λ_{max}=305	C,Mx,r	

CONT

FD81 (CONT.) $C_{19}H_{13}N_3O_2$

Code No.	Empirical Formula	Name and C.A. Number	Structural Formula	Solvent	Tech.	Medium		μ, M	pH	T, °C	Electrodes	App.	Experimental Parameters
FD81	$C_{19}H_{13}N_3O_2$	2-phenyl-3-(4-pyridylimino)-3H-indole 1,1'-dioxide	Table II	DMF	VA	Bu_4NClO_4	0.03	-	-	17.5 ±0.1	HMDE// SCE(NaCl, aq)	23AO	C=0.285, v=150, Hg Aux
FD82	$C_{19}H_{16}$	4-benzylbiphenyl C.A. 613-42-3	$4-C_6H_5C_6H_4CH_2-C_6H_5$	DMF	PY	Bu_4NI	0.25	-	-	-	DME// SCE(o)	2-O	C=2
						Bu_4NI PHEN	0.25 0.01						
FD83	$C_{19}H_{16}CrIO$	benzophenonebenzene-chromium(I) iodide C.A. 52346-47-1	$C_6H_5COC_6H_5Cr-C_6H_6^+ \, I^-$	H_2O	PY	BR GEL	- 0.04	-	1.6	-	DME//SCE	24-O	C=1, m=1.88, h=64, Pt Aux
									5				
									8.1				
									11.5				
				EtOH 50	PY	BR GEL	- 0.04	-	2.5	-	DME//SCE	24-O	C=1, m=1.88, h=64, Pt Aux
									4.25				
									12.5				
				DMF	PY	Et_4NClO_4	0.1	-	-	-	DME//SCE	24-O	C=1, m=1.88, h=64, Pt Aux
FD84	$C_{19}H_{16}CrO$	benzophenonebenzene-chromium(0) C.A. 52445-45-1	$C_6H_5COC_6H_5Cr-C_6H_6$	DMF	PY	Et_4NClO_4	1	-	-	-	DME//SCE	24-O	C=1, m=1.88, h=64, Pt Aux
					VY	Et_4NClO_4 GEL	1 0.04(?)	-	-	-	DME//SCE	3A-	C=1?, Pt Aux

$C_{19}H_{16}CrO$ FD84

Ref.	C/M	Charact. Potential		Response	Const. Value	n Tech.		Electrokinetic Data			Products and Identification	Description and Remarks	Code No.
		Value	vs.					Parameter	Value	From			
JE041 0429	427 a	E_p	-0.78F	SCE(NaCl, aq)	i_p	2.2F	QE	2	-	-	-	-	$C, E_p \neq f(v)$ for v=40-600,R,r	FD81
			-1.11F			2.1F							$C, E_p \neq f(v)$ for v=40-600,R	
			-1.03F			small							A,related to peak at -1.11V, $E_p \neq f(v)$ for v=40-600,R	
			-0.70F			2.4F							A,related to peak at -0.78V, $E_p \neq f(v)$ for v=40-600,R	
EA019 0629	-	$E_{\frac{1}{2}}$	-2.67	SCE(o)	i_ℓ	5.2	-	-	-	-	-	-	C,r	FD82
			-2.65			14.4	QE (-2.8V)	5.8	-	-	-	1-(4-cyclohexenyl)-4-(phenylmethyl)-benzene 95%	C,r	
JE050 0359	444 a	$E_{\frac{1}{2}}$	-0.7F	SCE	-	-	-	-	-	-	-	-	C,p	FD83
			-0.84F										C	
			-0.95F										C	
			-0.7F		-	-	-	-	-	-	-	-	C,p	
			-1.06F										C	
			-1.16F										C	
			-0.7F		-	-	QE	1	-	-	-	CP→benzophenone-benzenechromium(0), PY	C,p	
			-1.15F					2				CP→bisbenzene-chromium(0),UVS	C	
			-1.47F		-	-	-	-	-	-	-	-	C,p	
JE050 0359	444 a	$E_{\frac{1}{2}}$	-0.67F	SCE	-	-	-	-	-	-	-	-	C,p	
			-0.8F										C	
			-1.0F										C	
			-0.67F		-	-	-	-	-	-	-	-	C,p	
			-0.95F										C	
			-0.67F		-	-	-	-	-	-	-	-	C,p	
			-1.49F										C	
JE050 0359	444 a	$E_{\frac{1}{2}}$	-0.59	SCE	i_ℓ	1.5F	QE	1	-	-	-	CP→benzophenone-benzenechromium(0), PY	C,R,i_d,Tc=1.8 for T=20-50, $i_\ell/C \neq f(C)$,p	
			-1.91			1.5F		1					$C,B, i_\ell/C \neq f(C)$	
			-2.01			1.5F		1				CP→bisbenzenechromium(0),PY	$C,B, i_\ell/C \neq f(C)$	
JE050 0359	444 a	$E_{\frac{1}{2}}$	-0.59	SCE	-	-	-	-	Elog	59	-	-	A,R,i_d,p	FD84
			-1.91							-			C,B	
			-2.01										C,B	
JE051 0226	444 a	$E_{\frac{1}{2}}$	-0.59F	SCE	-	-	-	-	Elog	59	-	-	A,R,p	

FD85 $C_{19}H_{16}NO_6P$

Code No.	Empirical Formula	Name and C.A. Number	Structural Formula	Solvent	Tech.	Medium	μ, M	pH	T, °C	Electrodes	App.	Experimental Parameters
FD85 bm36	$C_{19}H_{16}NO_6P$	diphenyl 4-nitrobenzyl phosphate C.A. 10577-64-7	$4\text{-}O_2NC_6H_4CH_2O\text{-}P(:O)(OC_6H_5)_2$	MeCN	PY	Et_4NClO_4 0.1	-	-	22±2	DME//SCE	125A0	C=ns, Pt Aux
				DMF	PY	Et_4NClO_4 0.1	-	-	22±2	DME//SCE	125A0	C=ns, Pt Aux
					VR	Et_4NClO_4 0.1	-	-	22±2	Pt(Hg): xxwi// SCE	125A02	0.6→-2.0→ 0.6→-2.0V, A=0.026, Pt Aux
												C=1.04, E_T = -1.60 V, v=215, Pt Aux
FD86	$C_{19}H_{16}O$	(4-biphenylyl)benzyl alcohol C.A. 7598-80-3	$4\text{-}C_6H_5C_6H_4CHOH C_6H_5$	DMF	PY	Et_4NBr 0.1	-	-	-	DME// SCE(o)	2-0	C=2
						Et_4NBr 0.1 PHEN 0.01						
FD87 bm38	$C_{19}H_{16}O$	triphenylmethanol C.A. 76-84-6	$(C_6H_5)_3COH$	DMF	PY	Bu_4NI 0.1	-	-	-	DME// SCE(o)	2-0	C=2
FD88	$C_{19}H_{16}S_3$	2,5-bis(4-methylphenyl)-3,3a,4-trithiapentalene	Table II	MeCN 50 CH_2Cl_2 50	IL	$NaClO_4$ 0.1	-	-	-	Pt: xxbu// SCE	---	ns
FD89	$C_{19}H_{19}BrN_2S_2$	3-ethyl-2-[(3-ethyl-2(3H)-benzothiazolylidene)methyl]-benzothiazolium bromide C.A. 17095-31-7	Table II	MeOH 40	IL	KCl 1.2	-	-	-	g(Nujol): xxns// SCE	0--	A=0.126
				MeOH 50	PY	Me_4NOH 0.1	-	-	-	DME//SCE	0--	C=0.05, m=2.61, t=3, h=50
FD90 bm49	$C_{19}H_{19}N_7O_6$	folic acid C.A. 59-30-3	Table II	H_2O	PY	ACET 0.1	-	5.5	25± 0.1	DME// Ag/AgCl, KCl satd	2A0	C=0.005, W Aux
												C=0.050

CONT

TABLE I. Electrochemical Data

$C_{19}H_{19}N_7O_6$ (CONT.) FD90

Ref.	C/M	Charact. Potential		Response Const.		n Tech.	Electrokinetic Data			Products and Identification	Description and Remarks	Code No.
		Value	vs.		Value		Parameter	Value	From			
EA018 1025	187 c	$E_{\frac{1}{2}}$ -0.880 ±0.005	SCE	-	-	QE 1	Tomeš	72	-	CP→ 4,4'-dinitrobibenzyl 82-88%, GLC; diphenylphosphate, 94-100%, pptn as Fe^{2+} salt, IRS; nitrotoluene, ≥ 0.15%, GLC, VR($E_{p,C}$= -1.33 V, $E_{p,A}$= -1.25 V)	C,≠,W,i_ℓ=0.92$h^{\frac{1}{2}}$,i_d, ece,r	FD85 bm36
		-1.210 ±0.005				-		-		CP(-1.35 V) → 4,4'-dinitrobibenzyl radical anion, ESR	C,W,i_ℓ=0.84$h^{\frac{1}{2}}$,i_d	
EA018 1025	187 c	$E_{\frac{1}{2}}$ -0.810 ±0.005	SCE	-	-	QE 1	Tomeš	60	sttd	-	C,r	
		-1.160 ±0.005				-		-		CP(-1.35 V) → 4-nitrotoluene radical anion, ESR	C	
EA018 1025	187 c	E_p -1.01 ±0.005	SCE	-	-	-	-	-	-	-	C,r	
		-1.24 ±0.005									C	
		-1.14 ±0.005									A	
		-1.06 ±0.005		i_p	7.3F	QE 1	-	-	-	CP(-1.00 V) → 4,4'-dinitrobibenzyl, VR	C,i_p=steady-state values,r	
		-1.27 ±0.005			14.6F	-					C	
		-1.18 ±0.005			14.6F						A	
EA019 0629 (EA020 0143)	33e	$E_{\frac{1}{2}}$ -2.66	SCE(o)	i_ℓ	10.8	QE (-2.75V)	2Ap	-	-	-	QE(-2.75 V,[Bu_4NI] = 0.25) → 4-benzylbiphenyl 90%	C,r	FD86
		-2.58			21.8	-	-	-	-	-	C,r	
EA019 0629	-	$E_{\frac{1}{2}}$ -2.81	SCE(o)	-	-	QE (-2.90V)	1.95	-	-	-	triphenylmethane 95%	C,r	FD87 bm38
JE038 0479	-	E_p 0.69	SCE	-	-	QE 1	-	-	-	-	A,p	FD88
EA019 0215	-	$E_{p/2}$ 1.18	SCE	-	-	sttd 1	-	-	-	-	A,R,("$E_{\frac{1}{2}}$"=1.21c),p	FD89
		"$E_{\frac{1}{2}}$" -1.28				-					C,Mc($E_{\frac{1}{2}}$)	
EA019 0215	-	$E_{\frac{1}{2}}$ -1.48	SCE	I	2.54	sttd 2	-	-	-	-	C,i_d+i_a,p	
		-1.79			-	2					C	
AA096 0345	-	$E_{\frac{1}{2}}$ -0.646	Ag/AgCl, KCl satd	i_ℓ/C	5.40	-	-	-	-	-	C,r	FD90 bm49
		-0.620			6.00	QE 1.92	-	-	-	7,8-dihydrofolic acid	C,r	

CONT

FD90 (CONT.) $C_{19}H_{19}N_7O_6$

Code No.	Empirical Formula	Name and C.A. Number	Structural Formula	Solvent	Tech.	Medium		μ, M	pH	T, °C	Electrodes	App.	Experimental Parameters
FD90 bm49	$C_{19}H_{19}N_7O_6$	folic acid	Table II	H_2O	PY	ACET	0.1	-	5.5	25± 0.1	DME// Ag/AgCl, KCl satd	2AO	C=0.2, W Aux
					DI	ACET DTPA Fe(II) fumarate	0.1 0.15 ns	-	5.5	-	DME// Ag/AgCl, KCl satd	2AO	C=ns
					PV	ACET	0.1	-	5.5	25± 0.1	DME// Ag/AgCl, KCl satd	2AO	C=5E-4, Δe=10(rms), W Aux
													C=0.005
													C=0.2
					PVI	ACET DTPA Fe(II) fumarate	0.15 ns	-	5.5	25± 0.1	DME// Ag/AgCl, KCl satd	2AO	C=ns, Δe=10(rms)
					VA	ACET	0.1	-	5.5	-	HMDE// Ag/AgCl, KCl satd	25AO	-0.4→-1.2→ -0.4 V, C=0.02, v=100, Pt Aux
FD91	$C_{19}H_{23}ClN_2$	chlorimipramine C.A. 303-49-1	Table II	Me_2CO 10	PY	NaCl PHOS	0.5 0.006	-	3.2	-	DME//SCE	2AO	C=0.2, t(c)=1, h=66, Pt Aux
									3.7				
									4.6				
									6.0				
									7.0				
						NaCl PHOS	0.5 0.004		7.2				C=0.20
						NaCl PHOS	0.5 0.006		7.8				C=0.2
						NaCl BOR	0.5 0.003		8.3				C=0.1
					DI	NaCl PHOS	0.5 0.006	-	3.2	-	DME//SCE	2AO	C=0.2, t(c)=1, h=66, ΔE=5
									5.0				
									6.5				

CONT

$C_{19}H_{23}ClN_2$ (CONT.) FD91

Ref.	C/M	Charact. Potential		Response	Const. Value	n	Tech.	Electrokinetic Data			Products and Identification	Description and Remarks	Code No.
		Value	vs.					Parameter	Value	From			
AA096 0345	-	$E_{\frac{1}{2}}$	-0.600	Ag/AgCl, KCl satd	i_ℓ/C	2.61 6.00	-	-	$dE_{\frac{1}{2}}/dpH$	65	sttd	-	$C, i_\ell=kh^{\frac{1}{2}}, i_d, Tc=1.2$ for $T=20-50, i_\ell \downarrow$ as pH $\uparrow$ above 6,r	FD90 bm49
			-1.00F			5.60F	-		-				C	
AA096 0345	-	E_{su}	-0.53F	Ag/AgCl, KCl satd	i_{su}	0.2F	-	-	-	-	-	-	C,DTPA=diethylene- triaminepentaacetic acid,r	
AA096 0345	-	E_{su}	-0.638	Ag/AgCl, KCl satd	i_{su}/C	170	-	-	-	-	-	-	C,r	
			-0.622			162	-	-	-	-	-	-	C,r	
			-0.620			19.6	-	-	dE_{su}/dpH	65	sttd	-	C,r	
AA096 0345	-	E_{su}	-0.62F	Ag/AgCl, KCl satd	i_{su}	1.6F	-	-	-	-	-	-	C,DTPA=diethylene- triaminepentaacetic acid,r	
AA096 0345	-	E_p	-0.64F	Ag/AgCl, KCl satd	i_p	1.75F	-	-	-	-	-	-	$C, Mx, R, i_p/Cv^{\frac{1}{2}} \uparrow$ as $v \uparrow$ or $C \downarrow, i_a, r$	
			-1.00F			1.3F							C	
			-1.12F			0.5F							C	
			-0.62F			0.1F							A,R	
AA098 0093	-	$E_{\frac{1}{2}}$	-1.64F	SCE	i_ℓ	13.4F	-	-	-	-	-	-	C,r	FD91
			-1.63F			11.7F	-	-	-	-	-	-	C,r	
			-1.63F			11.2F	-	-	-	-	-	-	C,r	
			-1.62F			8.6F	-	-	-	-	-	-	C,r	
			-1.59F			4.5F	-	-	-	-	-	-	C,r	
			-			2.4F	-	-	-	-	-	-	$C, i_\ell=k_1C+k_2$ for $C>0.1, i_\ell=f[buffer]$ for $C>0.1$, $i_\ell \neq [buffer]$ for $C<0.02,r$	
			-1.57F			2.8F	-	-	-	-	-	-	C,r	
			-1.66F			1.5F	-	-	-	-	-	-	C,Mx at -1.72 V,r	
AA098 0093	-	E_{su}	-1.83F	SCE	-	-	-	-	-	-	-	-	C,r	
			-1.80F		i_{su}	25.5F	-	-	-	-	-	-	C,r	
			-1.74F			14.0F	-	-	-	-	-	-	C,r	CONT

FD91 (CONT.) $C_{19}H_{23}ClN_2$

Code No.	Empirical Formula	Name and C.A. Number	Structural Formula	Solvent	Tech.	Medium		μ, M	pH	T, °C	Electrodes	App.	Experimental Parameters
FD91	$C_{19}H_{23}ClN_2$	chlorimipramine	Table II	Me_2CO 10	DI	NaCl PHOS	0.5 0.006	-	7.5	-	DME//SCE	2AO	C=0.2, t(c)=1, h=66, ΔE=5, Pt Aux
									8.2				
FD92	$C_{19}H_{26}O_2$	17β-hydroxyandrosta-1,4-dien-3-one C.A. 846-48-0	Table II	EtOH 50	PY	BR		-	2.8	-	DME//SCE	O4AO	C=0.5, m=2.3, t=2.9, h=60
									4.3				
									6.0				
									8.0				
									9.9				
									11.5				
FD93	$C_{19}H_{27}FO_2$	6α-fluorotestosterone C.A. 1597-68-8	Table II	EtOH 50	PY	BR		-	3.1	-	DME//SCE	O4AO	C=0.5, m=2.3, t=2.9, h=60
									4.4				
									6.0				
									8.0				
FD94	$C_{19}H_{27}FO_2$	6β-fluorotestosterone C.A. 1852-58-0	Table II	EtOH 50	PY	BR		-	3.1	-	DME//SCE	O4AO	C=0.5, m=2.3, t=2.9, h=60
									4.4				
									6.0				
									8.0				
									9.3				
									11.3				
FD95 bm68	$C_{19}H_{28}O_2$	testosterone C.A. 58-22-0	Table II	EtOH 50	PY	BR		-	3.1	-	DME//SCE	O4AO	C=0.5, m=2.3, t=2.9, h=60
									4.4				
									6.0				
									8.0				
CONT													

TABLE I. Electrochemical Data   241

$C_{19}H_{28}O_2$ (CONT.) FD95

Ref.	C/M	Charact. Potential		Response Const.		n Tech.		Electrokinetic Data			Products and Identification	Description and Remarks	Code No.
		Value	vs.	Value	Value			Parameter	Value	From			
AA098 0093	–	E_{su}	–1.68F	SCE	i_{su}	4.2F	–	–	–	–	–	–	C,r	FD91
			–1.65F			2.6F	–	–	–	–	–	–	C,r	
C0027 2447	–	$E_{\frac{1}{2}}$	–1.06F	SCE	–	–	–	–	–	–	–	–	C,p	FD92
			–1.12F		–	–	–	–	$dE_{\frac{1}{2}}/dpH$	66F	–	–	C, $dE_{\frac{1}{2}}/dpH$ for pH=4.3-9.9,p	
			–1.235		i_d/C	0.55	–	–		66F		–	C,p	
			–1.36F		–	–	–	–		66F		–	C,p	
			–1.50F		–	–	–	–		20F		–	C, $dE_{\frac{1}{2}}/dpH$ for pH ≥ 9.9,p	
			–1.53F		–	–	–	–		20F		–	C,p	
C0027 2447	–	$E_{\frac{1}{2}}$	–0.92F	SCE	–	–	–	–	$dE_{\frac{1}{2}}/dpH$	63F	–	–	C,p	FD93
			–1.00F		–	–	–	–		63F		–	C,p	
			–1.090		i_d/C	1.05	–	–		63F		–	C,p	
			–1.23F		–	–	–	–		63F		–	C,p	
C0027 2447	–	$E_{\frac{1}{2}}$	–0.68F	SCE	–	–	–	–	$dE_{\frac{1}{2}}/dpH$	65F	–	–	C, $dE_{\frac{1}{2}}/dpH$ for pH=3.1-9.3,p	FD94
			–0.78F		–	–	–	–		65F		–	C,p	
			–0.860		i_d/C	0.94	–	–		65F		–	C,p	
			–1.01F		–	–	–	–		65F		–	C,p	
			–1.08F		–	–	–	–		0F		–	C, $dE_{\frac{1}{2}}/dpH$ for pH ≥ 9.3,p	
			–1.08F		–	–	–	–		0F		–	C,p	
C0027 2447	–	$E_{\frac{1}{2}}$	–1.17F	SCE	–	–	–	–	$dE_{\frac{1}{2}}/dpH$	55F	–	–	C, $dE_{\frac{1}{2}}/dpH$ for pH=3.1-9.3,p	FD95 bm68
			–1.28F		–	–	–	–		55F		–	C,p	
			–1.385		i_d/C	0.58	–	–		55F		–	C,p	
			–1.46F		–	–	–	–		55F		–	C,p	
			–1.71F				–	–		0F			C	

CONT

FD95 (CONT.) $C_{19}H_{28}O_2$

Code No.	Empirical Formula	Name and C.A. Number	Structural Formula	Solvent	Tech.	Medium	μ, M	pH	T, °C	Electrodes	App.	Experimental Parameters
FD95	$C_{19}H_{28}O_2$	testosterone	Table II	EtOH 50	PY	BR	–	–	9.3	–	DME//SCE	O4AO	C=0.5,m=2.3, t=2.9,h=60
								11.3				
FD96	$C_{20}H_6Br_4Na_2O_5$	eosin C.A. 20266-43-7	Table II	H_2O	PY	BR	–	–	4.0	–	DME//SCE		C=0.44, m=2.02, t=4.24,h=49
								6.0				
								8.08				
								9.0				
								10				
						BR	– 0.8	5.0±0.05	30.0 ±0.1	DME//SCE	5AO	C=0.865, m=2.0,t=4.1, h=80
								6.4±0.05				
								7.0±0.05				
								7.3±0.05				

CONT

TABLE I. Electrochemical Data

$C_{20}H_6Br_4Na_2O_5$ (CONT.) FD96

Ref.	C/M	Charact. Potential		Response Const.	n Tech.		Electrokinetic Data			Products and Identification	Description and Remarks	Code No.
		Value	vs.	Value			Parameter	Value	From			
C0027 2447	-	$E_{\frac{1}{2}}$	-1.51F	SCE	-	-	-	$dE_{\frac{1}{2}}/dpH$	55F	-	-	C,0 for pH > 9.3,p	FD95
			-1.71F						0F			C,$dE_{\frac{1}{2}}/dpH$ for pH ≥ 9.3	
			-1.71F		-	-	-		0F		-	C,p	
EA018 0265	356 a	$E_{\frac{1}{2}}$	-0.50F	SCE	i_ℓ	0.5F	-	-	$dE_{\frac{1}{2}}/dpH$	80F	-	-	C,i_a,$dE_{\frac{1}{2}}/dpH$ for pH=4-8	FD96
			-0.61F			1.4F				133F			C,i_d,$dE_{\frac{1}{2}}/dpH$ for pH=4-6	
			-1.10F			0.9F				-			C	
			-0.69F			0.4F	-	-		80F		-	C,i_a,i_ℓ ↓ as pH ↑,p	
			-0.93F			1.3F				133F			C,i_d,i_ℓ ↑ as pH ↑	
			-1.35F			0.4F				-			C,0 for C < 0.19, i_ℓ ↓ as pH ↑	
			-0.9			-	-	-		80F			C,0 for pH > 8.1,p	
			1.01			1.7F	i:i	2		0F			C,$dE_{\frac{1}{2}}/dpH$ for pH=7-11	
			-			-	-	-		-			C(?)	
	356 b		-1.01F			1.5F	-	-		0F			C,p	
			-1.45F			0.3F				50F			C,$dE_{\frac{1}{2}}/dpH$ for pH=9-10	
			-1.03F			1.1F	-	-		0F			C,p	
			-1.50F			0.6F				0F			C,$dE_{\frac{1}{2}}/dpH$ for pH ≥ 10	
EA018 0335	-	$E_{\frac{1}{2}}$	-0.63F	SCE	i_ℓ	0.6F	QE }	2	$dE_{\frac{1}{2}}/dpH$	64F	-	-	C,⊬,i_a,$dE_{\frac{1}{2}}/dpH$ for pH=5-6.1,r	
			-0.80F			2.7F				72F			C,⊬,Σi≠f(pH),i_a + i_d,$dE_{\frac{1}{2}}/dpH$ for pH=5-6.7	
			-1.38F			6.4F	i:i	4		0F			C,$dE_{\frac{1}{2}}/dpH$ for pH=5-7	
			-0.78F			0.6F	QE }	2		133F		-	C,⊬,i_a,$dE_{\frac{1}{2}}/dpH$ for pH=6-7.7,r	
			-0.93F			2.7F				72F			C,⊬,i_a + i_d	
			-1.38F			6.4F	i:i	4		0F			C,0 for pH > 6.8	
			-0.82F			3.3F		2		133F		-	C,⊬,i_a + i_d,r	
			-1.40F			6.4F		4		27±3F			C,⊬,M,X,$dE_{\frac{1}{2}}/dpH$ for pH=6.7-9	
			-0.88F			-	-	-		133F			C,i_a,r	
			-0.96			3.2F		2		27±3F			C,i_a + i_d	

CONT

FD96 (CONT.) $C_{20}H_6Br_4Na_2O_5$

Code No.	Empirical Formula	Name and C.A. Number	Structural Formula	Solvent	Tech.	Medium		μ, M	pH	T, °C	Electrodes	App.	Experimental Parameters
FD96	$C_{20}H_6Br_4Na_2O_5$	eosin	Table II	H_2O	PY	BR	-	0.8	7.9±0.05	30.0 ±0.1	DME//SCE	5AO	C=0.865, m=2.0, t=4.1, h=80
									8.4±0.05				
									9.0±0.05				
									10.4±0.05				
									11.4±0.05				
					IL	KCl	2(?)	-	-	-	g(Nujol): xxdi // SCE	O--	C=ns, A=0.126
					PK	ACET	ns	0.2	5.7±0.05	30.0 ±0.1	DME//SCE	5AO	C=1.01, m=2.0, t=4.1, h=80, f=6, E_{app}=-1.5V
				MeOH 40	IL	KCl	1.2	-	-	-	g(Nujol): xxdi // SCE	O--	C=ns, A=0.126
				MeOH 50	PY	Me_4NOH	0.1	-	-	-	DME//SCE	O--	C=0.05, m=2.61, t=3.0, h=50
FD97 bm73	$C_{20}H_6I_4Na_2O_5$	erythrosine C.A. 20266-45-9	Table II	H_2O	PY	BR TX100	- 6E-4	-	4.0	-	DME//SCE	---	C=0.58, m=2.02, t=4.24, h=49
									5.15				
FD96 CONT	$C_{20}H_6Br_4Na_2O_5$												

TABLE I. Electrochemical Data

$C_{20}H_6I_4Na_2O_5$ (CONT.) FD97

Ref.	C/M	Charact. Potential		Response Const.		n Tech.		Electrokinetic Data			Products and Identification	Description and Remarks	Code No.
		Value	vs.		Value			Parameter	Value	From			
EA018 0335	-	$E_{\frac{1}{2}}$	-0.97F	SCE	i_ℓ	3.2F	i:i	2	$dE_{\frac{1}{2}}/dpH$	27±3F	-	-	$C, i_a+i_d, E_{\frac{1}{2}}=$ f[buffer],r	FD96
			-0.99F			3.0F	}	2		27±3F		-	C,r	
			-			0.3F							C,QI	
			-0.99F			2.4F	}	2			0F	-	-	$C, dE_{\frac{1}{2}}/dpH$ for pH=9-12	
			-1.226			0.9F			α	35F 0.36	sttd		$C, dE_{\frac{1}{2}}/dpH$ for pH=9-12	
			-1.0F			1.8F	}	2	$dE_{\frac{1}{2}}/dpH$	0F	-	-	C, i_a+i_d, r	
			-1.274			1.4F			α	35F 0.57	sttd		C	
			-1.0F			1.8F	}	2	$dE_{\frac{1}{2}}/dpH$ $dE_{\frac{1}{2}}/d\log$[buffer]	0F 55	- sttd	-	$C, i_a+i_d, E_{\frac{1}{2}}\neq$ f[buffer],r	
			1.332			1.5F			$dE_{\frac{1}{2}}/dpH$ α $dE_{\frac{1}{2}}/d\log\mu$	35F 0.57 328	- sttd			
EA019 0215	-	$E_{p/2}$	0.85	SCE	-	-	sttd	2	-	-	-	-	A,≠,r	
EA018 0335	-	$E_{\frac{1}{2}}$	-0.10c	SCE	-	-	-	-	-	-	-	-	A,r	
EA019 0215	-	E_p	0.99	SCE	-	-	sttd	2	$E_{p/2}-E_p$	60	sttd	-	A,≠	
			1.10					-		-			A, i_a	
EA019 0215	-	$E_{\frac{1}{2}}$	-1.05	SCE	-	-	sttd	1	Tomeš	60		-	$C, R, Mc(E_{\frac{1}{2}}), p$	
			-		I	2.72		1		-			C	
EA018 0265	356 a	$E_{\frac{1}{2}}$	-0.52F	SCE	i_ℓ	0.3F	-	-	-	-	-	-	C,p	FD97 bm73
			-0.82F			0.5F							C	
			-1.31F			1.0F							C	
			-			0.3F	-	-	$dE_{\frac{1}{2}}/dpH$	-	-		$C, X, i_\ell=kh, i_a, i_\ell \downarrow$ as pH ↑, $\log i=k_1+k_2 \log C, p$	
			-0.82F			0.9F							C,≠,redn of furan, $i_\ell \downarrow$ as pH ↑, $i_\ell=kh^{\frac{1}{2}}, i_d$	
			-1.09F			2.1F				48F			C,≠,redn of furan, $i_\ell=kh^{\frac{1}{2}}, i_d, dE_{\frac{1}{2}}/dpH$ for pH=5-8	
			-1.36F			1.54F				11F			$C, \neq, i_\ell=kh^{\frac{1}{2}}, i_d$, $dE_{\frac{1}{2}}/dpH$ for pH=5-8	
			-			-				-			C,M(H)	

CONT

FD97 (CONT.) $C_{20}H_6I_4Na_2O_5$

Code No.	Empirical Formula	Name and C.A. Number	Structural Formula	Solvent	Tech.	Medium		μ, M	pH	T, °C	Electrodes	App.	Experimental Parameters
FD97 bm73	$C_{20}H_6I_4Na_2O_5$	erythrosine	Table II	H_2O	PY	BR TX100	6E-4	-	6.0	-	DME//SCE	---	C=0.58, m=2.02, t=4.24, h=49
									7.00				
									8.0				
									9.0				
									9.5				
									10.0				
									11.0				
					IL	KCl	2(?)	-	-	-	g(Nujol): xxdi// SCE	0--	C=ns, A=0.126
				MeOH 40	IL	KCl	1.2	-	-	-	g(Nujol): xxdi// SCE	0--	A=0.126
				MeOH 50	PY	Me_4NOH	0.1	-	-	-	DME//SCE	0--	C=0.05, m=2.61, t=3.0, h=50
FD98 bm88	$C_{20}H_{10}Na_2O_5$	fluorescein disodium salt C.A. 20266-41-5	Table II	H_2O	IL	KCl	2(?)	-	-	-	g(Nujol): xxdi// SCE	0--	C=ns, A=0.13
				MeOH 40	IL	KCl	1.2	-	-	-	g(Nujol): xxdi// SCE	0--	A=0.13
				MeOH 50	PY	Me_4NOH	0.1	-	-	-	DME//SCE	0--	C=0.05, m=2.61, t=3.0, h=50

TABLE I. Electrochemical Data

$C_{20}H_{10}Na_2O_5$ FD98

Ref.	C/M	Charact. Potential		Response Const.		n Tech.	Electrokinetic Data			Products and Identification	Description and Remarks	Code No.
		Value	vs.		Value		Parameter	Value	From			
EA018 0265	356 a	$E_{\frac{1}{2}}$	-	i_ℓ	0.3F	-	-	-	-	-	C, i_a, p	FD97 bm73
		-			0.9F			-			C	
		-1.05F			2.2F		$dE_{\frac{1}{2}}/dpH$	+48F	-		C	
		-1.35F			1.6F			+11F			C	
		-1.56F			0.7F			-20F			$C, dE_{\frac{1}{2}}/dpH$ for pH=6-9.5	
	356 c	-			0.1F	-	-	-		-	$C, i_a, 0$ for pH > 7, p	
		-			0.7F			-			C	
		-1.1F			2.6F			+48F			C	
		-1.34F			1.4F			+11F			C	
		-1.51F			0.86F			-20F			C	
		-			0.1F	-	-		-		$C, 0$ for pH > 8, p	
		-0.95F			3.4F			+48F			C	
		-1.33F			0.9F			+11F			$C, 0$ for pH > 8	
		-1.61F			1.0F			-20F			C	
		-0.99F			3.3F	-	-		-60F		-	C, redn of furan, p	
		-1.62F			1.40F			-20F			C	
		-1.04F			3.0F	-	-		-60	sttd	-	$C, dE_{\frac{1}{2}}/dpH$ for pH=8.5-9.5Ap, p	
		-1.63F			1.9F			-20			C, X	
		-1.05F			1.8F	-	-		-40F	-	-	$C, dE_{\frac{1}{2}}/dpH$ for pH > 10	
		-1.70F			2.6F			0F			$C, i_\ell \uparrow$ as pH $\uparrow$, $dE_{\frac{1}{2}}/dpH$ for pH > 10	
		-1.09F			1.36F	-	-		-40F		-	C, p	
		-1.70F			2.6F			0F			C	
EA019 0215	-	$E_{p/2}$	0.80	SCE	-	-	sttd	2	-	-	-	-	$A, \not=, E_{\frac{1}{2}}=0.80c, p$	
EA019 0215	-	$E_{p/2}$	0.92	SCE	-	-	sttd	2	$E_{p/2}-E_p$	48	sttd	-	$A, \not=, E_{\frac{1}{2}}=0.92c, p$	
		-			-			-			A, i_a, P	
EA019 0215	-	$E_{\frac{1}{2}}$	-1.05	SCE	I	10.23	sttd	4	Tomeš	56	sttd	-	$C, \not=, Mc(E_{\frac{1}{2}}), p$	
		-			-		4		-			C	
EA019 0215	-	$E_{p/2}$	0.75	SCE	-	-	sttd	1	-	-	-	-	$A, R, E_{\frac{1}{2}}=0.78c, p$	FD98 bm88
EA019 0215	-	E_p	0.88	SCE	-	-	sttd	1	$E_p - E_{p/2}$	58	sttd	-	$A, R, E_{\frac{1}{2}}=0.91c, p$	
		-			-			-			A, i_a, P	
EA019 0215	-	$E_{\frac{1}{2}}$	-1.22	SCE	-	-	sttd	1	Tomeš	58	sttd	-	$C, R, Mc(E_{\frac{1}{2}}), p$	
		-			I	2.26		1		-			C	

FD99 $C_{20}H_{12}$

Code No.	Empirical Formula	Name and C.A. Number	Structural Formula	Solvent	Tech.	Medium		μ, M	pH	T, °C	Electrodes	App.	Experimental Parameters
FD99 bm82 c182	$C_{20}H_{12}$	perylene C.A. 198-55-0	Table II	MeCN	VY	$AlCl_3$	1	-	-	-	Pt: rodi// Ag/AgCl, Me_4NCl satd	2A0	ns
				HF	VY	KF	0.1	-	-	0	glC: rodi// Cu/CuF_2, KF 0.2	2-0	C=1, A=0.07, v=100, ω=1, Pt Aux
				$MeNO_2$	VY	$AlCl_3$	1	-	-	-	Pt: rodi// Ag/AgCl, Me_4NCl satd	2A0	C=ns, A=0.01
						Et_4NClO_4	0.1						
FE00	$C_{20}H_{12}N_3NaO_7S$	Solochrome Black WDFA C.A. 1787-61-7	Table II	H_2O	PY	buffer	ns	-	1.8	35± 0.1	DME//SCE	OA0	C=0.5, EC=3.75, h=50
									4.8				
									8.0				
									9.15				
									11.50				
FE01	$C_{20}H_{13}BrN_2$	3-(4-bromophenyimino)-2-phenyl-3H-indole C.A. 31054-09-8	Table II	DMF	PY	$KClO_4$	0.02	-	-	20± 0.1	DME//SCE (NaCl, aq)	23A0	C=ns, m=4.84, t=2.9(at -1.15 V), h=80, Hg Aux
					VA	$KClO_4$	0.05	-	-	20± 0.1	HMDE// SCE(NaCl, aq)	23A0	-0.33 → -1.43 → -0.33 V, C=0.15, v=250, Hg Aux

$C_{20}H_{13}BrN_2$ FE01

Ref.	C/M	Charact.	Potential	Response	Const.	Tech.	n	Electrokinetic Data			Products and	Description and	Code
		Value	vs.		Value			Parameter	Value	From	Identification	Remarks	No.
JE039 0385	-	$E_{\frac{1}{2}}$	0.70	Fc	-	-	-	-	-	-	-	-	A,Σ,p	FD99 bm82 c182
JE054 0232	-	$E_{\frac{1}{2}}$	0.125	NHE	i_ℓ	-	-	-	Elog	-	-	radical cation	A,p	
			0.695			185F				54		dication	A,oxidn of radical cation	
JE039 0385	-	$E_{\frac{1}{2}}$	0.65	Fc	-	-	-	-	-	-	-	-	A,r	
			0.65		-	-	-	-	-	-	-	-	A,r	
JE054 0417	-	$E_{\frac{1}{2}}$	-0.12	SCE	-	-	PQ	2	αn_a, $k_{s,h}$, $dE_{\frac{1}{2}}/dpH$	0.86, 5.2 E-6, 31	sttd	-	C,$\not\equiv$,i_d,$E_{\frac{1}{2}}$=0.068-0.031pH±0.003, $i_\ell = kh^{\frac{1}{2}}$ and kC, $i_\ell \neq f(pH)$,p	FE00
			-0.210		-	-	-	-		0.86, 5.0 E-6, 31		-	C,$\not\equiv$,i_d,p	
			-0.300		-	-	-	-		0.86, 4.8 E-6, 31		-	C,$\not\equiv$,i_d,p	
			-0.360		-	-	-	-		0.86, 4.9 E-6, 31		-	C,$\not\equiv$,i_d,p	
			-1.19							0.43, 4.8 E-6, -			C,Mc($E_{\frac{1}{2}}$)	
			-		-	-	-	-		0.86, 4.9 E-6, 31		-	C,$\not\equiv$,i_d,p	
			-1.19							0.43, 3.4 E-6, -			C	
JE036 0147	427 ce	$E_{\frac{1}{2}}$	-0.908	SCE(NaCl,aq)	I	0.88	QE	-	Tomeš	60	-	3-(4-bromophenyl-amino)-2-phenyl-3H-indole,λ_{max}=308	C,R,i_ℓ=kC for C=0.1-1,i_d,Tc=1 for T=15-30,$\Delta E_{\frac{1}{2}}$=0.195σ,r	FE01
			-1.285			0.89		2± 0.1		89		-	C,i_d,i_ℓ=kC for C=0.1-1,$\not\equiv$	
JE036 0147	427 ce	E_p	-0.89F	SCE(NaCl,aq)	i_p	16F	-	-	-	-	-	-	C,$E_p \neq f(v)$,r?	
			-1.24F			10.7F							C	
			-0.71F			17.8F							A	

FE02 $C_{20}H_{13}BrN_2O$

Code No.	Empirical Formula	Name and C.A. Number	Structural Formula	Solvent	Tech.	Medium		μ, M	pH	T, °C	Electrodes	App.	Experimental Parameters
FE02	$C_{20}H_{13}BrN_2O$	3-(4-bromophenyl-imino)-2-phenyl-3H-indole 1-oxide C.A. 31083-67-7	Table II	DMF	PY	$KClO_4$	0.02	-	-	17.5 ±0.1	DME//SCE (NaCl, aq)	23A0	C=0.4Ap, m=1.71, t=2.9 (at -1.15 V), h=80, Hg Aux
FE03	$C_{20}H_{13}ClN_2$	3-(4-chlorophenyl-imino)-2-phenyl-3H-indole C.A. 31054-08-7	Table II	DMF	PY	$KClO_4$	0.02	-	-	20± 0.1	DME//SCE (NaCl, aq)	23A0	C=ns, m=4.84, t=2.9(at -1.15 V), h=80, Hg Aux
FE04	$C_{20}H_{13}ClN_2O$	3-(4-chlorophenyl-imino)-2-phenyl-3H-indole 1-oxide C.A. 31054-04-3	Table II	DMF	PY	$KClO_4$	0.02	-	-	17.5 ±0.1	DME//SCE (NaCl, aq)	23A0	C=0.4Ap, m=1.71, t=2.9(at -1.15 V), h=80, Hg Aux
					VA	$KClO_4$	0.05	-	-	17.5 ±0.1	HMDE// SCE(NaCl, aq)	23A0	-0.4→-1.4→-0.4 V, C=0.16, v=400, Hg Aux
FE05 bn03 c185	$C_{20}H_{14}$	9-phenylanthracene C.A. 602-55-1	Table II-2	MeCN	VA	$NaClO_4$	0.1	-	-	-	Pt: xxbu// SCE	---	0.95→1.30→0.95 V, C=2.0, v=156
						$NaClO_4$ PYR	0.1 1.0E-3						C=2.0, v=15.6
													v=156
													v=310
						$NaClO_4$ PYR	0.1 2.0E-3						v=15.6
													v=156
													v=600
						$NaClO_4$ PYR	0.1 4.0E-3						v=156
						$NaClO_4$ PYR	0.1 0.01						v=156
						$NaClO_4$ PYR	0.1 0.03						

TABLE I. Electrochemical Data   251

$C_{20}H_{14}$ FE05

Ref.	C/M	Charact. Potential		Response Const.		Tech.	n	Electrokinetic Data			Products and Identification	Description and Remarks	Code No.
		Value	vs.		Value			Parameter	Value	From			
JE041 0067	427 bf	$E_{\frac{1}{2}}$	-0.943	SCE(NaCl, aq)	I	1.76	QE	2Ap	Tomeš	60	sttd	CP→3-(4-bromophenylimino)-1-hydroxy-2-phenyl-3H-indole, PY	C, i_ℓ=kC for C=0.1-1, i_d, Tc=1 for T=15-30, $\Delta E_{\frac{1}{2}}$=0.2σ, r	FE02
			-1.234			1.61		2Ap		74		CP→3-(4-bromophenylimino)2-phenyl-indole, PY	C, i_ℓ=kC for C=0.1-1, i_d, Tc=1 for T=15-30	
JE036 0147	427 ce	$E_{\frac{1}{2}}$	-0.917	SCE(NaCl, aq)	I	0.84	QE	-	Tomeš	60	-	3-(4-chlorophenyl-amino)-2-phenyl-indole, λ_{max}=308	C,R, i_ℓ=kC for C=0.1-1, i_d, Tc=1 for T=15-30, $\Delta E_{\frac{1}{2}}$=0.19σ, r	FE03
			-1.330			0.83		2±0.1		89		-	C, ≠, i_ℓ=kC for C=0.1-1	
JE041 0067	427 bf	$E_{\frac{1}{2}}$	-0.947	SCE(NaCl, aq)	I	1.69	QE	2Ap	Tomeš	64	-	CP→3-(4-chlorophenylimino)-1-hydroxy-2-phenylindole, PY	C, i_ℓ=kC for C=0.1-1, i_d, Tc=1 for T=15-30, $\Delta E_{\frac{1}{2}}$=0.2σ, r	FE04
			-1.285			1.49		2Ap		68		CP→3-(4-chlorophenylimino)-2-phenyl-indole, PY	C, i_ℓ=kC for C=0.1-1, i_d, Tc=1 for T=15-30	
JE041 0067	427 bf	E_p	-0.91F	SCE(NaCl, aq)	i_p	2.2F	-	-	-	-	-	-	C, E_p≠f(v), R	
			-1.24F			1.6F							C, E_p→more neg as v ↑	
			-0.84F			0.9F							A	
			-0.72F			0.8F							A	
EA018 0665	-	E_p	1.15	SCE	i_p	180	-	-	$E_p-E_{p/2}$	75	sttd	-	A, r	FE05 bnU3 c185
			1.07			50				-			C	
			1.00			-							A, Pr, r	
			1.135										A	
			-										C	
			1.03		$i_p(u)$	27F	-	-	-	-		-	A, Pr, r	
			1.13			8F							A	
			1.07			2F							C	
			1.05		-	-	-	-	-	-		-	A, Pr, r	
			1.135										A	
			-										C	
			1.02		-	-	-	-	-	-		-	A, Pr, r	
			1.05		$i_p(u)$	225	-	-	$E_p-E_{p/2}$	65	sttd	-	A, Pr, r	
			-			-				-			A, 0	
			-			-				-			C, 0	
			1.08		-	-	-	-	-	-		-	A, Pr, r	
			1.025		i_p	296	-	-	$E_p-E_{p/2}$	55	sttd	-	A, Pr, r	
			1.003			298	-	-		55		-	A, Pr, r	
			0.99			298	-	-		55		-	A, Pr, r	

FE06 $C_{20}H_{14}N_2$

Code No.	Empirical Formula	Name and C.A. Number	Structural Formula	Solvent	Tech.	Medium		μ, M	pH	T, °C	Electrodes	App.	Experimental Parameters
FE06	$C_{20}H_{14}N_2$	2-phenyl-3-(phenyl-imino)-3H-indole C.A. 23073-34-9	Table II	DMF	PY	$KClO_4$	0.02	-	-	20±0.1	DME//SCE (NaCl, aq)	23AO	C=ns, m=4.84, t=2.9(at -1.15 V), h=80, Hg Aux
						$KClO_4$ $ClCH_2COOH$	0.02 1.4E-4						C=0.14, h=50, Hg Aux
						$KClO_4$ $ClCH_2COOH$ PVC	0.02 3.64E-4 5E-3						
FE07	$C_{20}H_{14}N_2O$	2-phenyl-3-phenyl-imino-3H-indole 1-oxide C.A. 5165-73-1	Table II	DMF	PY	$KClO_4$	0.02	-	-	17.5±0.1	DME//SCE (NaCl, aq)	23AO	C=0.4Ap, m=1.71, t=2.9(at -1.15 V), h=80, Hg Aux
						$KClO_4$ $CH_2ClCOOH$	0.02 1.4E-3						
						$KClO_4$ $CH_2ClCOOH$	0.02 0.046						
FE08	$C_{20}H_{14}O_4$	diphenyl phthalate C.A. 84-62-8	2-$C_6H_5OCOC_6H_4$-$C(O)OC_6H_5$	EtOH 40	PY	BOR	0.08	-	10	-	DME//SCE	04AO	C=ns, m=1.77, t=5, h=60
FE09	$C_{20}H_{16}$	triphenylethene C.A. 58-72-0	$(C_6H_5)_2C:CHC_6H_5$	DMF	VA	Me_4NBF_4	satd	-	-	-	Hg: xxsd//SCE	5AO	C=ns, v=300
													v=1E3
				HMP	VA	Bu_4NBF_4	satd	-	-	-	Hg: xxsd//SCE	5AO	v=300
						LiCl	satd						

TABLE I. Electrochemical Data   253

$C_{20}H_{16}$ FE09

Ref.	C/M	Charact. Potential		Response	Const. Value	n Tech.	Electrokinetic Data			Products and Identification	Description and Remarks	Code No.
		Value	vs.				Parameter	Value	From			
JE036 0147	427 ce	$E_{\frac{1}{2}}$	-0.953	SCE(NaCl, aq)	I	0.87	QE —	Tomeš	60	—	2-phenyl-3-phenyl-aminoindole, λ_{max}=308	C,R,i_ℓ=kC for C=0.1-1, i_d,Tc=1 for T=15-30, $\Delta E_{\frac{1}{2}}$=0.195σ,r	FE06
			-1.378			0.89	2± 0.1		88		—	$C,\not=,i_d,i_\ell$=kC for C=0.1-1	
			-0.32F		i_ℓ	0.40F	—				—	C,Mx,r	
			-0.88F			0.18F						C	
			-1.25F			0.14F						C,additional redn occurs for $[ClCH_2COOH]$ >8E-5	
			-0.3F			0.68F	— —	—	—	—	—	C,r	
			-0.8F			0.05F						C	
			—			—						C	
JE041 0067	427 bf	$E_{\frac{1}{2}}$	-0.987	SCE(NaCl, aq)	I	1.69	QE 2Ap	Tomeš	64	—	CP→1-hydroxy-2-phenyl-3-phenyl-imino indole,PY	C,i_ℓ=kC for C=0.1-1, i_d,Tc=1 for T=15-30, $\Delta E_{\frac{1}{2}}$=0.2σ,r	FE07
			-1.308			1.58	2Ap		72		CP→2-phenyl-3-phenyiimino indole, PY	C,i_ℓ=kC for C=0.1-1, i_d,Tc=1 for T=15-30	
			-0.4F			0.78F	—	—	—	—	—	C,Mx,r	
			-0.8F			0.42F						C	
			-0.95F			0.18F						C	
			-0.2F		i_ℓ	0.96F	—	—	—	—	—	C,Mx,r	
			-0.75F			0.6F						C	
C0028 1985	438 a	—	-1.68	SCE	—	—	sttd	4	—	—	—	—	$C,\Delta E_{\frac{1}{2}}$=0.39σ*,p	FE08
EA019 0951	—	XE	-2.04	SCE	$i_p(u)$	22	—	—	—	—	—	$C,A,XE=E_{p,A}+E_{p,C}/2$, R,p	FE09
		E_p	-2.30			1						$C,\not=$	
		—	—	SCE		8	—	—	—	—	—	C,A,p	
						1						C	
												C	
EA019 0951	—	XE	-2.05	SCE	—	—	—	—	—	—	—	$C,A,XE=(E_{p,A}+E_{p,C})/2$, R,p	
		E_p	-2.43									$C,\not=$	
		XE	-1.95	SCE	—	—	—	—	—	—	—	$C,A,XE=(E_{p,A}+E_{p,C})/2$, R,eec,p	
		E_p	-2.3Ap									$C,\not=$	
			-1.0Ap									A,present only after a hold with E more neg than -2.3 V and if $v_{reverse}$ ≥3E4	

FE10 $C_{20}H_{16}N_4O_2$

Code No.	Empirical Formula	Name and C.A. Number	Structural Formula	Solvent	Tech.	Medium		μ, M	pH	T, °C	Electrodes	App.	Experimental Parameters
FE10	$C_{20}H_{16}N_4O_2$	Solochrome Red ERS	Table II	H_2O	PY	buffer	ns	–	1.8	35±0.1	DME//SCE	OAO	C=0.5, EC=3.75, h=50
									4.80				
									5.8				
									8.0				
									11.50				
FE11	$C_{20}H_{16}O_2$	2-methyl-1-phenyl-3-(5-phenyl-2-furyl)-2-propen-1-one C.A. 51930-45-1	Table II	DMF	PY	Bu_4NClO_4	0.1	–	–	–	DME//SCE	12-0	C=1
						$LiClO_4$	0.1						
					PK	Bu_4NClO_4	0.1	–	–	–	DME//SCE	12-0	C=ns, E_{app}=–1.5 V, t=3, f=12.5
FE12	$C_{20}H_{18}N_2O_2$	1,4-diacetyl-1,4-dihydro-2,3-diphenylpyrazine C.A. 32174-84-8	Table II	DMF	VA	$LiClO_4$	0.1	–	–	–	Pt// Ag/AgI, I⁻ 0.1	2-0	C=ns, A=0.015, v=100
FE13	$C_{20}H_{18}O$	methyltriphenylmethyl ether C.A. 596-31-6	$(C_6H_5)_3COCH_3$	DMF	PY	Bu_4NI	0.1	–	–	–	DME// Ag/AgI, I⁻ 0.1	3-0	C=1, Pt Aux
						Bu_4NI PHEN	0.1 0.01						
FE14	$C_{20}H_{27}N$	4,4'-di(t-butyl)-diphenylamine C.A. 32367-67-2	$[4-(CH_3)_3C-C_6H_4]_2NH$	MeCN	VY	Et_4NClO_4	0.1	–	–	–	Pt: rodi// Ag/ $AgClO_4$ 0.01, Et_4NClO_4 0.1	2-0	C=ns, d=0.1, ω=10, Pt Aux

$C_{20}H_{27}N$ FE14

Ref.	C/M	Charact. Value	Potential vs.	Response	Const. Value	Tech.	n	Electrokinetic Data Parameter	Value	From	Products and Identification	Description and Remarks	Code No.
JE054 0417	-	$E_{\frac{1}{2}}$	-0.2	SCE	-	-	PQ	2	αn_a $k_{s,h}$ $dE_{\frac{1}{2}}/dpH$	0.86 5.1E-6 31	-	-	$C, \not=, i_d, E_{\frac{1}{2}} = 0.068 - 0.031pH \pm 0.003$ for pH 1.8-3.3, $i_\ell = kh^{\frac{1}{2}}, i_\ell = kC, p$	FE10
			-0.280	-	-	-	-		0.86 1.7E-6 -	-		$C, \not=, p$	
			-0.280	-	-	-	-		0.43 6.1E-6 0	-		$C, \not=, p$	
			-0.280	-	-	-	-		0.43 4.9E-6 0	-		$C, \not=, p$	
			-0.465	-	-	-	-		0.26 7.8E-6 0	-		C	
			-0.280	-	-	-	-		0.43 2.7E-6 0	-		$C, \not=, p$	
			-0.465	-	-	-	-		0.26 5.3E-6 0	-		C	
JE050 0351	-	$E_{\frac{1}{2}}$	-1.47	SCE	-	-	i:i	1	Elog	62	-	CP → radical anion, ESR	C, W, i_d, p	FE11
			-1.8							80			C, W, i_d	
			-2.7							-			C	
			-1.39	-	-		1		49		-	C, W, i_d, p	
			-1.69						62			C, W, i_d	
JE050 0351	-	$E_{\frac{1}{2}}$	-1.47	SCE	XR	2.44	-	-	-	-	-	-	$A, XR = i_\ell, A, PK/i_\ell, C, PY, p$	
EA021 0345	434 a	E_p	1.29	Ag/AgI, I⁻ 0.1	i_p	12.0	QE (1.4 V)	1.87	-	-	-	2,3-diphenylpyrazine 75%, NMR	$A, \not=, r$	FE12
EA020 0853	33k	$E_{\frac{1}{2}}$	-2.31	Ag/AgI, I⁻ 0.1	i_ℓ	3.0	QE (-2.3V)	2.2	-	-	-	triphenylmethane 100%	C, r	FE13
			-2.33			4.9		2.2	-	-	-	triphenylmethane 100%	C, r	
EA021 0557	147 n	$E_{\frac{1}{2}}$	0.51	Ag/ AgClO₄ 0.01, Et₄NClO₄ 0.1	i_ℓ/C	5.7	sttd	1	-	-	-	radical cation, VIS	$A, \Delta E_{\frac{1}{2}} = 0.46\sigma, r$	FE14
			1.30			-		1				-	A	

FE15 $C_{20}H_{30}O_2$

Code No.	Empirical Formula	Name and C.A. Number	Structural Formula	Solvent	Tech.	Medium		μ, M	pH	T, °C	Electrodes	App.	Experimental Parameters
FE15	$C_{20}H_{30}O_2$	6α-methyltestosterone C.A. 13251-86-0	Table II	EtOH 50	PY	BR	–	–	4.4	–	DME//SCE	04A0	C=0.5, m=2.3, t=2.9, h=60
									6.0				
									9.0				
									11.3				
FE16	$C_{20}H_{30}O_2$	6β-methyltestosterone C.A. 13252-06-7	Table II	EtOH 50	PY	BR	–	–	3.1	–	DME//SCE	04A0	C=0.5, m=2.3, t=2.9, h=60
									6.0				
									9.0				
FE17	$C_{20}H_{34}O_4$	2,5-bis(2-isopropoxy-2-methylpropyl)-hydroquinone C.A. 17208-12-7	Table II	EtOH 50	PY	ACET	0.2	–	→→0	25±0.1	DME//SCE	04A0	ns
FE18	$C_{21}H_{16}N_2$	2-phenyl-3-(p-tolylimino)-3H-indole C.A. 31054-05-4	Table II	DMF	PY	$KClO_4$	0.02	–	–	20±0.1	DME//SCE (NaCl, aq)	23A0	C=ns, m=4.84, t=2.9, h=80, Hg Aux
FE19	$C_{21}H_{16}N_2O$	3-(4-methoxyphenylimino)-2-phenyl-3H-indole C.A. 31054-06-5	Table II	DMF	PY	$KClO_4$	0.02	–	–	20±0.1	DME//SCE (NaCl, aq)	23A0	C=ns, m=4.84, t=2.9, h=80, Hg Aux
FE20	$C_{21}H_{16}N_2O$	2-phenyl-3-(p-tolylimino)-3H-indole 1-oxide C.A. 18852-46-5	Table II	DMF	PY	$KClO_4$	0.02	–	–	17.5±0.1	DME//SCE (NaCl, aq)	23A0	C=0.4Ap, m=1.71, t=2.9, h=80, Hg Aux
						$KClO_4$ 3,4-dimethylphenol	0.05 4.27E-4						C=0.43
						$KClO_4$ 3,4-dimethylphenol	0.05 1.28E-3						
					VA	$KClO_4$	0.05	–	–	17.5±0.1	HMDE//SCE(NaCl, aq)	23A0	-0.6→-1.2→-0.6 V, C=0.16, v=400, Hg Aux

TABLE I. Electrochemical Data

$C_{21}H_{16}N_2O$ FE20

Ref.	C/M	Charact. Potential		Response	Const. Value	n Tech.		Electrokinetic Data			Products and Identification	Description and Remarks	Code No.
		Value	vs.					Parameter	Value	From			
C0027 2447	-	$E_{\frac{1}{2}}$	-1.28F	SCE	-	-	-	-	$dE_{\frac{1}{2}}/dpH$	52F	-	-	C,p	FE15
			-1.375		i_d/C	0.61	-	-		52F		-	C,p	
			-1.52F							52F		-	C,0 for pH > 9.0,p	
			-1.73F		-	-	-	-		0F		-	C,0 for pH ≤ 8.0	
			-1.73F		-	-	-	-		0F		-	C,p	
C0027 2447	-	$E_{\frac{1}{2}}$	-1.17F	SCE	-	-	-	-	$dE_{\frac{1}{2}}/dpH$	54F	-	-	C,p	FE16
			-1.355		i_d/C	0.54	-	-		54F		-	C,p	
			-1.49F		-	-	-	-		54F		-	C,p	
C0032 2140	-	$E_{\frac{1}{2}}^{o}$	0.671	NHE	-	-	-	-	-	-	-	-	$A, E_{\frac{1}{2}} \rightarrow \rightarrow$ from data at pH=3.8-5.6,r	FE17
JE036 0147	427 ce	$E_{\frac{1}{2}}$	-0.991	SCE(NaCl, aq)	I	0.86	QE	-	Tomeš	63	-	3-(4-methylphenyl-amino)-2-phenyl-indole,λ_{max}=308nm	C,R,i_ℓ=kC for C=0.1-1,i_d,Tc=1 for T=15-30,$\Delta E_{\frac{1}{2}}$=0.19σ,r	FE18
			-1.458			0.99	2± 0.1			89		-	$C,\not{\tau},i_\ell$=kC for C=0.1-1,i_d	
JE036 0147	427 ce	$E_{\frac{1}{2}}$	-1.017	SCE(NaCl, aq)	I	0.82	QE	-	Tomeš	60	-	3-(4-methoxyphenyl-amino)-2-phenylin-dole,λ_{max}=308nm	C,R,i_ℓ=kC for C=0.1-1,i_d,Tc=1 for T=15-30,$\Delta E_{\frac{1}{2}}$=0.19σ,r	FE19
			-1.504			0.86	2± 0.1			94		-	$C,\not{\tau},i_\ell$=kC for C=0.1-1,i_d	
JE041 0067	427 bf	$E_{\frac{1}{2}}$	-1.020	SCE(NaCl, aq)	I	1.63	QE	2Ap	Tomeš	60	-	CP→ 1-hydroxy-3-(4-methylphenylimino)-2-phenylindole,PY	C,i_ℓ=kC for C=0.1-1,i_d,Tc=1 for T=15-30,$\Delta E_{\frac{1}{2}}$=0.2σ,r	FE20
			-1.365			1.56		2Ap		77		CP→ 3-(4-methyl-phenylimino)-2-phenylindole,PY	C,i_ℓ=kC for C=0.1-1,i_d,Tc=1 for T=15-30	
			-		i_ℓ	0.9F 0.16F	-	-		-	-	-	C,r C	
			-			1.02F 0.04F	-	-		-	-	-	C,r C	
JE041 0067	427 bf	E_p	-1.05F -0.94F	SCE(NaCl, aq)	i_p	1.76F 1.12F	-	-	-	-	-	-	C,r A	

FE21 $C_{21}H_{16}N_2O_2$

Code No.	Empirical Formula	Name and C.A. Number	Structural Formula	Solvent	Tech.	Medium		μ, M	pH	T, °C	Electrodes	App.	Experimental Parameters
FE21	$C_{21}H_{16}N_2O_2$	3-(4-methoxyphenyl-imino)-2-phenyl-3H-indole 1-oxide C.A. 18852-45-4	Table II	DMF	PY	$KClO_4$	0.02	-	-	17.5 ±0.1	DME//SCE (NaCl, aq)	23AO	C=0.4Ap, m=1.71, t=2.9, h=80, Hg Aux
FE22	$C_{21}H_{17}N_3O_2$	1-(4-nitrophenyl)-3,5-diphenyl-Δ^2-pyrazoline C.A. 10252-45-6	Table II	MeCN	IL	$LiClO_4$	0.1	-	-	-	Pt: xxns// SCE	2--	C=1, v=300
					IR	$LiClO_4$	0.1	-	-	-	Pt: xxns// SCE	2--	C=1, E_{app}=1.0, t=1-10
FE23	$C_{21}H_{18}N_2$	1,3-diphenyl-2-propen-1-one phenyl-hydrazone C.A. 37799-62-5	$C_6H_5CH:CHC-(C_6H_5):NNHC_6H_5$	MeCN	IL	$LiClO_4$	0.1	-	-	-	Pt: xxns// SCE	2--	0→1.0→0 V, C=1, v=300
					IR	$LiClO_4$	0.1	-	-	-	Pt: xxns// SCE	2--	C=1, E_{app}=1.0, i=4-10
					QE	$LiClO_4$	0.1	-	-	-	Pt: nsns// SCE	2--	C=5.5, E_{app}=0.97± 0.03
													C=21
FE24	$C_{21}H_{18}N_2$	1,3,5-triphenyl-Δ^2-pyrazoline C.A. 742-01-8	Table II	MeCN	IR	$LiClO_4$	0.1	-	-	-	Pt: xxns// SCE	2--	C=1, E_{app}=1.0, t=4-10
					QE	$LiClO_4$	0.1	-	-	-	Pt: nsns// SCE	2--	C=5.5, E_{app}=0.97± 0.03
CONT													

TABLE I. Electrochemical Data

$C_{21}H_{18}N_2$ (CONT.) FE24

Ref.	C/M	Charact. Potential		Response Const.		n Tech.	Electrokinetic Data			Products and Identification	Description and Remarks	Code No.
		Value	vs.		Value		Parameter	Value	From			
JE041 0067	427 bf	$E_{\frac{1}{2}}$ -1.042	SCE(NaCl, aq)	I	1.57	2Ap QE	Tomeš	64	-	CP→1-hydroxy-3-(4-methoxyphenylimino)-2-phenylindole,PY	$C, i_\ell = kC$ for $C=0.1-1$, i_d, Tc=1 for T=15-30, $\Delta E_{\frac{1}{2}} = 0.2\sigma$, r	FE21
		-1.396			1.49	2Ap		77	-	CP→3-(4-methoxyphenylimino)-2-phenylindole,PY	$C, i_\ell = kC$ for $C=0.1-1$, i_d, Tc=1 for T=15-30	
EA021 0621	-	E_p 1.210	SCE	$i_p/v^{\frac{1}{2}}$	1.08	sttd 1	-	-	-	-	A,r	FE22
EA021 0621	-	-	-	$it^{\frac{1}{2}}$	15.3	sttd 1	-	-	-	-	A,R,r	
EA021 0621	-	E_p 0.790	SCE	$i_p/v^{\frac{1}{2}}$	1.04 ±0.04	sttd 1	$dE_p/d\log v$	20.2 ±0.4	sttd	-	A,Mc($i_p/v^{\frac{1}{2}}$), $i_p/v^{\frac{1}{2}}$ ↑ as v ↑ for v < 100, r	FE23
							$dE_p/d\log C$	19.1 ±1.9				
							$E_p-E_{p/2}$	55				
		0.9F			0.5F	-	-	-			A	
EA021 0621	-	-	-	$it^{\frac{1}{2}}$	12±1	sttd 1	-	-	-	-	A,r	
EA021 0621	-	-	-	-	-	QE 1.67	-	-	-	1,3,5-triphenylpyrazole 28%,MPT,IRS; 1,3,5-triphenyl-Δ^2-pyrazolium perchlorate 2%,MPT,IRS;4,4'-bis[3,5-diphenyl-Δ^2-pyrazolynil-(1)]biphenyl diperchlorate 25%,MAS,IRS;4,4'-bis[3,5-diphenylpyrazolyl-(1)]biphenyl 27%,MAS	A,r	
	-	-	-	-	-	QE 1.8	-	-	-	1,3,5-triphenylpyrazole 15%,MPT,IRS; 1,3,5-triphenyl-Δ^2-pyrazolinium perchlorate 4%,MPT,IRS; 4,4'-bis[3,5-diphenyl-Δ^2-pyrazolynil-(1)]-biphenyl diperchlorate 30%,MAS,IRS;4,4'-bis[3,5-diphenylpyrazolyl-(1)]biphenyl 35%,MAS	A,r	
EA021 0621	419 f	-	-	$it^{\frac{1}{2}}$	12±1	sttd 1	-	-	-	-	A,r	FE24
EA021 0621	419 f	-	-	-	-	QE 1.47	-	-	-	1,3,5-triphenylpyrazole 1%,MPT,IRS; 1,3,5-triphenyl-Δ^2-pyrazolinium perchlorate 6%,MPT,IRS; 4,4'-bis[3,5-diphenyl-Δ^2-pyrazolynil-(1-)biphenyl diperchlorate 25%,MAS,IRS; 4,4'-bis[3,5-diphenylpyrazolyl-(1)]biphenyl 35%,MAS	A,r	CONT

FE24 (CONT.) $C_{21}H_{18}N_2$

Code No.	Empirical Formula	Name and C.A. Number	Structural Formula	Solvent	Tech.	Medium		μ, M	pH	T, °C	Electrodes	App.	Experimental Parameters
FE24	$C_{21}H_{18}N_2$	1,3,5-triphenyl-Δ^2-pyrazoline	Table II	MeCN	QE	$LiClO_4$	0.1	-	-	-	Pt: nsns// SCE	2--	C=21, E_{app}=0.97± 0.03
					VA	$LiClO_4$	0.1	-	-	-	Pt: xxns// SCE	2--	0→1.0→0 V, C=1, v=300
					VR	$LiClO_4$	0.1	-	-	-	Pt: xxns// SCE	2--	0→1.0→0→ 1.0 V, C=1, v=300
FE25	$C_{21}H_{21}BrN_2S_2$	3-ethyl-2-[3-(3-ethyl-2(3H)-benzo-thiazolylidene)-1-propenyl]benzothia-zolium bromide C.A. 17389-14-9	Table II	MeOH 50	PY	Me_4NOH	0.1	-	-	-	DME//SCE	2--	C=0.05, m=2.61, t=3.0, h=50
FE26	$C_{21}H_{25}N_6NaO_{15}P_2$	sodium deamino nicotinamide adenine dinucleotide C.A. 1851-07-6	Table II	DMSO	PY	Et_4NClO_4	0.1	-	-	-	DME//SCE //Rb^+	35AF0	C=ns, m=1.02, t=5.0, h=70
					VR	Et_4NClO_4	0.1	-	-	-	HMDE// SCE//Rb^+	35AF0	0→-2.2→ 0→-2.2 V, C=ns, v=100
													0→-1.5→ 0→-1.5 V, v=150
													v=200
FE27	$C_{21}H_{27}N_7NaO_{17}P_3$	sodium nicotinamide adenine dinucleotide phosphate C.A. 53-59-8	Table II	DMSO	PY	Et_4NClO_4	0.1	-	-	-	DME//SCE //Rb^+	35AF0	C=ns, m=1.02, t=5.0, h=70
CONT					VR	Et_4NClO_4	0.1	-	-	-	HMDE// SCE//Rb^+	35AF0	0→-2.2→ 0→-2.2 V, C=ns, v=100

TABLE I. Electrochemical Data

$C_{21}H_{27}N_7NaO_{17}P_3$ (CONT.) FE27

Ref.	C/M	Charact. Potential		Response Const.		n Tech.	n	Electrokinetic Data			Products and Identification	Description and Remarks	Code No.
		Value	vs.		Value			Parameter	Value	From			
EA021 0621	419 f	-	-	-	-	QE	1.5	-	-	-	1,3,5-triphenylpyrazole,trace,MPT,IRS; 1,3,5-triphenyl-Δ^2-pyrazolinium perchlorate 3%,MPT,IRS IRS;4,4'-bis[3,5-diphenyl-Δ^2-pyrazolynil-(1)]biphenyl diperchlorate 32%,MAS,IRS; 4,4'-bis[3,5-diphenylpyrazolyl-(1)]biphenyl 40%, MAS	A,r	FE24
EA021 0621	419 f	E_p 0.770	SCE	$i_p/v^{\frac{1}{2}}$	1.10±0.02	sttd	1	$dE_p/d\log v$ $dE_p/d\log C$ $E_p-E_{p/2}$	19.3 ±0.7 21.0 ±1.6 40	sttd	-	A,≠,r	
		0.62			-	-		-				C	
		0.40										C	
EA021 0621	419 f	E_p 0.45F	SCE	i_p	5F	-	-	-	-	-	-	A,0 on first scan	
		0.76F			29F	sttd	1					A	
		0.60F			18F		-					C	
		0.41F			6F							C	
EA019 0215	-	$E_{\frac{1}{2}}$ -1.05	SCE	I	1.47	sttd	1	-	-	-	-	$C,i_d,Mc(E_{\frac{1}{2}}),p$	FE25
		-1.25			-		1					C,i_a	
		-1.57					-					C	
JA095 5482	42e f	$E_{\frac{1}{2}}$ -1.00	SCE	I	0.65	QE	1	Tomeš	50	sttd	dimer,UVS	$C,R($from PV$),E_{\frac{1}{2}}$ corr for E_j	FE26
		-1.96			1.4		-	-				C,redn of Na(I)	
JA095 5482	42e f	E_p -1.10	SCE	$i_p/v^{\frac{1}{2}}$	4.4	-	-	-	-	-	-	C,r	
		-1.96			-							C,redn of Na(I),R	
		-1.90			-							A,R	
		-0.22			1.9							C	
		-1.10			4.1	-	-	-	-	-	-	C,r	
		-0.22			2.4							A	
		-1.10			4.3	-	-	-	-	-	-	C,r	
		-0.22			2.4							A	
JA095 5482	42e f	$E_{\frac{1}{2}}$ -1.06	SCE	I	0.91	-	-	Tomeš	48	sttd	dimer,UVS	$C,R($from PV$),E_{\frac{1}{2}}$ corr for E_j,$i_\ell=kh^{0.55}$,r	FE27
		-1.96			2				-			C,redn of Na(I)	
JA095 5482	42e f	E_p -1.16	SCE	$i_p/v^{\frac{1}{2}}$	2.8	-	-	-	-	-	-	C,R for v > 1E4,r	
		-1.96			-							C,redn of Na(I),R	
		-1.90			-							A,R	
		-0.28			1.26							A	

CONT

FE27 (CONT.) $C_{21}H_{27}N_7NaO_{17}P_3$

Code No.	Empirical Formula	Name and C.A. Number	Structural Formula	Solvent	Tech.	Medium	μ, M	pH	T, °C	Electrodes	App.	Experimental Parameters
FE27	$C_{21}H_{27}N_7NaO_{17}P_3$	sodium nicotinamide adenine dinucleotide phosphate	Table II	DMSO	VR	Et_4NClO_4 0.1	-	-	-	HMDE// SCE//Rb^+	35AFO	$0 \to -1.5 \to 0 \to -1.5$ V, C=ns, v=150
												v=200
FE28	$C_{21}H_{27}N_7O_{14}P_2$	nicotinamide adenine dinucleotide C.A. 53-84-9	Table II	DMSO	PY	Et_4NClO_4 0.1	-	-	-	DME//SCE //Rb^+	35AFO	C=ns, m=1.02, t=5.0, h=70
						Et_4NClO_4 0.1 BENZ						
					PV	Et_4NClO_4 0.1	-	-	-	DME//SCE //Rb^+	35AFO	C=0.82, f=50, Δe=10
						Et_4NClO_4 0.1 BENZ 2E-3						
					VR	Et_4NClO_4 0.1	-	-	-	HMDE// SCE//Rb^+	35AFO	$0 \to -2.2 \to 0 \to -2.2$ V, C=ns, v=100
												v=480
												v=2.5E3
						Et_4NClO_4 0.1 BENZ (2-3)E-3						C=1.03
FE29 bo12	$C_{21}H_{30}O_2$	progesterone C.A. 57-83-0	Table II-2	EtOH 50	PY	BR	-	6.0	-	DME//SCE	OAO	C=0.5, m=2.3, t=2.9, h=60
FE30	$C_{21}H_{30}O_4$	Δ^4-pregnen-17α,21-diol-3,20-dione C.A. 152-58-9	Table II	EtOH 50	PY	BR	-	4.6	-	DME//SCE	O4AO	C=0.5, m=2.3, t=2.9, h=60
								6.0				
								8.2				
								9.9				
								11.4				
FE31	$C_{22}H_{14}O_2$	2,3-diphenyl-1,4-naphthoquinone C.A. 33753-12-7	Table II	DMF	VR	Bu_4NClO_4 0.5	-	-	20	Pt: xxdi// Ag/AgCl	125A2	C=ns, Pt Aux

TABLE I. Electrochemical Data

$C_{22}H_{14}O_2$ FE31

| Ref. | C/M | Charact. Potential | | Response Const. | | n | | Electrokinetic Data | | | Products and | Description and | Code |
		Value	vs.		Value	Tech.		Parameter	Value	From	Identification	Remarks	No.
JA095 5482	42e f	E_p	-1.16	SCE	$i_p/v^{1/2}$	2.5	-	-	-	-	-	-	C,r	FE27
			-0.28			1.55							A	
			-1.16			2.5	-	-	-	-	-	-	C,r	
			-0.28			1.34							A	
JA095 5482	42e f	$E_{1/2}$	-0.98	SCE	I	1.20	QE	1	Tomeš	46	sttd	dimer, UVS	C,R(from PV), $i_\ell k h^{0.55}$, r	FE28
			-0.98		-	-	sttd	1	-	-	-	-	C, i_d, r	
			-1.99					-					C, $i_{\ell,2}/i_{\ell,1}$ ↑ and → 1 as [BENZ]/[NAD⁺] → 8 for pH > 12.2	
JA095 5482	42e f	E_{su}	-1.02	SCE	-	-	sttd	1	$\Delta E_{su/2}$	88	sttd	-	C,W,R,r	
			-1.02		$i_{su}(u)$	1	-	-	-	-	-	-	C,W,R,r	
			-2.15			1							C	
JA095 5482	42e f	E_p	-0.33	SCE	$i_p/v^{1/2}$	-	-	-	-	-	-	-	C,r	
			-1.03			12.9							C	
			-0.23			3.9							A	
			-0.33			-							C,r	
			-1.05			13.0							C	
			-0.23			4.4							A	
			-0.33			-							C,r	
			-1.06			12.0							C,R for v > 1E4	
			-0.23			7.0							A	
			-1.03		$i_p(u)$	1	-	-	-	-	-	-	C,r	
			-1.99			3							C	
C0027 2447	-	$E_{1/2}$	-1.365	SCE	i_d/C	0.60	-	-	-	-	-	-	C,p	FE29 bo12
C0027 2447	-	$E_{1/2}$	-1.29	SCE	-	-	-	-	$dE_{1/2}/dpH$	58F	-	-	C,p	FE30
			-1.380		i_d/C	0.54	-	-		58F		-	C,p	
			-1.50F		-	-	-	-		58F		-	C,0 for pH > 8.2, p	
			-1.71F							0F			C	
			-1.71F		-	-				20F			C,p	
			-1.74F		-	-	-	-		-	-	-	C,p	
JA095 6688	-	$E_{1/2}$	-0.73	Ag/AgCl	-	-	-	-	-	-	-	-	C,p	FE31
			-1.52										C	

FE32 C$_{22}$H$_{16}$BrN$_5$S

Code No.	Empirical Formula	Name and C.A. Number	Structural Formula	Solvent	Tech.	Medium		μ, M	pH	T, °C	Electrodes	App.	Experimental Parameters
FE32	C$_{22}$H$_{16}$BrN$_5$S	4-(4-bromophenylazo)-3,5-diphenyl-1-thiocarbamoylpyrazole C.A. 24743-48-4	Table II	DMF	PY	KCl	1	-	-	-	DME//SCE	OAO	C=0.1,EC=3.75
FE33	C$_{22}$H$_{16}$ClN$_5$S	4-(3-chlorophenylazo)-3,5-diphenyl-1-thiocarbamoylpyrazole C.A. 24749-16-4	Table II	DMF	PY	KCl	1	-	-	-	DME//SCE	OAO	C=0.1,EC=3.75
FE34	C$_{22}$H$_{16}$ClN$_5$S	4-(4-chlorophenylazo)-3,5-diphenyl-5-thiocarbamoylpyrazole C.A. 24743-47-3	Table II	DMF	PY	KCl	1	-	-	-	DME//SCE	OAO	C=0.1,EC=3.75
FE35	C$_{22}$H$_{17}$N$_5$S	3,5-diphenyl-4-phenylazo-4-thiocarbamoylpyrazole C.A. 24749-13-1	Table II	DMF	PY	KCl	1	-	-	-	DME//SCE	OAO	C=0.1,EC=3.75
FE36	C$_{22}$H$_{19}$N$_3$	3-(4-dimethylaminophenylimino)-2-phenylindole C.A. 24279-81-0	Table II	DMF	PY	KClO$_4$	0.02	-	-	20±0.1	DME//SCE (NaCl, aq)	23AO	C=ns,m=4.84, t=2.9,h=80, Hg Aux
						KClO$_4$ 3,4-dimethylphenol	0.05 3.4E-3						C=0.45
						KClO$_4$ 3,4-dimethylphenol	0.05 2.3E-4						
FE37	C$_{22}$H$_{19}$N$_3$O	3-(4-dimethylaminophenylimino)-2-phenylindole 1-oxide C.A. 5169-66-4	Table II	DMF	PY	KClO$_4$	0.02	-	-	17.5±0.1	DME//SCE (NaCl, aq)	23AO	C=0.4Ap, m=1.71, t=2.9,h=80, Hg Aux
FE38	C$_{22}$H$_{23}$BrN$_2$S$_2$	3-ethyl-2-[(3-ethyl-2(3H)-benzothiazoylidene)-2-methyl-1-propenyl]benzothiazolium bromide C.A. 1745-32-0	Table II	MeOH 40	IL	KCl	1.2	-	-	-	g(Nujol): xxdi// SCE	O--	A=0.126
				MeOH 50	PY	Me$_4$NOH	0.1	-	-	-	DME//SCE	O--	C=0.05, m=2.61, t=3.0,h=50
FE39	C$_{22}$H$_{31}$FO$_4$	6α-fluoro-17,21-dihydroxy-16α-methylpregn-4-ene-3,20-dione C.A. 378-59-6	Table II	EtOH 50	PY	BR	-	-	3.1	-	DME//SCE	O4AO	C=0.5,m=2.3, t=2.9,h=60
									6.0				
									8.2				
									9.2				
									10.6				

TABLE I. Electrochemical Data

$C_{22}H_{31}FO_4$ FE39

Ref.	C/M	Charact. Potential		Response Const.		n Tech.		Electrokinetic Data			Products and Identification	Description and Remarks	Code No.
		Value	vs.		Value			Parameter	Value	From			
JE054 0411	198 1	$E_{\frac{1}{2}}$	-0.61	SCE	-	-	-	-	-	-	-	-	C,p	FE32
JE054 0411	198 1	$E_{\frac{1}{2}}$	-0.6	SCE	-	-	-	-	-	-	-	-	C,p	FE33
JE054 0411	198 1	$E_{\frac{1}{2}}$	-0.65	SCE	-	-	-	-	-	-	-	-	C,p	FE34
JE054 0411	198 1	$E_{\frac{1}{2}}$	-0.72	SCE	-	-	-	-	-	-	-	-	C,p	FE35
JE036 0147	427 ce	$E_{\frac{1}{2}}$	-1.054	SCE(NaCl, aq)	I	0.76	QE	-	Tomeš	64	-	3-(4-dimethylamino-phenylamino)-2-phenylindole, λ_{max}=308 nm	C,R,i_ℓ=kC for C=0.1-1,i_d,Tc=1 for T=15-30,$\Delta E_{\frac{1}{2}}$=0.19σ,r	FE36
			-1.515			0.81		2± 0.1		95	-	-	C,↕,i_d,i_ℓ=kC for C=0.1-1	
			-0.94F		i_ℓ	2.34F		2	-	-	-	-	C,r	
			-1.0F			1.6F	-	-	-	-	-	-	C,r	
			-1.44F			0.74F							C	
JE041 0067	427 bf	$E_{\frac{1}{2}}$	-1.089	SCE(NaCl, aq)	I	1.69	QE	2Ap	Tomeš	64	-	CP→3-(4-dimethyl-aminophenylimino)-1-hydroxy-2-phenyl-indole,PY	C,i_ℓ=kC for C=0.1-1,i_d,Tc=1 for T=15-30,$\Delta E_{\frac{1}{2}}$=0.2σ,r	FE37
			-1.413			1.44		2Ap		79	-	CP→3-(4-dimethyl-aminophenylimino)-2-phenylindole,PY	C,i_ℓ=kC for C=0.1-1,i_d,Tc=1 for T=15-30	
EA019 0215	-	$E_{p/2}$	0.63	SCE	D	5.10	sttd	1	$E_p-E_{p/2}$	60	sttd	-	A,R,$E_{\frac{1}{2}}$=0.66c,p	FE38
			0.73			-		-		-			A,i_a	
EA019 0215	-	$E_{\frac{1}{2}}$	-1.07	SCE	I	1.42	sttd	1	-	-	-	-	C,i_d+i_a,Mc($E_{\frac{1}{2}}$),p	
			-1.64			-		1					C	
C0027 2447	-	$E_{\frac{1}{2}}$	-0.87	SCE	-	-	-	-	$dE_{\frac{1}{2}}/dpH$	67F	-	-	C,Mc(name),p	FE39
			-1.060		i_d/C	0.54	-	-		67F		-	C,p	
			-1.21F			-	-	-		67F		-	C,p	
			-1.37F			-	-	-		0F		-	C,p	
		$E_{\frac{1}{2}}$	-1.37F			-	-	-		0F		-	C,p	

FE40 $C_{22}H_{31}FO_4$

Code No.	Empirical Formula	Name and C.A. Number	Structural Formula	Solvent	Tech.	Medium		μ, M	pH	T, °C	Electrodes	App.	Experimental Parameters
FE40	$C_{22}H_{31}FO_4$	6β-fluoro-17,21-dihydroxy-16α-methyl-pregn-4-ene-3,20-dione	Table II	EtOH 50	PY	BR		-	3.1	-	DME//SCE	O4AO	C=0.5,m=2.3, t=2.9,h=60
									6.0				
									8.4				
									11.4				
FE41	$C_{23}H_{16}S_3$	2,3,5-triphenyl-1,6,6a-trithiopentalene C.A. 16094-76-1	Table II	MeCN 50 CH_2Cl_2 50	IL	$NaClO_4$	0.1	-	-	-	Pt: xxbu// SCE	---	C=ns
FE42	$C_{23}H_{17}ClO_5$	2,4,6-triphenyl-pyrylium perchlorate C.A. 1484-88-4	Table II	MeCN	VY	Et_4NClO_4	0.1	-	-	20±1	Pt: rodi// SCE	25AO	C=0.2,ω=78, A=7.8E-3
													C=2,ω=78
													C=2,ω=330
				DMF	PY	Et_4NClO_4	0.1	-	-	20±1	DME//SCE	25AO	C=ns
FE43	$C_{23}H_{19}N_5OS$	3,5-diphenyl-4-(2-methoxyphenyl-1-thiocarbamoyl-pyrazole C.A. 24743-49-5	Table II	DMF	PY	KCl	1	-	-	-	DME//SCE	OAO	C=0.1, EC=3.75
FE44	$C_{23}H_{19}N_5OS$	3,5-diphenyl-4-(4-methoxyphenyl)-1-thiocarbamoyl-pyrazole C.A. 24743-51-9	Table II	DMF	PY	KCl	1	-	-	-	DME//SCE	OAO	C=0.1, EC=3.75
FE45	$C_{23}H_{19}N_5S$	3,5-diphenyl-4-(2-methylphenylazo)-1-thiocarbamoyl-pyrazole C.A. 24749-14-2	Table II	DMF	PY	KCl	1	-	-	-	DME//SCE	OAO	C=0.1, EC=3.75
FE46	$C_{23}H_{19}N_5S$	3,5-diphenyl-4-(3-methylphenylazo)-thiocarbamoyl-pyrazole C.A. 53428-42-5	Table II	DMF	PY	KCl	1	-	-	-	DME//SCE	OAO	C=0.1, EC=3.75
FE47	$C_{23}H_{19}N_5S$	3,5-diphenyl-4-(4-methylphenylazo)-1-thiocarbamoyl-pyrazole C.A. 24749-15-3	Table II	DMF	PY	KCl	1	-	-	-	DME//SCE	OAO	C=0.1, EC=3.75
FE48	$C_{23}H_{23}BrN_2S$	1-ethyl-2-[3-(3-ethyl-2(3H)benzothiazolylidene)-1-propenyl]quinolium bromide C.A. 52886-84-7	Table II	MeOH 50	PY	Me_4NOH	0.1	-	-	-	DME//SCE	O--	C=0.05, m=2.61, t=3.0,h=50

TABLE I. Electrochemical Data

$C_{23}H_{23}BrN_2S$ FE48

Ref.	C/M	Charact. Potential		Response Const.		n Tech.		Electrokinetic Data			Products and Identification	Description and Remarks	Code No.
		Value	vs.		Value			Parameter	Value	From			
C0027 2447	-	$E_{\frac{1}{2}}$	-0.64F	SCE	-	-	-	-	$dE_{\frac{1}{2}}/dpH$	63F	-	-	C,Mc(name),p	FE40
			-0.820		-	-	-	-		63F	-	-	C,slightly soluble,p	
			-1.02F		-	-	-	-		16F	-	-	C,p	
			-1.07F		-	-	-	-		16F	-	-	C,p	
JE038 0479	-	E_p	1.16	SCE	-	-	QE	1	-	-	-	-	A,p	FE41
EA021 0497	-	$E_{\frac{1}{2}}$	-	-	-	-	-	-	-	-	-	-	A,O,p	FE42
			-0.410	SCE			QE (-0.5V)	0.98				red radical,ESR	C,R	
			-0.392	-	-	-	-	-	-	-	-	-	C,R,p	
			-0.404	-	-	-	-	-	-	-	-	-	C,R,p	
EA021 0497	-	$E_{\frac{1}{2}}$	-0.32	SCE	$i_\ell(u)$	1	-	-	$E\log i^{2/3}$	62	sttd	-	$C, i_\ell=kC, i_\ell=kh^{\frac{1}{2}},p$	
			-1.22			1				-			C,R	
JE054 0411	198 1	$E_{\frac{1}{2}}$	-0.58	SCE	-	-	-	-	-	-	-	-	C,p	FE43
JE054 0411	198 1	$E_{\frac{1}{2}}$	-0.73	SCE	-	-	-	-	-	-	-	-	C,p	FE44
JE054 0411	198 1	$E_{\frac{1}{2}}$	-0.640	SCE	-	-	-	-	-	-	-	-	C,p	FE45
JE054 0411	198 1	$E_{\frac{1}{2}}$	-0.660	SCE	-	-	-	-	-	-	-	-	C,p	FE46
JE054 0411	198 1	$E_{\frac{1}{2}}$	-0.715	SCE	-	-	-	-	-	-	-	-	C,p	FE47
EA019 0215	-	$E_{\frac{1}{2}}$	-1.07	SCE	I	2.66	sttd	2	-	-	-	-	$C, i_d + i_a, Mc(E_{\frac{1}{2}}),p$	FE48
			-1.66			-		2					C	

FE49 $C_{23}H_{23}IN_2$

Code No.	Empirical Formula	Name and C.A. Number	Structural Formula	Solvent	Tech.	Medium		μ, M	pH	T, °C	Electrodes	App.	Experimental Parameters
FE49	$C_{23}H_{23}IN_2$	1-ethyl-2-[(1-ethyl-2(1H)-quinolinylidene)methyl]quinolinium iodide C.A. 977-96-8	Table II	MeOH 50	PY	Me$_4$NOH	0.1	-	-	-	DME//SCE	O--	C=0.05, m=2.61, t=3.0, h=50
FE50	$C_{23}H_{23}IN_2$	1-ethyl-2-[(1-ethyl-4(1H)-quinolinylidene)methyl]quinolinium iodide C.A. 634-21-9	Table II	MeOH 50	PY	Me$_4$NOH	0.1	-	-	-	DME//SCE	O--	C=0.05, m=2.61, t=3.0, h=50
FE51	$C_{23}H_{23}IN_2S_2$	3-ethyl-2-[5-(3-ethyl-2(3H)benzothiazolylindene)-1,3-pentadienylbenzothiazolium iodide C.A. 514-73-8	Table II	MeOH 50	PY	Me$_4$NOH	0.1	-	-	-	DME//SCE	O--	C=0.05, m=2.61, t=3.0, h=50
FE52	$C_{23}H_{25}BrN_2S_2$	3-ethyl-2-[2-[(3-ethyl-2(3H)benzothiazolylidene)methyl]-1-butenyl]-benzothiazolium bromide C.A. 3028-95-3	Table II	MeOH 50	PY	Me$_4$NOH	0.1	-	-	-	DME//SCE	O--	C=0.05, m=2.61, t=3.0, h=50
FE53	$C_{23}H_{29}ClO_4$	17α-acetoxy-6α-chloropregna-1,4-diene-3,20-dione C.A. 151-69-6	Table II	EtOH 50	PY	BR		-	-	2.8	-	DME//SCE	O4AO	C=0.5, m=2.3, t=2.9, h=60
										6.0				
										9.9				
										11.4				
FE54	$C_{23}H_{29}ClO$	17α-acetoxy-6-chloropregna-4,6-diene-3,20-dione C.A. 302-22-7	Table II	EtOH 50	PY	BR		-	-	2.8	-	DME//SCE	O4AO	C=0.5, m=2.3, t=2.9, h=60
										6.0				
										8.8				
										9.9				
										11.4				
FE55	$C_{23}H_{30}O_4$	17α-acetoxypregna-1,4-diene-3,20-dione	Table II	EtOH 50	PY	BR		-	-	2.8	-	DME//SCE	O4AO	C=0.5, m=2.3, t=2.9, h=60
										4.3				
										6.0				
										11.0				

$C_{23}H_{30}O_4$ FE55

TABLE I. Electrochemical Data

Ref.	C/M	Charact. Potential Value	vs.	Response Const. Value		n Tech.		Electrokinetic Data Parameter	Value	From	Products and Identification	Description and Remarks	Code No.
EA019 0215	-	$E_{\frac{1}{2}}$ -1.13	SCE	I	1.39	sttd	1	-	-	-	-	$C, i_d + i_a, Mc(E_{\frac{1}{2}}), p$	FE49
		-1.68		-			1					C	
EA019 0215	-	$E_{\frac{1}{2}}$ -1.12	SCE	I	1.22	sttd	1	-	-	-	-	$C, i_d + i_a, Mc(E_{\frac{1}{2}}), p$	FE50
		-1.72		-			1					C	
EA019 0215	-	$E_{\frac{1}{2}}$ -0.95	SCE	I	1.33	sttd	1	-	-	-	-	$C, i_d + i_a, Mc(E_{\frac{1}{2}}), p$	FE51
		-1.42		-			1					C	
EA019 0215	-	$E_{\frac{1}{2}}$ -1.08	-	I	1.42	sttd	1	-	-	-	-	$C, i_d + i_a, Mc(E_{\frac{1}{2}}), p$	FE52
		-		-			1					C	
C0027 2447	-	$E_{\frac{1}{2}}$ -0.74F	SCE	-	-	-	-	$dE_{\frac{1}{2}}/dpH$	68F	-	-	$C, dE_{\frac{1}{2}}/dpH$ for pH=2.8-8.8, p	FE53
		-0.965		i_d/C	1.01	-	-		68F		-	C, p	
		-1.22F		-	-	-	-		20F		-	$C, dE_{\frac{1}{2}}/dpH$ for pH > 9.9, p	
		-1.25F		-	-	-	-		20F		-	C, p	
C0027 2447	-	$E_{\frac{1}{2}}$ -0.76F	SCE	-	-	-	-	$dE_{\frac{1}{2}}/dpH$	61F	-	-	$C, dE_{\frac{1}{2}}/dpH$ for pH=2.8-8.8, p	FE54
		-0.970		i_d/C	0.51	-	-		61F		-	C, p	
		-1.430			0.49				-			C	
		-1.125F			-	-	-		61F		-	C, p	
		-1.27F		-	-	-	-		13F		-	$C, dE_{\frac{1}{2}}/dpH$ for pH > 9.9, p	
		-1.29F		-	-	-	-		13F		-	C, p	
C0027 2447	-	$E_{\frac{1}{2}}$ -1.06F	SCE	-	-	-	-	-	-	-	-	C, p	FE55
		-1.11F		-	-	-	-	$dE_{\frac{1}{2}}/dpH$	58F	-	-	$C, dE_{\frac{1}{2}}/dpH$ for pH=4.3-11, p	
		-1.220		i_d/C	0.52	-	-		58F		-	C, p	
		-1.50F			-	-	-		58F		-	C, p	

FE56 $C_{23}H_{30}O_4$

Code No.	Empirical Formula	Name and C.A. Number	Structural Formula	Solvent	Tech.	Medium	μ, M	pH	T, °C	Electrodes	App.	Experimental Parameters
FE56	$C_{23}H_{30}O_4$	17α-acetoxypregna-4,6-diene-3,20-dione C.A. 425-51-4	Table II	EtOH 50	PY	BR	—	—	2.8	—	DME//SCE	O4AO	C=0.5,m=2.3, t=2.9,h=60
									6.0				
									9.9				
									11.4				
FE57	$C_{23}H_{31}ClO_4$	17α-acetoxy-6α-chloropregn-4-ene-3,20-dione C.A. 2477-73-8	Table II	EtOH 50	PY	BR	—	—	3.0	—	DME//SCE	O4AO	C=0.5,m=2.3, t=2.9,h=60
									4.4				
									6.0				
									7.9				
									8.9				
									11.4				
FE58	$C_{23}H_{31}ClO_4$	17α-acetoxy-6β-chloropregn-4-ene-3,20-dione C.A. 2658-74-4	Table II	EtOH 50	PY	BR	—	—	6.0	—	DME//SCE	O4AO	C=0.5,m=2.3, t=2.9,h=60
FE59	$C_{23}H_{31}FO_4$	17α-acetoxy-6α-fluoropregn-4-ene-3,20-dione	Table II	EtOH 50	PY	BR	—	—	3.0	—	DME//SCE	O4AO	C=0.5,m=2.3, t=2.9,h=60
									4.6				
									6.0				
									8.3				
									9.5				
									11.6				
FE60	$C_{23}H_{31}FO_4$	17α-acetoxy-6β-fluoropregn-4-ene-3,20-dione C.A. 336-79-8	Table II	EtOH 50	PY	BR	—	—	3.0	—	DME//SCE	O4AO	C=0.5,m=2.3, t=2.9,h=60
									4.6				
									6.0				
									8.3				
									9.2				
									11.3				

TABLE I. Electrochemical Data

$C_{23}H_{31}FO_4$ FE60

Ref.	C/M	Charact. Potential		Response Const.		n	Tech.	Electrokinetic Data			Products and Identification	Description and Remarks	Code No.
		Value	vs.		Value			Parameter	Value	From			
C0027 2447	-	$E_{\frac{1}{2}}$	-0.84F	SCE	-	-	-	-	$dE_{\frac{1}{2}}/dpH$	72F	-	-	C,$dE_{\frac{1}{2}}/dpH$ for pH=2.8-9.9,p	FE56
			-1.065		i_d/C	0.56	-	-		72F	-	-	C,p	
			-1.35F		-	-	-	-		20F	-	-	C,$dE_{\frac{1}{2}}/dpH$ for pH > 9.9,p	
			-1.38F		-	-	-	-		20F	-	-	C,p	
C0027 2447	-	$E_{\frac{1}{2}}$	-0.70F	SCE	-	-	-	-	$dE_{\frac{1}{2}}/dpH$	80F	-	-	C,$dE_{\frac{1}{2}}/dpH$ for pH=3.0-4.4,p	FE57
			-0.81F		-	-	-	-		95F	-	-	C,$dE_{\frac{1}{2}}/dpH$ for pH=4.4-8.9,p	
			-0.940		i_d/C	1.06	-	-		95F	-	-	C,p	
			-1.12F		-	-	-	-		95F	-	-	C,p	
			-1.18F		-	-	-	-		0F	-	-	C,$dE_{\frac{1}{2}}/dpH$ for pH > 8.9,p	
			-1.18F		-	-	-	-		0F	-	-	C,p	
C0027 2447	-	$E_{\frac{1}{2}}$	-	SCE	i_d/C	1.00	-	-	-	-	-	-	C,M,p	FE58
C0027 2447	-	$E_{\frac{1}{2}}$	-0.90F	SCE	-	-	-	-	$dE_{\frac{1}{2}}/dpH$	50F	-	-	C,$dE_{\frac{1}{2}}/dpH$ for pH=3-4.6,p	FE59
			-0.98F		-	-	-	-		65F	-	-	C,$dE_{\frac{1}{2}}/dpH$ for pH=4.6-8.3,p	
			-1.040		i_d/C	0.84	-	-		65F	-	-	C,p	
			-1.22F		-	-	-	-		65F	-	-	C,p	
			-1.38F		-	-	-	-		5F	-	-	C,$dE_{\frac{1}{2}}/dpH$ for pH=9.5-11.6,p	
			-1.39F		-	-	-	-		5F	-	-	C,p	
C0027 2447	-	$E_{\frac{1}{2}}$	-0.69F	SCE	-	-	-	-	$dE_{\frac{1}{2}}/dpH$	56F	-	-	C,$dE_{\frac{1}{2}}/dpH$ for pH=3.0-4.6,p	FE60
			-0.78F		-	-	-	-		89F	-	-	C,$dE_{\frac{1}{2}}/dpH$ for pH=4.6-8.3,p	
			-0.895		i_d/C	0.85	-	-		89F	-	-	C,p	
			-1.11F		-	-	-	-		89F	-	-	C,p	
			-1.15F		-	-	-	-		0F	-	-	C,$dE_{\frac{1}{2}}/dpH$ for pH > 9.2,p	
			-1.15F		-	-	-	-		0F	-	-	C,p	

FE61 $C_{23}H_{32}O_4$

Code No.	Empirical Formula	Name and C.A. Number	Structural Formula	Solvent	Tech.	Medium	μ, M	pH	T, °C	Electrodes	App.	Experimental Parameters
FE61	$C_{23}H_{32}O_4$	17α-acetoxypregn-4-ene-3,20-dione C.A. 302-23-8	Table II	EtOH 50	PY	BR	–	4.5	–	DME//SCE	O4AO	C=0.5,m=2.3, t=2.9,h=60
								6.0				
								7.9				
								8.8				
								11.4				
FE62	$C_{24}H_{14}Cl_2N_4O_4$	2,7-dichloro-5,10-di(4-nitrophenyl)-5,10-dihydrophenazine C.A. 61228-23-7	Table II	MeCN	VY	Et_4NClO_4 0.1(?) 4-cyanopyridine 0.002	–	–	–	Pt: rodi// Ag/ AgClO$_4$ 0.01, Et$_4$NClO$_4$ 0.1	2-O	C=1?,d=0.1, ω=10,Pt Aux
FE63	$C_{24}H_{14}Cl_4N_2$	2,7-dichloro-5,10-di(4-chlorophenyl)-5,10-dihydrophenazine C.A. 31438-41-2	Table II	MeCN	VY	Et_4NClO_4 0.1	–	–	–	Pt: rodi// Ag/ AgClO$_4$ 0.01, Et$_4$NClO$_4$ 0.1	2-O	C=1?,d=0.1, ω=10,Pt Aux
FE64 CONT	$C_{24}H_{14}K_2O_2$	1,2-diphenylcyclobuta[b]naphthalene-3,8-diol dipotassium salt C.A. 50982-49-5	Table II	DMF	VR	Bu_4NClO_4 1.0	–	–	20	Pt: xxdi// Ag/AgCl	125A2	C=0.3,v=240, Pt Aux
												v=480
												v=940
												v=4.8E3
												v=1E4
												v=4.8E4
						Bu_4NClO_4 0.5				Pt: xxdi// Ag/AgCl// naphthoquinone		C$\leq$0.5, v=1E4, Pt Aux
												$-0.7\rightarrow-1.1$V, C=0.8Ap, v=1E3, Pt Aux
												$-0.15\rightarrow-1.1$ $\rightarrow-0.15\rightarrow-1.1$, v=4E3

TABLE I. Electrochemical Data

$C_{24}H_{14}K_2O_2$ (CONT.) FE64

Ref.	C/M	Charact. Potential		Response	Const. Value	n	Tech.	Electrokinetic Data			Products and Identification	Description and Remarks	Code No.
		Value	vs.					Parameter	Value	From			
C0027 2447	-	$E_{1/2}$	-1.27F	SCE	-	-	-	-	$dE_{1/2}/dpH$	63F	-	-	C,$dE_{1/2}/dpH$ for pH=4.5-7.9,p	FE61
			-1.375		i_d/C	0.60	-	-		63F		-	C,p	
			-1.48_5F		-	-	-	-		63F		-	C,p	
			-1.52F		-	-	-	-		-		-	C,p	
			1.70F										C,$dE_{1/2}/dpH$ for pH > 8.8	
			-1.72F		-	-	-	-		8F		-	C,p	
EA021 0557	-	$E_{1/2}$	1.10	Ag/ AgClO$_4$ 0.01, Et$_4$NClO$_4$ 0.1	-	-	-	-		-	-	radical cation,UVS, VIS,ESR	A,r	FE62
EA020 1019	428 ab	$E_{1/2}$	0.09	Ag/ AgClO$_4$ 0.01, Et$_4$NClO$_4$ 0.1	-	-	sttd	1		-	-	radical cation,ESR, UVS,VIS	A,R,r	FE63
			0.75					1				dication,UVS,VIS	A,R	
JA095 6688	-	E_p	-0.81	Ag/AgCl	$i_{p,2}/i_{p,1}$	-	-	-	$dE_p/dlogv$	-	-	-	A,r	FE64
			-0.33			1.00				44F			A	
			-0.32F		-	-				-			A,r	
			-			1.13F				44F			A	
			-		-	-				-			A,r	
			-0.30F			1.29F				44F			A	
			-		-	-				-			A,r	
			-0.27F			1.29F				44F			A	
			-		-	-				-			A,r	
			-0.26F			1.23F				25F			A	
			-		-	-				-			A,r	
			-0.243F			1.00F				25F			A	
			-0.25		-	-	sttd	1	$E_p-E_{p/2}$	41±8	sttd	-	A,≠,r	
			-0.90					1		57±2			A,R	
			-0.81F		i_p	140F	-	-		-	-	-	A,ec,QI,$E_p\neq f(v)$ for v=(1-9)E3,r	
			-0.90F			140F							C,QI	
			-0.25F			160F	-	-		-	-	-	A,≠,r	
			-0.4F			70F							C,X,≠	
			-0.81F			180F							A,QI	
			-0.90F			55F							C,QI	CONT

FE64 (CONT.) $C_{24}H_{14}K_2O_2$

Code No.	Empirical Formula	Name and C.A. Number	Structural Formula	Solvent	Tech.	Medium		μ, M	pH	T, °C	Electrodes	App.	Experimental Parameters
FE64	$C_{24}H_{14}K_2O_2$	1,2-diphenylcyclobuta[b]naphthalene-3,8-diol dipotassium salt	Table II	DMF	VR	Bu_4NClO_4	1.0	-	-	20	Pt: xxdi// Ag/AgCl// naphthoquinone	125A 2	C=1.5, v=200, Pt Aux
													v=6E4
FE65	$C_{24}H_{16}Cl_4N_2$	tetrakis(4-chlorophenyl)hydrazine C.A. 31438-34-3	$[(4-ClC_6H_4)_2N]_2$	MeCN	VY	Et_4NClO_4	0.1	-	-	-	Pt: rodi// Ag/AgClO$_4$ 0.01, Et$_4$NClO$_4$ 0.1	2-O	C=ns, d=0.1, ω=10, Pt Aux
FE66	$C_{24}H_{16}N_6O_8$	tetrakis(4-nitrophenyl)hydrazine C.A. 31438-32-1	$[(4-O_2NC_6H_4)_2N]_2$	MeCN	VY	Et_4NClO_4	0.1	-	-	-	Pt: rodi// Ag/AgClO$_4$ 0.01, Et$_4$NClO$_4$ 0.1	2-O	C=ns, d=0.1, ω=10, Pt Aux
FE67	$C_{24}H_{16}O_2$	1,2-diphenylcyclobuta[b]naphthalene-3,8-diol C.A. 29510-58-5	Table II	EtOH 50	PY	ACET	0.1	-	5.6	-	DME//SCE	---	C=ns
FE68	$C_{24}H_{18}Br_2N_2$	N,N'-di(4-bromophenyl)-N-phenyl-phenylenediamine C.A. 62958-80-9	Table II	MeCN	VY	Et_4NClO_4	0.1	-	-	-	Pt: rodi// Ag/AgClO$_4$ 0.01, Et$_4$NClO$_4$ 0.1	2-O	C=ns, d=0.1, ω=10, Pt Aux
FE69	$C_{24}H_{18}N_2$	5,10-dihydro-5,10-diphenylphenazine C.A. 3665-72-3	Table II	MeCN	VY	Et_4NClO_4	0.1	-	-	-	Pt: rodi// Ag/AgClO$_4$ 0.01, Et$_4$NClO$_4$ 0.1	2-O	C=ns, d=0.1, ω=10, Pt Aux
FE70	$C_{24}H_{18}O_4$	2,3-bis(4-methoxyphenyl)-1,4-naphthoquinone C.A. 50982-55-3	Table II	DMF	IL	Bu_4NClO_4	0.5	-	-	20	Pt: xxdi// Ag/AgCl	125A 2	C=ns
FE71	$C_{24}H_{20}ClO_4P$	tetraphenylphosphonium perchlorate C.A. 19859-51-9	$(C_6H_5)_4P^+\ ClO_4^-$	MeCN	QE	Et_4NClO_4	0.1	-	-	-	Hg: nsns// Ag/AgClO$_4$ 0.01, Et$_4$NClO$_4$ 0.1	2AO	C=34, E$_{app}$ on pl of first wave, Pt Aux
				DMF	PY	Et_4NClO_4	0.1	-	-	-28	DME// Ag/AgClO$_4$ 0.01, Et$_4$NClO$_4$ 0.1	2AO	C=1.5, Hg Aux

CONT

TABLE I. Electrochemical Data 275

$C_{24}H_{20}ClO_4P$ (CONT.) FE71

| Ref. | C/M | Charact. Potential | | Response | Const. | n | | Electrokinetic Data | | | Products and | Description and | Code |
		Value	vs.		Value	Tech.		Parameter	Value	From	Identification	Remarks	No.
JA095 6688	-	E_p	-0.8	Ag/AgCl	$i_{p,2}/i_{p,1}$	-	-	-	$E_p-E_{p/2}$	40	sttd	-	A,r	FE64
			-0.3			0.06F				-			A	
			-			-	-	-		-	-	-	A,r	
			-			1.23F							A	
EA020 1019	-	$E_{\frac{1}{2}}$	0.665	Ag/AgClO$_4$ 0.01, Et$_4$NClO$_4$ 0.1	-	-	sttd	1	-	-	-	-	A,R,r	FE65
			1.43					-					A	
EA020 1019	-	$E_{\frac{1}{2}}$	1.20	Ag/AgClO$_4$ 0.01, Et$_4$NClO$_4$ 0.1	-	-	-	-	-	-	-	-	A,r	FE66
JA092 4139	-	$E_{\frac{1}{2}}$	0.163	SCE	-	-	sttd	2	-	-	-	-	A,⊭,i_ℓ ↓ as $\underline{t}$ ↑,Σ,p	FE67
EA021 1171	147 jkl	$E_{\frac{1}{2}}$	0.31	Ag/AgClO$_4$ 0.01, Et$_4$NClO$_4$ 0.1	-	-	-	-	-	-	-	radical cation	A,r	FE68
			0.65									diimine monocation	A	
EA020 1019	428 ab	$E_{\frac{1}{2}}$	-0.11	Ag/AgClO$_4$ 0.01, Et$_4$NClO$_4$ 0.1	-	-	sttd	1	-	-	-	radical cation,ESR, UVS,VIS	A,R,r	FE69
			0.64					1				dication,UVS,VIS	A,R	
JA095 6688	-	$E_{\frac{1}{2}}$	-0.78	Ag/AgCl	-	-	-	-	-	-	-	-	C,p	FE70
			-1.55										C	
EA020 0021	-	-	-	-	-	-	sttd	1	-	-	-	triphenylphosphine 50%,NMR,MAS,PY;tri-phenylphosphine oxide 50%,NMR,MAS, PY;benzene 100%, GSC ;biphenyl trace	C,similar results in DMF,r	FE71
EA020 0021	-	-	-	-	i_ℓ/C	0.92	-	-	-	-	-	-	C,r	

CONT

FE71 (CONT.) $C_{24}H_{20}ClO_4P$

Code No.	Empirical Formula	Name and C.A. Number	Structural Formula	Solvent	Tech.	Medium		μ, M	pH	T, °C	Electrodes	App.	Experimental Parameters
FE71	$C_{24}H_{20}ClO_4P$	tetraphenylphosphonium perchlorate	$(C_6H_5)_4P^+ ClO_4^-$	DMF	PY	Bu_4NI	0.1	—	—	20	DME// Ag/ AgClO_4 0.01, Et_4NClO_4 0.1	2AO	C=1.5, Hg Aux
						Bu_4NI OH^-	0.1 1.5E-3						
					PR	Et_4NClO_4	0.1	—	—	-28	DME// Ag/ AgClO_4 0.01, Et_4NClO_4 0.1	2AO	C=0.87, v=50, Hg Aux
													v=500
													v=5E3
													v=5E4
													v=2.5E5
						Bu_4NI	0.1			20			C=1.5, v=?, MP Aux
FE72	$C_{24}H_{20}Cr$	bis(η^6-biphenyl)-chromium(0) C.A. 33085-81-3	$(\eta-C_6H_5C_6H_5)_2Cr$	DMF	VY	Et_4NClO_4 GEL	1 0.04?	—	—	—	Pt: pdhb// SCE	3A-	C=1?, Pt Aux
FE73 bp22	$C_{24}H_{20}CrI$	bis(η^6-biphenyl)-chromium(I) iodide C.A. 12099-17-1	$(\eta-C_6H_5C_6H_5)_2Cr^+$ I^-	DMF	PY	Bu_4NClO_4	0.1	—	—		DME//SCE	12-O	C=1
						$LiClO_4$	0.1						
FE74	$C_{24}H_{20}N_2$	N,N'-diphenyl-benzidine C.A. 531-91-9	$(4-C_6H_5NHC_6H_4)_2$	MeCN	VA	Et_4NClO_4	0.1	—	—	—	Pt: xxdi// Ag/Ag+ 0.01	2-O	-0.1→0.7→ -0.1V, C=0.3, d=0.1, v=18, Pt Aux
CONT						Et_4NClO_4 $HClO_4$	0.1 4.5E-4						

TABLE I. Electrochemical Data

$C_{24}H_{20}N_2$ (CONT.) FE74

Ref.	C/M	Charact. Potential		Response Const.		n Tech.		Electrokinetic Data			Products and Identification	Description and Remarks	Code No.
		Value	vs.		Value			Parameter	Value	From			
EA020 0021	-	$E_{\frac{1}{2}}$	-1.8F	SCE	i_ℓ	3F	-	-	-	-	-	-	C, DMF <u>must</u> be OH^- free, $i_\ell/C=1.5$ in Et_4NClO_4 0.1, i_d, r	FE71
			-2.5F			2F							C, i_k	
			-2.7F			3F							C	
			-2.5F			4.2F	-	-	-	-	-	-	C, r	
			-2.8F			4.3F							C	
EA020 0021	-	-	-	-	$i_p/v^{\frac{1}{2}}$	10.3	-	-	-	-	-	-	C, r	
			-			-							C	
			-	-		10.2	-	-	-	-	-	-	C, r	
			-			-							C, i_{p_2}/i_{p_1} ↓ as v ↑	
			-	-		9.5	-	-	-	-	-	-	C, r	
			-			-							C, i_{p_2}/i_{p_1} ↓ as v ↑	
			-	-		9.6	-	-	-	-	-	-	C, r	
			-			-							C, i_{p_2}/i_{p_1} ↓ as v ↑	
			-			-							C	
			-	-		9.1	-	-	-	-	-	-	C, r	
			-			-							C, 0 for v > 1.8E5 if C=1 or for v > 9E4 if C=0.53	
			-			-							C	
		E_p	-1.8F	SCE	i_p	2.7F	-	-	-	-	-	-	C, ⇌, r	
			-2.5F			1.6F							C	
			-2.8F			2.7F							C	
			-2.7F			2.7F							A	
			-2.5F			1.5F							A	
JE051 0226	443 a	$E_{\frac{1}{2}}$	-0.74	SCE	-	-	-	-	Elog	58	-	-	A, R, p	FE72
			0.93							140			A, ⇌	
									βn_b	0.42	Elog			
JE050 0351	-	$E_{\frac{1}{2}}$	-0.625	SCE	-	-	sttd	1	Elog	58	sttd	-	C, R(PK), p	FE73 bp22
			-0.625		-	-		1		58		-	C, p	
EA020 1011	419 d	E_p	0.36F	Ag/Ag$^+$ 0.01	i_p	9.6F	-	-	-	-	-	-	A, r	FE74
			0.50F			6.8F							A	
			0.44F			6.8F							C	
			0.32F			6.8F							C	
			0.52F			14.8F	-	-	-	-	-	-	A, r	
	419 c		0.45F			14.0F							C	

CONT

FE74 (CONT.) $C_{24}H_{20}N_2$

Code No.	Empirical Formula	Name and C.A. Number	Structural Formula	Solvent	Tech.	Medium	μ, M	pH	T, °C	Electrodes	App.	Experimental Parameters
FE74	$C_{24}H_{20}N_2$	N,N'-diphenyl-benzidine	$(4-C_6H_5NHC_6H_4)_2$	MeCN	VA	Et_4NClO_4 0.1 2,6-lutidine 1.2E-3	-	-	-	Pt: xxdi// Ag/Ag+ 0.01	2-0	$-0.4 \to 0.6 \to -0.4$ V, C=0.6, v=23, Pt Aux
					VY	Et_4NClO_4 0.1	-	-	-	Pt: rodi// Ag/Ag+ 0.01	2-0	C=0.5, d=0.1, ω=10, Pt Aux
FE75 dg38	$C_{24}H_{20}N_2$	1,1,2,2-tetraphenyl-diazane C.A. 632-52-0	$[(C_6H_5)_2N]_2$	MeCN	VY	Et_4NClO_4 0.1	-	-	-	Pt: rodi// Ag/ AgClO4 0.01, Et_4NClO_4 0.1	2-0	C=ns, d=0.1, ω=10, Pt Aux
FE76	$C_{24}H_{20}N_2$	N,N,N'-triphenyl-phenylenediamine C.A. 19606-98-5	$4-(C_6H_5)_2N-C_6H_4NHC_6H_5$	MeCN	VY	Et_4NClO_4 0.1	-	-	-	Pt: rodi// Ag/ AgClO4 0.01, Et_4NClO_4 0.1	2-0	C=ns, d=0.1, ω=10, Pt Aux
FE77	$C_{24}H_{21}N_5S$	4-(2,4-dimethylphen-ylazo)-3,5-diphenyl-1-thiocarbamoyl-pyrazole C.A. 24743-54-2	Table II	DMF	PY	KCl 1	-	-	-	DME//SCE	OAO	C=0.1, EC=3.75
FE78	$C_{24}H_{21}N_5S$	4-(2,6-dimethylphen-ylazo)-3,5-diphenyl-1-thiocarbamoyl-pyrazole C.A. 24743-56-4	Table II	DMF	PY	KCl 1	-	-	-	DME//SCE	OAO	C=0.1, EC=3.75
FE79	$C_{24}H_{24}N_2O_4$	N,N,N',N'-tetra-acetyl-1,4-diamino-1,4-diphenyl-1,3-butadiene C.A. 60389-45-9	$[(CH_3CO)_2NC-(C_6H_5):CH]_2$	DMF	VA	$LiClO_4$ 0.1	-	-	-	Pt: nsns// Ag/AgI, I- 0.1	2-0	C=ns, A=0.015, v=100
FE80	$C_{24}H_{26}N_4$	1-diphenylmethyl-4-[(6-methyl-2-pyridyl)methylene-amino]piperazine C.A. 3601-19-2	Table II	MeOH 10	PT	BR	0.12	0	-	DME//SCE	2AO	C=ns, m=2.57, t=3.46, h=55, v=1-2, Pt Aux
								1.0				
								3.6				
								9.1				
								10				
								14.0				

TABLE I. Electrochemical Data

$C_{24}H_{26}N_4$ FE80

Ref.	C/M	Charact. Potential		Response Const.		n Tech.	n	Electrokinetic Data			Products and Identification	Description and Remarks	Code No.
		Value	vs.		Value			Parameter	Value	From			
EA020 1011	-	E_p	0.30F	Ag/Ag^+ 0.01	i_p	41F	-	-	-	-	-	-	A,r	FE74
			0.22F			25F							C	
			-0.04F			3F							C	
EA020 1011	419 d	$E_{\frac{1}{2}}$	0.352	Ag/Ag^+ 0.01	$i_\ell(u)$	1	sttd	1	-	-	-	radical cation,ESR, VIS,UVS	A,i_d,r	
			0.485			1		1				dication,"relatively stable",VIS,UVS	A,i_d	
EA020 1019	-	$E_{\frac{1}{2}}$	0.49	$Ag/AgClO_4$ 0.01, Et_4NClO_4 0.1	-	-	-	-	-	-	-	-	A,r	FE75 dg38
			1.33										A	
EA021 1171	147 jkl	$E_{\frac{1}{2}}$	0.195	$Ag/AgClO_4$ 0.01, Et_4NClO_4 0.1	-	-	-	-	-	-	-	radical cation	A,r	FE76
			0.63									diimine monocation	A	
JE054 0411	198 1	$E_{\frac{1}{2}}$	-0.620	SCE	-	-	-	-	-	-	-	-	C,p	FE77
JE054 0411	198 1	$E_{\frac{1}{2}}$	-0.600	SCE	-	-	-	-	-	-	-	-	C,p	FE78
EA021 0345	435 a	E_p	>2.2	Ag/AgI, I^- 0.1	-	-	QE (2.5 V)	4	-	-	-	3,6-diphenylpyrid-azine 72%	A,≠,r	FE79
AA094 0119	431 a	$E_{\frac{1}{2}}$	-0.27	SCE	-	-	sttd	3	$dE_{\frac{1}{2}}/dpH$	110F	-	-	C,pK_1'=1.0,$dE_{\frac{1}{2}}/dpH$ for pH=0-1,r	FE80
			-0.38F		-	-		3		58F	-	-	C,$dE_{\frac{1}{2}}/dpH$ for pH=1-3.6,r	
			-0.53		-	-		3		91F	-	-	C,pK_2=3.6,$dE_{\frac{1}{2}}/dpH$ for pH=3.6-9.1,r	
			-1.03F		-	-	-	-		44F	-	-	C,pK_2'=9.1,$dE_{\frac{1}{2}}/dpH$ for pH > 9.1	
			-1.07F		-	-	-	-		44F	-	-	C,r	
			-1.23		-	-	-	-		44F	-	-	C,r	

FE81 $C_{24}H_{32}Cl_2P_2$

Code No.	Empirical Formula	Name and C.A. Number	Structural Formula	Solvent	Tech.	Medium		μ, M	pH	T, °C	Electrodes	App.	Experimental Parameters
FE81	$C_{24}H_{32}Cl_2P_2$	1,1,4,4-tetraethyl-1,4-dihydro-2,5-diphenyl-1,4-diphosphorinium dichloride C.A. 20439-99-0	Table II	DMF	PY	Et_4NClO_4	0.1	-	-	-	DME//SCE	2--	ns
FE82	$C_{24}H_{32}O_4$	17α-acetoxy-6α-methylpregna-1,4-diene-3,20-dione C.A. 151-68-8	Table II	EtOH 50	PY	BR	-	-	2.8	-	DME//SCE	O4AO	C=0.5,m=2.3, t=2.9,h=60
									6.0				
									10.0				
									11.4				
FE83	$C_{24}H_{32}O_4$	17α-acetoxy-6-methylpregna-4,6-diene-3,20-dione C.A. 595-33-5	Table II	EtOH 50	PY	BR	-	-	2.8	-	DME//SCE	O4AO	C=0.5,m=2.3, t=2.9,h=60
									4.3				
									6.0				
									9.9				
									11.4				
FE84	$C_{24}H_{34}O_4$	17α-acetoxy-6α-methylpregn-4-ene-3,20-dione C.A. 71-58-9	Table II	EtOH 50	PY	BR	-	-	4.5	-	DME//SCE	O4AO	C=0.5,m=2.3, t=2.9,h=60
									6.0				
									7.9				
									8.8				
									11.4				
FE85	$C_{24}H_{38}B_2F_8N_2$	N,N'-diheptyl-4,4'-bipyridylium tetrafluoroborate C.A. 36530-85-5	Table II	MeCN	PY	NH_4BF_4	0.02	-	-	-	DME// Ag/AgCl	123A O	C=1, Ag/AgCl Aux
						NH_4BF_4	0.1						
						KCl	0.01						
						KCl	0.1						
						KCl	4						

TABLE I. Electrochemical Data

$C_{24}H_{38}B_2F_8N_2$ FE85

Ref.	C/M	Charact. Potential		Response Const.		n Tech.		Electrokinetic Data			Products and Identification	Description and Remarks	Code No.
		Value	vs.	Response	Value			Parameter	Value	From			
JE042 0309	-	$E_{\frac{1}{2}}$	-0.774	SCE	-	-	sttd	1	-	-	-	-	C,p	FE81
C0027 2447	-	$E_{\frac{1}{2}}$	-1.04F	SCE	-	-	-	-	$dE_{\frac{1}{2}}/dpH$	63F	-	-	C,$dE_{\frac{1}{2}}/dpH$ for pH=2.8-10,p	FE82
			-1.220		i_ℓ/C	0.64	-	-		63F		-	C,p	
			-1.49F		-	-	-	-		43F		-	C,$dE_{\frac{1}{2}}/dpH$ for pH > 10,p	
			-1.55F		-	-	-	-		43F		-	C,p	
C0027 2447	-	$E_{\frac{1}{2}}$	-0.91F	SCE	-	-	-	-	$dE_{\frac{1}{2}}/dpH$	47F	-	-	C,$dE_{\frac{1}{2}}/dpH$ for pH=2.8-4.3,p	FE83
			-0.98F		-	-	-	-		68F		-	C,$dE_{\frac{1}{2}}/dpH$ for pH=4.3-9.9	
			-1.095		i_ℓ/C	0.71	-	-		68F		-	C,p	
			-1.36F		-	-	-	-		40F		-	C,$dE_{\frac{1}{2}}/dpH$ for pH > 9.9,p	
			-1.42F		-	-	-	-		40F		-	C,p	
C0027 2447	-	$E_{\frac{1}{2}}$	-1.27F	SCE	-	-	-	-	$dE_{\frac{1}{2}}/dpH$	96F	-	-	C,$dE_{\frac{1}{2}}/dpH$ for pH=4.5-7.9,p	FE84
			-1.380		i_ℓ/C	0.64	-	-		96F		-	C,p	
			-1.50F		-	-	-	-		96F		-	C,p	
			-1.52F		-	-	-	-		-		-	C,0 for pH > 8.8,p	
			-1.75F							0F			C,$dE_{\frac{1}{2}}/dpH$ for pH > 8.8	
			-1.75F		-	-	-	-		0F		-	C,p	
JS121 1555	-	$E_{\frac{1}{2}}$	-0.50	Ag/AgCl	-	-	-	-	$dE_{\frac{1}{2}}/dpH$	+28F	-	-	C,p	FE85
			-0.81F							-43F			C	
			-0.48F		-	-	-	-		+28F		-	C,order of shift of $E_{\frac{1}{2}}$ with anion is $F^- > Cl^- > Br^- > BF_4^- > ClO_4^-$	
			-0.84F							-43F			C	
			-0.64F		-	-	-	-		-		-	C,p	
			-0.72F										C	
			-0.62F		-	-	-	-		-		-	C,p	
			-0.75F										C	
			-0.52F										C,p	
			-0.82F										C	

FE86 $C_{25}H_{24}N_3Na$

Code No.	Empirical Formula	Name and C.A. Number	Structural Formula	Solvent	Tech.	Medium	μ, M	pH	T, °C	Electrodes	App.	Experimental Parameters
FE86	$C_{25}H_{24}N_3Na$	2-(4-dimethylamino-phenyl)-4,5-bis-(4-methylphenyl)imidazole, sodium salt C.A. 31909-33-8	Table II	MeCN	VY	Bu_4NClO_4 0.1	-	-	-	Pt: rodi// Ag/AgCl// cobaltocinium	2AO	C=1.9, A=0.012, v=15.3
FE87	$C_{25}H_{24}N_3NaO_2$	4,5-bis(4-methoxy-phenyl)-2-(4-dimethylaminophenyl)imidazole, sodium salt C.A. 31909-34-9	Table II	MeCN	VY	Bu_4NClO_4 0.1	-	-	-	Pt: rodi// Ag/AgCl// cobaltocinium	2AO	C=1.9, A=0.012, v=15.3
FE88	$C_{25}H_{25}BrN_2$	1-ethyl-2-[3-(1-ethyl-2(1H)-quinolinylidene)-1-propenyl]quinolinium bromide C.A. 2670-67-9	Table II	MeOH 40	IL	KCl 1.2	-	-	-	g(Nujol): xxdi// SCE	0--	C=ns, A=0.126
				MeOH 50	PY	Me_4NOH 0.1	-	-	-	DME//SCE	0--	C=0.05, m=2.61, t=3.0, h=50
FE89	$C_{25}H_{25}ClN_2O_4S_2$	3-ethyl-2-[7-(3-ethyl-2(3H)-benzo-thiazolylidene)-1,3,5-heptatrienyl]-benzothiazolium perchlorate C.A. 22268-66-2	Table II	MeOH 50	PY	Me_4NOH 0.1	-	-	-	DME//SCE	0--	C=0.05, m=2.61, t=3.0, h=50
FE90 bp68	$C_{25}H_{31}N_3$	tris(4-dimethyl-aminophenyl)methane C.A. 603-48-5	$[4-(CH_3)_2N-C_6H_4]_3CH$	MeCN	VA	$LiClO_4$ ns H_2O 0.1Ap	-	-	-	Pt: xxns// Ag/ $AgClO_4$ 0.01	28AO	0→0.8→0V, C=0.2, A=0.05, v=2, Pt Aux
											2AO	C=1, v=100
					VY	$LiClO_4$ ns H_2O 0.1Ap	-	-	-	Pt: xxns// Ag/ $AgClO_4$ 0.01	2AO	C=1, Pt Aux
FE91	$C_{26}H_{16}K_2O_2$	1,2-dibenzylidene-cyclobutano[3,4-b]-3,8-naphthohydro-quinone dipotassium salt C.A. 50982-54-2	Table II	DMF	VR	Bu_4NClO_4 0.5	-	-	20	Pt: xxdi// Ag/AgCl (KCl?)// naphthoquinone	125A2	C≤0.5, Pt Aux
												v=54

TABLE I. Electrochemical Data   283

$C_{26}H_{16}K_2O_2$ FE91

| Ref. | C/M | Charact. Potential | | Response | Const. Value | Tech. | n | Electrokinetic Data | | | Products and Identification | Description and Remarks | Code No. |
		Value	vs.					Parameter	Value	From			
C0036 0575	397 a	$E_{\frac{1}{2}}$	-0.024	Ag/AgCl	$i_\ell(u)$	1.00	Elog(!)	1	αn_a	0.069	Elog	-	A,R,i_d,r	FE86
			0.469			1.10		1		0.066			A,R,i_d,r	
C0036 0575	397 a	$E_{\frac{1}{2}}$	-0.075	Ag/AgCl	$i_\ell(u)$	1.00	Elog(!)	1	αn_a	0.064	Elog	-	A,R,i_d,r	FE87
			0.422			1.12		1		0.063			A,R,i_d,r	
EA019 0215	-	$E_{p/2}$	0.57	SCE	D	-	-	-	-	-	-	-	A,i_a,p	FE88
			0.49			4.00	sttd	1	$E_p-E_{p/2}$	57	sttd		A,R	
			-0.94			-	-	-		-			C	
			-										C(?)	
EA019 0215	-	$E_{\frac{1}{2}}$	-1.10	-	I	2.76	sttd	2	-	-	-	-	C,i_d,Mc($E_{\frac{1}{2}}$),p	
			-1.30			-		-					C,i_a	
			-1.59					2					C	
EA019 0215	-	$E_{\frac{1}{2}}$	-0.77	SCE	I	1.12	sttd	1	-	-	-	-	C,i_d+i_a,Mc($E_{\frac{1}{2}}$),p	FE89
			-			-		1					C	
JE048 0081	-	E_p	0.32F	Ag/AgClO$_4$ 0.01	i_p	23.3F	-	-	-	-	-	-	A,p	FE90 bp68
			0.72F			12.6F							A	
			0.69F			5.2F							C	
			0.22F			5.2F							C	
			0.44F			0.74F	-	-	-	-	-	CP→crystal violet, UVS;bi-(4-dimethyl-iminium-2,5-cyclo-hexadien-1-ylidene), UVS,VA	A,⇌,p	
			0.62F			0.38F							A	
			0.48F			small							C	
JE048 0081	-	$E_{\frac{1}{2}}$	0.36F	Ag/AgClO$_4$ 0.01	i_ℓ	5F	i:i	2	Elog	50	-	-	A,⇌,i_ℓ=kC for C=0.1-1,i_d,p	
			0.66F			2.8F		-		-			A	
JA095 6688	-	"$E_{\frac{1}{2}}$"	-0.45	Ag/AgCl	-	-	sttd	1	-	-	-	-	A,r	FE91
			-1.22					1					A	
		E_p	-0.42F		i_p	51F		1	-	-	-	-	A,r	
			-0.45F			46F		1					C	
			-1.21F			50F		1					A	
			-1.27F			40F		1					C	

FE92 $C_{26}H_{16}N_4O_8$

Code No.	Empirical Formula	Name and C.A. Number	Structural Formula	Solvent	Tech.	Medium		μ, M	pH	T, °C	Electrodes	App.	Experimental Parameters
FE92	$C_{26}H_{16}N_4O_8$	tetrakis(4-nitrophenyl)ethylene C.A. 47797-98-8	$[(4-O_2NC_6H_4)_2C:]_2$	MeCN	XT	Et_4NClO_4	0.1	-	-	-30±1	HMDE// Ag/ AgClO$_4$ 0.01	2-2	C=ns,v= (1.5-5.5)E2
										0±1			
										20±0.5			
				DMF	IL	Et_4NClO_4	0.1	-	-	25±0.5	HMDE// SCE	2-2	C=ns, v=1.2E3
					XT	Et_4NClO_4	0.1	-	-	0±1	HMDE// SCE	2-2	C=ns,v= (1.5-5.5)E2
										30±0.5			
										60±0.5			
FE93	$C_{26}H_{18}$	9,9'-bifluorenyl C.A. 1530-12-7	Table II	DMF	PY	Bu_4NI PHEN	0.1 1E-3	-	-	-	DME// Ag/AgI, I⁻ 0.1	2AO	C=1
					IL	Bu_4NI PHEN	0.1 0.01	-	-	-	HMDE// Ag/AgI, I⁻ 0.1	2AO	C=ns, v=400
					VA	Bu_4NI	0.1	-	-	-	HMDE// Ag/AgI, I⁻ 0.1	2AO	-1.5→-2.5→ -1.5 V,C=1, v=400
						Bu_4NI PHEN	0.1						-1.5→-2.3→ -1.5 V,C=1, v=400
FE94 bp78 cm56 dg47	$C_{26}H_{18}$	9,10-diphenylanthracene C.A. 1499-10-1	Table II-2	CH_2Cl_2 66.65 CF_3COOH 33.35	VA	Bu_4NBF_4	0.1	-	-	-	Pt: nsns// SCE	---	C=ns,v=150
						Bu_4NBF_4 H_2O	0.1 0.6						
CONT													

TABLE I. Electrochemical Data

$C_{26}H_{18}$ (CONT.) FE94

Ref.	C/M	Charact. Potential		Response Const.		n Tech.	Electrokinetic Data			Products and Identification	Description and Remarks	Code No.
		Value	vs.		Value		Parameter	Value	From			
JE047 0115	-	XE	0.275 ±0.002	Ag/ AgClO₄ 0.01	-	-	-	-	-	-	-	C,XT=semiintegral chronoamperometry with linear potential sweep, XE= difference between potentials for the two steps,p	FE92
		-										C	
		0.295 ±0.002		-	-	-	-	-	-	-	-	C,p	
		-										C	
		0.310 ±0.002		-	-	-	-	-	-	-	-	C,p	
		-										C	
JE047 0215	-	E_p	-0.81F	SCE	i_p	6.8F	-	-	-	-	-	-	C, data for first wave only, p
		-										C(?)	
JE047 0115	-	XE	0.401 ±0.002	SCE	-	-	-	-	-	-	-	-	C,XT=semiintegral chronoamperometry with linear potential sweep, XE= difference between potentials for the two steps, p
		-										C	
		0.417 ±0.002		-	-	-	-	-	-	-	-	C,p	
		-										C	
		0.436 ±0.002		-	-	-	-	-	-	-	-	C,p	
		-										C	
EA020 0143	33 i	$E_{\frac{1}{2}}$	-1.93	Ag/AgI, I⁻ 0.1	-	-	-	-	-	-	-	-	C,r	FE93
		-2.07										C	
EA020 0143	33 i	E_p	-2.0F	Ag/AgI, I⁻ 0.1	$i_p(u)$	15F	-	-	-	-	-	C,S,r	
		-2.2F			23F							C,p	
EA020 0143	33 i	E_p	-2.0F	Ag/AgI, I⁻ 0.1	$i_p(u)$	9	-	-	-	-	-	C,r	
		-2.2F			1							C	
		-			-							A,O	
		-2.0F			13F	-	-	-	-	-	-	C,r	
		-2.2F			22F							C	
		-			-							A,O	
JE038 009A	420 a	E_p	0.88	SCE	i_p	67F	QP	1	-	-	-	-	A,p	FE94 bp78 cm56 dg47
		1.60			67F		1					A	
		0.78			30F		-					C	
		0.40			35F							C, peak broadens and disappears as H_2O is added	
		0.88		-	-	-	-	-	-	-	-	A,p	
		1.6										A	
		0.78										C	
		0.35										C,O in absence of H_2O	CONT

FE94 (CONT.) $C_{26}H_{18}$

Code No.	Empirical Formula	Name and C.A. Number	Structural Formula	Solvent	Tech.	Medium		μ, M	pH	T, °C	Electrodes	App.	Experimental Parameters
FE94 bp78 cm56 dg47	$C_{26}H_{18}$	9,10-diphenylanthracene	Table II-2	MeCN	QE	$LiClO_4$ PYR	0.01 2E-5	-	-	25±1	Pt: xfgz// SCE	26--	$C=0.01$, $E_{app}=1.35$ V, Pt Aux
					VY	$AlCl_3$	1	-	-	-	Pt: rodi// Ag/AgCl, Me_4NCl satd//Fc	2A0	ns
				CH_3NO_2	VY	$AlCl_3$	1	-	-	-	Pt: rodi// Ag/AgCl, Me_4NCl satd//Fc	2A0	$C=3, A=0.01$
						Et_4NClO_4	0.1						$C=ns, A=0.01$
				DMF	VA	Bu_4NI	0.1	-	-	-	Pt: xxdi//Ag /Bu_4NI 0.1	---	$C=0.85$, $v=21-511$
FE95	$C_{26}H_{18}K_2O_4$	1,2-bis(4-methoxyphenyl)cyclobuteno-[3,4-b]-3,8-naphthohydroquinone dipotassium salt	Table II	DMF	VR	Bu_4NClO_4	0.5	-	-	20	Pt: xxdi// Ag/AgCl// naphthoquinone	125A 2	$C \leq 0.5$, $v=1E4$, Pt Aux
						Bu_4NClO_4	1.0						$C=0.25$, $v=316$, Pt Aux
													$v=630$
													$v=1.6E3$
													$v=4.8E3$
													$C=0.5, v=690$
													$v=2.5E3$
													$v=3.8E4$
FE96 bp79	$C_{26}H_{18}O_2$	9,9'-bi(9-hydroxyfluorenyl) C.A. 3073-51-6	Table II-2	DMF	PY	Et_4NBr	0.1	-	-	-	DME// SCE(o)	2A0	$C=1$
						Et_4NBr PHEN	0.1 2.1E-2						
CONT													

$C_{26}H_{18}O_2$ (CONT.) FE96

Ref.	C/M	Charact. Potential Value	vs.	Response Const. Value		n Tech.	Electrokinetic Data Parameter	Value	From	Products and Identification	Description and Remarks	Code No.
JE054 0305	420 b	-	-	-	-	QE	2.0 ±0.1	-	-	-	9,10-diphenyl-9,10-pyridinium-9,10-dihydroanthracene diperchlorate 85%Ap,CHN,MPT	A,p	FE94 bp78 cm56 dg47
JE039 0385	-	$E_{\frac{1}{2}}$	0.9 1.28	Fc	-	-	-	-	-	-	-	A,Σ,p A	
JE039 0385	-	$E_{\frac{1}{2}}$	0.88 1.28	Fc	i_ℓ	3F 3F	QE	1 1	-	-	-	blue monocation dication	A,r A
			0.88 1.28		-	-	-	-	-	-	-	A,r A	
JE046 0141	-	E_p	-1.327 ±0.005 -	Ag	$i_p/v^{\frac{1}{2}}$	0.021 0.021	-	-	ΔE_p	61 -	sttd	-	A,p(reference electrode unpoised) C
JA095 6688	-	E_p	-1.00 -0.38	Ag/AgCl	-	-	sttd	1 1	$E_p-E_{p/2}$	57±2 46±6	sttd	-	A,R,r A,⋡	FE95
			- -0.429F		$i_{p,2}/i_{p,1}$	- 0.55F	-	-	-	-	-	-	A,r A	
			- -0.424F			- 1.13F	-	-	-	-	-	-	A,r A	
			- -0.395F			- 1.22F	-	-	-	-	-	-	A,r A	
			- -0.411F			- 1.34F	-	-	-	-	-	-	A,r A	
			- -0.420F			- 0.39F	-	-	-	-	-	-	A,r A	
			- -0.406F			- 1.05F	-	-	-	-	-	-	A,r A	
			- -0.380F			-	-	-	-	-	-	-	A,r A	
EA020 0143	33 i	$E_{\frac{1}{2}}$	-1.76 -1.98 -2.46	SCE(o)	i_ℓ	3.4 2.2 2.0	-	-	-	-	-	-	C,r C C	FE96 bp79
			-2.43 -2.73			14.8 7.2	-	-	-	-	-	-	C,r C	

CONT

FE96 (CONT.) $C_{26}H_{18}O_2$

Code No.	Empirical Formula	Name and C.A. Number	Structural Formula	Solvent	Tech.	Medium		μ, M	pH	T, °C	Electrodes	App.	Experimental Parameters
FE96 bp79	$C_{26}H_{18}O_2$	9,9'-bi(9-hydroxy-fluorenyl)	Table II-2	DMF	PY	Bu_4NI PHEN	0.1 1E-3	-	-	-	DME// Ag/AgI, I⁻ 0.1	2AO	C=1
					QE	Bu_4NI	0.25	-	-	-	Hg: srpo// Ag/AgI, I⁻ 0.1	5AO	C=ns, A=12, E_{app}= -2.3 V
					VA	Bu_4NI	0.1	-	-	-	HMDE// Ag/AgI, I⁻ 0.1	5AO	C=1, v=400
						Bu_4NI PHEN	0.1 0.01						
FE97 bp86	$C_{26}H_{20}$	tetraphenylethene C.A. 59856-51-8	$[(C_6H_5)_2C:]_2$	DMF	VR	Et_4NBF_4	satd	-	-	-	Hg: xxsd// SCE	5AO	C=ns, v=300
				HMP	VR	LiCl	satd	-	-	-	Hg: xxsd// SCE	5AO	v=300
						Bu_4NBF_4	satd						
FE98	$C_{26}H_{21}O_3S_3$	2,3,5-tris(4-methoxyphenyl)trithiapentalene C.A. 38755-22-5	Table II	MeCN 50 CH_2Cl_2 50	IL	$NaClO_4$	0.1	-	-	-	Pt: xxbu// SCE	---	C=ns
FE99	$C_{26}H_{22}N_2$	2,7-dimethyl-5,10-diphenyl-5,10-dihydrophenazine C.A. 59130-88-0	Table II	MeCN	VY	Et_4NClO_4	0.1	-	-	-	Pt: rodi// Ag/ $AgClO_4$ 0.01, Et_4NClO_4 0.1	2-0	C=ns, d=0.1, ω = 10, Pt Aux
FF00	$C_{26}H_{22}N_2O_2$	2,7-dimethoxy-5,10-diphenyl-5,10-dihydrophenazine C.A. 59130-87-9	Table II	MeCN	VY	Et_4NClO_4	0.1	-	-	-	Pt: rodi// Ag/ $AgClO_4$ 0.01, Et_4NClO_4 0.1	2-0	C=ns, d=0.1, ω = 10, Pt Aux
FF01	$C_{26}H_{22}O$	bis(diphenylmethyl) ether C.A. 574-42-5	$[(C_6H_5)_2CH]_2O$	DMF	PY	Bu_4NI	0.1	-	-	-	DME// Ag/AgI, I⁻ 0.1	3-0	C=1, Pt Aux
						Bu_4NI PHEN	0.1 0.01						

TABLE I. Electrochemical Data

$C_{26}H_{22}O$ FF01

Ref.	C/M	Charact. Potential		Response	Const. Value	n Tech.	n	Electrokinetic Data			Products and Identification	Description and Remarks	Code No.
		Value	vs.					Parameter	Value	From			
EA020 0143	33 i	$E_{\frac{1}{2}}$	-1.94	Ag/AgI, I⁻ 0.1	-	-	-	-	-	-	-	-	C,r	FE96 bp79
			-2.06										C	
EA020 0143	33 i	-	-	-	-	-	QE	4 (fixed)	-	-	-	9-fluorenol 45%, fluorene 45%	C,r	
EA020 0143	33 i	E_p	-1.25F	Ag/AgI, I⁻ 0.1	$i_p(u)$	4F	-	-	-	-	-	-	C,S,r	
			-1.35F			8F							C	
			-2.1F			1.5F							C	
			-1.54F			4F							A	
			-0.64F			27F							A	
			-2.0F			35F	-	-	-	-	-	-	C,≠,r	
			-2.25F			22F							C,≠	
EA019 0951	-	XE	-2.05	SCE	-	-	sttd	2	-	-	-	-	C,R,XE=$(E_{p,A}+E_{p,C})/2$, eec,p	FE97 bp86
			-					-					A	
EA019 0951	-	E_p	-1.94	SCE	-	-	sttd	1	-	-	-	-	C,R(?),p	
			-2.13					1	ΔE_p	60	sttd		C,R,M	
			-2.1Ap					1		-			A,R	
			-1.88					1					A,R(?)	
			-0.95					-					A,O unless E is held at a value more neg than -2.13 V and v ≥ 3E4	
			-2.00		-	-	-	-	-	-	-	-	C,R,p	
			-2.20										C,R	
			-2.16										A,R	
			-1.94										A,R	
JE038 0479	-	E_p	0.80	SCE	-	-	QE	1	-	-	-	-	A,p	FE98
EA020 1019	428 ab	$E_{\frac{1}{2}}$	-0.20	Ag/AgClO₄ 0.01, Et₄NClO₄ 0.1	-	-	sttd	1	-	-	-	radical cation,ESR, UVS,VIS	A,R,r	FE99
			0.54					1				dication,UVS,VIS	A,R	
EA020 1019	428 ab	$E_{\frac{1}{2}}$	-0.265	Ag/AgClO₄ 0.01, Et₄NClO₄ 0.1	-	-	sttd	1	-	-	-	radical cation,ESR, UVS,IRS	A,R,r	FF00
			0.395					1				dication,UVS,IRS	A,R	
EA020 0853	33 j	$E_{\frac{1}{2}}$	-2.22	Ag/AgI, I⁻ 0.1	i_ℓ	2.0	QE (-2.40V)	2.1	-	-	-	benzhydrol 50%, diphenylmethane 50%	C,r	FF01
			-2.22			3.2	-	-	-	-	-	-	C,r	

FF02 $C_{26}H_{23}ClO_8$

Code No.	Empirical Formula	Name and C.A. Number	Structural Formula	Solvent	Tech.	Medium	μ, M	pH	T, °C	Electrodes	App.	Experimental Parameters
FF02	$C_{26}H_{23}ClO_8$	2,4,6-tris(4-methoxyphenyl)pyrylium perchlorate C.A. 63373-56-8	Table II	MeCN	VY	Et_4NClO_4 0.1	-	-	20±1	Pt: rodi// SCE	25A0	C=ns, A=7.8E-3
FF03	$C_{26}H_{24}N_2$	1,2-di(4-methylphenyl)-N,N'-diphenylhydrazine C.A. 18440-53-4	Table II	MeCN	VY	Et_4NClO_4 0.1	-	-	-	Pt: rodi// Ag/ AgClO_4 0.01 Et_4NClO_4 0.1	2-0	C=ns, d=0.1, ω=10, Pt Aux
FF04	$C_{26}H_{24}N_2$	N,N'-bis(4-methylphenyl)-N-phenyl-1,4-phenylenediamine C.A. 62958-79-6	Table II	MeCN	VY	Et_4NClO_4 0.1	-	-	-	Pt: rodi// Ag/ AgClO_4 0.01, Et_4NClO_4 0.1	2-0	C=ns, d=0.1, ω=10, Pt Aux
FF05	$C_{26}H_{24}N_2O_2$	N,N'-bis(4-methoxyphenyl)benzidine C.A. 59131-00-9	Table II	MeCN	VY	Et_4NClO_4 0.1	-	-	-	Pt: rodi// Ag/ AgClO_4 0.01, Et_4NClO_4 0.1	2-0	C=0.5, d=0.1, ω=10, Pt Aux
FF06	$C_{26}H_{24}N_2O_2$	N,N'-bis(4-methoxyphenyl)-N-phenyl-phenylenediamine C.A. 62958-59-2	Table II	MeCN	VY	Et_4NClO_4 0.1	-	-	-	Pt: rodi// Ag/ AgClO_4 0.01, Et_4NClO_4 0.1	2-0	C=ns, d=0.1, ω=10, Pt Aux
FF07	$C_{26}H_{24}N_2O_2$	1,2-bis(4-methoxyphenyl)-N,N'-diphenylhydrazine C.A. 34839-21-9	Table II	MeCN	VY	Et_4NClO_4 0.1	-	-	-	Pt: rodi// Ag/ AgClO_4 0.01, Et_4NClO_4 0.1	2-0	C=ns, d=0.1, ω=10, Pt Aux
FF08	$C_{27}H_{30}N_5Na$	2,4,5-tris(4-dimethylaminophenyl)imidazole, sodium salt C.A. 31909-35-0	Table II	MeCN	VY	Bu_4NClO_4 0.1	-	-	-	Pt: rodi// Ag/AgCl	2A0	C=1.9, A=0.012, v=15.3
FF09	$C_{28}H_{12}N_4O_2$	3,3'-bi(6H,11H-6-oxopyrazolo[3,4,5-de]anthracene) dianion	Table II	EtOH 50	PY	NaOH 0.65 GEL 0.005	-	-	-	DME//SCE	04A0	C=1.0, t=0.87, h=44
FF10	$C_{28}H_{14}N_6S_4$	5,10-dihydro-2,7-dithiocyanato-5,10-bis(4-thiocyanatophenyl)phenazine C.A. 61228-19-1	Table II	MeCN	VY	Et_4NClO_4 0.1(?)	-	-	-	Pt: rodi// Ag/ AgClO_4 0.01, Et_4NClO_4 0.1	2-0	C=ns, d=0.1, ω=10, Pt Aux

TABLE I. Electrochemical Data

$C_{28}H_{14}N_6S_4$ FF10

Ref.	C/M	Charact. Potential		Response Const.		n Tech.	Electrokinetic Data			Products and Identification	Description and Remarks	Code No.
		Value	vs.		Value		Parameter	Value	From			
EA021 0497	–	$E_{\frac{1}{2}}$	-0.58	SCE	–	–	–	–	–	–	–	C,p	FF02
EA020 1019	–	$E_{\frac{1}{2}}$	0.43	Ag/ AgClO$_4$ 0.01, Et$_4$NClO$_4$ 0.1	–	–	–	–	–	–	–	A,r	FF03
			1.25									A	
EA021 1171	147 jkl	$E_{\frac{1}{2}}$	0.14	Ag/ AgClO$_4$ 0.01, Et$_4$NClO$_4$ 0.1	–	–	–	–	–	–	radical cation	A,r	FF04
			0.58								diimine cation	A	
EA020 1011	419 d	$E_{\frac{1}{2}}$	0.255	Ag/ AgClO$_4$ 0.01, Et$_4$NClO$_4$ 0.1	$i_\ell(u)$	1	–	–	–	–	radical cation	A,r	FF05
			0.400			1					dication	A	
			1.14			–						A	
EA021 1171	147 jkl	$E_{\frac{1}{2}}$	0.13	Ag/ AgClO$_4$ 0.01, Et$_4$NClO$_4$ 0.1	–	–	–	–	–	–	radical cation	A,r	FF06
			0.55								diimine cation	A	
EA020 1019	–	$E_{\frac{1}{2}}$	0.325	Ag/ AgClO$_4$ 0.01, Et$_4$NClO$_4$ 0.1	–	–	sttd	1	–	–	–	A,R,r	FF07
			0.97					–				A	
C0036 0575	397 a	$E_{\frac{1}{2}}$	-0.237	Ag/AgCl	$i_d(u)$	1.00	Elog	1	$2.3RT/\alpha n_a$	0.066	Elog	–	A,R,i_d,r	FF08
			0.230			1.18		1		0.063		A,R,i_d	
C0026 1763	–	$E_{\frac{1}{2}}$	-0.91	SCE	–	–	i:i	2	–	–	–	C,i_d,p	FF09
EA021 0557	–	$E_{\frac{1}{2}}$	0.26	Ag/ AgClO$_4$ 0.01, Et$_4$NClO$_4$ 0.1	–	–	–	–	–	–	radical cation, UVS, VIS, ESR	A,r	FF10
			0.85									A	

FF11 $C_{28}H_{18}N_2O_2$

Code No.	Empirical Formula	Name and C.A. Number	Structural Formula	Solvent	Tech.	Medium	μ, M	pH	T, °C	Electrodes	App.	Experimental Parameters
FF11	$C_{28}H_{18}N_2O_2$	2-phenyl-3-(2-phenyl-3H-indol-3-ylidene)-3H-indole 1,1'-dioxide C.A. 2196-95-4	Table II	DMF 99.995	PY	Et_4NClO_4 0.1	–	–	20±0.1	DME//SCE (NaCl)	235A0	C=0.19, m=1.17, t=5.94, h=80
						Et_4NClO_4 0.1 2-HOC_6H_4COOH 2E-3			23±0.1			C=0.5
						Et_4NClO_4 0.1 $(HOOCCH_2)_2$ 2E-3						
						Et_4NClO_4 0.1 $CH_2ClCOOH$ 2E-3						
						Et_4NClO_4 0.1 4-$NO_2C_6H_4$-COOH 2E-3						
						Et_4NClO_4 0.1 BENZ 2E-3						
						Et_4NClO_4 0.1 HOAc 2E-3						
					VA	Et_4NClO_4 0.1	–	–	22±0.1	HMDE//SCE(NaCl)	235A2	-0.03 → -1.13 → -0.03V, C=0.99, v=250
						Et_4NClO_4 0.1 $(HOOCCH_2)_2$ 5.9E-3						0.13 → -0.8 → 0.13V, C=0.994, v=250
CONT		2-phenyl-3-(2-phenyl-3H-indol-3-ylidene)-3H-indole 1,1'-dioxide	Table II		PY	Et_4NClO_4 0.1						

TABLE I. Electrochemical Data   293

$C_{27}H_{18}N_2O_2$ (CONT.) FF11

Ref.	C/M	Charact. Potential		Response Const.		n Tech.	Electrokinetic Data			Products and Identification	Description and Remarks	Code No.
		Value	vs.		Value		Parameter	Value	From			
JE051 0341	426 a	$E_{\frac{1}{2}}$ -0.395	SCE(NaCl)	I	1.33	QE 1	Tomeš	60	sttd	CP→ radical anion, ESR	C,R,W, i_ℓ=kC for C=0.03-1, i_ℓ=kh$^{\frac{1}{2}}$, i_d, r	FF11
		-0.75			1.32	1		55		dianion	C,R,W, i_ℓ=kC for C=0.03-1, i_d	
	426 b	0.046 ±0.005		-	-	2	$dE_{\frac{1}{2}}/dpK_{AH}$	66	sttd	CP→ 1,1'-dihydroxy-2,2'-biphenyl-3,3'-biindole,UVS	C,Mx; $E_{\frac{1}{2}}$=0.590- 0.066pK_{AH} for pK_{AH}=8-14 and [AH]/C=2-10, pK_{AH}=8.23 for 2-HOC$_6$H$_4$COOH;Tc= 0.5 for T=5-35,r	
		-			-		-	-			C	
		-			-		-	-			C	
		-0.070 ±0.005		-	-	-	-	66		-	C,Mx, pK_{AH}=10.05 for (HOOCCH$_2$)$_2$,Tc=0.5 for T=5-35,r	
		-			-		-	-			C	
		-			-		-	-			C	
		-0.095 ±0.005		-	-	-	-	66		-	C,Mx, pK_{AH}=10.4 for CH$_2$ClCOOH,Tc=0.5 for T=5-35,r	
		-			-		-	-			C	
		-			-		-	-			C	
		-0.100 ±0.005		-	-	-	-	66		-	C,Mx, pK_{AH}=10.6 for 4-NO$_2$C$_6$H$_4$COOH,Tc= 0.5 for T=5-35,r	
		-			-		-	-			C	
		-			-		-	-			C	
		-0.220 ±0.005		-	-	-	-	66		-	C,B,Mx, pK_{AH}=12.25 for BENZ,Tc=0.5 for T=5-35,r	
		-			-		-	-			C,B	
		-			-		-	-			C	
		-0.290 ±0.005		-	-	-	-	66		-	C,B,Mx, pK_{AH}=13.4 for HOAc,Tc=0.5 for T=5-35,r	
		-			-		-	-			C,B	
		-			-		-	-			C	
JE051 0341	426 a	E_p -0.42	SCE(NaCl)	i_p	9.8F	-	-	-	-	-	C; E_p and $i_p/v \neq f(v)$ for v=40-1500,r	
		-0.77			16F						C, E_p and $i_p/v \neq f(v)$ for v=40-1500	
		-0.71			9.8F						A	
		-0.35			12.4F						A	
	426 b	-0.11F			22F	-	-	-	-	-	C,R,r	
		-0.01F			27F						A,R	

CONT

FF11 (CONT.) $C_{26}H_{18}N_2O_2$

Code No.	Empirical Formula	Name and C.A. Number	Structural Formula	Solvent	Tech.	Medium	μ, M	pH	T, °C	Electrodes	App.	Experimental Parameters
FF11	$C_{26}H_{18}N_2O_2$	2-phenyl-3-(2-phenyl-3H-indol-3-ylidene)-3H-indole 1,1'-dioxide	Table II	DMF 99.995	VA	Et_4NClO_4 0.1	–	–	22± 0.1	Pt: xxdi// SCE(NaCl)	23A2	-0.1→-1.1→ -0.1V, C=0.83, v=250
						Et_4NClO_4 0.1 $(HOOCCH_2)_2$ 1.8E-3						0.8→-0.6→ 0.8V, C=0.83, v=250
					VY	Et_4NClO_4 0.1	–	–	22± 0.1	Pt: rodi// SCE(NaCl)	235A0	C=0.95, d=0.6, ω=33
						Et_4NClO_4 0.1 2-HOC_6H_4COOH 2E-4						C=0.1, d=0.6, ω=33
						Et_4NClO_4 0.1 $(HOOCCH_2)_2$ 2E-4						
						Et_4NClO_4 0.1 $CH_2ClCOOH$ 2E-4						
						Et_4NClO_4 0.1 4-$NO_2C_6H_4$-COOH 2E-4						
						Et_4NClO_4 0.1 BENZ 2E-4						
						Et_4NClO_4 0.1 HOAc 2E-4						
FF12	$C_{28}H_{22}Cl_2N_4O_2$	2,7-di(acetylamino)-5,10-di(4-chlorophenyl-5,10-dihydrophenazine C.A. 61228-21-5	Table II	MeCN	VY	Et_4NClO_4 0.1(?)	–	–	–	Pt: rodi// Ag/ $AgClO_4$ 0.01, Et_4NClO_4 0.1	2-0	C=ns, d=0.1, ω=10, Pt Aux
FF13 CONT	$C_{28}H_{22}O_2$	[9,9'-bi(9-hydroxy-9,10-dihydroanthryl)] C.A. 4393-30-0	Table II	DMF	PY	Et_4NBr 0.1	–	–	–	DME// SCE (o)	2A0	C=1
						Et_4NBr 0.1 PHEN 2.1E-2						

TABLE I. Electrochemical Data

$C_{28}H_{22}O_2$ (CONT.) FF13

Ref.	C/M	Charact. Potential		Response Const.		n Tech.		Electrokinetic Data			Products and Identification	Description and Remarks	Code No.
		Value	vs.		Value			Parameter	Value	From			
JE051 0341	426 a	E_p	-0.44F	SCE(NaCl)	i_p	70F	-	-	-	-	-	-	C,r	FF11
			-0.8F			70F							C	
			-0.7F			25F							A	
			-0.3F			75F							A	
	426 b		-0.25F			90F	-	-	-	-	-	-	C,≠,r	
			0.26F			40F							A	
JE051 0341	426 a	$E_{\frac{1}{2}}$	-0.39	SCE(NaCl)	i_ℓ	168F	-	-	-	-	-	-	C,$E_{\frac{1}{2}}\neq f(\omega)$,$i_d$,r	
			-0.79			160F							C,$E_{\frac{1}{2}}\neq f(\omega)$,$i_d$	
	426 b		-0.256 ±0.005		-	-	-	-	$dE_{\frac{1}{2}}/dpK_{HA}$	71	sttd	-	C,B,$E_{\frac{1}{2}}$= -0.071- 0.022pK_{AH} for pK_{AH}=8-14 and [AH]/C=2-10,Tc=1 for T=5-35,r	
			-							-			C	
			-							-			C	
			-0.290 ±0.005		-	-	-	-		71		-	C,B,Tc=1 for T=5-35, r	
			-							-			C	
			-							-			C	
			-0.306 ±0.005		-	-	-	-		71		-	C,B,Tc=1 for T=5-35, r	
			-							-			C	
			-							-			C	
			-0.306 ±0.005		-	-	-	-		71		-	C,B,Tc=1 for T=5-35, r	
			-							-			C	
			-							-			C	
			-0.345 ±0.005		-	-	-	-		71		-	C,B,Tc=1 for T=5-35, r	
			-							-			C	
			-							-			C	
			-0.375 ±0.005		-	-	-	-		71		-	C,B,Tc=1 for T=5-35, r	
			-							-			C	
			-							-			C	
EA021 0557	-	$E_{\frac{1}{2}}$	-0.18	Ag/ AgClO$_4$ 0.01, Et$_4$NClO$_4$ 0.1	-	-	-	-	-	-	-	radical cation,UVS, VIS,ESR	A,r	FF12
			0.46										A	
EA020 0143	33f	$E_{\frac{1}{2}}$	-1.74	SCE(o)	i_ℓ	2.6	-	-	-	-	-	-	C,r	FF13
			-2.18			6.0							C	
			-2.79			3.4							C	
	33g		-1.66			1.8	-	-	-	-	-	-	C,r	
			-2.18			6.6							C	
			-2.78			6.6							C	CONT

FF13 (CONT.) $C_{28}H_{22}O_2$

Code No.	Empirical Formula	Name and C.A. Number	Structural Formula	Solvent	Tech.	Medium	μ, M	pH	T, °C	Electrodes	App.	Experimental Parameters
FF13	$C_{28}H_{22}O_2$	[9,9'-bi(9-hydroxy-9,10-dihydroanthryl)	Table II	DMF	QE	Bu_4NI 0.25	-	-	-	Hg: srpo// Ag/AgI, I⁻	5AO	C=ns, A=12, E_{app}= -2.0 V
FF14	$C_{28}H_{22}O_2$	9,10-di-p-anisyl-anthracene C.A. 24672-76-2	Table II	MeCN	VA	Bu_4NBF_4 0.2 Al_2O_3 0.25g cm⁻³	-	-	-	Pt: xxbu// Ag/AgNO₃ 0.1, Bu_4NBF_4 0.1	25AFO	0→1.5→0 V, C=1.0, v=86
				C_6H_5CN	VA	Bu_4NBF_4 0.2 Al_2O_3 0.25g cm⁻³	-	-	-	Pt: xxbu// Ag/AgNO₃ 0.1, Bu_4NBF_4 0.1	25AFO	0→1.6→0 V, C=1.0, v=86
				2-PrCN	VA	Bu_4NBF_4 0.2 Al_2O_3 0.25g cm⁻³	-	-	-	Pt: xxbu// Ag/AgNO₃ 0.1, Bu_4NBF_4 0.1	25AFO	0→1.5→0 V, C=1.0, v=86
				CH_2Cl_2	VA	Bu_4NBF 0.2 Al_2O_3 0.25g cm⁻³	-	-	-	Pt: xxbu// Ag/AgNO₃ 0.1, Bu_4NBF_4 0.1	25AFO	0→1.6→0 V, C=1.0, v=86
				CH_2Cl_2 88 F_3CCOOH 2 $(F_3CCO)_2O$ 10	VA	Bu_4NBF_4 0.2	-	-	-	Pt: xxbu// Ag/AgNO₃ 0.1, Bu_4NBF_4 0.1	25AO	0→1.6→0 V, C=1.0, v=43
												v=86
												v=161
												v=210
				$MeNO_2$	VA	Bu_4NBF_4 0.2 Al_2O_3 0.25g cm⁻³	-	-	-	Pt: xxbu// Ag/AgNO₃ 0.1, Bu_4NBF_4 0.1	25AFO	0→1.5→0 V, C=1.0, v=86
CONT				$C_6H_5NO_2$	VA	Bu_4NBF 0.2 Al_2O_3 0.25g cm⁻³	-	-	-	Pt: xxbu// Ag/AgNO₃ 0.1, Bu_4NBF_4 0.1	25AFO	0→1.6→0 V, C=1.0, v=86

TABLE I. Electrochemical Data

$C_{28}H_{22}O_2$ (CONT.) FF14

Ref.	C/M	Charact. Potential		Response Const.		n Tech.	Electrokinetic Data			Products and Identification	Description and Remarks	Code No.
		Value	vs.		Value		Parameter	Value	From			
EA020 0143	33f	-	-	-	-	QE	2.8	-	-	-	9,10-dihydroanthracene 90%	C,r	FF13
EA018 0537	419 e	E_p	1.15	SCE	-	-	sttd	1	-	-	-	radical cation	A,R,Al_2O_3 added to remove H_2O,r	FF14
			1.37					1				dication	A,R	
			1.31					1					C,R	
			1.09					1					C,R	
EA018 0537	419 e	E_p	1.19	SCE	-	-	sttd	1	-	-	-	radical cation	A,R,r	
			1.42					1				dication	A,R	
			1.36					1					C,R	
			1.13					1					C,R	
EA018 0537	419 e	E_p	1.23	SCE	-	-	sttd	1	-	-	-	radical cation	A,R,r	
			1.39					1				dication	A,R	
			1.33					1					C,R	
			1.16					1					C,R	
EA018 0537	419 e	E_p	1.18	SCE	-	-	sttd	1	-	-	-	radical cation	A,R,r	
			1.48					1				dication	A,R	
			1.41					1					C,R	
			1.12					1					C,R	
EA018 0537	419 e	-	-	-	$i_p/Cv^{\frac{1}{2}}$	5.6	sttd	1	-	-	-	radical cation	A,R,r	
						4.8		1				dication	A,R	
						4.8		1					C,R	
						5.6		1					C,R	
		E_p	1.19	SCE		5.6	QE	1.0	-	-	-	radical cation,ESR,IRS	A,R,r	
			1.51			5.2		1.0				dication,VIS	A,R	
			1.42			5.2	sttd	1					C,R	
			1.12			5.6		1					C,R	
			-			5.4		1	-	-	-	radical cation	A,R,r	
			-			5.4		1				dication	A,R	
			-			5.4		1					C,R	
			-			5.4		1					C,R	
			-			5.4		1	-	-	-	radical cation	A,R,r	
			-			5.2		1				dication	A,R	
			-			5.2		1					C,R	
			-			5.4		1					C,R	
EA018 0537	419 e	E_p	1.12	SCE	-	-	sttd	1	-	-	-	radical cation	A,R,r	
			1.32					1				dication	A,R	
			1.25					1					C,R	
			1.05					1					C,R	
EA018 0537	419 e	E_p	1.16	SCE	-	-	sttd	1	-	-	-	radical cation	A,R,r	
			1.42					1				dication	A,R	
			1.35					1					C,R	
			1.10					1					C,R	

CONT

FF14 (CONT.) $C_{28}H_{22}O_2$

Code No.	Empirical Formula	Name and C.A. Number	Structural Formula	Solvent	Tech.	Medium	μ, M	pH	T, °C	Electrodes	App.	Experimental Parameters
FF14	$C_{28}H_{22}O_2$	9,10-di-p-anisyl-anthracene	Table II	EtCN	VA	Bu_4NBF_4 0.2 Al_2O_3 0.25g cm^{-3}	-	-	-	Pt: xxbu// Ag/AgNO$_3$ 0.1, Bu$_4$NBF$_4$ 0.1	25AF0	$0 \to 1.5 \to 0$ V, C=1.0, v=86
				F$_3$CCOOH 90 (F$_3$CCO)$_2$O 0.2	VA	Bu$_4$NBF$_4$ 0.2	-	-	-	Pt: xxbu// Ag/AgNO$_3$ 0.1, Bu$_4$NBF$_4$ 0.1	25AO	$0 \to 1.6 \to 0$ V, C=1.0, v=86
FF15	$C_{28}H_{24}Br_2P_2$	1,1,4,4-tetraphenyl-diphosphoniacyclo-hexa-2,5-diene dibromide C.A. 15924-61-5	Table II	DMF	PY	Et$_4$NClO$_4$ 0.1	-	-	-	DME//SCE	2--	ns
FF16	$C_{28}H_{24}N_2O_2$	N,N'-di(4-acetyl-phenyl)-N-phenyl-phenylenediamine	Table II	MeCN	VY	Et$_4$NClO$_4$ 0.1	-	-	-	Pt: rodi// Ag/AgClO$_4$ 0.01, Et$_4$NClO$_4$ 0.1	2-0	C=ns, d=0.1, ω=10, Pt Aux
FF17	$C_{28}H_{24}N_4O_2$	2,7-bis(acetylamino)-5,10-dihydro-5,10-diphenylphenazine	Table II	MeCN	VY	Et$_4$NClO$_4$ 0.1	-	-	-	Pt: rodi// Ag/AgClO$_4$ 0.01, Et$_4$NClO$_4$ 0.1	2-0	C=ns, d=0.1, ω=10, Pt Aux
FF18	$C_{28}H_{26}Br_2P_2$	1,1,4,4-tetraphen-yldiphosphoniacyclo-hexa-2-ene dibromide C.A. 13274-97-0	Table II	DMF	PY	Et$_4$NClO$_4$ 0.1	-	-	-	DME//SCE	2--	ns
FF19	$C_{28}H_{26}N_2$	2,7-dimethyl-5,10-bis(4-methylphenyl)-5,10-dihydrophena-zine C.A. 31438-42-3	Table II	MeCN	VY	Et$_4$NClO$_4$ 0.1	-	-	-	Pt: rodi// Ag/AgClO$_4$ 0.01, Et$_4$NClO$_4$ 0.1	2-0	C=ns, d=0.1, ω=10, Pt Aux
FF20	$C_{28}H_{26}N_2O_4$	2,7-dimethoxy-5,10-bis(4-methoxyphenyl)-5,10-dihydrophena-zine C.A. 31438-43-4	Table II	MeCN	VA	Et$_4$NClO$_4$ 0.1	-	-	-	Pt: xxdi// Ag/AgClO$_4$ 0.01, Et$_4$NClO$_4$ 0.1	2-0	$-0.5 \to 0.6 \to -0.5$ V, C=0.6, v=2E3, Pt Aux
						Et$_4$NClO$_4$ 0.1 Et$_4$NCl 2.7E-3						
					VY	Et$_4$NClO$_4$ 0.1	-	-	-	Pt: rodi// Ag/AgClO$_4$ 0.01, Et$_4$NClO$_4$ 0.1	2-0	C=ns, d=0.1, ω=10, Pt Aux

TABLE I. Electrochemical Data

$C_{28}H_{26}N_2O_4$ FF20

Ref.	C/M	Charact. Potential		Response	Const. Value	Tech.	n	Electrokinetic Data			Products and Identification	Description and Remarks	Code No.
		Value	vs.					Parameter	Value	From			
EA018 0537	419 e	E_p	1.20	SCE	-	-	sttd	1	-	-	-	radical cation	A,R,r	FF14
			1.39					1				dication	A,R	
			1.32					1					C,R	
			1.14					1					C,R	
EA018 0537	419 e	E_p	1.13	SCE	-	-	sttd	1	-	-	-	radical cation	A,R,r	
			1.46					1				dication	A,R	
			1.40					1					C,R	
			1.07					1					C,R	
JE042 0309	-	$E_{\frac{1}{2}}$	-0.56	SCE	-	-	sttd	1	-	-	-	-	C,"$E_{\frac{1}{2}}$" on Pt (VA?)= -0.61V,p	FF15
EA021 1171	147 jkl	$E_{\frac{1}{2}}$	0.42	Ag/ AgClO₄ 0.01, Et₄NClO₄ 0.1	-	-	-	-	-	-	-	radical cation	A,r	FF16
			0.77									diimine cation	A	
EA021 0557	-	$E_{\frac{1}{2}}$	-0.22	Ag/ AgClO₄ 0.01, Et₄NClO₄ 0.1	-	-	-	-	-	-	-	radical cation, UVS, VIS, ESR	A,r	FF17
			0.40										A	
JE042 0309	-	$E_{\frac{1}{2}}$	-0.75	SCE	-	-	sttd	1	-	-	-	-	C,⊨,$i_a + i_d$, "$E_{\frac{1}{2}}$" on Pt (VA?)= -0.79 V,p	FF18
EA020 1019	428 ab	$E_{\frac{1}{2}}$	-0.23	Ag/ AgClO₄ 0.01, Et₄NClO₄ 0.1	-	-	sttd	1	-	-	-	radical cation, ESR, UVS, VIS	A,R,r	FF19
			0.51					1				dication, UVS, VIS	A,R	
EA020 1019	428 ab	E_p	-0.26F	Ag/ AgClO₄ 0.01, Et₄NClO₄ 0.1	i_p	10F	-	-	-	-	-	-	A,r	FF20
			0.41F			12F							A	
			0.35F			11F							C	
			-0.33F			13F							C	
			-0.26F			10F		-	-	-	-	-	A,R,r	
			-0.47F			35F							A,⊭	
			-0.33F			14F							C,R	
EA020 1019	428 ab	$E_{\frac{1}{2}}$	-0.28	Ag/ AgClO₄ 0.01, Et₄NClO₄ 0.1	-	-	sttd	1	-	-	-	radical cation, ESR, UVS, VIS	A,R,r	
			0.37					1				dication, UVS, VIS	A,R	

FF21 $C_{28}H_{26}O_2$

Code No.	Empirical Formula	Name and C.A. Number	Structural Formula	Solvent	Tech.	Medium		μ, M	pH	T, °C	Electrodes	App.	Experimental Parameters
FF21	$C_{28}H_{26}O_2$	2,3-bis[(4-biphenyl-yl)]-2,3-butanediol C.A. 10426-00-3	Table II	DMF	PY	Et_4NBr	0.1	-	-	-	DME// SCE(o)	2AO	C=1
						Et_4NBr PHEN	0.1 2.1E-2						
					QE	Bu_4NI PHEN	0.25 ns	-	-	-	Hg: srpo// Ag/AgI, I-	5AO	C=ns, A=12, E_{app}=-2.3 V
FF22	$C_{28}H_{28}Br_2P_2$	1,1,4,4-tetraphenyl-diphosphoniacyclo-hexane dibromide C.A. 2316-28-1	Table II	DMF	PY	Et_4NClO_4	0.1	-	-	-	DME//SCE	2--	ns
FF23	$C_{28}H_{28}N_2$	tetrakis(4-methyl-phenyl)hydrazine C.A. 1807-53-0	$[(4-CH_3C_6H_4)_2N]_2$	MeCN	VY	Et_4NClO_4	0.1	-	-	-	Pt: rodi// Ag/ $AgClO_4$ 0.01, Et_4NClO_4 0.1	2-0	C=ns, d=0.1, ω=10, Pt Aux
FF24	$C_{29}H_{20}N_6O_{12}S_3$	5-[(4-amino-3-sulfo-anthraquinon-1-yl)-amino]-2-{[3-hyd-roxy-5-(3-sulfophen-yl)]-s-triazinyl-amino}benzenesulfon-ic acid	Table II	H_2O	PY	BR	-	-	2.05	25	DME//NCE	0-0	C=0.134, m=2.5, t=4±0.2, h=65±16
									4				
									5.1				
						PHOS	ns		7				
						BR	-		8				
									11.1				
FF25 bq83	$C_{29}H_{20}O$	tetraphenylcyclo-pentadienone C.A. 479-33-4	Table II-2	EtOH 96(?)	PY	LiCl HCl	0.1 0.001	-	-	25	DME//NCE (LiCl, o)	OAO	C=0.36, m=1.12, t=3.2, h=40
						LiCl HCl	0.1 6.5E-4						
						LiCl HCl	0.1 4.5E-4						
CONT													

$C_{29}H_{20}O$ (CONT.) FF25

Ref.	C/M	Charact. Potential		Response	Const. Value	Tech.	n	Electrokinetic Data			Products and Identification	Description and Remarks	Code No.
		Value	vs.					Parameter	Value	From			
EA020 0143	33f	$E_{\frac{1}{2}}$	-2.72	SCE(o)	i_ℓ	10.7	-	-	-	-	-	-	C,r	FF21
	33g		-2.70			18.1	-	-	-	-	-	-	C,r	
EA020 0143	33g	-	-	-	-	-	sttd	21	-	-	-	4-cyclohexyl-1-ethylbenzene 90%	C, no redn in absence of PHEN; in its presence redn is incomplete for n=4	
JE042 0309	-	$E_{\frac{1}{2}}$	-1.84	SCE	-	-	sttd	2	-	-	-	-	C,$\mp$,$i_a + i_d$,"$E_{\frac{1}{2}}$" on Pt(VA?)= -1.70 V,p	FF22
EA020 1019	-	$E_{\frac{1}{2}}$	0.34	Ag/ AgClO$_4$ 0.01, Et$_4$NClO$_4$ 0.1	-	-	-	-	-	-	-	-	A,r	FF23
			1.15										A	
JE036 0167	-	$E_{\frac{1}{2}}$	-0.4F	NCE	$i_\ell(u)$	7F	-	-	-	-	-	-	C,p	FF24
			-1.2F			14F							C	
			-0.5F			7F	-	-	-	-	-	-	C,p	
			-1.3F			13F							C	
			-0.7F			7F	-	-	-	-	-	-	C,p	
			-1.4F			9F							C,Mx	
			-0.7F		-	-	-	-	-	-	-	-	C,i_ℓ=kC,D=5±0.3, i_d for T=20-30, i_ℓ=kh$^{\frac{1}{2}}$,Tc=3 for T>30,p	
			-1.25F										C	
			-0.85F			7F							C,p	
			-1.45F			4F							C,split wave	
			-1.0F			-	-	-	-	-	-	-	C,X,split wave,Ap,p	
			-1.4F										C,X,split wave,Ap	
C0030 4143	436 a	$E_{\frac{1}{2}}$	-0.82F	NCE (LiCl,o)	i_ℓ/i_d	1.0F	-	-	-	-	-	-	C,p	FF25 bq83
			-1.4F			0.4F(?)							C,H	
			-0.39F			0.9F							C,p	
			-0.84F			<0.1F							C	
			-			-							C,H	
			-0.39F			0.6F	-	-	-	-	-	-	C,p	
			-0.84F			0.4F							C	
			-			-							C,H	

CONT

FF25 (CONT.) $C_{29}H_{20}O$

Code No.	Empirical Formula	Name and WLN Code	Structural Formula	Solvent	Tech.	Medium		μ, M	pH	T, °C	Electrodes	App.	Experimental Parameters
FF25 bq83	$C_{29}H_{20}O$	tetraphenylcyclopentadienone	Table II-2	EtOH 96(?)	PY	LiCl HCl	0.1 2.5E-4	-	-	25± 0.1	DME//NCE (LiCl, o)	OAO	C=0.36, m=1.12, t=3.2, h=4.0
						LiCl	0.1						
FF26	$C_{29}H_{21}IN_2S_2$	3-phenyl-2-[3-(3-phenyl-2(3H)-benzothiazolylidene)-1-propenyl]benzothiazolium iodide C.A. 52886-85-8	Table II	MeOH 40	IL	KCl	1.2	-	-	-	g(Nujol): xxns// SCE	0--	C=ns, A=0.126
				MeOH 50	PY	Me_4NOH	0.1	-	-	-	DME//SCE	0--	C=0.05, m=2.61, t=3.0, h=50
FF27	$C_{29}H_{22}O$	2,3,4,5-tetraphenyl-2-cyclopenten-1-one C.A. 7317-52-4	Table II	EtOH 96(?)	PY	LiCl	0.1	-	-	-	DME//NCE (LiCl,o)	OAO	C=0.33, m=1.12, t=3.2, h=40
FF28 CONT	$C_{30}H_{22}Cl_2N_6O_8Ru$	bis[2,6-di(3-pyridyl)pyridine]ruthenium(2+) bisperchlorate C.A. 15746-82-4	Table II	MeCN	VA	Bu_4NBF_4	0.1	-	-	-	Pt: xxdi// Ag:xxwi	2F-	$0 \rightarrow 1.4 \rightarrow -2 \rightarrow 0$ V, C=1.05, v=20, Pt Aux
													v=50
													v=100
													v=200
													v=500

TABLE I. Electrochemical Data   303

$C_{30}H_{22}Cl_2N_6O_8Ru$ (CONT.) FF28

Ref.	C/M	Charact. Potential		Response	Const. Value	n Tech.		Electrokinetic Data			Products and Identification	Description and Remarks	Code No.
		Value	vs.					Parameter	Value	From			
C0030 4143	436 a	$E_{\frac{1}{2}}$	-0.39F -0.84F -1.4F	NCE (LiCl,o)	i_ℓ/i_d	0.25F 0.75F 0.75F	-	-	-	-	-	-	C,p C C,H	FF25 bq83
	436 b		-0.85 -1.65			1.00F 0.2F	-	-	-	-	-	-	C,p C	
EA019 0215	-	$E_{p/2}$	0.72 0.84	SCE	D	4.90 -	sttd	2	$E_p/E_{p/2}$	80 -	sttd	-	A,≠,p A,i_a	FF26
EA019 0215	-	$E_{\frac{1}{2}}$	-1.14 -1.56	SCE	I	1.28 -	sttd	1 1	-	-	-	-	C,$i_d + i_a$,Mc($E_{\frac{1}{2}}$),p C,Mc($E_{\frac{1}{2}}$)	
C0030 4143	436 b	$E_{\frac{1}{2}}$	-1.65F	NCE (LiCl,o)	i_ℓ/i_d	0.5F	-	-	-	-	-	-	C,i_d, redn of tetraphenyldienone?	FF27
JA095 6582	374 c	E_p	1.28 - -1.43 -1.70 - -	SCE	$i_p/Cv^{\frac{1}{2}}$	3.2 - 3.2 3.0 - -	-	-	ΔE_p	60 - 64 94 - -	sttd	-	A,$i_{p,C}/i_{p,A}$=1.0,R,r C,R C,$i_{p,A}/i_{p,C}$=0.89 C,$i_{p,A}/i_{p,C}$=0.82 A A	FF28
			1.28 - -1.43 -1.70 - -			3.3 - 3.2 3.4 - -				65 - 61 85 - -			A,r C C C A A	
			1.28 - -1.43 -1.70 - -			3.3 - 3.3 3.6 - -				64 - 59 80 - -			A,$i_{p,C}/i_{p,A}$=1.1,r C C,$i_{p,A}/i_{p,C}$=0.93 C,$i_{p,A}/i_{p,C}$=0.80 A A	
			1.28 - -1.43 -1.70 - -			3.3 - 3.5 3.7 - -				65 - 50 82 - -			A,$i_{p,C}/i_{p,A}$=1.1,r C C,$i_{p,A}/i_{p,C}$=0.90 C,$i_{p,A}/i_{p,C}$=0.92 A A	
			1.28 - -1.43 -1.70 - -			3.2 - 3.4 3.7 - -				59 - 50 81 - -			A,$i_{p,C}/i_{p,A}$=1.1,r C C,$i_{p,A}/i_{p,C}$=0.91 C,$i_{p,A}/i_{p,C}$=0.90 A A	
			- -	NCE (LiCl,o)	i_ℓ/i_d	- -				- -			A A	CONT

FF28 (CONT.) $C_{30}H_{22}Cl_2N_6O_8Ru$

Code No.	Empirical Formula	Name and C.A. Number	Structural Formula	Solvent	Tech.	Medium		μ, M	pH	T, °C	Electrodes	App.	Experimental Parameters
FF28 bg09	$C_{30}H_{22}Cl_2N_6O_8Ru$	bis[2,6-di(3-pyridyl)pyridine]ruthenium(2+) bisperchlorate	Table II	MeCN	VA	Bu_4NBF_4	0.1	–	–	–	Pt: xxdi//Ag	2F-	0 → -1.8 → 0 V, C=1, v=200, Pt Aux
													0 → 1.4 – 0 V
FF29 see DG64	$C_{30}H_{24}Cl_2N_6O_8Ru$	tris(2,2'-bipyridine-N,N')ruthenium-(2+) diperchlorate C.A. 15635-95-7	Table II	MeCN	VA	Bu_4NBF_4	0.1	–	–	–	Pt: xxdi//Ag	2F-	0 → 1.4 → -2.6 → 0 V, C=ns, Pt Aux
													0 → -2.0 → 0 V
													0 → -2.6 → 0 V
													0 → 1.4 → 0 V
CONT		bis[2,6-di(3-pyridyl)pyridine]ruthenium		MeCN		Bu_4NBF_4	0.1					2F-	

TABLE I. Electrochemical Data

$C_{30}H_{24}Cl_2N_6O_8Ru$ (CONT.) FF29

Ref.	C/M	Charact. Potential		Response Const.		n Tech.	Electrokinetic Data			Products and Identification	Description and Remarks	Code No.
		Value	vs.		Value		Parameter	Value	From			
JA095 6582	374 c	E_p	-1.35F	Ag	i_p	5.8F	-	-	-	-	-	-	C,r	FF28 bg09
		-1.62F			4.5						C	
		-1.54F			4F						A	
		-1.30F			2F						A	
		1.26F			10.6F	-	-	-	-	-	-	A,r	
		1.18F			10F						C	
JA095 6582	374 c	E_p	1.354	SCE	-	-	QE (1.40V)	1.04	-	-	-	trication,ESR,bright green,unstable to air overnight, stable several days in absence of air, exposure to air gives $Ru(bipy)_3^{2+}$ (VIS), reverse electrolysis at 1 hr gives $Q_C/Q_A=0.92$	A,r	FF29 see DG64
		-						-				C	
		-1.322					i:i	1			monocation,ESR	C	
		-1.517						1			neutral molecule	C	
		-1.764						1			monoanion,ESR	C	
		-2.4						2.7			ESR signal	C,≠ with v < 2E4, R and n=1 with v > 2E4, R with v > 2E5 in DMF	
		-						-				A	
		-						-				A	
		-						-				A	
		-						-				A	
		-1.43F		Ag	i_p	17F	-	-	-	-	-	monocation,ESR	C,r	
		-1.63F			18F						neutral molecule	C	
		-1.88F			16F						monoanion,ESR	C	
		-1.82F			17F							A	
		-1.57			17F							A	
		-1.37			15F							A	
		-1.43F			17	-	-	-	-	-	-	C,r	
		-1.63F			18							C	
		-1.88F			16							C	
		-2.56F			45							C	
		-2.33F			2							A	
		-1.82F			14.5F							A	
		-1.57F			10F							A	
		-1.38F			6F							A	
		-0.94F			6F							A	
		-0.71F			4F							A	
		-0.52			2F							A	
		1.32F			20F	-	-	-	-	-	trication,ESR,bright green,unstable to air overnight, stable several days in absence of air, exposure to air gives $Ru(bipy)_3^{2+}$ (VIS), reverse electrolysis at 1 hr gives $Q_C/Q_A=0.92$	A,r	
		1.28F			20F							C	CONT

FF29 (CONT.) $C_{30}H_{24}Cl_2N_6O_8Ru$

Code No.	Empirical Formula	Name and C.A. Number	Structural Formula	Solvent	Tech.	Medium		μ, M	pH	T, °C	Electrodes	App.	Experimental Parameters
FF29 see DG64	$C_{30}H_{24}Cl_2N_6O_8Ru$	tris(2,2'-bipyridine-N,N')ruthenium-(2+) diperchlorate	Table II	MeCN	VA	Bu_4NBF_4	0.1	-	-	25	Pt: xxdi//Ag	2F-	C=1.27, v=20, Pt Aux
													v=50
													v=100
													v=200
													v=500
													v=2E3
													v=2E4
CONT DG64		tris(2,2'-bipyridine-N,N')ruthenium-(2+) diperchlorate	Table II	MeCN	VA	Bu_4NBF_4	0.1	-	-	25	Pt:	2F-	

$C_{30}H_{24}Cl_2N_6O_8Ru$ (CONT.) FF29

Ref.	C/M	Charact. Potential		Response Const.		n Tech.	n	Electrokinetic Data			Products and Identification	Description and Remarks	Code No.
		Value	vs.		Value			Parameter	Value	From			
JA095 6582	374 c	E_p	1.354	Ag	$i_p/Cv^{\frac{1}{2}}$	3.2	-	-	ΔE_p	60	sttd	-	$A,R,i_{p,C}/i_{p,A}=1.0,r$	FF29 see DG64
			-			-				-			C	
			-1.332			3.3				59			$C,i_{p,A}/i_{p,C}=0.96$	
			-1.517			3.4				65			$C,i_{p,A}/i_{p,C}=0.96$	
			-1.764			3.7				68			$C,i_{p,A}/i_{p,C}=0.98$	
			-			-				-			A	
			-			-				-			A	
			-			-				-			A	
			1.354			3.4	-	-		62		-	$A,R,i_{p,C}/i_{p,A}=0.99,r$	
			-			-				-			C	
			-1.332			3.8				56			$C,R,i_{p,A}/i_{p,C}=0.98$	
			-1.517			3.4				70			$C,R,i_{p,A}/i_{p,C}=0.98$	
			-1.764			3.6				66			$C,i_{p,A}/i_{p,C}=0.97$	
			-			-				-			A	
			-			-				-			A	
			-			-				-			A	
			1.354			3.1	-	-		61		-	$A,R,i_{p,C}/i_{p,A}=1.0,r$	
			-			-				-			C	
			-1.332			4.3				62			$C,R,i_{p,A}/i_{p,C}=0.98$	
			-1.517			3.5				64			$C,R,i_{p,A}/i_{p,C}=0.93$	
			-1.764			3.3				65			$C,R,i_{p,A}/i_{p,C}=1.0$	
			-			-				-			A	
			-			-				-			A	
			-			-				-			A	
			1.354			3.4	-	-		64		-	$A,R,i_{p,C}/i_{p,A}=1.0,r$	
			-			-				-			C	
			-1.332			4.8				60			$C,R,i_{p,A}/i_{p,C}=0.98$	
			-1.517			3.5				66			$C,R,i_{p,A}/i_{p,C}=0.93$	
			-1.764			3.7				60			$C,R,i_{p,A}/i_{p,C}=0.96$	
			-			-				-			A	
			-			-				-			A	
			-			-				-			A	
			1.354			3.2	-	-		55		-	$A,R,i_{p,C}/i_{p,A}=0.98,r$	
			-			-				-			C	
			-1.332			3.9				61			$C,R,i_{p,A}/i_{p,C}=0.97$	
			-			-				-			A	
			1.354			4.1	-	-		68		-	$A,R,i_{p,C}/i_{p,A}=1.0,r$	
			-			-				-			C	
			-1.332			4.2				65			C	
			-			-				-			A	
			1.354			3.3	-	-		62		-	A,r	
			-			-				-			C	
			-1.332			4.1				78			C	
			-			-				-			A	
														CONT

FF29 (CONT.) $C_{30}H_{24}Cl_2N_6O_8Ru$ (1)

Code No.	Empirical Formula	Name and C.A. Number	Structural Formula	Solvent	Tech.	Medium	μ, M	pH	T, °C	Electrodes	App.	Experimental Parameters
FF29	$C_{30}H_{24}Cl_2N_6O_8Ru$	tris(2,2'-bipyridine-N,N')ruthenium-(2+) diperchlorate	Table II	MeCN	VY	Bu_4NBF_4 0.1	-	-	-	Pt: rord// Ag:xxwi	2F-	C=ns, E_{ring}=1.48 V, Pt Aux
FF30	$C_{30}H_{25}P$	pentaphenylphosphorane C.A. 2588-88-7	$(C_6H_5)_5P$	DMF	PA	Bu_4NI 0.1	-	-	20	DME//SCE	2AO	C=ns, Hg Aux
FF31	$C_{30}H_{26}N_2O_4$	5,10-di(4-acetylphenyl)-2,7-dimethoxy-5,10-dihydrophenazine C.A. 61228-20-4	Table II	MeCN	VY	Et_4NClO_4 0.1(?)	-	-	-	Pt: rodi// Ag/ $AgClO_4$ 0.01, Et_4NClO_4 0.1	2-0	C=ns, d=0.1, ω=10, Pt Aux
FF32	$C_{30}H_{28}Br_2P_2$	2,5-dimethyl-1,1,4,4-tetraphenyldiphosphoniacyclohexa-2,5-diene dibromide C.A. 21557-88-0	Table II	DMF	PY	Et_4NClO_4 0.1	-	-	-	DME//SCE	2--	ns
FF33	$C_{30}H_{28}N_4O_4$	5,10-bis(4-acetylaminophenyl)-2,7-dimethoxy-5,10-dihydrophenazine C.A. 61255-13-8	Table II	MeCN	VY	Et_4NClO_4 0.1(?)	-	-	-	Pt: rodi// Ag/ $AgClO_4$ 0.01, Et_4NClO_4 0.1	2-0	C=ns, d=0.1, ω=10, Pt Aux
FF34	$C_{31}H_{40}O_2$	vitamin K_2	Table II	MeOH 70	PY	ACET ns $NaClO_4$ 0.2	-	4.8	25	DME//SCE	2-0	C=0.1, m=0.93, t=7.32, Pt Aux
					PV	ACET ns $NaClO_4$ 0.2	-	4.8	25	DME//SCE	2-0	C=0.1, m=0.93, t=7.32, Pt Aux
					VA	ACET ns $NaClO_4$ 0.2	-	4.8	25	HMDE// SCE	2-0	0→-1.2→ 0.2→0 V, C=0.167, A=0.052, v=100, Pt Aux
FF35	$C_{32}H_{18}K_2O_2$	6,11-dioxido-2,4-diphenyldibenzo[b,h]-biphenylene dipotassium salt	Table II	DMF	VR	Bu_4NClO_4 0.5	-	-	20	Pt: xxdi// Ag/AgCl// naphthoquinone	125A2	C ≤ 0.5, Pt Aux
										Pt: xxdi// Ag/AgCl		-1.8→-0.2→ -1.8→-0.2 V, v=1.1E4

CONT

TABLE I. Electrochemical Data

$C_{32}H_{18}K_2O_2$ (CONT.) FF35

Ref.	C/M	Charact. Potential		Response	Const. Value	n Tech.		Electrokinetic Data			Products and Identification	Description and Remarks	Code No.
		Value	vs.					Parameter	Value	From			
JA095 6582	374 c	E_{disc}	-1.37 SCE	i_ℓ	40F	-	-	-	-	-	-	C,r	FF29
			-1.60		40F							C	
			-1.90		38F							C	
EA020 0021	-	E_p	-2.75 SCE	-	-	-	-	-	-	-	-	C,≠,r	FF30
EA021 0557	-	$E_{\frac{1}{2}}$	-0.19 Ag/AgClO$_4$ 0.01, Et$_4$NClO$_4$ 0.1	-	-	-	-	-	-	-	radical cation,UVS, VIS,ESR	A,r	FF31
JE042 0309	-	$E_{\frac{1}{2}}$	-0.806 SCE	-	-	sttd	1	-	-	-	-	C,p	FF32
EA021 0557	-	$E_{\frac{1}{2}}$	-0.26 Ag/AgClO$_4$ 0.01, Et$_4$NClO$_4$ 0.1	-	-	-	-	-	-	-	radical cation,UVS, VIS,ESR	A,p	FF33
			0.40									A	
JE049 0133	-	$E_{\frac{1}{2}}$	-0.26 SCE	i_ℓ	0.4F	sttd	2	-	-	-	-	C,i_d,r	FF34
JE049 0133	-	E_{su}	0.12 SCE	XR	80F	-	-	-	-	-	-	C,i_a;XR=1/(impedance at summit),$\mu(\Omega^{-1})$;r	
			-0.26		100F							C	
			-1.05		6F							C,i_a	
JE049 0133	-	E_p	-0.24F SCE	i_p	7.3F	-	-	-	-	-	-	C,i_p,$Cv^{\frac{1}{2}}$ ↑ as v ↑ or C ↓,r	
			-0.23F		4.7F							A,$i_p/Cv^{\frac{1}{2}}$ ↑ as v ↑ or C ↓	
			0.12F		1.3F							A,i_a	
			0.10F		1.3F							C,i_a	
JA095 6688	-	"$E_{\frac{1}{2}}$"	-0.36 Ag/AgCl	-	-	sttd	1	-	-	-	-	A,r	FF35
			-1.17				1					A	
			-				-					C	
			-									C	
		E_p	-1.00F Ag/AgCl	i_p	170F	-	-	-	-	-	-	A,r	
			-0.29F		185F							A	
			-0.42F		170F							C	
			-1.13F		143F							C,Mx	CONT

FF35 (CONT.) $C_{32}H_{18}K_2O_2$

Code No.	Empirical Formula	Name and C.A. Number	Structural Formula	Solvent	Tech.	Medium	μ, M	pH	T, °C	Electrodes	App.	Experimental Parameters
FF35	$C_{32}H_{18}K_2O_2$	6,11-dioxido-2,4-diphenyldibenzo[b,h]-biphenylene dipotassium salt	Table II	DMF	VR	Bu_4NClO_4 0.5	-	-	20	Pt: xxdi// Ag/AgCl	125A 2	$-0.2 \to -2.0 \to -0.2 \to -2.0$ V, $C > 1, v=108$, Pt Aux
												$v=1080$
FF36	$C_{32}H_{22}N_6Na_2O_6S_2$	Congo Red	Table II	MeOH 50	PY	Me_4NOH 0.1	-	-	-	DME//SCE	0--	$C=0.05$, $m=2.61$, $t=3.0, h=50$
FF37	$C_{32}H_{24}$	1,2,4,7-tetraphenyl-cyclooctatetraene C.A. 19099-38-8	Table II	MeCN	PY	Et_4NClO_4 ns	-	-	0	DME//SCE (NaCl)	26A0	$C=ns, m=2.19$, $h=65, EC=2.05$, Pt Aux
									22			
FF38	$C_{32}H_{24}$	1,3,5,7-tetraphenyl-cyclooctatetraene C.A. 35087-43-5	Table II	MeCN	PY	Et_4NClO_4 ns	-	-	22	DME//SCE (NaCl)	26A0	$C=ns, m=2.19$, $h=65, EC=2.05$, Pt Aux
				PrCN	PY	Et_4NClO_4 ns	-	-	0	DME//SCE (NaCl)	26A0	$C=ns, m=2.19$, $h=65, EC=2.05$, Pt Aux
									22			
					VA	Et_4NClO_4 ns	-	-	22	HMDE// SCE(NaCl)	2A0	ns → ns → ns V, $C=ns, v=200$, Pt Aux
										Pt: xxns// SCE(NaCl)		
FF39	$C_{32}H_{30}N_6O_4$	2,7-bis(acetylamino)-5,10-bis(4-acetyl-aminophenyl-5,-10-dihydrophenazine C.A. 61228-17-9	Table II	MeCN	VY	Et_4NClO_4 0.1(?)	-	-	-	Pt: rodi// Ag/ $AgClO_4$ 0.01, Et_4NClO_4 0.1	2-0	$C=ns, d=0.1$, $\omega=10$, Pt Aux
FF40	$C_{32}H_{32}Br_2P_2$	2,5-diethyl-1,1,4,4-tetraphenyldiphos-phoniacyclohexa-2,5-diene dibromide C.A. 21557-91-5	Table II	DMF	PY	Et_4NClO_4 0.1	-	-	-	DME//SCE	2--	$C=ns$
FF41	$C_{32}H_{34}N_2$	N,N'-di(2,3,5,6-tetramethylphenyl)-benzidine C.A. 59245-29-3	Table II	MeCN	VY	Et_4NClO_4 0.1	-	-	-	Pt: rodi// Ag/ $AgClO_4$ 0.01, Et_4NClO_4 0.1	2-0	$C=0.5, d=0.1$, $\omega=10$, Pt Aux

TABLE I. Electrochemical Data

$C_{32}H_{34}N_2$ FF41

Ref.	C/M	Charact. Potential		Response	Const. Value	n		Electrokinetic Data			Products and Identification	Description and Remarks	Code No.
		Value	vs.			Tech.		Parameter	Value	From			
JA095 6688	–	E_p	–1.00F	Ag/AgCl	i_p	2.2F	–	–	–	–	–	–	A,r	FF35
			–0.33F			2.5F							A	
			–1.11F			1.3F							C	
			–1.77F			1.0F							C	
			–1.00F			5F	–	–	–	–	–	–	A,r	
			–0.29F			8F							A	
			–1.17F			3F							C	
			–1.89F			3F							C	
EA019 0215	–	$E_{\frac{1}{2}}$	–0.77	SCE	I	5.84	sttd	4	Tomeš	15	sttd	–	C,R,Mc($E_\frac{1}{2}$),p	FF36
JE056 0409	–	$E_{\frac{1}{2}}$	–1.85	SCE(NaCl)	–	–	–	–	–	–	–	–	C,p	FF37
			–1.873		I	3.37	–	–	–	–	–	dianion	C, $i_l=kh^{\frac{1}{2}}, i_d$, p	
JE056 0409	–	$E_{\frac{1}{2}}$	–1.67	SCE(NaCl)	I	3.11	1:1	2	Elog Tomeš	28.4 28	sttd	dianion	C, $i_l=kh^{\frac{1}{2}}, i_d$, p	FF38
JE056 0409	–	$E_{\frac{1}{2}}$	–1.706	–	–	–	–	–	–	–	–	–	C,p	
			–1.718	SCE(NaCl)	I	3.56	sttd	2	Tomeš	28	–	orange dianion (stable)	C, $i_l=kh^{\frac{1}{2}}, i_d$, p	
JE056 0409	–	E_p	–1.74F	SCE(NaCl)	i_p	13F	–	–	–	–	–	–	C,R,Ql for v > 800, p	
			–1.70F			12F			$E_p-E_{p/2}$	27.5			A,R,Ql for v > 800	
			–1.67F			18F	–	–	–	–	–	–	C,p	
			–1.58F			18F							A	
EA021 0557	–	$E_{\frac{1}{2}}$	–0.24	Ag/ AgClO$_4$ 0.01, Et$_4$NClO$_4$ 0.1	–	–	–	–	–	–	–	radical cation, UVS, VIS, ESR	A,r	FF39
			0.415										A	
JE042 0309	–	$E_{\frac{1}{2}}$	–0.969	SCE	–	–	sttd	1	–	–	–	–	C,p	FF40
EA020 1011	419 d	$E_{\frac{1}{2}}$	0.33	Ag/ AgClO$_4$ 0.01, Et$_4$NClO$_4$ 0.1	$i_l(u)$	1	–	–	–	–	–	–	A,r	FF41
			0.48			1							A	

FF42 $C_{32}H_{38}N_4$

Code No.	Empirical Formula	Name and C.A. Number	Structural Formula	Solvent	Tech.	Medium	μ, M	pH	T, °C	Electrodes	App.	Experimental Parameters
FF42 br50	$C_{32}H_{38}N_4$	etioporphyrin I C.A. 448-71-5	Table 11-2	CH_2Cl_2	VA	Bu_4NPF_6 0.1	-	-	25±2	Pt: xxbe// SCE(NaCl)	2AO	ns→ns→nsV, C=1
FF43	$C_{33}H_{28}O_4S_3$	2,3,4,5-tetrakis(4-methoxyphenyl)tri-thiapentalene C.A. 38755-23-6	Table 11	MeCN 50 CH_2Cl_2 50	IL	$NaClO_4$ 0.1	-	-	-	Pt: xxbu// SCE	---	C=ns
FF44	$C_{33}H_{36}N_4ORu$	carbonyletio-porphyrin(I)ruthen-ium(II) C.A. 43145-31-9	Table 11	CH_2Cl_2	VR	Bu_4NPF_6 0.1	-	-	25±2	Pt: xxbe// SCE(NaCl)	2AO	C=1
FF45 cm84	$C_{33}H_{36}N_4O_6$	bilirubin C.A. 635-65-4	Table 11-3	DMSO	PK	$LiClO_4$ 0.1	-	-	-	DME// Ag/AgCl	--0	-1.0→ -1.55 V, C=1.37, f=20
												-1.5→-2.0 V
										glC: rodi// Ag/AgCl		-1.0→-1.7 V
				VA		Et_4NClO_4 0.1	-	-	-	Au: xxns// SCE	--0	-0.6→-1.6→ -0.6 V, C=0.95, A=0.2, v=100
												v=1000
										Au: xxns// Ag/AgCl	18-0	0→0.6→ -0.6→0 V, C=1.81, v=2
										HMDE// SCE	--2	-1.0→-2.3 →-1.0 V, C=0.9, v=2E3
				VR		$LiClO_4$ 0.1	-	-	-	Au: xxns// SCE	1-0	-0.8→0.8→ -0.8→0.8 V, C=3.44, v=20
						Et_4NClO_4 0.1				Au: xxns// Ag/AgCl	8-0	0→-1.7→ 0→-1.7 V, C=4.12, v=2

CONT

TABLE I. Electrochemical Data

$C_{33}H_{36}N_4O_6$ (CONT.) FF45

Ref.	C/M	Charact.	Potential Value	vs.	Response Const.	Value	n Tech.		Electrokinetic Data Parameter	Value	From	Products and Identification	Description and Remarks	Code No.
JA095 5939	-	"$E_{\frac{1}{2}}$"	0.78	SCE(NaCl)	-	-	QE	0.97				-	A,R	FF42 br50
			1.34				-	-					A,R	
			-										C	
			-										C	
JE038 0479	-	E_p	0.84	SCE	-	-	QE	1	-	-	-	-	A,p	FF43
JA095 5939	329 1	"$E_{\frac{1}{2}}$"	0.61	SCE(NaCl)	-	-	QE	0.88				-	A,R	FF44
			1.11				-	-					A,R	
			-										C	
			-										C	
JE048 0447	343 c	E_τ	-1.43F	Ag/AgCl	i_ℓ	0.3F	-	-	-	-	-	-	A,R,$i \downarrow$ as $f \downarrow$,ec,p	FF45 cm84
						0.3F							C,R,$i \neq f(f)$	
			-1.77F			0.4F	-	-	-	-	-	-	A,R,p	
						0.5F							C,R	
			-1.42F			1.2F	-	-	-	-	-	-	A,0 at low f,p	
						7F							C	
JE048 0447	343 c	E_p	-1.4F	SCE	i_p	28F	-	-	-	-	-	-	C,$i_p=f(C)$,p	
													A,0	
			-1.45F			100F	QE	1.03	-	-	-	-	C,p	
			-1.35F			small		-					A	
			0.50F	Ag/AgCl		32.5F		1.97	βn_b	0.74	$E_p - E_{p/2}$	-	A,p	
			-0.41F			20.6F				-			C	
			-1.36	SCE		8F		1.9	-	-	-		C,⟂,$i_p \uparrow$ on subsequent scans,$E_p \rightarrow$ more neg as v $\uparrow$,r	
			-1.90			11F		-					C,$E_p \neq f(v)$	
			-2.20			9F							C	
			-2.05F			7ApF							A	
			-1.81F			4F							A	
JE048 0447	343 c	E_p	0.7F	SCE	$i_p(u)$	29F	-	-	-	-	-	CP→biliverdine,UVS,VIS	A,⟂,$i_p=k_1+k_2C$ for v=20-100 and C=0.36-3.5,$E_p \rightarrow$ more pos as v $\uparrow$,p	
			-0.3F			10F							C	
			-0.7F	Ag/AgCl	i_p	13F	i:i QE:QE	0.7 0.94 ± 0.1	-	-	-	-	C,$i_p \downarrow$ on subsequent scans,p	
			-1.34F			30F		-					C,$i_p \downarrow$ on subsequent scans	
			-0.8F			3F							A	
			-0.5F			5F							A	CONT

FF45 (CONT.) $C_{33}H_{36}N_4O_6$

Code No.	Empirical Formula	Name and C.A. Number	Structural Formula	Solvent	Tech.	Medium		μ, M	pH	T, °C	Electrodes	App.	Experimental Parameters
FF45 cm84	$C_{33}H_{36}N_4O_6$	bilirubin	Table II-3	DMSO	VR	Et_4NClO_4	0.1	-	-	-	HMDE// SCE	--0	$-0.7 \to -1.7 \to -0.7 \to -1.7$ V, C=0.89, v=200
													v=5E4
FF46	$C_{33}H_{40}N_4O_6$	2,17-diethyl-1,10,-19,22,23,24-hexahydro-3,7,13,18-tetramethyl-1,19-dioxo-21H-biline-8,-12-dipropanoic acid C.A. 16568-56-2	Table II	H_2O	PY	BR	-	-	8.0	-	DME//SCE	O4A0	C=0.014, m=1.32, t=3.11
													C=0.084
									8.25				
									10.1				C=0.014
													C=0.084
									12.0				C=0.014
													C=0.084
													C=0.177
									12.9				C=0.014
													C=0.084
													C=0.177
FF47	$C_{34}H_{24}Cl_2O_8S_6$	5,5'-diphenyl-3,3'-(2,5-diphenyl-3,4-dithia-1,6-hexamethylene)dithiolium diperchlorate	Table II	MeCN 50 CH_2Cl_2 50	IL	$NaClO_4$	0.1	-	-	-	Pt: xxbu// SCE	---	C=ns
FF48	$C_{34}H_{38}O_6$	2,2'-bis(3-methoxyphenylethyl)-4,4',5,5'-tetramethoxybiphenyl C.A. 51095-50-2	Table II	MeCN	VA	$NaClO_4$	0.1	-	-	-	Pt: xxbe// SCE	2A2	$0.4 \to 1.5 \to 0.4$ V, C=ns, v=86
FF49	$C_{35}H_{28}N_3Na$	4,5-di(biphenylyl)-2-(dimethylaminophenyl)imidazole sodium salt C.A. 31909-32-7	Table II	MeCN	VY	Bu_4NClO_4	0.1	-	-	-	Pt: rodi// Ag/AgCl// cobaltocinium	2A0	C=1.9, A=0.012, v=15.3

$C_{35}H_{28}N_3Na$ FF49

Ref.	C/M	Charact. Potential		Response Const.		n		Electrokinetic Data			Products and Identification	Description and Remarks	Code No.
		Value	vs.		Value	Tech.		Parameter	Value	From			
JE048 0447	343 c	E_p	-1.38F	SCE	i_p	6F	-	-	-	-	-	-	C,r A,	FF45 cm84
			-1.5F			60F	-	-	-	-	-	-	C,r	
			-1.27F			30F	-	-	-	-	-	-	A	
C0026 2271	-	$E_{\frac{1}{2}}$	0.03	SCE	-	-	-	-	-	-	-	-	A,p	FF46
			0.08		-	-	-	-	-	-	-	-	A,p	
			-1.62F			3.57	-	-	-	-	-	-	C,p	
			-0.11		-	-	-	-	-	-	-	-	A,p	
			-0.08		-	-	-	-	-	-	-	-	A,p	
			-0.20		-	-	-	-	-	-	-	-	A,p	
			-0.15		-	-	-	-	-	-	-	-	A,p	
			-1.62			3.57	-	-	-	-	-	-	C,p	
			-0.27			3.57	-	-	-	-	-	Hg salt	A,$i_\ell=kC,i_d$,p	
			-0.22		-	-	-	-	-	-	-	-	A,i_a,$i_\ell=kh$,$i \neq f(C)$,p	
			-1.62			3.57	-	-	-	-	-	CP(-1.7 V) → mesobilirubinogen (urobilinogen)	C,p	
JE038 0479	-	E_p	0.32	SCE	-	-	QE	0.8	-	-	-	-	C,p	FF47
JA095 7132	437 c	E_p	1.22F	SCE	$i_p(u)$	4F	-	-	-	-	-	-	A,R,r	FF48
			1.36F			4F	-	-	-	-	-	-	A,R	
			1.31F			4F	-	-	-	-	-	-	C,R	
			1.15F			2F	-	-	-	-	-	-	C,R	
C0036 0575	397 a	$E_{\frac{1}{2}}$	0.012	Ag/AgCl	$i_\ell(u)$	1.00	Elog!	1	αn_a	0.85	Elog	-	A,R,i_d,r	FF49
			0.472			1.03		1		0.87			A,R,i_d	

FF50 $C_{36}H_{24}Cl_2N_6O_8Ru$

Code No.	Empirical Formula	Name and C.A. Number	Structural Formula	Solvent	Tech.	Medium	μ, M	pH	T, °C	Electrodes	App.	Experimental Parameters
FF50	$C_{36}H_{24}Cl_2N_6O_8Ru$	tris(1,10-phenanthroline)ruthenium(II) perchlorate C.A. 22873-66-1	Table II	MeCN	VA	Bu_4NBF_4 0.1	-	-	-	Pt: xxdi//Ag	2F-	ns→ns→ns V, C=ns, Pt Aux
												0→ 1.65→ 0 V, C=1, v=200, Pt Aux
												0→-1.7→0 V
												0→-2.5→0 V
FF51	$C_{36}H_{24}Cl_3N_{12}O_{12}Ru$	bis(2,4,6-tri-2-pyridyl-s-triazine)-ruthenium(III) perchlorate C.A. 21791-68-4	Table II	MeCN	VA	Bu_4NBF_4 0.1	-	-	-	Pt: xxdi//Ag	2F-	ns→ns→ns V, C=ns, Pt Aux
												0→ 1.8→ 0 V, C=1, v=200, Pt Aux
												0→-2.0→ 0 V
												0→-2.6→ 0 V
CONT			Table II	MeCN	VA	Bu_4NBF_4 0.1				Pt: xxdi//Ag		

$C_{36}H_{24}Cl_3N_{12}O_{12}Ru$ (CONT.) FF51

Ref.	C/M	Charact. Potential Value	vs.	Response Const. Value		n Tech.		Electrokinetic Data Parameter	Value	From	Products and Identification	Description and Remarks	Code No.
JA095 6582	374 c	E_p	1.40	SCE	-	-	i:i	1	-	-	-	-	A,r	FF50
			-					-					C	
			-1.41					1					C	
			-1.54					1					C, desorption spike on reversal	
			-1.84					1					C, i_p small, broad peak	
			-2.24					1					C,⇌	
			-					-					A	
			-					-					A	
			-					-					A	
			-					-					A	
			1.52F			65F	-	-	-	-	-	-	A,r	
			1.46F			60F							C	
			-1.42			54F	-	-	-	-	-	-	C,r	
			-1.56			61F							C	
			-1.37			150F							A,Mx	
			-1.42F			54F	-	-	-	-	-	-	C,r	
			-1.56F			61F							C	
			-1.90			7F							C	
			-2.28F			300F							C,Mx	
			-2.14F			11F							A	
			-1.78F			35F							A	
			-1.37F			65F							A	
JA095 6582	374 c	E_p	1.52	SCE	-	-	i:i	1	-	-	-	-	A,r	FF51
			-					-					C	
			-0.84					1					C	
			-1.00					1					C	
			-1.66					1					C	
			-1.91					1					C	
			-					-					A	
			-										A	
			-										A	
			-										A	
			1.70F	Ag	i_p	18F	-	-	-	-	-	-	A,r	
			1.60F			13F							C	
			-0.80F			12	-	-	-	-	-	-	C,r	
			-1.00F			11							C	
			-1.68F			12							C	
			-1.95F			8							C	
			-1.83F			6F							A	
			-1.60F			8F							A	
			-0.92F			13F							A	
			-0.75F			6F							A	
			-0.80F			12F							C,r	
			-1.00F			11F							C	
			-1.68F			12F							C	
			-1.95F			8F							C	
			-2.4F			108F							C	
			-1.83F			6F							A	
			-1.60F			4F							A	
			-0.92F			10F							A	
			-0.73F			3F							A	CONT

FF51 (CONT.) $C_{36}H_{24}Cl_3N_{12}O_{12}Ru$

Code No.	Empirical Formula	Name and C.A. Number	Structural Formula	Solvent	Tech.	Medium		μ, M	pH	T, °C	Electrodes	App.	Experimental Parameters
FF51 CONT	$C_{36}H_{24}Cl_3N_{12}O_{12}Ru$	bis(2,4,6-tri-2-pyridyl-s-triazine)-ruthenium(III) perchlorate	Table II	MeCN	VA	Bu_4NBF_4	0.1	-	-	-	Pt: xxdi//Ag	2F-	C=1.44, v=20, Pt Aux
													v=200
													v=500
													v=1E3
													v=2E3
													v=2E4
FF52	$C_{36}H_{28}Cl_2O_8S_6$	5,5'-diphenyl-3,3'-[2,5-bis(4-methoxyphenyl)-3,4-dithia-1,6-hexamethyleno]-dithiolium perchlorate	Table II	MeCN 50 CH_2Cl_2 50	IL	$NaClO_4$	0.1	-	-	-	Pt: xxbu// SCE	---	C=ns
FF53	$C_{36}H_{36}MgN_4O_4$	magnesium(II) protoporphyrin IX dimethylester C.A. 14724-63-1	Table II	DMF	PY	Et_4NClO_4	0.1	-	-	-	DME//SCE	25--	C=ns

$C_{36}H_{36}MgN_4O_4$ (CONT.) FF53

Ref.	C/M	Charact. Potential		Response Const.		n		Electrokinetic Data			Products and Identification	Description and Remarks	Code No.
		Value	vs.		Value	Tech.		Parameter	Value	From			
JA095 6582	374 c	E_p		$i_p/Cv^{\frac{1}{2}}$	2.3	-	-	ΔE_p	70	sttd	-	A, r	FF51
		1.52	SCE		-				-			$C, i_{p,A}/i_{p,C}=1.37$	
		-0.84			2.4				98			$C, R, i_{p,A}/i_{p,C}=1.0$	
		-1.00			2.7				94			$C, R, i_{p,A}/i_{p,C}=1.0$	
		-1.66			2.9				80			$C, R, i_{p,A}/i_{p,C}=1.0$	
		-1.91			2.8				94			$C, R, i_{p,A}/i_{p,C}=1.0$	
		-			-				-			A	
		-										A	
		-										A	
		-										A	
		1.54			2.0	-	-		68		-	$A, i_{p,C}/i_{p,A}=0.83, r$	
		-			-				-			C	
		-0.84			2.2				70			$C, i_{p,A}/i_{p,C}=0.9$	
		-1.00			2.8				70			$C, R, i_{p,A}/i_{p,C}=1.00$	
		-1.66			2.9				72			$C, R, i_{p,A}/i_{p,C}=1.0$	
		-1.91			3.1				83			$C, R, i_{p,A}/i_{p,C}=1.0$	
		-			-				-			A	
		-			-				-			A	
		-			-				-			A	
		-			-				-			A	
		-0.84			24	-	-		83		-	C, r	
		-1.00			2.9				89			$C, i_{p,A}/i_{p,C}=0.89$	
		-1.66			-				78			C	
		-1.91							110			C	
		-			-				-			A	
		-										A	
		-										A	
		-										A	
		-0.84			2.2	-	-		91		-	C, r	
		-			-				-			A	
		-0.84			2.3	-	-		74		-	C, r	
		-			-				-			A	
		-0.84			2.0	-	-		71		-	C, r	
		-			-				-			A	
JE038 0479	-	E_p	0.42	SCE	-	-	QE	0.7	-	-	-	C, p	FF52
EA021 1149	328 b	$E_{\frac{1}{2}}$	-1.54 ±0.02	SCE	D	4.52± 0.08	QE (-1.75V)	1.0	Elog $k_{s,h}$	58 (7.5± 0.5)E-2	-	-	$C, QI, E_{\frac{1}{2}} \neq f(C), i_\ell=kh^{\frac{1}{2}}, i_d, r$	FF53
		-1.86 ±0.02			-	(-2.10V)	1.0		67 (4.6± 0.4)E-2			$C, R, E_{\frac{1}{2}} \neq f(C), i_\ell=kh^{\frac{1}{2}}, i_d$	
		-2.40 ±0.02				(-2.55V)	1.03		67 (3.5± 0.5)E-2			$C, R, E_{\frac{1}{2}} \neq f(C), i_\ell=kh^{\frac{1}{2}}, i_d$	CONT

FF53 (CONT.) $C_{36}H_{36}MgN_4O_4$

Code No.	Empirical Formula	Name and C.A. Number	Structural Formula	Solvent	Tech.	Medium	μ, M	pH	T, °C	Electrodes	App.	Experimental Parameters
FF53	$C_{36}H_{36}MgN_4O_4$	magnesium(II) protoporphyrin IX dimethyl ester	Table II	DMF	PR	Et_4NClO_4 0.1	-	-	-	DME//SCE	35--	C=0.34, v=2E4
FF54	$C_{36}H_{40}Br_2P_2$	2,5-di-tert-butyl-1,1,4,4-tetraphenyl-diphosphoniacyclohexa-2,5-diene dibromide C.A. 41480-68-6	Table II	DMF	PY	Et_4NClO_4 0.1	-	-	-	DME//SCE	2--	C=ns
FF55	$C_{36}H_{40}MgN_4O_4$	magnesium(II)mesoporphyrin IX dimethyl ester C.A. 53513-44-3	Table II	DMF	PY	Et_4NClO_4 0.1	-	-	-	DME//SCE	25--	C=ns
					PR	Et_4NClO_4 0.1	-	-	-	DME//SCE	35--	C=0.324, v=2E4
FF56 dh04	$C_{36}H_{46}N_4$	2,3,7,8,12,13,17,18-octaethylporphine C.A. 2683-82-1	Table II-4	CH_2Cl_2	VR	Bu_4NPF_6 0.1	-	-	25±2	Pt: xxbe// SCE(NaCl)	2AO	C=1
FF57	$C_{37}H_{30}ClOP_2Rh$	carbonyl(chloro)bis-(triphenylphosphine)rhodium C.A. 13938-94-8	trans-$[(C_6H_5)_3P]_2$-RhCl(CO)	MeCN 50 $CH_3C_6H_5$ 50	PY	Et_4NClO_4 0.1 $(C_6H_5)_3P$ 1.73E-2	-	-	25± 0.1	DME//SCE	-6AO	C=1.73, m=0.235, t(c)=1, h=68.0
FF58	$C_{37}H_{44}N_4ORu$	carbonyl(octaethylporphyrinyl)ruthenium C.A. 41636-35-5	Table II	CH_2Cl_2	VR	Bu_4NPF_6 0.1	-	-	25±2	Pt: xxbe// SCE(NaCl)	2AO	C=1

TABLE I. Electrochemical Data

$C_{37}H_{44}N_3ORu$ FF58

Ref.	C/M	Charact. Potential		Response Const.		n Tech.		Electrokinetic Data			Products and Identification	Description and Remarks	Code No.
		Value	vs.		Value			Parameter	Value	From			
EA021 1149	328 b	E_p	−1.475	SCE	i_p	4F	−	−	−	−	−	−	$C, \Delta E_p \uparrow$ as $v \uparrow$, $E_p \neq f(C), i_p = kv^{\frac{1}{2}}$ for $v=(5-100)E3, i_d, r$	FF53
			−1.935			4F							$C, \Delta E_p \uparrow$ as $v \uparrow$, $E_p \neq f(C), i_p = kv^{\frac{1}{2}}$ for $v=(5-100)E3, i_d$	
			−2.615			5.8F							$C, \Delta E_p \uparrow$ as $v \uparrow$, $E_p \neq f(C), i_p = kv^{\frac{1}{2}}$ for $v=(5-100)E3, i_d$	
			−2.4			2F							A	
			−1.84			2.5F							A	
			−1.349			2.5F							A	
JE042 0309	−	$E_{\frac{1}{2}}$	−0.723	SCE	−	−	sttd	1	−	−	−	−	C,p	FF54
			−1.08					1					C	
EA021 1149	328 c	$E_{\frac{1}{2}}$	−1.64 ±0.02	SCE	D	4.5± 0.3	QE (−1.72V)	1.0	E log k_s	57 (5.7± 0.1)E-2	−	−	$C, R, E_{\frac{1}{2}} \neq f(C), i_\ell = kh^{\frac{1}{2}}, i_d, r$	FF55
			−1.97 ±0.02				(−2.10V)	1.2		53 (2.7± 0.2)E-2			$C, R, E_{\frac{1}{2}} \neq f(C), i_\ell = kh^{\frac{1}{2}}, i_d$	
			−2.25 ±0.02				(2.45V)	0.95		85 −			$C, E_{\frac{1}{2}} \neq f(C), i_\ell = kh^{\frac{1}{2}}$	
EA021 1149	328 c	E_p	−1.628	SCE	i_p	0.5F	−	−	$dE_p/d\log v$	0	sttd	−	$C, \Delta E_p \uparrow$ as $v \uparrow$, $i_p = kv^{\frac{1}{2}}$ for $v=(5-100)E3, i_d, r$	
			−2.033			1.0F				0			$C, \Delta E_p \uparrow$ as $v \uparrow$, $i_p = kv^{\frac{1}{2}}$ for $v=(5-100)E3, i_k$	
			−2.385			0.8F			α	60 0.50			$C, \Delta E_p \uparrow$ as $v \uparrow$, $i_p \neq kv^{\frac{1}{2}}$ for $v=(5-100)E3, i_k$	
			−2.298			0.6F				−			A	
			−1.957			0.5F							A	
			−1.563			0.1F							A	
JA095 5939	−	"$E_{\frac{1}{2}}$"	0.83	SCE(NaCl)	−	−	QE	1.19	−	−	−	−	A,R	FF56 dh04
			1.39				−	−					A,R	
			−										C	
			−										C	
JE040 0063	−	$E_{\frac{1}{2}}$	−1.75	SCE	−	−	QE, IR	2	−	−	−	−	$C, \neq, i_d, r$	FF57
JA095 5939	329 1	"$E_{\frac{1}{2}}$"	0.64	SCE(NaCl)	−	−	QE	0.88	−	−	−	−	A,R	FF58
			1.21				−	−					A,R	
			−										C	
			−										C	

FF59 $C_{38}H_{30}O_2$

Code No.	Empirical Formula	Name and C.A. Number	Structural Formula	Solvent	Tech.	Medium		μ, M	pH	T, °C	Electrodes	App.	Experimental Parameters
FF59	$C_{38}H_{30}O_2$	1,2-bis(4-biphenyl-yl)-1,2-diphenyl-ethanediol C.A. 13224-48-1	[4-$C_6H_5C_6H_4$C-(OH)(C_6H_5)-]$_2$	DMF	PY	Et_4NBr	0.1	-	-	-	DME// SCE(o)	2A0	C=1
						Et_4NBr PHEN	0.1 2.10E-2						
					CP	Bu_4NI	0.25	-	-	-	Hg: srpo// Ag/AgI, Bu_4NI 0.25	59A0	C=ns, A=12, E_{app} = -2.3 V, Q=6F mol^{-1}
						Bu_4NI PHEN	0.25 ns						Q=4F mol^{-1}
FF60	$C_{38}H_{32}Cl_2O_{10}S_6$	5,5'-bis(4-methoxy-phenyl)-3,3'-[2,4-bis(4-methylphenyl)-3,4-dithia-1,6-hexa-methylene]dithiolium perchlorate	Table II	MeCN 50 CH_2Cl_2 50	IL	$NaClO_4$	0.1	-	-	-	Pt: xxbu// SCE	---	C=ns
FF61	$C_{40}H_{28}Cl_2F_4P_2$	2,5-diphenyl-1,1,4,4-tetra(4-fluoro-phenyl)diphosphonia-cyclohexa-2,5-diene dichloride C.A. 39927-31-6	Table II	DMF	PY	Et_4NClO_4	0.1	-	-	-	DME//SCE	2--	C=ns
FF62	$C_{40}H_{32}Br_2P_2$	1,1,2,4,4,5-hexa-phenyldiphosphonia-cyclohexa-2,5-diene dibromide C.A. 15362-58-0	Table II	DMF	PY	Et_4NClO_4	0.1	-	-	-	DME//SCE	2--	C=ns
FF63	$C_{40}H_{32}Cl_2P_2$	1,1,2,4,4,5-hexa-phenyldiphosphonia-cyclohexa-2,5-diene dichloride C.A. 41480-71-1	Table II	DMF	PY	Et_4NClO_4	0.1	-	-	-	DME//SCE	2--	C=ns
FF64	$C_{40}H_{50}N_2$	2,7-di-tert-butyl-5,10-bis(4-tert-butylphenyl)-5,10-dihydrophenazine	Table II	MeCN	VY	Et_4NClO_4	0.1(?)	-	-	-	Pt: rodi// Ag/ $AgClO_4$ 0.01, Et_4NClO_4 0.1	2-0	C=ns, d=0.1, ω=10, Pt Aux
FF65	$C_{42}H_{34}N_4$	4,4'-bis[3,5-di-phenyl-Δ^2-pyrazo-lin-1-yl]biphenyl C.A. 43040-07-9	Table II	MeCN	VA	$LiClO_4$	0.1	-	-	-	Pt: xxns// SCE	25--	0→1.0→0 V, C=0.5, v=300, Pt Aux
FF66	$C_{43}H_{52}N_5Ru$	acetonitrilopyridi-nooctaethylporphy-rinruthenium(III) ion C.A. 43070-16-2	Table II	MeCN	VR	Bu_4NPF_6	0.1	-	-	25±2	Pt: xxbe// SCE(NaCl)	2A0	C=1

TABLE I. Electrochemical Data   323

$C_{43}H_{52}N_5Ru$ FF66

Ref.	C/M	Charact. Value	Potential vs.	Response	Const. Value	Tech.	n	Electrokinetic Parameter	Value	From	Products and Identification	Description and Remarks	Code No.
EA020 0143	33f	$E_{\frac{1}{2}}$	-1.82	SCE(o)	i_ℓ	3.0	-	-	-	-	-	-	C,r	FF59
			-2.09			2.4							C	
			-2.69			9.2							C	
	33g		-			1.0	-	-	-	-	-	-	C,r	
			-2.64			21.4							C	
EA020 0143	33f	-	-	-	-	-	-	-	-	-	-	(4-biphenylyl)phen- ylmethane 90%	C,r	
	33g	-	-	-	-	-	-	-	-	-	-	(4-biphenylyl)phen- ylmethane 60%, starting cmpd 25%, other products 15%	C,r	
JE038 0479	-	E_p	0.3	SCE	-	-	QE	0.7	-	-	-	-	C,p	FF60
JE042 0309	-	$E_{\frac{1}{2}}$	-0.457	SCE	-	-	sttd	1	-	-	-	-	C,"$E_{\frac{1}{2}}$" on Pt (VA?)= -0.46 V,p	FF61
JE042 0309	-	$E_{\frac{1}{2}}$	-0.515	SCE	-	-	sttd	1	-	-	-	-	C,p	FF62
JE042 0309	-	$E_{\frac{1}{2}}$	-0.488	SCE	-	-	sttd	1	-	-	-	-	C,p	FF63
EA021 0557	-	$E_{\frac{1}{2}}$	-0.235	Ag/ AgClO$_4$ 0.01, Et$_4$NClO$_4$ 0.1	-	-	-	-	-	-	-	radical cation, UVS, VIS, ESR	A,r	FF64
			0.52										A	
EA021 0621	419 g	E_p	0.46F	SCE	i_p	12F	sttd	1	ΔE_p	60	sttd	radical cation	A,R	FF65
			0.63F			12F		-		70		dication	A,R	
			0.57F			12F				-			C,R	
			0.41F			12F							C,R	
JA095 5939	329 m	"$E_{\frac{1}{2}}$"	0.08	SSCE	-	-	-	-	-	-	-	-	C,R	FF66
			1.05										C,R	
			-										A	
			-										A	

FF67 $C_{44}H_{30}N_4$

Code No.	Empirical Formula	Name and C.A. Number	Structural Formula	Solvent	Tech.	Medium		μ, M	pH	T, °C	Electrodes	App.	Experimental Parameters
FF67 bs31 dh18	$C_{44}H_{30}N_4$	5,10,15,20-tetraphenylporphine C.A. 917-23-7	Table II-2	CH_2Cl_2	VR	Bu_4NPF_6	0.1	-	-	25±2	Pt: xxbe// SCE(NaCl)	2AO	C=1
FF68	$C_{44}H_{40}Br_2O_4P_2$	2,5-diphenyl-1,1,4,4-tetrakis(4-methoxyphenyl)diphosphoniacyclohexa-2,5-diene dibromide C.A. 39927-30-5	Table II	DMF	PY	Et_4NClO_4	0.1	-	-	-	DME//SCE	2--	C=ns
FF69	$C_{45}H_{28}N_4ORu$	carbonyl(tetraphenylporphyrinyl)-ruthenium C.A. 32073-84-0	Table II	CH_2Cl_2	VR	Bu_4NPF_6	0.1	-	-	25±2	Pt: xxbe// SCE	2AO	C=1
FF70	$C_{46}H_{32}Cl_2O_8S_6$	4,4',5,5'-tetraphenyl-3,3'-(2,5-diphenyl-3,4-dithia-1,6-hexamethylene) dithiolium perchlorate	Table II	MeCN 50 CH_2Cl_2 50	IL	$NaClO_4$	0.1	-	-	-	Pt: xxbu// SCE	---	C=ns
FF71	$C_{50}H_{33}N_5ORu$	carbonyl(pyridinotetraphenylporphyrinyl)ruthenium(II) C.A. 41751-82-0	Table II	CH_2Cl_2	VA	Bu_4NPF_6	0.1	-	-	25±2	Pt: xxbe// SCE(NaCl)	2AO	C=1 v=200
FF72	$C_{52}H_{44}ClP_4Rh$	bis-[cis-1,2-bis(diphenylphosphino)-ethylene]rhodium(I) chloride C.A. 22754-44-5	[Rh{$(C_6H_5)_2$-PCH:CH-$P(C_6H_5)_2$}$_2$] Cl	MeCN	PY	Bu_4NClO_4	0.1	-	-	25± 0.1	DME//SCE	-6AO	C=ns,m=0.23, t(c)=1,h=68
FF73	$C_{52}H_{44}Cl_2O_{12}S_6$	4,4',5'5-tetrakis(4-methoxyphenyl)-3,3'-[2,5-bis(4-methylphenyl)-3,4-dithia-1,6-hexylene] dithiolium perchlorate	Table II	MeCN 50 CH_2Cl_2 50	IL	$NaClO_4$	0.1	-	-	-	Pt: xxbu// SCE	---	C=ns
FF74	$C_{52}H_{48}ClCoO_4P_4$	bis(1,2-bisdiphenylphosphinoethane)-cobalt(I) perchlorate C.A. 34881-50-0	[Co{$(C_6H_5)_2$-PCH_2CH_2P-$(C_6H_5)_2$}$_2$] ClO_4	MeCN	PY	Bu_4NClO_4	0.1	-	-	25	DME//SCE	-A-	C=0.85, m=0.238,t=1
FF75	$C_{52}H_{48}ClIrP_4$	bis-[1,2-bis(diphenylphosphino)ethane]-iridium(I) chloride C.A. 15390-38-2	[Ir{$(C_6H_5)_2PCH_2$-$CH_2P(C_6H_5)_2$}$_2$]-Cl	MeCN	PY	Bu_4NClO_4	0.1	-	-	25± 0.1	DME//SCE	-6AO	C=0.98, m=0.308, h=45

TABLE I. Electrochemical Data

$C_{52}H_{48}ClP_4Ir$ FF75

| Ref. | C/M | Charact. | Potential | | Response Const. | | n | Electrokinetic Data | | | Products and | Description and | Code |
		Value	vs.		Value	Tech.		Parameter	Value	From	Identification	Remarks	No.
JA095 5939	-	"$E_{\frac{1}{2}}$"	0.95	SCE(NaCl)	-	-	QE	1.15	-	-	-	-	A,R,p	FF67 bs31 dh18
			1.28					-					A,R	
			-										C	
			-										C	
JE042 0309	-	$E_{\frac{1}{2}}$	-0.622	SCE	-	-	sttd	1	-	-	-	-	C,"$E_{\frac{1}{2}}$" at Pt(VA?)= 0.61V,p	FF68
JA095 5939	329 1	"$E_{\frac{1}{2}}$"	0.82	SSCE	-	-	QE	0.90	-	-	-	-	A,R,p	FF69
			1.21				-	-					A,R	
			-										C	
			-										C	
JE038 0479	-	E_p	-0.03	SCE	-	-	QE	1.0	-	-	-	-	C,p	FF70
JA095 5939	329 1	"$E_{\frac{1}{2}}$"	0.81	SCE(NaCl)	-	-	QE	1.01	-	-	-	-	A,R,p	FF71
			1.36				-	-					A,R	
			-										C	
			-										C	
			0.83F			33F	sttd	1	-	-	-	-	A,p	
			1.39F			31F		1					A	
			1.29F			33F		1					C	
			0.73F			31F		1					C	
JE047 0089	50v	$E_{\frac{1}{2}}$	-1.60	SCE	-	-	sttd	2	-	-	-	CP(-1.75 V,C=1.5) → [(C$_6$H$_5$)$_2$PCH:CH-P(C$_6$H$_5$)$_2$]$_2$RhH, IRS	C,R,p	FF72
JE038 0479	-	E_p	0.29	SCE	-	-	QE	0.8	-	-	-	-	C,p	FF73
JE050 0295	-	$E_{\frac{1}{2}}$	-1.12	SCE	I	2.2	QE	1	-	-	-	CP → bis(1,2-bisdiphenylphosphino-ethane)cobalt(0), IRS	C,R,i_d,r	FF74
			-1.6			2.2		1				-	C,R,i_d	
JE045 0483	-	$E_{\frac{1}{2}}$	-1.70	SCE	I	4.0	-	-	Elog	33	-	IrH[(C$_6$H$_5$)$_2$PCH$_2$CH$_2$-P(C$_6$H$_5$)$_2$]$_2$, IRS	C,R,W,i_d,r	FF75

FF76 $C_{52}H_{48}ClP_4Rh$

Code No.	Empirical Formula	Name and C.A. Number	Structural Formula	Solvent	Tech.	Medium	μ, M	pH	T, °C	Electrodes	App.	Experimental Parameters
FF76	$C_{52}H_{48}ClP_4Rh$	bis[1,2-bis(diphenylphosphino)ethane]-rhodium(I) chloride C.A. 15043-47-7	[Rh{$(C_6H_5)_2PCH_2$-$CH_2P(C_6H_5)_2$}$_2$]-Cl	MeCN	PY	Bu_4NClO_4 0.1	–	–	25±0.1	DME//SCE	-6AO	C=1, m=0.308, h=45
					PA	Bu_4NClO_4 0.1	–	–	25±0.1	DME//SCE	-6AO	C=1, v=250
FF77	$C_{54}H_{38}N_6Ru$	bis(pyridino)-α,β,γ,δ-tetraphenylporphinylruthenium C.A. 34690-41-0	Table II	CH_2Cl_2	VR	Bu_4NPF_6 0.1	–	–	25±2	Pt: xxbe// SCE(NaCl)	2AO	-0.2→1.6→ -0.2→1.6 V, C=1 v=200
FF78	$C_{55}H_{46}IrOP_3$	hydridomonocarbonyl-tris(triphenylphosphine)iridium(I) C.A. 17250-25-8	[IrH(CO)-{$P(C_6H_5)_3$}$_3$]	MeCN 70 $CH_3C_6H_5$ 30	VY	Et_4NClO_4 ns $(C_6H_5)_3P$ ns	–	–	25	Pt: pdns// SCE	---	C=1
FF79	$C_{55}H_{46}OP_3Rh$	hydridomonocarbonyl-tris(triphenylphosphine)rhodium(I) C.A. 17185-29-4	[RhH(CO)-{$P(C_6H_5)_3$}$_3$]	MeCN 70 $CH_3C_6H_5$ 30	VY	Et_4NClO_4 ns $(C_6H_5)_3P$ ns	–	–	25	Pt: pdns// SCE	---	C=1
FF80	$C_{66}H_{56}Cl_2O_{14}S_6$	4,4',5,5'-tetrakis-(4-methoxyphenyl)-3,3'-[2,5-bis(4-methylphenyl)-1,6-bis(4-methoxyphenyl)-3,4-dithia-1,6-hexylene]dithiolium perchlorate	Table II	MeCN 50 CH_2Cl_2 50	IL	$NaClO_4$ 0.1	–	–	–	Pt: xxbu// SCE	---	C=ns

TABLE I. Electrochemical Data

$C_{66}H_{56}Cl_2O_{14}S_6$ FF80

Ref.	C/M	Charact. Potential		Response Const.		n Tech.	Electrokinetic Data			Products and Identification	Description and Remarks	Code No.
		Value	vs.		Value		Parameter	Value	From			
JE045 0483	-	$E_{\frac{1}{2}}$	-1.71	SCE	i_ℓ	1.7F	QE,IR	2	Elog	33	-	$RhH[(C_6H_5)_2PCH_2CH_2-P(C_6H_5)_2]_2$, IRS	C,R,W,i_d,r	FF76
JE045 0483	-	E_p	-1.73F -1.67F	SCE	i_p	10F 10F	-	-	-	-	-	-	C,r A	
JA095 5939	329 m	"$E_{\frac{1}{2}}$"	1.26 0.21 - -	SCE(NaCl)	-	-	- QE -	- 0.94 				-	A,R,p A,R C,R C,R	FF77
			0.22 1.30 1.19 0.12F			34F 31F 33F 33F	sttd	1 1 1 1	-	-	-	-	A,p A C C	
JE042 005A	-	$E_{\frac{1}{2}}$	0.00 -	SCE	-	-	-	-	-	-	$CP \to IrH(CO)-[P(C_6H_5)_3]_3^+$, PY	A,QI,i_d $E_{\frac{1}{2}} \neq f[(C_6H_5)_3P]$,p A	FF78
JE042 005A	-	$E_{\frac{1}{2}}$	0.06 -	SCE	-	-	QE -	1	-	-	-	$CP \to RhH(CO)-[P(C_6H_5)_3]_3^+$, PY	A,≠,i_d, $E_{\frac{1}{2}} \neq f[(C_6H_5)_3P]$,p A	FF79
JE038 0479	-	E_p	-0.35	SCE	-	-	QE	1.1	-	-	-	-	C,p	FF80

FF81 $C_{10}H_8$

Code No.	Compound oxidized (=R), empirical formula	Compound reduced (=0), empirical formula	Solvent	Medium	μ	pH	T	Electrodes	App.
FF81 ed27	naphthalene $C_{10}H_8$	same	MeCN	Bu_4NClO_4 0.1 tetramethyl-p-phenylenediamine (Wurster's Blue) ns	-	-	-	-/SCE	-A-
FF82 fb87 ff29	10-methylphenothiazine $C_{13}H_{11}NS$	tris(2,2'-bipyridine)-ruthenium(2+) diperchlorate $C_{30}H_{24}Cl_2N_6O_8Ru$	MeCN	Bu_4NBF_4 0.1	-	-	-	Pt:xxwi//Ag	2FO
FF83 de99	fluoranthene $C_{16}H_{10}$	same	MeCN	Bu_4NClO_4 0.1 Wurster's Blue ns	-	-	-	-/SCE	-A-
FF84 ck07	pyrene $C_{16}H_{10}$	same	MeCN	Bu_4NClO_4 0.1 Wurster's Blue ns	-	-	-	-/SCE	-A-
FF85 bj69 fe94	tetrabutylammonium bromide $C_{16}H_{36}BrN$	9,10-diphenylanthracene $C_{26}H_{18}$	MeCN	Bu_4NClO_4 0.1	-	-	-	Pt:nsco//SCE	235-0
FF86 - fe94	tetrabutylammonium iodide $C_{16}H_{36}IN$	9,10-diphenylanthracene $C_{26}H_{18}$	MeCN	Bu_4NClO_4	-	-	-	Pt:nsco//SCE	235-0

TABLE I. Electrochemical Data

$C_{16}H_{36}IN$ FF86

Exptl. parameters	Ref.	C/M	Excitation					Emission			Remarks	Code No.
			Type	f	E_1	E_2	vs.	Species	Type	Characteristics		
$C_R=0.1$	JA095 7164	351g	-	-	1.54	-2.47	SCE	-	-	$\lambda_{max}=316$ nm	quenching by Wurster's Blue (which is oxidized at +0.20 V vs. SCE),p	FF81 ed27
C_R=ns,C_O=ns,Pt Aux	JA095 6582	374b	□	ns	1.1	-1.4 to -1.8	Ag	-	-	-	similar to $Ru(bipy)_3^{2+}$ alone in intensity and λ_{max}, larger peaks on oxidn part of cycle,p	FF82 fb87 ff29
$C_R=0.1$	JA095 7164	351g	-	-	1.45	-1.81	SCE	-	-	$\lambda_{max}=358$ nm	quenching by Wurster's Blue (which is oxidized at +0.20 V vs. SCE),p	FF83 de99
$C_R=0.1$	JA095 7164	351g	-	-	1.26	-2.07	SCE	-	-	$\lambda_{max}=357$ nm	quenching by Wurster's Blue (which is oxidized at +0.20 V vs. SCE),p	FF84 ck07
$C_R=0.1,C_O=1,A=1.9$	EA018 0639	351f	□	5	0.9	-2.3	SCE	$^1O^*$	-	$\lambda_{max}=430$ nm	r	FF85 bj69 fe94
$C_R=1$			□	5	0.5→3.0	-2.0	SCE	$^1O^*$	-	$\lambda_{max}=430$ nm	π begins at $E_1=0.7$ V,π =max at 1.0 and 1.5 V $π_{max}$ at 1.5 V↓ as C_R ↑,π = 0 for $E_1 \geq 2.0$ V,r	
$C_R=100$			□	5	0→2.0	-2.0	SCE	$^1O^*$	-	$\lambda_{max}=430$ nm	π begins at $E_1=0.7$ V(oxidn of Br^-),π=max at $E_1=0.9$ V, π related to presence of Br_3^-, π ≈ 3(π in absence of Br^-), π=0 for $E_1 >$ 1.3 V,r	
$C_R=1,C_O=1,A=1.9$	EA018 0639	351f	□	5	0→3.0	-2.0	SCE	$^1O^*$	-	$\lambda_{max}=430$ nm	π begins at $E_1=0.3$ V,π= max at $E_1=0.6(I_3^- + O_2^-)$ and 1.6 V ($O_2^+ + O_2^-$),π=0 for $E_1=1.0$ and $E_1 > 2.0$ V,r	FF86 - fe94
$C_R=5,C_O=1$			□	5	0→3.0	-2.0	SCE	$^1O^*$	-	$\lambda_{max}=430$ nm	π=max at $E_1=0.6$ and 1.6 V,π=0 for $E_1=0.8$-1.2 V and $E_1 >$ 2 V,r	
$C_R=10,C_O=1$			□	5	0→3.0	-2.0	SCE	$^1O^*$	-	$\lambda_{max}=430$ nm	π begins at $E_1=0.3$ V(oxidn of I^-),π = max at $E_1=0.5$ V, π=0 for $E_1 >$ 1.0 V,r	

FF87 $C_{16}H_{36}I_3N$

Code No.	Compound oxidized (=R), empirical formula	Compound reduced (=O), empirical formula	Solvent	Medium	μ	pH	T	Electrodes	App.
FF87 – fe94	tetrabutylammonium triiodide $C_{16}H_{36}I_3N$	9,10-diphenylanthracene $C_{26}H_{18}$	MeCN	Bu_4NClO_4	–	–	–	Pt:nsco//SCE	235-0
FF88 c105	1,2-benzanthracene $C_{18}H_{12}$	same	MeCN	Bu_4NClO_4 0.1 tetramethyl-p-phenylenediamine (Wurster's Blue) ns	–	–	–	–/SCE	–A–
				Bu_4NClO_4 0.1 Wurster's Blue ns					
FF89 – fe94	octadecyldimethyl-benzylammonium chloride $C_{27}H_{50}ClN$	9,10-diphenylanthracene $C_{26}H_{18}$	MeCN	Bu_4NClO_4	–	–	–	Pt:nsco//SCE	235-0
FF90 ff28	bis[2,6-di(3-pyridyl)-pyridine]ruthenium(2+) bisperchlorate $C_{30}H_{22}Cl_2N_6O_8Ru$	same	MeCN	Bu_4NClO_4	–	–	–	Pt:xxwi//Ag	2FO
FF91 (see also ff29)	tris(2,2'-bipyridine-N,N')ruthenium(2+) dichloride $C_{30}H_{24}Cl_2N_6Ru$	same	MeCN	Bu_4NClO_4 0.1	–	–	–	Pt:xxwi//Ag	2FO
								Pt:rord//Ag	
			C_6H_6 50 MeCN 50	Bu_4NBF_4 0.2	–	–	–	Pt:xxwi//Ag	2FO
			DMF	Bu_4NBF_4 0.1	–	–	–	Pt:xxwi//Ag	2FO

TABLE I. Electrochemical Data   331

$C_{30}H_{24}Cl_2N_6Ru$ FF91

Exptl. parameters	Ref.	C/M	Excitation					Emission			Remarks	Code No.
			Type	f	E_1	E_2	vs.	Species	Type	Characteristics		
A=1.9	EA018 0639	351f	□	5		-2.2	SCE	$^1O^*$	-	λ_{max}=430 nm	r	FF87 - fe94
C_R=0.1	JA095 7164	351g	-		1.18	-2.00	SCE			λ_{max}=358 nm	quenching by Wurster's Blue (which is oxidized by +0.20 V vs. SCE),r	FF88 c105
					1.36	-2.25				λ_{max}=316 nm	quenching by Wurster's Blue (which is oxidized by +0.20 V vs. SCE),r	
C_R=10,C_O=1,A=1.9	EA018 0639	351f	□	5	0→3.0	-2.0	SCE	$^1O^*$		λ_{max}=430 nm	π begins at E_1= +1.0 V(oxidn of Cl$^-$),π =max at E_1=1.46 V,π=0 for E_1 > 2.0 ,r	FF89 - fe94
					1.0	-0.5→ -3.0		$^1O^*$		λ_{max}=430 nm	π begins at E_2= -1.8 V(first redn of O),π = max at E_2= -2.0 V,π=0 for E_2 more neg than 2.5 V	
C_R= -,Pt Aux	JA095 6582	374c	□	50	1.3	-1.45	Ag			emission 550-750 nm, λ_{max}=660 nm	larger peaks on redn part of cycle for f= 50 but on oxidn part of cycle when f=0.2, π ↑ on later cycles,r	FF90 ff28
C_R=1,Pt Aux	JA095 6582	374c	□	0.2	1.4	-1.4	Ag	R^3	se	emission at 520-770 nm, λ_{max}=610 nm, $\lambda_{shoulder}$=630 nm	larger peaks on oxidn part of cycle,π depends on reduced species produced as follows: πRu(bipy)$_3^{+1}$: πRu(bipy)$_3^0$: πRu(bipy)$_3^{-1}$=1: 2.5:2.9,π lower for Cl$^-$ salt than ClO$_4^-$ salt,π ↓ as [O_2] ↑,r	FF91 (see also ff29)
E_{disc}=1.48					1.48					∅=5-6%	π ≠f(E_{ring}),r	
C_R=1.0,Pt Aux			□	0.2	-	-	-	-	-	-	π greater than in MeCN but is only 40% of that for ClO$_4^-$ salt,r	
C_R=1,Pt Aux	JA095 6582	374c	□	0.2	1.4	-1.7	Ag	-	-	-	λ_{max}=610 nm,π depends on reduced species produced as follows: πRu(bipy)$_3^+$: πRu(bipy)$_3^0$: πRu(bipy)$_3^{-1}$= 1:4:11	

FF92 $C_{36}H_{24}Cl_2N_6O_8Ru$

Code No.	Compound oxidized (=R), empirical formula	Compound reduced (=O), empirical formula	Solvent	Medium		u	pH	T	Electrodes	App.
FF92 ff50	tris(1,10-phenanthroline)ruthenium(II) perchlorate $C_{36}H_{24}Cl_2N_6O_8Ru$	same	MeCN	Bu_4NBF_4	0.1	-	-	-	Pt:xxwi∥Ag	2FO
FF93 ff51	bis(2,4,6-tripyridyl-s-triazine)ruthenium-(III) perchlorate $C_{36}H_{24}Cl_3N_{12}O_{12}Ru$	same	MeCN	Bu_4NBF_4	0.1	-	-	-	Pt:xxwi∥Ag	2FO
FF94 dh13	rubrene $C_{42}H_{28}$	same	C_6H_5CN	Et_4NClO_4	0.1	-	-	-	ZnO:xxfl∥Ag	2F-
FF95 dh13 fe42	rubrene $C_{42}H_{28}$	2,4,6-triphenyl-pyrylium perchlorate $C_{23}H_{17}ClO_5$	DMF	Et_4NClO_4	0.1	-	-	20±1	Pt:nsns∥SCE	2AO

TABLE I. Electrochemical Data

$C_{42}H_{28}$ FF95

Exptl. parameters	Ref.	C/M	Excitation					Emission			Remarks	Code No.
			Type	f	E_1	E_2	vs.	Species	Type	Characteristics		
C_R=1, Pt Aux	JA095 6582	374c	□	0.2	1.6	-1.5	Ag	-	-	λ_{max}=590 nm	π↑ on oxidn part of cycle, p	FF92 ff50
C_R=1.0, Pt Aux	JA095 6582	374c	-	-	-	-	-	-	-	-	no emission, p	FF93 ff51
C_R=2.4, A=0.15	EA020 0007	-	-	-	2.5	-	Pt	-	-	-	no emission, p	FF94 dh13
C_R=0.17, C_O=0.34	EA021 0497	-	□	50	0	>1.2	SCE	'R*	-	λ_{max}=455 nm F	p	FF95 dh13 fe42

TABLE II.
STRUCTURAL FORMULAS

This table is a supplement to Table I. It gives the structural formulas of the 337 compounds included in Table I. The compounds included here are those whose structural formulas could not be unambiguously and clearly represented in line form in the fourth column of Table I. Compounds are listed in the order of their code numbers.

TABLE II. Structural Formulas

FA20	$C_5H_4O_6$
FA34	$C_6H_2O_6$
FA35	$C_6H_3F_3$
FA39	$C_6H_4O_2$
FA46	$C_6H_8N_2$
FA57	$C_7H_9ClN_2O$
FA63	$C_8H_6N_2$
FA73	$C_9H_6O_3$
FA82	C_9H_{12}
FA84	$C_9H_{16}NO_2$
FA85	$C_9H_{18}NO$
FA86	$C_9H_{18}NO$
FA88	$C_{10}H_6N_2$
FA90	$C_{10}H_8O_4$
FA94	$C_{10}H_{10}O$
FB01	$C_{10}H_{12}F_3NO_2$
FB09	$C_{11}H_{10}$
FB12	$C_{11}H_{14}O_2$
FB15	$C_{12}H_4N_4$
FB22	$C_{12}H_8OS_2$
FB23	$C_{12}H_8O_2S_2$
FB24	$C_{12}H_8O_2S_2$
FB26	$C_{12}H_9BrN_2O$

FB27 C₁₂H₉ClN₂O	FB28 C₁₂H₉ClN₂O₂	FB29 C₁₂H₉ClN₂O₂
FB31 C₁₂H₉IN₂O	FB32 C₁₂H₉N₃O₃	FB33 C₁₂H₉N₃O₃
FB34 C₁₂H₉N₃O₃	FB39 C₁₂H₁₀ClN₃S	FB45 C₁₂H₁₀N₂O₄S
FB54 C₁₂H₁₂N₂O₂	FB55 C₁₂H₁₂O	FB56 C₁₂H₁₂O	FB57 C₁₂H₁₃N₃
FB59 C₁₂H₁₃N₅S	FB62 C₁₂H₂₀Cl₂N₂O	FB63 C₁₂H₂₂N₂	FB64 C₁₂H₂₂N₂O
FB65 C₁₂H₂₂N₂O	FB66 C₁₂H₂₂N₂O₂	FB67 C₁₃H₈Cl₂OS

TABLE II. Structural Formulas 339

| FB68 | $C_{13}H_8N_2Na_2O_6S$ | FB69 | $C_{13}H_8N_4O_7$ | FB70 | $C_{13}H_8N_4O_8$ |

| FB72 | $C_{13}H_9ClN_2O_3$ | FB73 | $C_{13}H_9ClN_2O_3$ | FB74 | $C_{13}H_9FOS$ |

| FB75 | $C_{13}H_9N_3O_5$ | FB78 | $C_{13}H_{10}N_2O$ | FB79 | $C_{13}H_{10}N_2O_3$ |

| FB80 | $C_{13}H_{10}N_2O_3$ | FB84 | $C_{13}H_{10}O$ | FB85 | $C_{13}H_{10}OS$ |

| FB88 | $C_{13}H_{11}N_3O$ | FB89 | | | $C_{14}H_{13}ClN_6O_6$ |

| FB90 | $C_{13}H_{12}N_2O$ | FB91 | $C_{13}H_{12}N_2O$ | FB92 | $C_{13}H_{12}N_4O_2$ |

FB93	$C_{13}H_{12}N_4O_2$	FB94	$C_{13}H_{12}N_4O_2$	FC00	$C_{13}H_{14}N_2O$
FC01	$C_{13}H_{14}N_2O_2$	FC02	$C_{13}H_{14}N_4O_2$	FC03	$C_{13}H_{15}N_5OS$
FC04	$C_{13}H_{15}N_5OS$	FC06	$C_{14}H_7N_2O$	FC12	$C_{14}H_{10}N_2$
FC17	$C_{14}H_{11}NO_2S$	FC18	$C_{14}H_{11}NO_3$	FC20	$C_{14}H_{12}$
FC25	$C_{14}H_{12}O$	FC26	$C_{14}H_{12}O$	FC27	$C_{14}H_{12}OS$	FC28	$C_{14}H_{12}OS$
FC29	$C_{14}H_{13}ClN_2O$	FC31	$C_{14}H_{13}N_3O_3$	FC33	$C_{14}H_{14}BrN_3$

TABLE II. Structural Formulas

FC34 $C_{14}H_{14}BrN_3$ FC35 $C_{14}H_{14}ClN_3$ FC36 $C_{14}H_{13}ClN_3$

FC37 $C_{14}H_{14}ClN_3$ FC38 $C_{14}H_{14}FN_3$ FC39 $C_{14}H_{14}FN_3$

FC40 $C_{14}H_{14}IN_3$ FC41 $C_{14}H_{14}IN_3$ FC45 $C_{14}H_{14}N_2O$

FC47 $C_{14}H_{14}N_2O_2$ FC48 $C_{14}H_{14}N_4O_2$ FC49 $C_{14}H_{14}N_4O_2$

FC53 $C_{14}H_{14}O_2$ FC58 $C_{14}H_{15}N_3O$ FC59 $C_{14}H_{15}N_3O$

FC60 $C_{14}H_{15}N_3O_3$ FC61 $C_{14}H_{15}N_3O_3$ FC62 $C_{14}H_{15}N_3O_3S$

FC63 $C_{14}H_{15}N_3O_3S$ FC66 $C_{14}H_{17}N_5OS$

FC67 $C_{14}H_{17}N_5OS$ FC68 $C_{15}H_9Br_3O_2$ FC69 $C_{15}H_9BrO_3$

FC70 $C_{15}H_{10}O_2$ FC71 $C_{15}H_{11}N_3$ FC79 $C_{15}H_{13}NO_3$

FC80 $C_{15}H_{13}NO_3$ FC81 $C_{15}H_{13}NO_3$ FC86 $C_{15}H_{14}O$

FC87 $C_{15}H_{14}O_2S$ FC88 $C_{15}H_{15}NOS$ FC89 $C_{15}H_{15}NO_2$

FC90 $C_{15}H_{15}N_3O_2$ FC91 $C_{15}H_{15}N_3O_2$ FC93 $C_{15}H_{15}N_3O_4$

TABLE II. Structural Formulas 343

FC94 $C_{15}H_{16}N_2O_2$

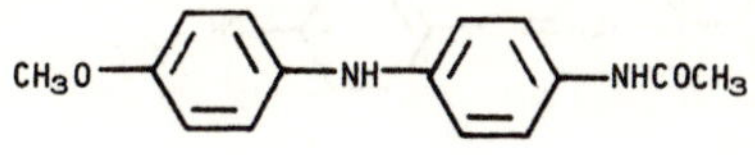

FC95 $C_{15}H_{16}N_2O_4$

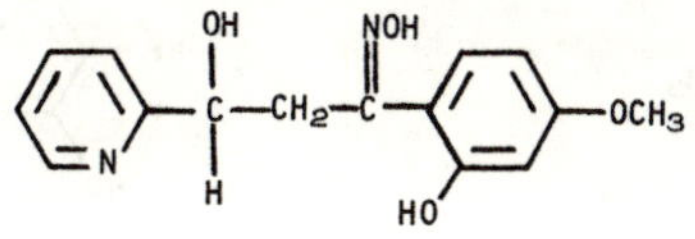

FC96 $C_{15}H_{16}N_2O_4$

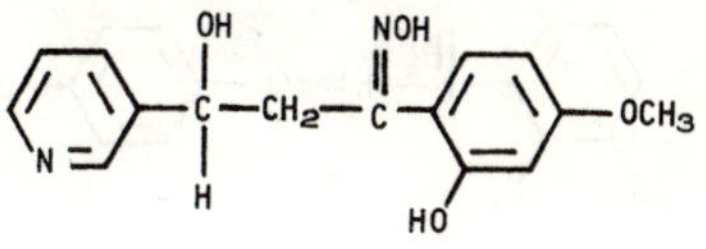

FC97 $C_{15}H_{16}N_2O_4$

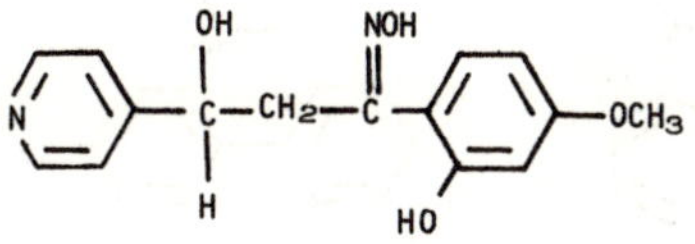

FD00 $C_{15}H_{17}N_3$

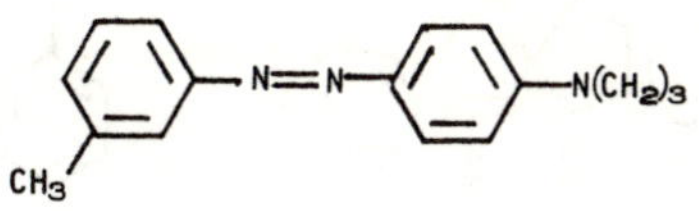

FD01 $C_{15}H_{17}N_3$

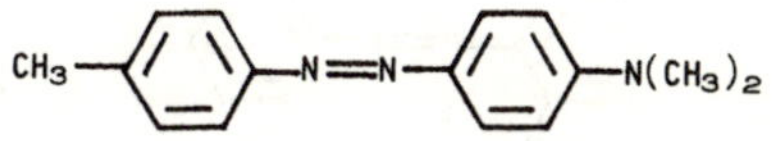

FD02 $C_{15}H_{17}N_3O$

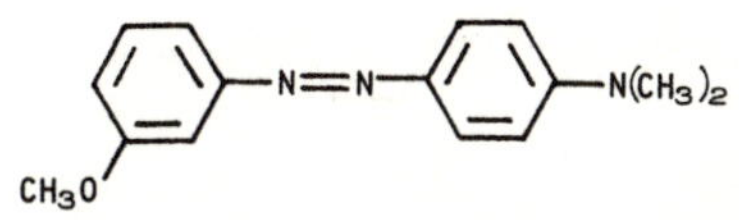

FD03 $C_{15}H_{17}N_3O$

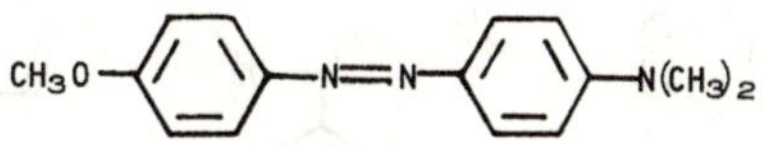

FD04 $C_{15}H_{17}N_3O_4$

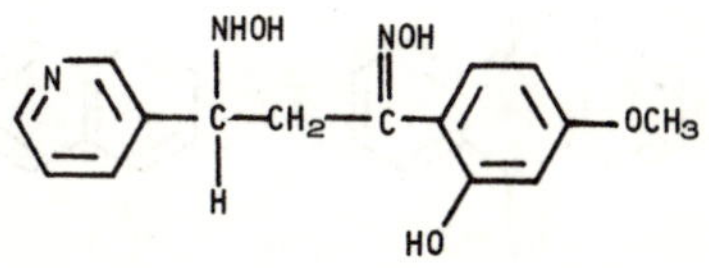

FD05 $C_{15}H_{24}O_3$

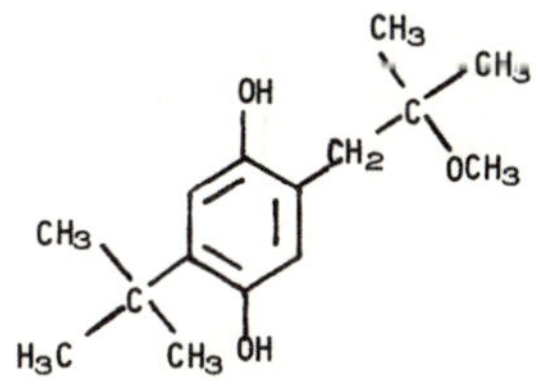

FD06 $C_{16}H_4N_6O_8$

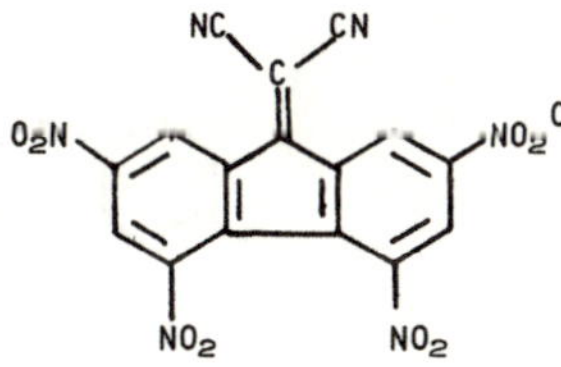

FD07 $C_{16}H_5N_5O_6$

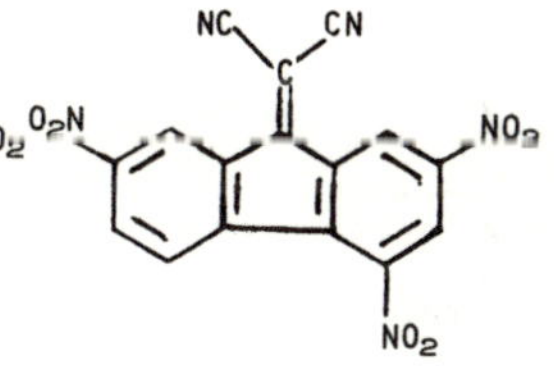

FD08 $C_{16}H_9ClO_2$

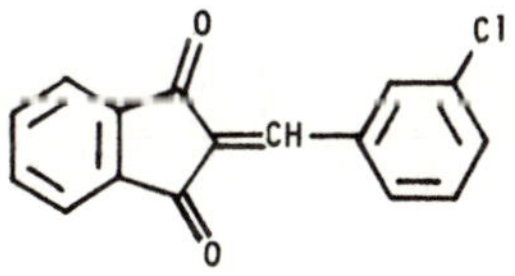

FD09 $C_{16}H_9ClO_2$

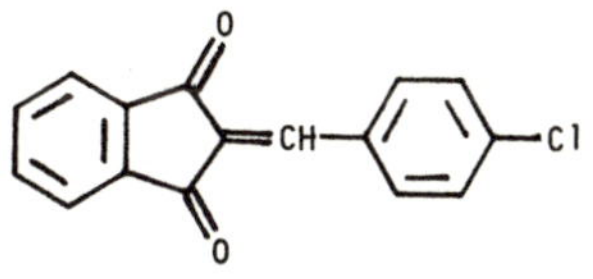

FD10 $C_{16}H_{16}O_2$

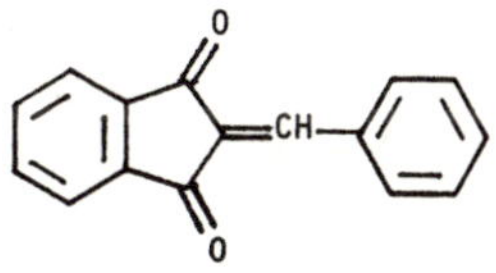

FD11 $C_{16}H_{10}O_3$

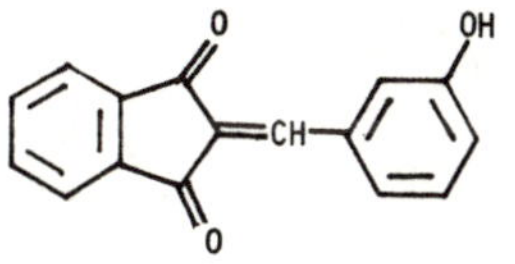

FD12 $C_{16}H_{10}O_4$

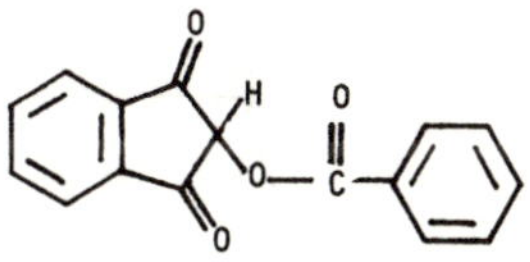

FD13 $C_{16}H_{11}Br_3O_2$

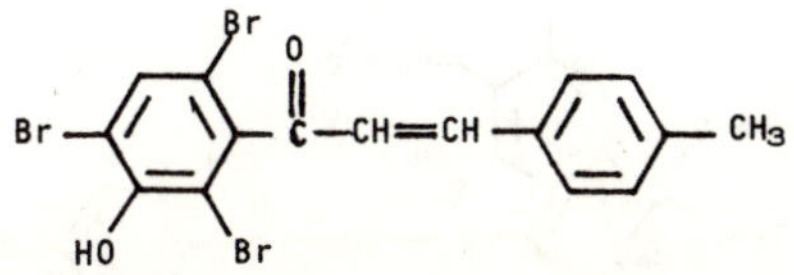

FD14 $C_{16}H_{11}Br_3O_3$

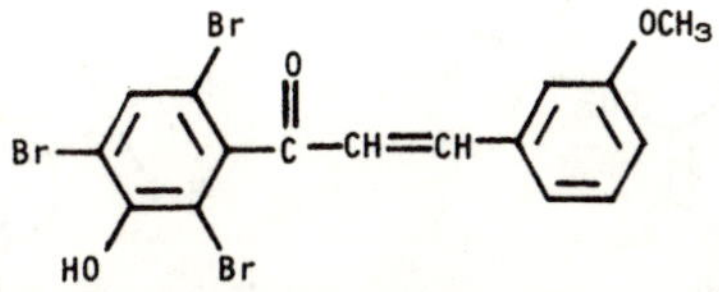

FD15	$C_{16}H_{11}Br_3O_3$
FD16	$C_{16}H_{11}Br_3O_4$
FD17	$C_{16}H_{11}N_2NaO_6S$
FD21	$C_{16}H_{12}O_4$
FD23	$C_{16}H_{14}$
FD24	$C_{16}H_{14}N_2O_2$
FD25	$C_{16}H_{14}N_2O_2$
FD26	$C_{16}H_{14}O$
FD28	$C_{16}H_{15}ClN_2O$
FD34	$C_{16}H_{16}O_4S$
FD35	$C_{16}H_{16}O_4S$
FD37	$C_{16}H_{18}ClN_3O_4S$
FD39	$C_{16}H_{18}N_2O_4S$
FD41	$C_{16}H_{19}NO_4$
FD43	$C_{16}H_{22}O_4$
FD44	$C_{16}H_{26}O_3$
FD45	$C_{16}H_{26}O_3$
FD46	$C_{16}H_{26}O_3$

TABLE II. Structural Formulas

FD47 $C_{16}H_{26}O_4$	FD48 $C_{17}H_{10}O_4$	FD49 $C_{17}H_{11}ClO_3$
FD50 $C_{17}H_{11}NO_5$	FD51 $C_{17}H_{12}OS$	FD52 $C_{17}H_{12}O_2$
FD53 $C_{17}H_{12}O_3$	FD54 $C_{17}H_{12}O_3$	FD55 $C_{17}H_{12}O_4$
FD56 $C_{17}H_{12}S_3$	FD58 $C_{17}H_{18}O_3$	FD59 $C_{17}H_{19}N_3O$
FD60 $C_{17}H_{20}O_3$	FD61 $C_{17}H_{28}O_3$	FD62 $C_{17}H_{28}O_3$
FD63 $C_{18}H_{10}O_6$	FD64 $C_{18}H_{12}N_6$	FD65 $C_{18}H_{13}ClN_2$

FD66 $C_{18}H_{14}O_3$ 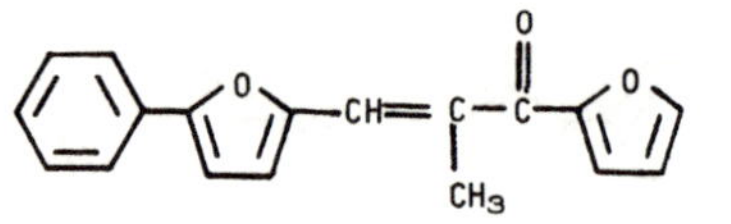
FD67 $C_{18}H_{14}S_3$ 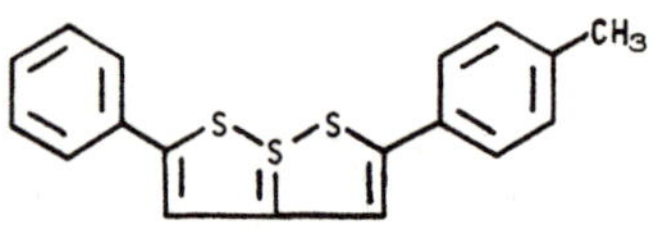
FD68 $C_{18}H_{15}ClN_4$

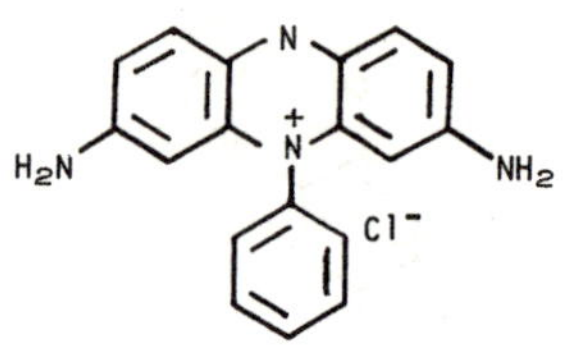

FD73 $C_{18}H_{20}Cl_2N_2O$ 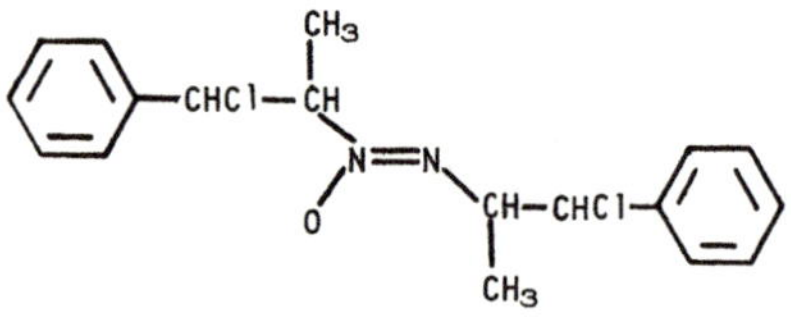
FD74 $C_{18}H_{22}O_4$ 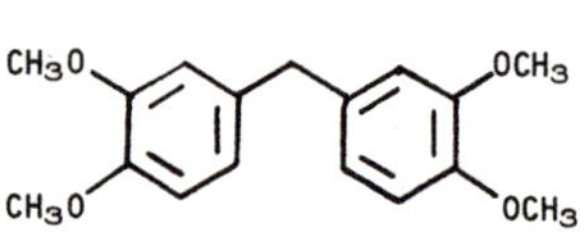
FD75 $C_{18}H_{24}Cr$

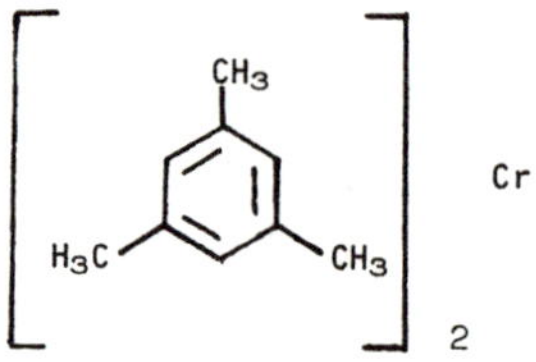

FD76 $C_{18}H_{30}O_3$ 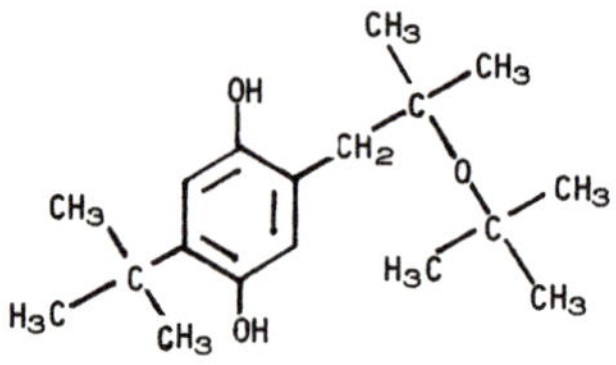
FD77 $C_{18}H_{30}O_4$ 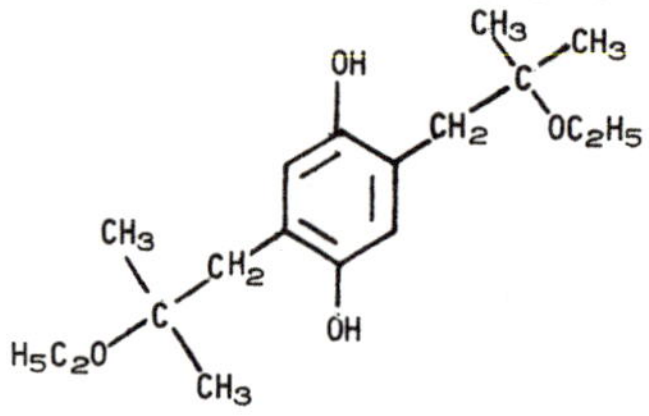
FD78 $C_{18}H_{30}O_4$

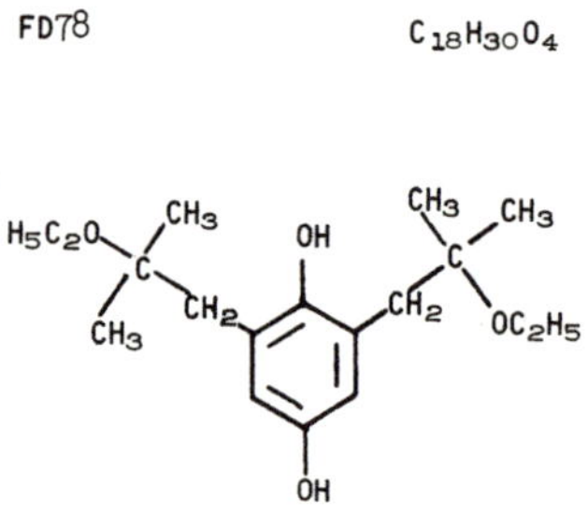

FD80 $C_{19}H_{13}N_3O$ 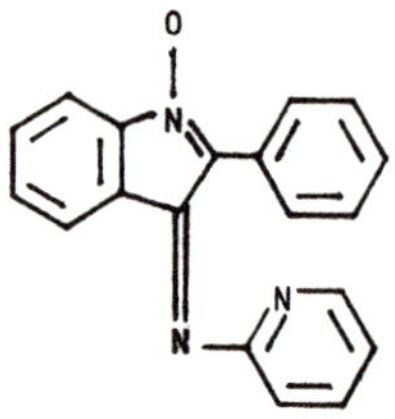
FD81 $C_{19}H_{13}N_3O_2$ 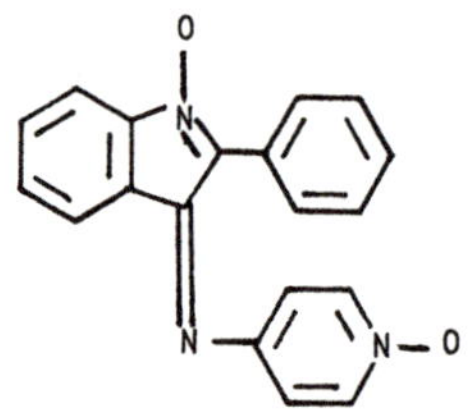
FD88 $C_{19}H_{16}S_3$

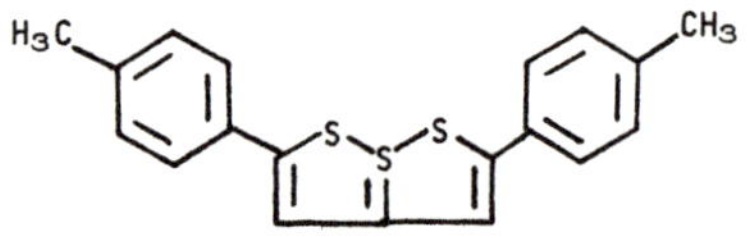

FD89 $C_{19}H_{19}BrN_2S_2$ 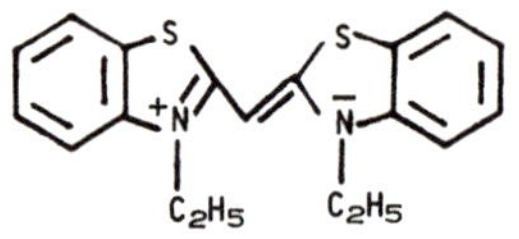
FD90 $C_{19}H_{19}N_7O_6$

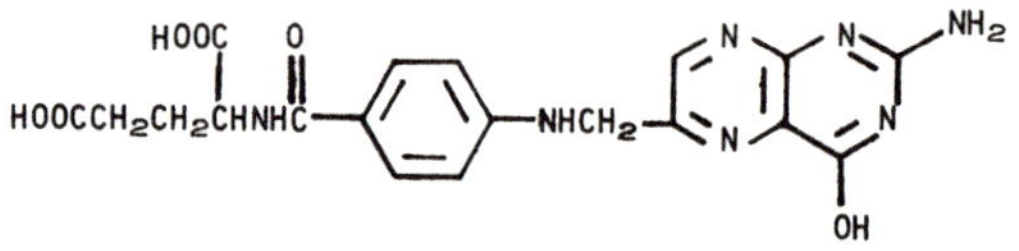

FD91 $C_{19}H_{23}ClN_2$ 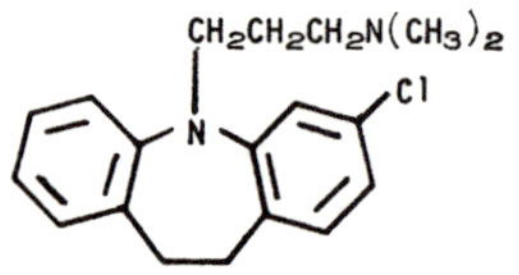
FD92 $C_{19}H_{26}O_2$ 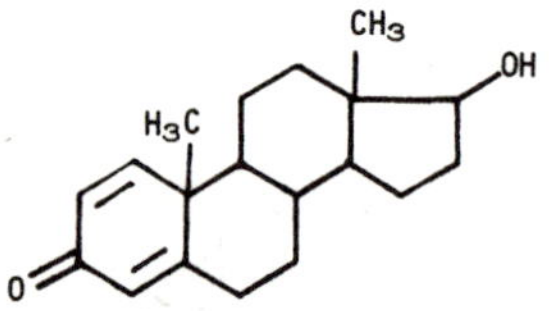
FD93 $C_{19}H_{27}FO_2$

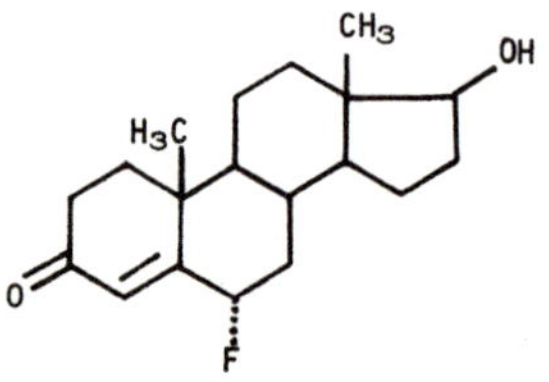

TABLE II. Structural Formulas 347

FD94	$C_{19}H_{27}FO_2$	FD95	$C_{19}H_{28}O_2$	FD96	$C_{20}H_6Br_4Na_2O_5$
FD97	$C_{20}H_6I_4Na_2O_5$	FD98	$C_{20}H_{10}Na_2O_5$	FD99	$C_{20}H_{12}$
FE00	$C_{20}H_{12}N_3NaO_7S$	FE01	$C_{20}H_{13}BrN_2$	FE02	$C_{20}H_{13}BrN_2O$
FE03	$C_{20}H_{13}ClN_2$	FE04	$C_{20}H_{13}ClN_2O$	FE06	$C_{20}H_{14}N_2$	FE07	$C_{20}H_{14}N_2O$
FE10	$C_{20}H_{16}N_4O_2$	FE11	$C_{20}H_{16}O_2$
FE12	$C_{20}H_{18}N_2O_2$	FE15	$C_{20}H_{30}O_2$	FE16	$C_{26}H_{30}O_2$

FE17 $C_{20}H_{34}O_4$ 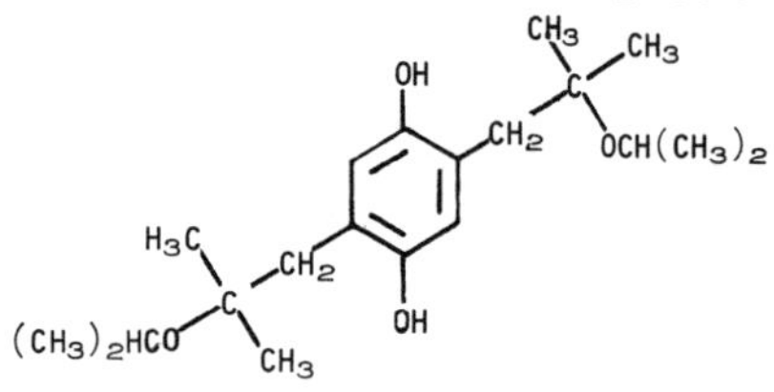
FE18 $C_{21}H_{16}N_2$ 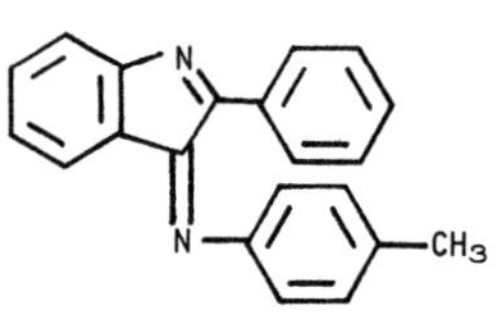
FE19 $C_{21}H_{16}N_2O$ 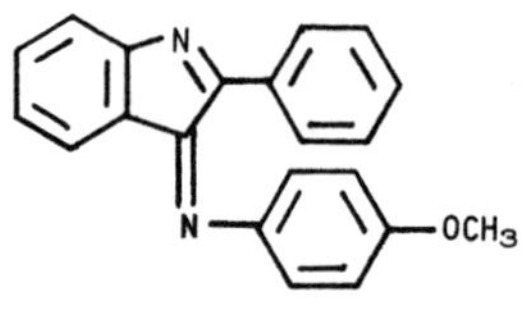
FE20 $C_{21}H_{16}N_2O$ 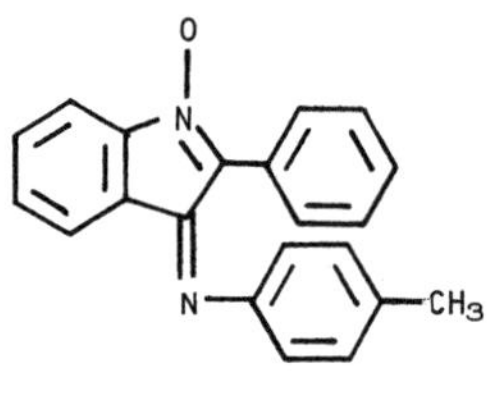
FE21 $C_{21}H_{16}N_2O_2$ 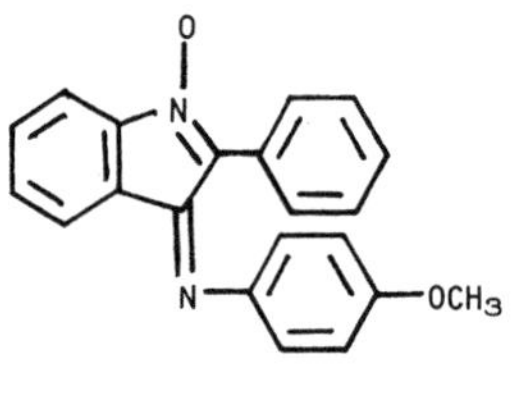
FE22 $C_{21}H_{17}N_3O_2$ 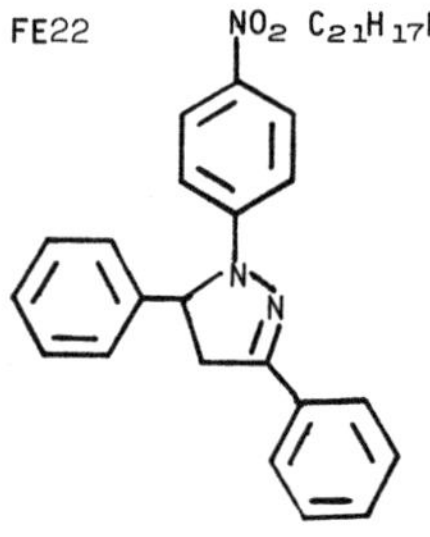
FE24 $C_{21}H_{18}N_2$ 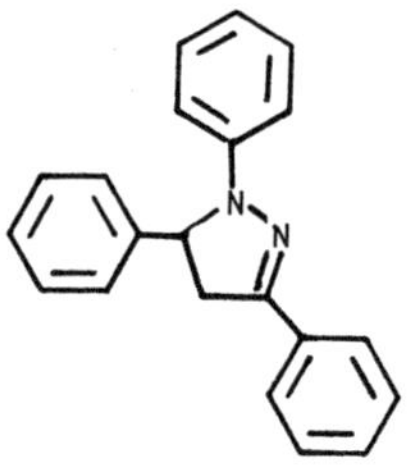
FE25 $C_{21}H_{21}BrN_2S_2$ 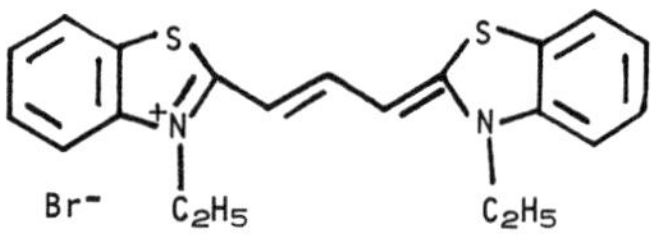
FE26 $C_{21}H_{25}N_6NaO_{15}P_2$ 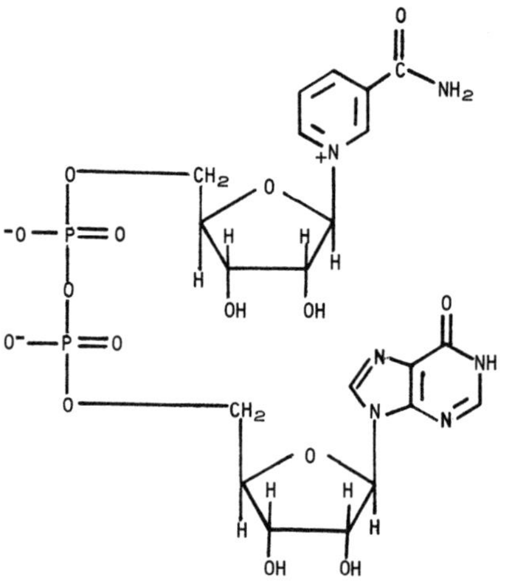
FE27 $C_{21}H_{27}N_7NaO_{17}P_3$ 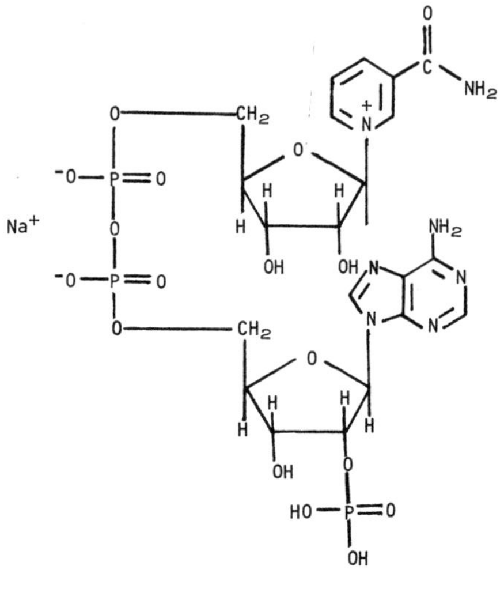
FE28 $C_{21}H_{27}N_7O_{14}P_2$

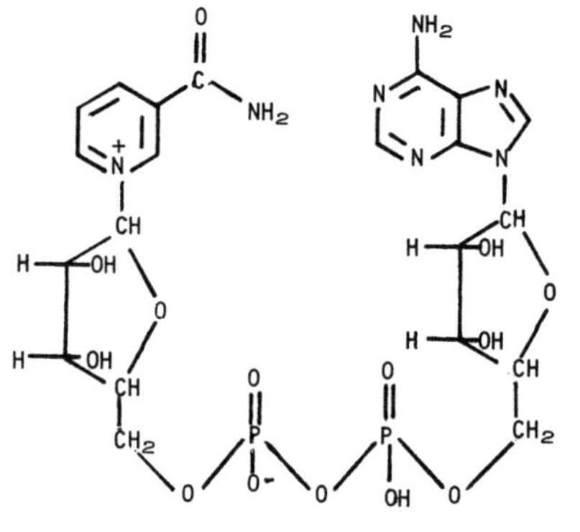

TABLE II. Structural Formulas

FE30	$C_{21}H_{30}O_4$	FE31	$C_{22}H_{14}O_2$	FE32	$C_{22}H_{16}BrN_5S$
FE33	$C_{22}H_{16}ClN_5S$	FE34	$C_{22}H_{16}ClN_5S$	FE35	$C_{22}H_{17}N_5S$
FE36	$C_{22}H_{19}N_3$	FE37	$C_{22}H_{19}N_3O$	FE38	$C_{22}H_{23}BrN_2S_2$
FE39	$C_{22}H_{31}FO_4$	FE43	$C_{22}H_{31}FO_4$	FF41	$C_{23}H_{16}S_3$
FE42	$C_{23}H_{17}ClO_5$	FF43	$C_{23}H_{19}N_5OS$	FE44	$C_{23}H_{19}N_5OS$
FE45	$C_{23}H_{19}N_5S$	FE46	$C_{23}H_{19}N_5S$	FE47	$C_{23}H_{19}N_5S$

FE48 C₂₃H₂₃BrN₂S   FE49 C₂₃H₂₃IN₂   FE50 C₂₃H₂₃IN₂

FE51 C₂₃H₂₅BrN₂S₂   FE52 C₂₃H₂₅BrN₂S₂   FE53 C₂₃H₂₉ClO₄

FE54 C₂₃H₂₉ClO₄   FE55 C₂₃H₃₀O₄   FE56 C₂₃H₃₀O₄

FE57 C₂₃H₃₁ClO₄   FE58 C₂₃H₃₁ClO₄   FE59 C₂₃H₃₁FO₄

FE60 C₂₃H₃₁FO₄   FE61 C₂₃H₃₂O₄   FE62 C₂₄H₁₄Cl₂N₄O₄

FE63 C₂₄H₁₄Cl₄N₂   FE64 C₂₄H₁₄K₂O₂   FE67 C₂₄H₁₆O₂

TABLE II. Structural Formulas   351

FE68	$C_{24}H_{18}Br_2N_2$	FE69	$C_{24}H_{18}N_2$	FE70	$C_{24}H_{18}O_4$
FE77	$C_{24}H_{21}N_5S$	FE78	$C_{24}H_{21}N_5S$	FE80	$C_{24}H_{26}N_4$
FE81	$C_{24}H_{32}Cl_2P_2$	FE82	$C_{24}H_{32}O_4$	FE83	$C_{24}H_{32}O_4$
FE84	$C_{24}H_{34}O_4$	FE85	$C_{24}H_{38}B_2F_8N_2$	FE86	$C_{25}H_{24}N_3Na$
FE87	$C_{25}H_{24}N_3NaO_2$	FE88	$C_{25}H_{25}BrN_2$		
FE89	$C_{25}H_{25}ClN_2O_4S_2$	FE91	$C_{26}H_{18}K_2O_2$		

FE93 $C_{26}H_{18}$	FE95 $C_{26}H_{18}K_2O_4$	FE98 $C_{21}H_{21}O_3S_3$
FE99 $C_{26}H_{22}N_2$	FF00 $C_{26}H_{22}N_2O_2$	FF02 $C_{26}H_{23}ClO_2$
FF03 $C_{26}H_{24}N_2$	FF04 $C_{26}H_{24}N_2$	FF05 $C_{26}H_{24}N_2O_2$
FF06 $C_{26}H_{24}N_2O_2$	FF07 $C_{26}H_{24}N_2O_2$	
FF08 $C_{27}H_{30}N_5Na$	FF09 $C_{28}H_{12}N_4O_2$	
FF10 $C_{28}H_{14}N_6S_4$	FF11 $C_{28}H_{18}N_2O_2$	FF12 $C_{28}H_{22}Cl_2N_4O_2$

TABLE II. Structural Formulas

FF13	$C_{28}H_{22}O_2$
FF14	$C_{28}H_{22}O_2$
FF15	$C_{28}H_{24}Br_2P_2$
FF16	$C_{28}H_{24}N_2O_2$
FF17	$C_{28}H_{24}N_4O_2$
FF18	$C_{28}H_{26}Br_2P_2$
FF19	$C_{28}H_{26}N_2$
FF20	$C_{28}H_{26}N_2O_4$
FF21	$C_{28}H_{26}O_2$
FF22	$C_{28}H_{28}Br_2P_2$
FF24	$C_{29}H_{21}N_6O_{12}S_3$
FF26	$C_{29}H_{21}IN_2S_2$
FF27	$C_{29}H_{22}O$
FF28	$C_{30}H_{22}Cl_2N_6O_8Ru$
FF29	$C_{30}H_{24}Cl_2N_6O_8Ru$
FF31	$C_{30}H_{26}N_2O_4$
FF32	$C_{30}H_{28}Br_2P_2$

FF33 $C_{30}H_{28}N_4O_4$	FF34 $C_{31}H_{40}O_2$	
FF35 $C_{32}H_{18}K_2O_2$	FF36 $C_{32}H_{22}N_6Na_2O_6S_2$	FF37 $C_{32}H_{24}$
FF38 $C_{32}H_{24}$	FF39 $C_{32}H_{30}N_6O_4$	FF40 $C_{32}H_{32}Br_2P_2$
FF41 $C_{32}H_{34}N_2$	FF43 $C_{33}H_{28}O_4S_3$	
FF44 $C_{33}H_{36}N_4ORu$	FF46 $C_{33}H_{40}N_3O_6$	
FF47 $C_{34}H_{24}Cl_2O_8S_6$	FF48 $C_{34}H_{38}O_6$	

TABLE II. Structural Formulas

FF49 C₃₅H₂₈N₃Na

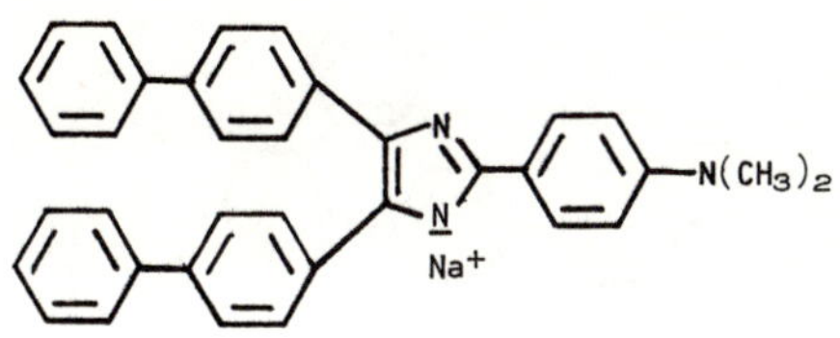

FF50 C₃₆H₂₄Cl₂N₆O₈Ru

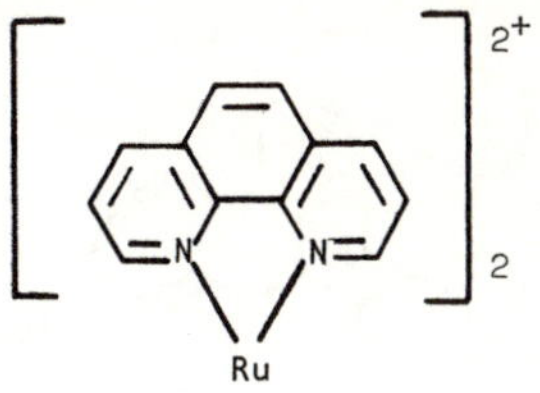

FF51 C₃₆H₂₄Cl₃N₁₂O₁₂Ru

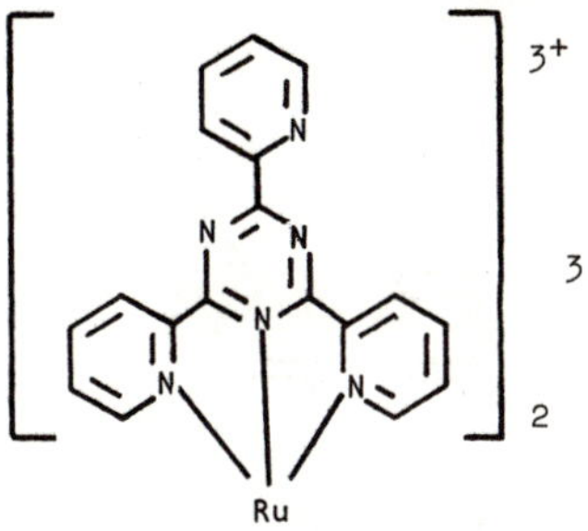

FF52 C₃₆H₂₈Cl₂O₈S₆

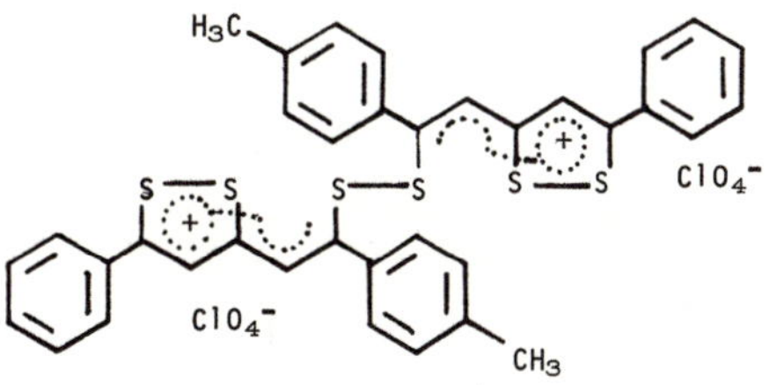

FF53 C₃₆H₃₆MgN₄O₄

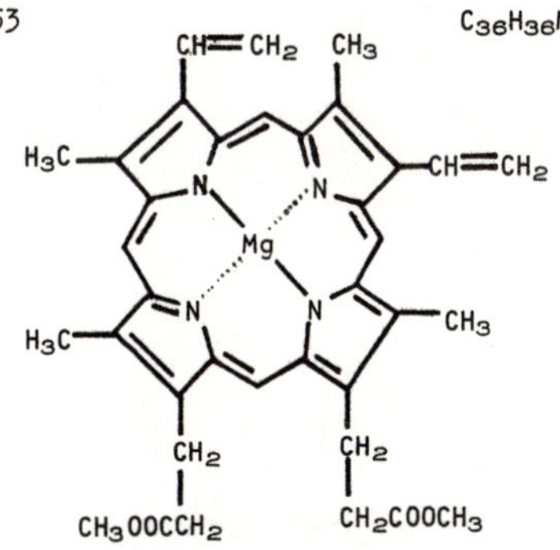

FF54 C₃₆H₄₀Br₂P₂

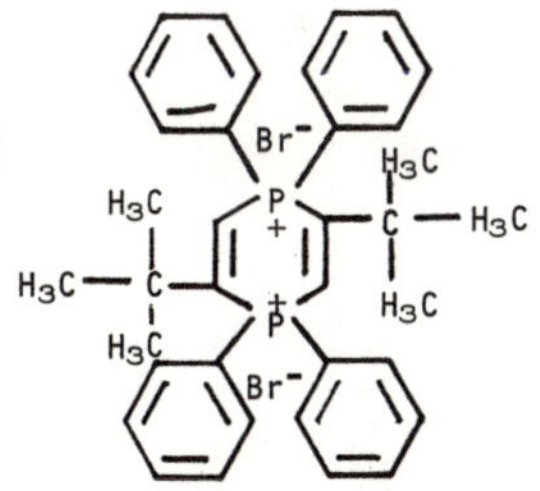

FF55 C₃₆H₄₀MgN₄O₄

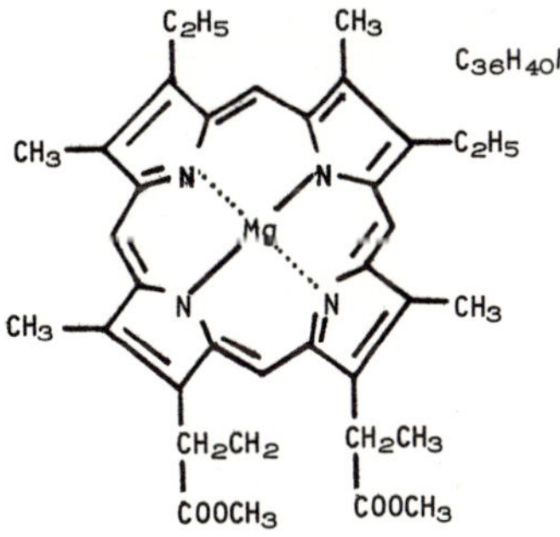

FF58 C₃₇H₄₄N₄ORu

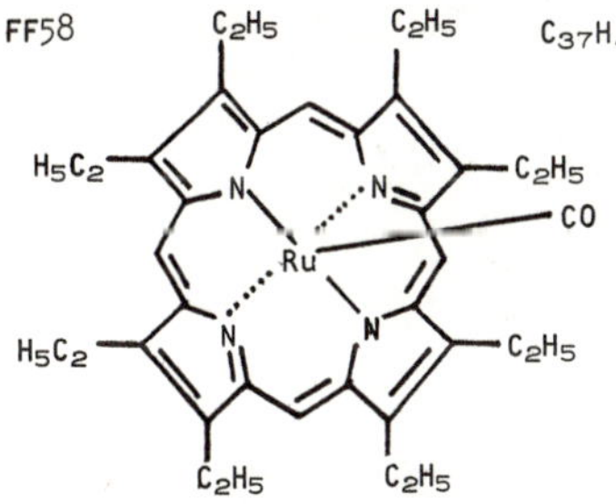

FF60 C₃₈H₃₂Cl₂O₁₀S₆

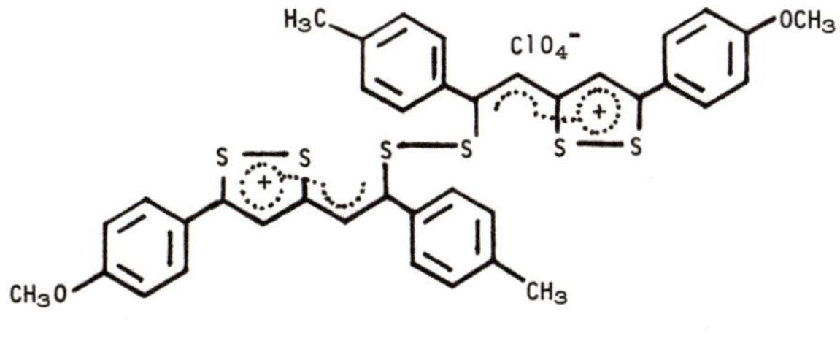

FF61 C₄₀H₂₈Cl₂F₄P₂

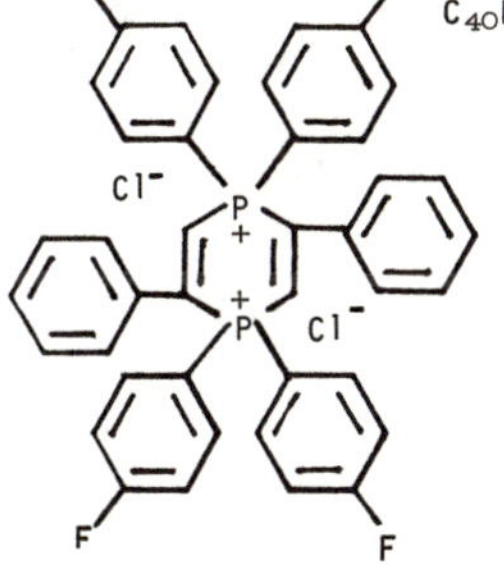

FF62 C₄₀H₃₂Br₂P₂

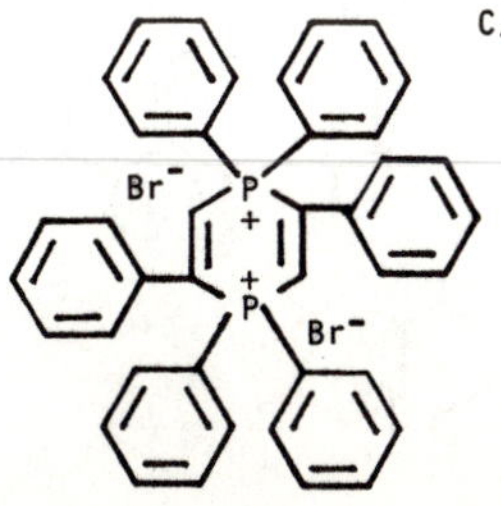

FF63 C₄₀H₃₂Cl₂P₂

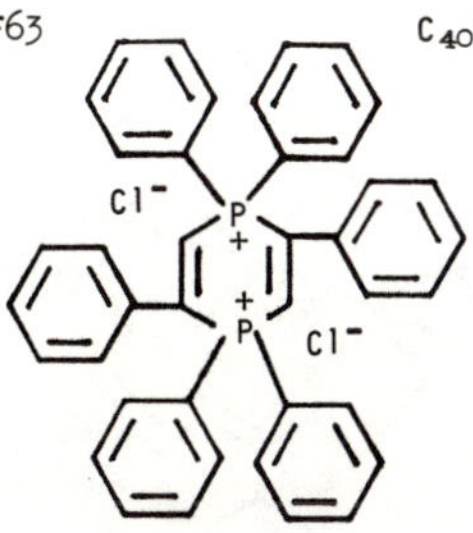

FF64 $C_{40}H_{50}N_2$

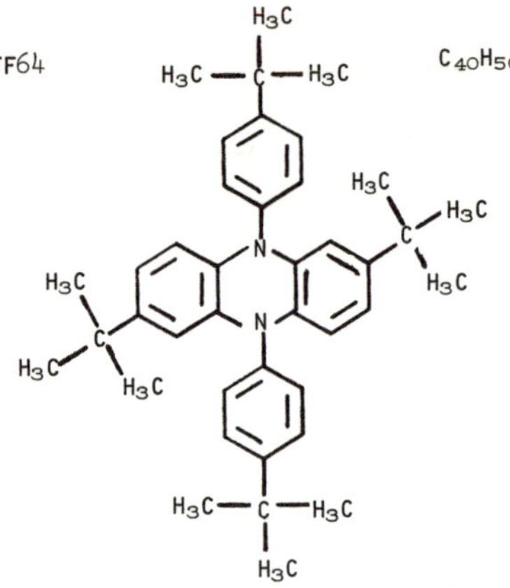

FF65 $C_{42}H_{34}N_4$

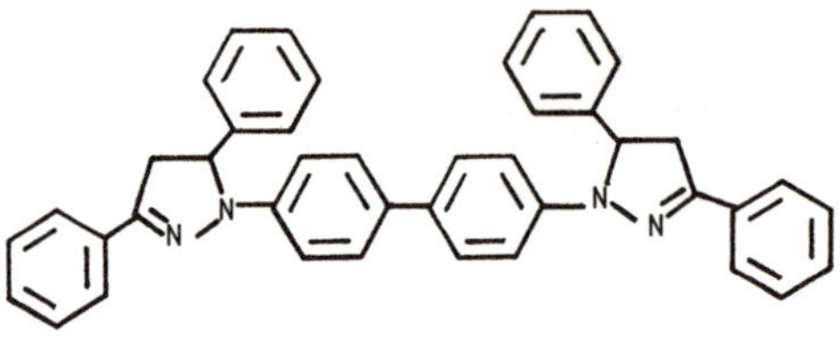

FF66 $C_{43}H_{52}N_5Ru$

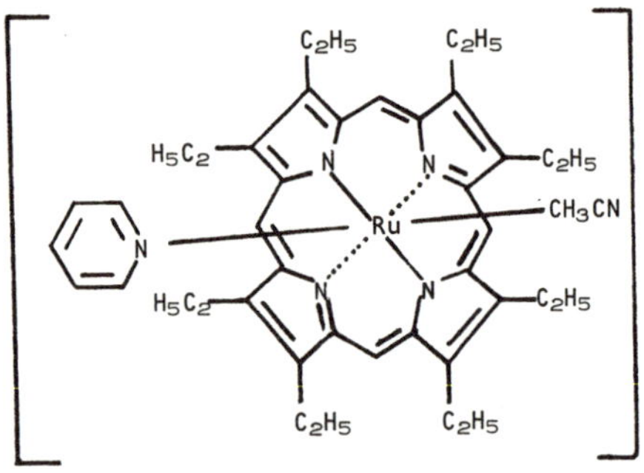

FF68 $C_{44}H_{40}Br_2O_4P_2$

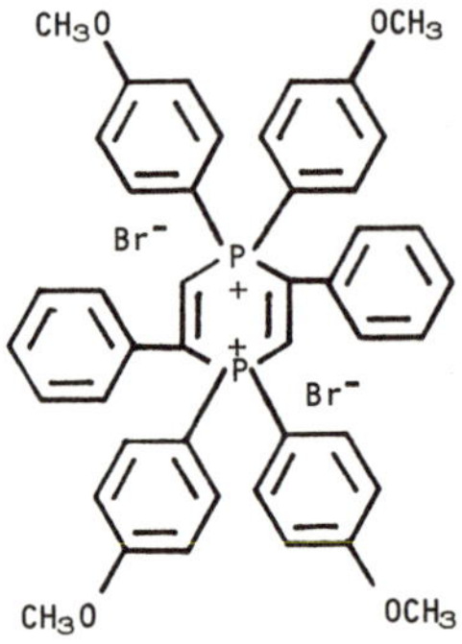

FF69 $C_{45}H_{28}N_4ORu$

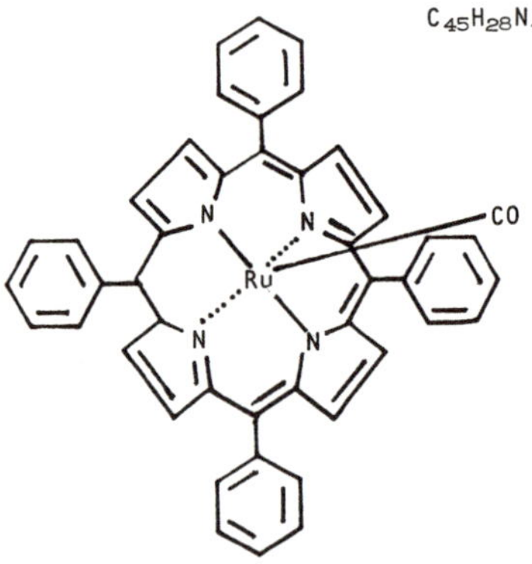

FF70 $C_{46}H_{32}Cl_2O_8S_6$

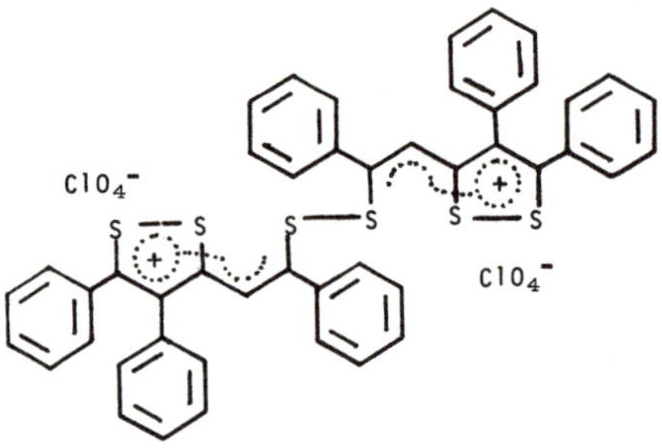

FF71 $C_{50}H_{33}N_5ORu$

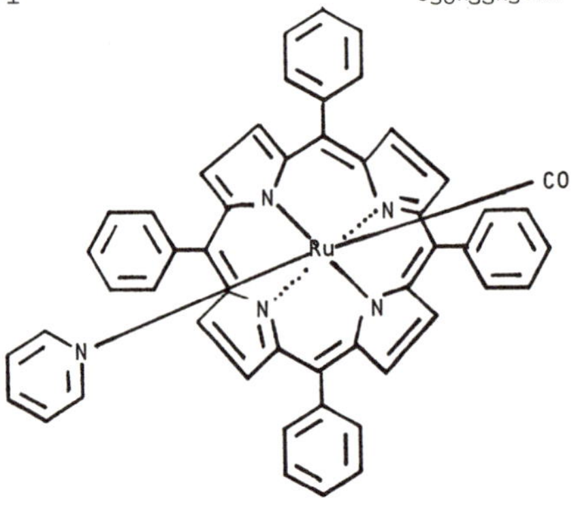

FF73 $C_{52}H_{44}Cl_2O_{12}S_6$

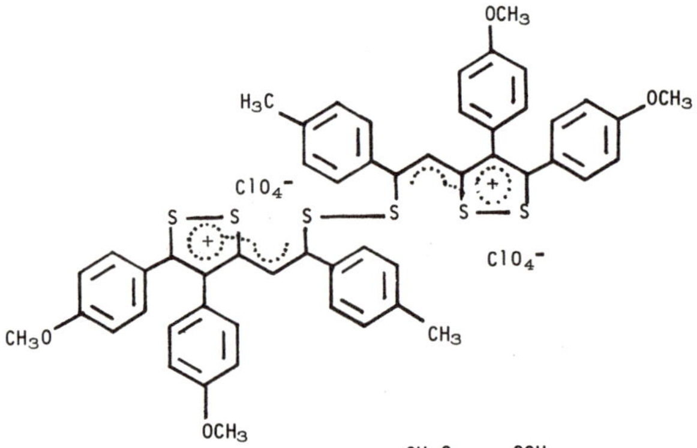

FF77 $C_{54}H_{38}N_6Ru$

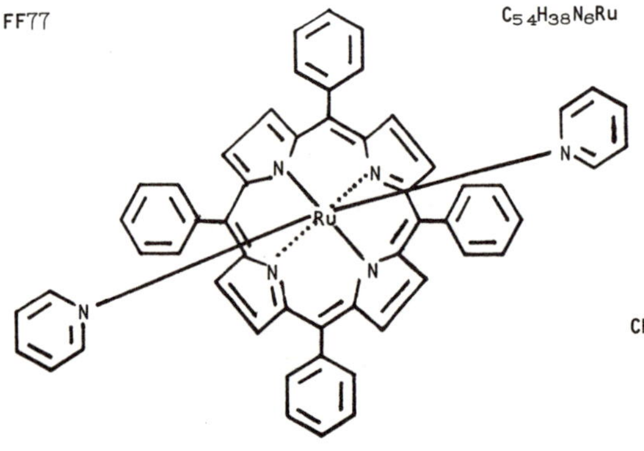

FF80 $C_{66}H_{56}Cl_2O_{14}S_6$

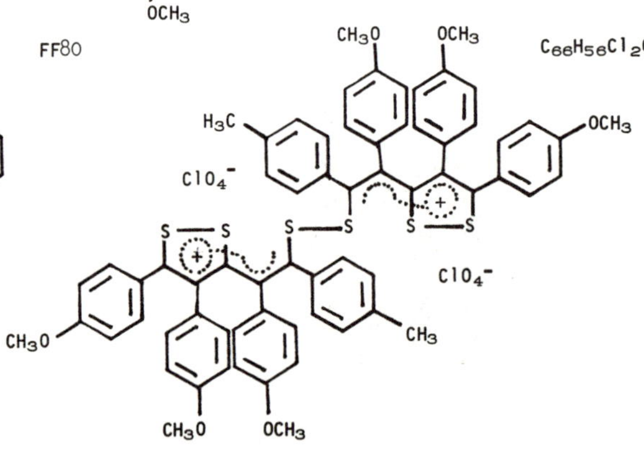

TABLE III.
COURSES AND MECHANISMS OF HALF-REACTIONS

This table is a supplement to Table 1. It gives equations for the electron-transfer steps, and for the chemical reactions associated with them, that are involved in the half-reactions undergone by the compounds listed in Table 1. It also gives values of the rate and equilibrium constants of these processes whenever those values appear in the original article.

References in this table appear in the fifteenth column ("C/M") of Table I in the form "50s". Such an entry signifies that this table provides equations for the steps by which the half-reaction proceeds, and that it may also contain the values of one or more physical constants associated with these steps. If the fifteenth column of Table I contains a dash, there is no additional information to be found in Table III.

The general nature of the mechanism is indicated by the Arabic number in the reference in Column 15 of Table I. For acrylonitrile (code number FA07) the entry in the colmn is "50s" and hence the Arabic number is "50." Under this number in Table III it is possible to find the mechanism for the reduction of acrylonitrile. Neither the Arabic number nor any letter following it (such as the "s" in this example has any electrochemical or chemical significance. The Arabic numbers appear consecutively in Table III and are centered on its pages.

To prevent repetition, related mechanism are frequently grouped together. Each individual mechanism is given by a set of several equations and is denoted by a reference number, which is a combination of the Arabic number and a letter, e.g., "50s". These reference numbers appear immediately below the corresponding Arabic numbers; for example, under Arabic number 50 the reference numbers are 50r, 50s, 50t, 50u, and 50v. Following each reference number, which is also given in Column 15 of Table I, are the numbers of the equations that were proposed for the corresponding mechanism. For example, "50s: 50-3, and -16" means that the individual steps in that mechanism are given by the equations number 50-3 and 50-16. The chemical equations for individual steps follow. In another example "53e: 53-1 to -4, -10 to -15, and -17" means that the individual steps in the mechanism denoted as 53e are given by the equations number 53-1, 53-2, 53-3, 53-4, 53-10, 53-11, 53-12, 53-13, 53-14, 53-15, and 53-17.

On the same line in Table III, opposite the number and letter that define each particular mechanism (e.g. 50s), the code numbers (referring to Table I) are given for the compounds for which this mechanism was proposed. For mechanism 50u the code number FA85 is the only one given: this means that 2,2,6,6-tetramethylpiperidinenitroxide cation (FA85) is the only substance included in this volume for which this mechanism was proposed. For 50t, four code numbers (FA31, FB15, FD06, and FD07) are given: this means that mechanism 50t was proposed for the four compounds having these code numbers. If an entry like "product of 283 (FA21, FA62)" appears among the code numbers, it means that the compounds with code number FA 21 and FA 62 are first reduced according to the mechanism identified by the Arabic number 283 and that the resulting products are further reduced according to mechanism 65c.

For some compounds the proposed course or mechanism of the electrode process is affected by changes of the composition of the supporting electrolyte, the nature of the electrode used, or other experimental conditions. Such reaction conditions are indicated in parentheses following the number of the last reaction for a given reference number. For acetonitrile (code number FA57) in Table I, the data obtained in a non-aqueous solvent are accompanied by the entry "42e" in Column 15. The corresponding entry "42e: 42-2, -3, and -8 (non-aqueous)" in the present table, together with the equations identified by these numbers, indicates that, in a non-aqueous solvent without added proton donor, the unprotonated 1-methylpyridinium-3-carbamoyl ion is reduced by one electron to produce a radical which dimerizes. The dimer undergoes a two-electron oxidation to produce the original species. For the same compound in Table I the entry "42f" appears in Column 15 opposite data obtained with benzoic acid added. This indicates that the mechanism is different from the one that is applicable without added benzoic acid. In Table III the reference number 42f is followed by "42-2 and -6 (non-aqueous with proton donor added)" indicating that the unprotonated species is again reduced to a radical, but in the prescence of a proton donor the radical is further reduced, with the addition of one proton and one electron, to 1-methyl-3-carbamoyl-1,2-dihydropyridine.

To indicate that all mechanisms following one Arabic number are limited to a certain type of solvent (e.g., aqueous, non-aqueous, protic, aprotic, etc.), a note is given in parentheses following this Arabic number.

To illustrate the situation when related mechanisms are collected under a single number, examples appearing under Arabic number 33 in Table III can be used. For 3-hydroxy-3-phenyl-1-propyne (code number FA78) the entry 33c appears in Column 15 of Table I. Inspection of the page in Table III containing the Arabic number 33 indicates that 3-hydroxy-3-phenyl-1-propyne is reduced by the addition of two electrons and one proton with the elimination of water and formation of 3-phenyl-1-propyne monocation. The cation is rearranged and a proton is added, forming 1-phenylpropadiene, which is reduced in two steps to 1-phenylpropene and then 1-phenylpropane. In a similar manner, 9-fluorenol (code number FB84) is reduced by the addition of two electrons and one proton to form water and a monocation, fluorenyl monocation. The cation is protonated to form fluorene, which can be reduced by the addition of two electrons and two protons. Mechanisms for several alcohols and ethers in which the elimination of water or an alcohol resulted from the addition of two electrons and two protons were thus combined under Arabic number 33 in order to save space, because these mechanism have several steps in common.

For most compounds having two or more electoactive groups, the order of reduction (or oxidation) of these groups is known. In such cases Column 15 of Table I gives only a single reference number, corresponding to the electrode reactions that occur first. The subsequent process is indicated by another reference number, which appears in this Table after either the reference number describing the initial process or the equation for the last reaction involved in the electrolysis of the first group.

For example, for the reduction of ethyl 3-bromo-2-oxopropanoate (code number FA21), Column 15 in Table I gives reference number "283e". Under the Arabic number 283, there appears "283e: 283-9 and -4 to -7, followed by 65c or k". This indicates that reduction of the unhydrated form occurs first with the elimination of the bromo group, in the manner described by equations 283-4 to -7 and that this reduction is followed by the reduction of the carbonyl group, for which the reference number 65c gives steps 65-1, -2, -8 and -9, corresponding to a two-step reduction of the protonated form while reference number 65k gives steps 65-10, -12, and -13, corresponding to a two-step reduction of the unprotonated form.

Many of the mechanisms cited in Table I have already appeared in Volumes II, III, IV, and V. If the equations given in an earlier volume are applicable without change to a compound appearing in this volume, the mechanism appears here without its equations, and there is a note in parentheses giving the number of the volume in which the equations may be found. This is exemplified by the material appearing under Arabic number 8. However, if new equations are needed, they are given here together with those that appeared in earlier volumes. This is exemplified by the material appearing under Arabic number 33, where step 33-1 was given in Volume II, and step 33-1 was expanded to 33-1 to -4 and steps 33-5 to -29 were added in this volume. A few mechanism or groups of mechanisms that appeared in previous volumes have been replaced by modified versions. In such cases there is an explanatory note under the new Arabic number; for example under "215" there is the note "To replace old 215". This means that a reference number found in Column 15 of Table I of this volume may not be possible to decode by referring to Table III in any earlier volume.

Some mechanisms (like those denoted by the Arabic numbers 1 through 7) that appeared in Volumes II, III, and IV are not applicable to any of the compounds in this volume, and therefore do not appear below.

Since closely related compounds are often (though not always) reduced or oxidized in similar ways, there are many cases in which groups or substituents that do not affect the course or mechanism of a half-reaction are represented by symbols like "R", "R^1", "R^2", "R^3" (mostly for alkyl groups), Ar (for aryl groups), and Pyr (for pyridyl groups). The symbol "X" is often used to denote a halogen substituent.

Roman numerals are often used to denote many reactants, intermediates, and products. In the first equation in which it appears in a mechanism, each Roman numeral follows the line or structural formula of the substance it denotes. The number of that equation follows the Roman numeral at each reappearance, to facilitate finding the formula of the substance and following its course through the mechanism.

We have attempted to show enough of each molecular structure to enable the reader to decipher the chemical significance of each equation without undue difficulty. In cases where certain reaction steps are restricted to compounds with certain groups R, Ar, or X, this is indicated by a statement like the one under Arabic number 33:

"In 33-7 through 33-10, R^1 = $C_6H_5C(Ar):CH-$ and $R^2 = R^3 = R^4 = H$"

which precedes the equation for those steps. This means that the rearrangements and reductions represented by equations 33-7 through 33-10 occur only with at least one aromatic group in the 1-position to the double bond in a propenyl cation.

The chemical equations are often followed by values of the rate and equibrium constants that pertain to mechanism and that appear in the chemical equations (e.g., the equilibrium and rate constants in the case of mechanism 50t). Such values are given here only for compounds that appear in Table I of the present volume (but not for compounds dealt with in Volumes I through V). Most of them were deduced from the electrochemical data; the original literature must of course be consulted for details. A few values that were obtained by spectrophotometric or other nonelectrochemical techniques are also included to permit comparison with, or to aid in the interpretation of, the accompanying electrochemical results.

8
(Vol IV)

Mechanism:
8g: 8-2, -6 and -7

RHgX

R = $4\text{-}NH_2C_6H_4\text{-}$

X = acetate

k_7, $cm^2 s^{-1} mol^{-1}$ = 4.33E13

Proposed for Compound:
FA68

33

Mechanism:

33a: 33-1, -2, -5 and -6 as one step

33b: 33-1, -2, and -5 to -10

33c: 33-1, -3 and -5 as one step, and -11 to -15

33d: 33-1, -3 and -5 as one step, -16 and -15

33e: 33-1, -3, -5 and -6 as one step, and -17

33f: 33-1, -2, and -5 as one step, -6 and -18

33g: 33-1, -2, -5, -6 and -19 to -22 (protic solvent)

33h: 33-1, -2, -5, -19, -20 and -23 to -26 (aprotic solvent)

33i: 33-1, -2, -5, -19, -20, (-23 to -25 as one step), and -26

33j: 33-1, -2, and -5 as one step, -3 and -6

33k: 33-1, -2, -5 and -6

33l: 33-1, -2, -5, -6 and -27 to -29

Proposed for Compound:
AH75, AH76, AL14, A006, AR85, AT49, BJ62, BN69, B063, DG28

FA80, FC85

FA78, FA94, FC77

FC84

FC52, FD86

Product of 65z (FB71), FB77, FB84

FF13, FF21, FF59

FF13, FF21, FF59

FE93, FE96

FC51, FF01

FB03, FC50, FC99, FE13

FC86

$$R^1 - \underset{\underset{OR^4}{|}}{\overset{\overset{R^2}{|}}{C}} - R^3 + e \rightarrow \left[R^1 - \underset{\underset{OR^4}{|}}{\overset{\overset{R^2}{|}}{C}} - R^3 \right]^{\tau} \quad (I) \qquad 33\text{-}1$$

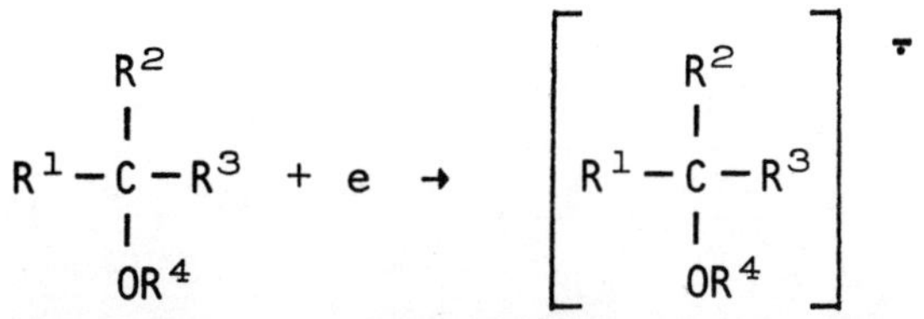

$$I(33\text{-}1) \rightarrow R^1-\underset{\cdot}{\overset{\overset{R^2}{|}}{C}}-R^3 \ (II) \ + \ R^4O^- \qquad 33\text{-}2$$

$$I(33\text{-}1) \ + \ H^+ \rightarrow II(33\text{-}2) \ + \ R^4OH \qquad 33\text{-}3$$

$$R^4OH \rightleftarrows R^4O^- \ + \ H^+ \qquad 33\text{-}4$$

$$II(33\text{-}2) \ + \ e \rightarrow R^1-\underset{-}{\overset{\overset{R^2}{|}}{C}}-R^3 \ (III) \qquad 33\text{-}5$$

$$III(33\text{-}5) \ + \ H^+ \rightarrow R^1-\underset{\underset{H}{|}}{\overset{\overset{R^2}{|}}{C}}-R^3 \ (IV) \qquad 33\text{-}6$$

In 33-7 through 33-10,

$$R^1 = C_6H_5\overset{\overset{Ar}{|}}{C}=CH-\quad \text{and} \quad R^2 = R^3 = R^4 = H$$

$$III(33\text{-}5) \leftrightarrow \ -\!\!\left\langle\!\!\!\bigcirc\!\!\!\right\rangle\!\!=\!\overset{\overset{Ar}{|}}{C}-CH=CH_2 \qquad 33\text{-}7$$

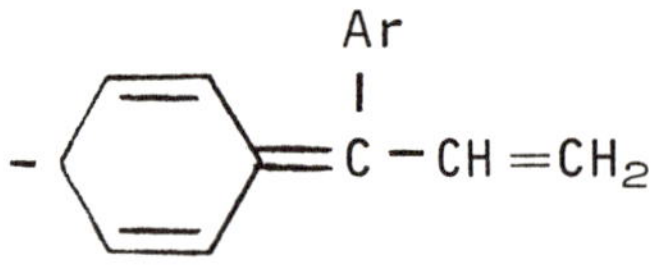

$$I(33\text{-}1) \ + \ H^+ \rightarrow \cdot\!\!\left\langle\!\!\!\bigcirc\!\!\!\right\rangle\!\!=\!\overset{\overset{Ar}{|}}{C}-CH_2CH_2OH \ (V) \qquad 33\text{-}8$$

$$V(33\text{-}8) \ + \ e \rightarrow \ -\!\!\left\langle\!\!\!\bigcirc\!\!\!\right\rangle\!\!=\!\overset{\overset{Ar}{|}}{C}-CH_2CH_2OH \ (VI) \qquad 33\text{-}9$$

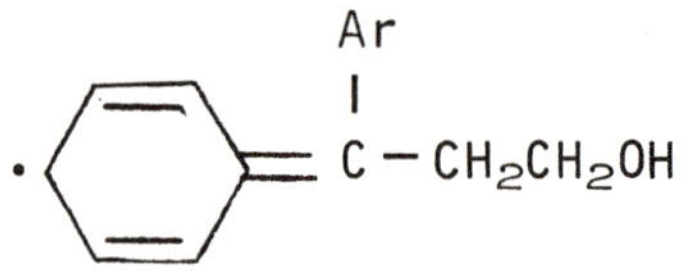

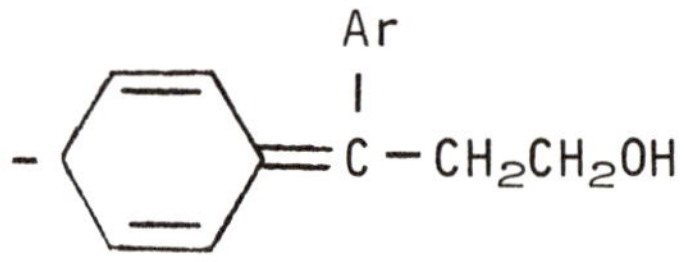

$$VI(33\text{-}9) \ + \ H^+ \rightarrow \left\langle\!\!\!\bigcirc\!\!\!\right\rangle\!\!-\!\overset{\overset{Ar}{|}}{C}HCH_2CH_2OH \qquad 33\text{-}10$$

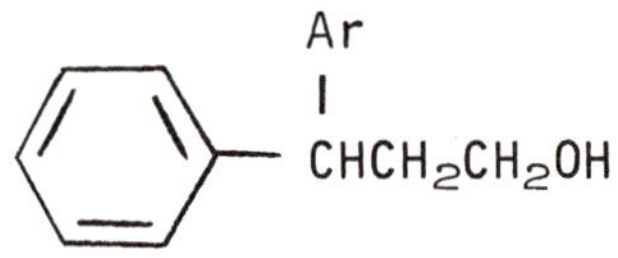

In 33-11 through 33-15,

$$R^1 = C_6H_5, \ R^3 = HC\equiv C\text{-}, \ \text{and} \ R^4 = H$$

$$III(33\text{-}5) \leftrightarrow C_6H_5-\overset{\overset{R^2}{|}}{C}=C=\bar{C}H \quad (VII) \qquad 33\text{-}11$$

$$\text{VII}(33\text{-}11) + H^+ \rightarrow C_6H_5-\underset{\underset{R^2}{|}}{C}=C=CH_2 \quad (\text{VIII}) \qquad 33\text{-}12$$

$$\text{VIII}(33\text{-}12) + 2e + H^+ \rightarrow C_6H_5-\underset{\underset{R^2}{|}}{\underline{C}}-CH=CH_2 \quad (\text{IX}) \qquad 33\text{-}13$$

$$\text{IX}(33\text{-}13) + H^+ \rightarrow C_6H_5-\underset{\underset{R^2}{|}}{C}=CH-CH_3 \quad (\text{X}) \qquad 33\text{-}14$$

$$\text{X}(33\text{-}14) + 2e + 2H^+ \rightarrow C_6H_5-\underset{\underset{R^2}{|}}{C}HCH_2CH_3 \qquad 33\text{-}15$$

In 33-16,

$R^1 = R^2 = C_6H_5$, $R^3 = CH=CH-$, and $R^4 = H$

$$\text{III}(33\text{-}5) + H^+ \rightarrow \text{X}(33\text{-}14) \qquad 33\text{-}16$$

In 33-17,

$R^1 = 4\text{-}C_6H_5C_6H_4\text{-}$ and $R^2 = R^4 = H$

$$\text{IV}(33\text{-}6) + 4e + 4H^+ \rightarrow \text{[cyclohexenyl-phenyl]}-CH_2R^3 \qquad 33\text{-}17$$

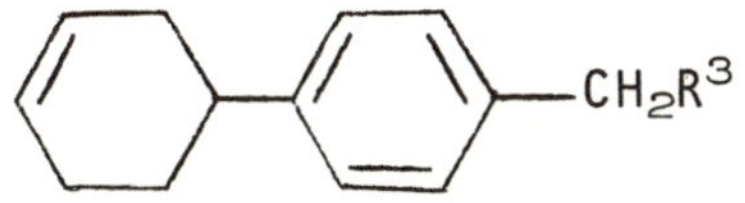

In 33-18,

$R^1-\underset{\underset{OR^4}{|}}{\overset{\overset{R^2}{|}}{C}}-R^3$ = [9-hydroxy-9H-fluorene structure with HO and H on the 9-position]

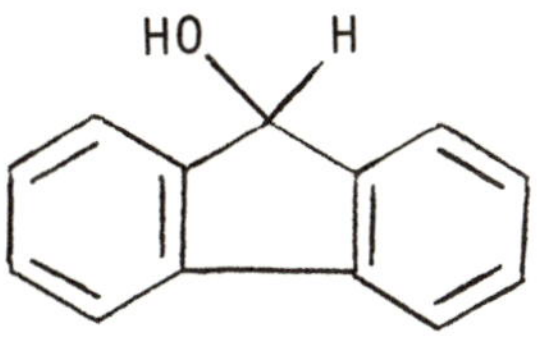

$$\text{IV}(33\text{-}6) + 2e + 2H^+ \rightarrow \text{products} \qquad 33\text{-}18$$

In 33-19 and 33-20,

$R^1 = R^3-\underset{\underset{OH}{|}}{\overset{\overset{R^2}{|}}{C}}-$ and $R^4 = H$

$$\text{III}(33\text{-}5) \rightarrow \underset{\underset{R^3}{|}}{R^3}-\overset{\overset{R^2}{|}}{C}=\overset{\overset{R^2}{|}}{C}-R^3 \quad (\text{XI}) \;+\; OH^- \qquad 33\text{-}19$$

$$\text{XI}(33\text{-}19) \;+\; 2e \;+\; 2H^+ \rightarrow R^3-\overset{\overset{R^2}{|}}{\underset{\underset{H}{|}}{C}}-\overset{\overset{R^2}{|}}{\underset{\underset{H}{|}}{C}}-R^3 \quad (\text{XII}) \qquad 33\text{-}20$$

$$\text{IV}(33\text{-}6) \;+\; 2e \;+\; 2H^+ \rightarrow \text{XII}(33\text{-}20) \qquad 33\text{-}21$$

$$\text{XII}(33\text{-}20) \;+\; 2e \;+\; 2H^+ \rightarrow 2\, R^3-\overset{\overset{R^2}{|}}{\underset{\underset{H}{|}}{C}}-H \qquad 33\text{-}22$$

$$\text{XII}(33\text{-}20) \;+\; e \rightarrow \left[R^3-\overset{\overset{R^2}{|}}{\underset{\underset{H}{|}}{C}}-\overset{\overset{R^2}{|}}{\underset{\underset{H}{|}}{C}}-R^3 \right]^{\overline{\cdot}} \quad (\text{XIII}) \qquad 33\text{-}23$$

$$\text{XIII}(33\text{-}23) \overset{k}{\rightarrow} R^3-\overset{\overset{R^2}{|}}{\underset{\underset{H}{|}}{C}}\cdot \;+\; R^3-\overset{\overset{R^2}{|}}{\underset{\underset{H}{|}}{C}}- \qquad 33\text{-}24$$

$$R^3-\overset{\overset{R^2}{|}}{\underset{\underset{H}{|}}{C}}\cdot \;+\; e \rightarrow R^3-\overset{\overset{R^2}{|}}{\underset{\underset{H}{|}}{C}}- \qquad 33\text{-}25$$

$$R^3-\overset{\overset{R^2}{|}}{\underset{\underset{H}{|}}{C}}- \;+\; H^+ \rightarrow R^3-\overset{\overset{R^2}{|}}{\underset{\underset{H}{|}}{C}}-H \qquad 33\text{-}26$$

TABLE III. Courses and Mechanisms of Half-Reactions 363

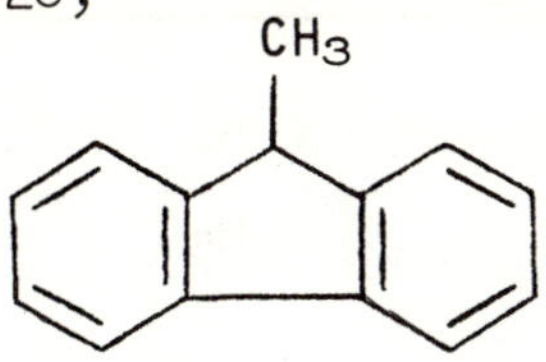

In 33-27 and 33-28,

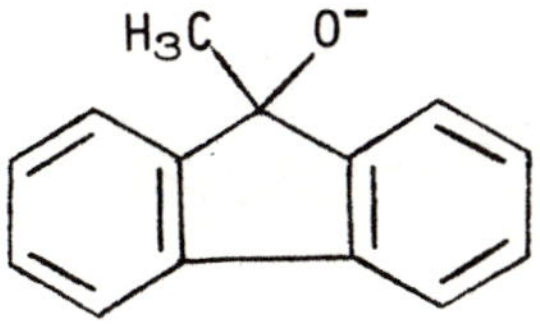

$R^1-\underset{\underset{R^3}{|}}{\overset{\overset{R^2}{|}}{C}}-R^3$ = [9-methylfluorene structure] and $R^4 = CH_3$

I(33-1) → [9-methyl-9-fluorenoxide structure] (XIV) + $CH_3\cdot$ 33-27

XIV(33-27) + H^+ → [9-methyl-9-fluorenol structure]   33-28

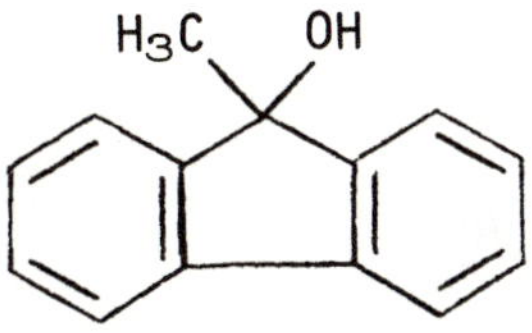

2 II(33-2) → dimer   33-29

Code No.	R^1	R^2	R^3	R^4
FA78	C_6H_5	H	$HC\equiv C-$	H
FA80	$C_6H_5CH=CH-$	H	H	H
FA94	C_6H_5	CH_3	$HC\equiv C-$	H
FB03	$C_6H_5CH-CH=$	H	H	CH_3
FB71, FB77, FB84	$R^1-\underset{\underset{}{|}}{\overset{\overset{R^2}{|}}{C}}$ = [fluorenyl structure]		H	H
FC50	C_6H_5	C_6H_5	H	CH_3
FC51	C_6H_5	H	H	$C_6H_5CH_2-$
FC52	$4-C_6H_5C_6H_4$	H	CH_3	H
FC77	C_6H_5	C_6H_5	$HC\equiv C-$	H
FC84	C_6H_5	C_6H_5	$CH_2=CH-$	H
FC85	$(C_6H_5)_2C=CH-$	H	H	H
FC86	$R^1-\overset{\overset{R^2}{|}}{C}$ = [9,9-dimethylfluorenyl structure]		CH_3	CH_3
FC99	$R^1-\underset{\underset{R^3}{|}}{\overset{\overset{R^2}{|}}{C}}-R^3$ = $CH_3O-C_6H_4-C_6H_4-$			CH_3

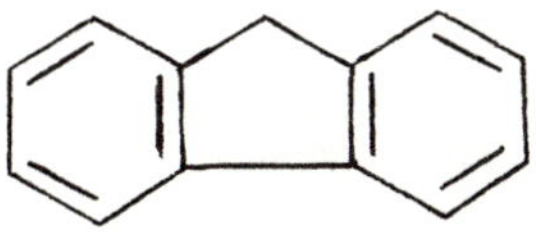

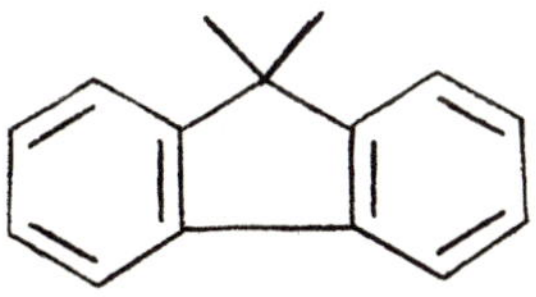

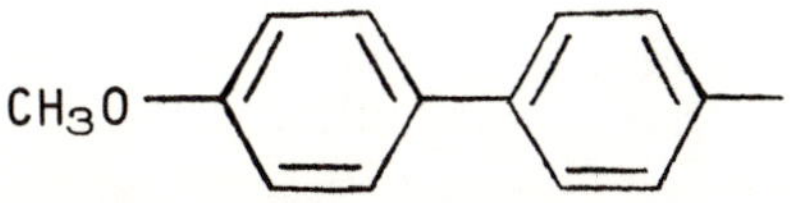

Code No.	R^1	R^2	R^3	R^4
FD86	4-$C_6H_5C_6H_4$-	H	C_6H_5	H
FE13	C_6H_5	C_6H_5	C_6H_5	CH_3
FE93, FE96	$R^2-\overset{R^3}{\underset{OH}{C}}-$	$R^3-C=$ (9,9-dimethylfluorenyl)		H
FF01	C_6H_5	C_6H_5	H	-$CH(C_6H_5)_2$
FF13	$R^2-\overset{R^3}{\underset{OH}{C}}-$	$R^2-\overset{R^3}{C}=$ (9,9-dimethyl-9,10-dihydroanthracenyl)		H
FF21	$R^2-\overset{R^3}{\underset{OH}{C}}-$	4-$C_6H_5C_6H_4$-	CH_3	H
FF59	$R^2-\overset{R^3}{\underset{OH}{C}}-$	4-$C_6H_5C_6H_4$	C_6H_5	H

42

(Vol V)

Mechanism:

42e: 42-2, -3 and -8 (nonaqueous)   Proposed for Compound: FA57, FB13, FE26, FE27, FE28

42f: 42-2 and -6 (nonaqueous with proton donor added)   FA57, FB13, FE26, FE27, FE28

Code No.	Solvent	k_3, $dm^3 mol^{-1} s^{-1}$ (at 40°C)
FA57	MeCN	1E6
FE26	DMSO	1E6
FE27	DMSO	5E6
FE28	DMSO	9E5

50

Mechanism:
50r: 50-3 and -16 (see below)

Proposed for Compound:
FA48, FA83, FA96, FA98, FB10, FB49, FC76

50s: 50-3, -4 and -8 (H^+ available)

FA07, FA14, FA23, FA47, FA48, FA49, FA69, FA74, FA83, FA96, FA98, FB08, FB10, FB49, FC76, FD20

50t: 50-3 and -17

FA31, FB15, FD06, FD07

50u: 50-18 and -19

FA85

50v: 50-20 and -21

FF72

$RH^+ \rightleftarrows R + H^+$	50-1
$RH^+ + e \rightarrow RH\cdot$	50-2
$R + e \rightarrow R^{\cdot -}$	50-3
$R^{\cdot -} + H^+ \rightleftarrows RH\cdot$	50-4
$R^{\cdot -} + e \rightarrow R^{2-}$	50-5
$RH\cdot + e \rightarrow RH^-$	50-6
$R^{2-} + H^+ \rightleftarrows RH^-$	50-7
$2RH\cdot \rightarrow HRRH$	50-8
$RH\cdot + H^+ \rightarrow RH_2^{\cdot +}$	50-9
$R^{2-} + R \rightarrow 2R^{\cdot -}$	50-10
$RH_2^{\cdot +} + e \rightarrow RH_2$	50-11
$RH^- + H^+ \text{ (or } HX) \rightarrow RH_2 + (X^-)$	50-12
$RH^- \rightarrow R + H^+ + 2e$	50-13
$RH\cdot + 5e + 5HQ \rightarrow$ ring opened products	50-14
$RH^- + ne \rightarrow$ products	50-15
$2R^{\cdot -} \rightarrow RR^{2-}$	50-16
$R^{\cdot -} \rightarrow$ products	50-17
$R^+ + e \rightarrow R\cdot$	50-18
$R\cdot + H^+ \xrightarrow[19]{k} RH^{\cdot +}$	50-19
$R^+ + 2e \rightleftarrows R^-$	50-20
$R^- + H^+ \rightarrow RH$	50-21

$$R = \;\diagup\!\!\!\!C=C-COO-,\; \diagup\!\!\!\!C=C-CN,\; \diagup\!\!\!\!C=C-CO-,$$

$$R^+ = \left[\begin{array}{c}\text{2,2,6,6-tetramethylpiperidine-1-oxyl cation}\end{array}\right]^+ \quad \text{or bis[}\underline{cis}\text{-1,2-bis(diphenylphosphino)-ethylene]rhodium(I)}$$

mechanism 50r applies under the following conditions

Code No.	
FA48	DMF with $H_2O < 1\%$
FA83	DMF with $H_2O < 1\%$
FA96	DMF with $H_2O < 20\%$
FA98	MeCN
FB10	MeCN
FB49	MeCN with $H_2O < 5\%$ with PHEN < 50mM
FC76	DMF with $H_2O < 1\%$

Code No.	k_3, cm s^{-1}	k_{17}, s^{-1}	k_{19}, dm^{3}mol^{-1} s^{-1}
FA31	1.9E-3	(4.5 ± 0.9)E-2	
FA85			5.9E3
FB15	(3.5 ± 0.3)E-3	(1.7 ± 0.5)E-3	
FD06	(2.2 ± 0.6)E-3	< 1E-3	
FD07	(3.9 ± 0.3)E-3	< 1E-3	

53

Mechanism:
53d: 53-1 to 4, -7, -8, -9 and -16 (acidic)

Proposed for Compound:
FA15, FA25, product of 421 (FA06, FA26, FA99)

53e: 53-1 to -4, -10 to -15 and -17 (basic)

FA15, FA25, product of 421 (FA06, FA26, FA99)

RCOCOOH $\rightleftarrows$ RCOCOO⁻ + H⁺ 53-1

RCOCOO⁻ + H₂O $\rightleftarrows$ RC(OH)₂COO⁻ 53-2

RCOCOOH + H₂O $\rightleftarrows$ RC(OH)₂COOH 53-3

RC(OH)₂COOH $\rightleftarrows$ RC(OH)₂COO⁻ + H⁺ 53-4

RCOCOOH + ne → product 53-5

RCOCOO⁻ + ne → product 53-6

In 53-7 to -15, R = XCH₂

XCH₂C(:OH⁺)COOH $\rightleftarrows$ XCH₂COCOOH + H⁺ 53-7

XCH₂C(:OH⁺)COOH $\longleftrightarrow$ XC⁺H₂=C(OH)COOH 53-8

XC⁺H₂=C(OH)COOH $\rightleftarrows$ XCH=C(OH)COOH + H⁺ 53-9

XCH₂COCOOH + B $\rightleftarrows$ XC̄HC(O)COOH + BH⁺ 53-10

XC̄HC(O)COOH $\longleftrightarrow$ XCH=C(O⁻)COOH 53-11

XCH=C(OH)COOH $\rightleftarrows$ XCH=C(O⁻)COOH + H⁺ 53-12

XCH₂COCOO⁻ + B $\rightleftarrows$ XC̄HCOCOO⁻ + BH 53-13

XC̄HCOCOO⁻ $\longleftrightarrow$ XCH=C(O⁻)COO⁻ 53-14

XCH=C(OH)COO⁻ $\rightleftarrows$ XCH=C(O⁻)COO⁻ + H⁺ 53-15

RCOCOOH + 2e + 2H⁺ → RCHOHCOOH 53-16

RCOCOO⁻ + 2e + 2H⁺ → RCHOHCOO⁻ 53-17

R = CH₃SCH₂, C₂H₅SCH₂, or CH₃

65
(Vol V)

Mechanism:
65c: 65-1, -2, -8 and -9 (acidic)

Proposed for Compound:
FA24, FA61, product of 283 (FA21, FA62), product of 156k (FB67, FB74), FB85, FC28), FC72, FC73, FC74, FC75, FC78, FD19, FD29, FD30, FD31, FD32

65e: 65-10, -11, -4 and -5   BC85, BD15, BD16, BF50, BF51, BM09

65k: 65-10, -12, and -13 (basic)   BC67, BJ72, FA24, FA61, product of 283 (FA21, FA62)

65y: 65-10, -11, -3, -8 and -9   FA67, FA81, FA97, FB04, FB11

65z: 65d followed by 33f   FB71

82

(Vol IV)

Mechanism:
82f: 82-2, -5, -7 and -8 (pH < 6.4)   Proposed for Compound: FD59

82g: 82-2 and -5 (pH > 6.4)   FD59

$$R^1R^2NNO = (C_6H_5)_2CH-N\overset{\frown}{\underset{\smile}{}}N-NO$$

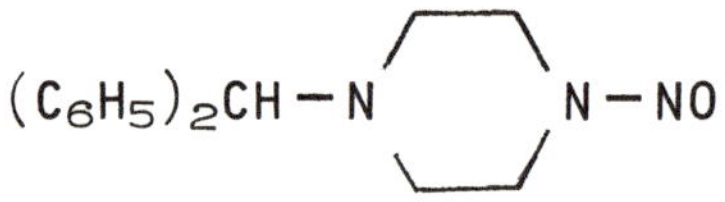

94

Mechanism:
94h: 94-1 (n=0) and -24   Proposed for Compound: FC22, FC23

$$RCHBr(CH_2)_nCHBrR + 2e \rightarrow RCHBr(CH_2)_n\overline{C}HR + Br^- \qquad 94\text{-}1$$

$$RCHBr(CH_2)_n\overline{C}HR + H^+ \xrightarrow{slow} RCHBr(CH_2)_nCH_2R \qquad 94\text{-}2$$

$$RCHBr(CH_2)_nCH_2R + 2e \rightarrow R\overline{C}H(CH_2)_nCH_2R + Br^- \qquad 94\text{-}3$$

$$R\overline{C}H(CH_2)_nCH_2R + H^+ \rightarrow RCH_2(CH_2)_nCH_2R \qquad 94\text{-}4$$

In 94-5, n = 1

$$RCHBrCH_2CHBrR + 2e \rightarrow \triangle_{R}^{R} + 2Br^- \qquad 94\text{-}5$$

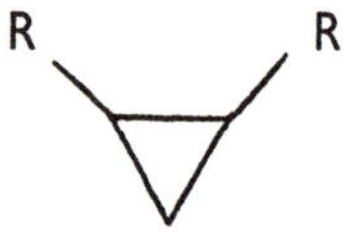

(dibromodecalin) + 2e → (decalin) + 2Br$^-$	94-6

In 94-7 through -13, n=1 and R=CH$_3$

CH$_3$CHBrCH$_2\overline{\text{C}}$HCH$_3$(94-1) + CH$_3$CHBrCH$_2$CHBrCH$_3$ → CH$_3$CHBrCH$_2$CH$_2$CH$_3$ + CH$_3$CHBr$\overline{\text{C}}$HCHBrCH$_3$ or $\overline{\text{C}}$H$_2$CHBrCH$_2$CHBrCH$_3$	94-7
CH$_3$CHBr$\overline{\text{C}}$HCHBrCH$_3$(94-7) → CH$_3$CH:CHCHBrCH$_3$ + Br$^-$	94-8
$\overline{\text{C}}$H$_2$CHBrCH$_2$CHBrCH$_3$(94-7) → CH$_2$:CHCH$_2$CHBrCH$_3$ + Br$^-$	94-9
CH$_3$CH:CHCHBrCH$_3$(94-8) + 2e → CH$_3$CH:CH$\overline{\text{C}}$HCH$_3$ + Br$^-$	94-10
CH$_3$CH:CHCH$_2$CH$_3$ ⇌ CH$_3$CH:CH$\overline{\text{C}}$HCH$_3$(94-10) + H$^+$	94-11
CH$_2$:CHCH$_2$CHBrCH$_3$(94-9) + 2e → CH$_2$:CHCH$_2\overline{\text{C}}$HCH$_3$ + Br$^-$	94-12
CH$_2$:CHCH$_2$CH$_2$CH$_3$ ⇌ CH$_2$:CHCH$_2\overline{\text{C}}$HCH$_3$(94-12) + H$^+$	94-13

In 94-14 through -23, R=H

CH$_2$Br(CH$_2$)$_n$CH$_2$Br + e → [CH$_2$Br(CH$_2$)$_n$CH$_2$]$^\cdot_{ads}$ + Br$^-$	94-14
[CH$_2$Br(CH$_2$)$_n$CH$_2$]$^\cdot_{ads}$ + e → CH$_2$Br(CH$_2$)$_n\overline{\text{C}}$H$_2$	94-15
[CH$_2$Br(CH$_2$)$_n$CH$_2$]$^\cdot_{ads}$ + e → [$\dot{\text{C}}$H$_2$(CH$_2$)$_n\dot{\text{C}}$H$_2$]$_{ads}$ + Br$^-$	94-16
[$\dot{\text{C}}$H$_2$(CH$_2$)$_n$CH$_2$]$^\cdot_{ads}$ + Hg → polymeric products	94-17
[$\dot{\text{C}}$H$_2$(CH$_2$)$_n\dot{\text{C}}$H$_2$]$_{ads}$ + e → [$\dot{\text{C}}$H$_2$(CH$_2$)$_n\overline{\text{C}}$H$_2$]$_{ads}$	94-18

In 94-19 and -20, n=2

[$\dot{\text{C}}$H$_2$(CH$_2$)$_2\overline{\text{C}}$H$_2$]$_{ads}$ + 2H$^+$ + e → CH$_3$(CH$_2$)$_2$CH$_3$	94-19
2[$\dot{\text{C}}$H$_2$(CH$_2$)$_2\overline{\text{C}}$H$_2$]$_{ads}$ + Hg + 2H$^+$ → [CH$_3$(CH$_2$)$_2$CH$_2$]$_2$Hg	94-20

In 94-21 through -23, n=1

CH$_2$BrCH$_2\overline{\text{C}}$H$_2$ → cyclopropane (CH$_2$–CH$_2$–CH$_2$) + Br$^-$	94-21
[$\dot{\text{C}}$H$_2$CH$_2\overline{\text{C}}$H$_2$]$_{ads}$ + H$^+$ → [CH$_3$CH$_2\dot{\text{C}}$H$_2$]$_{ads}$	94-22
[CH$_3$CH$_2\dot{\text{C}}$H$_2$]$_{ads}$ + H$^+$ + e → CH$_3$CH$_2$CH$_3$	94-23

In 94-24, n= 0

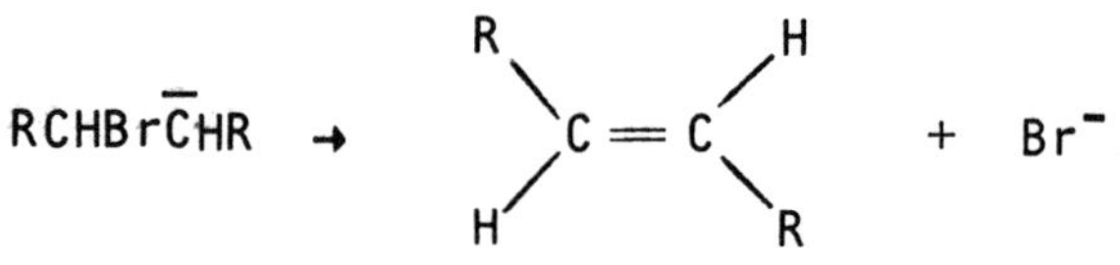

RCHBrC̄HR → $\underset{H}{\overset{R}{>}}C=C\underset{R}{\overset{H}{<}}$ + Br⁻

94-24

R=C₆H₅

103
(Vol V)

Mechanism:
103g: 103-1 and -20

Proposed for Compound:
Product of 124j (FA29, FB05, FB06)

109
(Vol V)

Mechanism:
109d: 109-1, -10, and -11

Proposed for Compound:
FC02, FC60, FC61, FC95, FD04

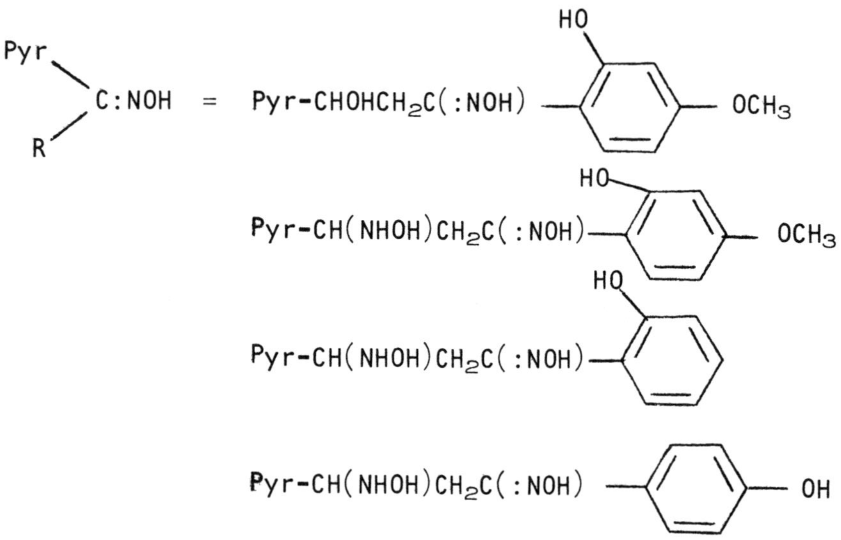

$\underset{R}{\overset{Pyr}{>}}$C:NOH = Pyr-CHOHCH₂C(:NOH)—[3-HO, 4-OCH₃-C₆H₃]

Pyr-CH(NHOH)CH₂C(:NOH)—[3-HO, 4-OCH₃-C₆H₃]

Pyr-CH(NHOH)CH₂C(:NOH)—[2-HO-C₆H₄]

Pyr-CH(NHOH)CH₂C(:NOH)—[4-HO-C₆H₄]

Pyr-C(NHOH)CH₂C(:NOH)-Pyr

119

Mechanism:
119j: 119-7, -8, -3 and -2

Proposed for Compound:
FA39, FA45

$HOC_6H_4OH \rightleftarrows 2H^+ + {}^-OC_6H_4O^-$ 119-1

(I) [p-benzoquinone] $+ H^+ \rightleftarrows$ (II) [protonated p-benzoquinone, OH$^+$ / O]   119-2

II(119-2) $+ e \rightleftarrows$ (III) [semiquinone radical anion with OH and O·]   119-3

III(119-3) $+ e \rightarrow$ (IV) [hydroquinone mono-anion with OH and O$^-$]   119-4

IV(119-4) $+ H^+ \rightleftarrows$ (V) [hydroquinone, OH / OH]   119-5

I(119-2) $+ e \rightarrow$ (VI) [dianion diradical, O· / O$^-$]   119-6

V(119-5) $\rightleftarrows$ (VII) [OH$^+$ / OH] $+ e$ 119-7

VII(119-7) ⇄ III(119-3) + H⁺   119-8

2 III (119-3) —fast→ hemiketal, reaction partly heterogeneous   119-9

hemiketal —slow→ I(119-1) + V(119-5)   119-10

VI (119-6) + e → [benzene ring with O⁻ at top and O⁻ at bottom] (VIII)   119-11

VII (119-7) → [benzene ring with OH⁺ at top and OH⁺ at bottom] (IX) + e   119-12

II (119-2) + 2e + H⁺ → V(119-5)   119-13
I (119-2) + 2e + 2H⁺ → V(119-5)   119-14
VIII(119-11) + H⁺ ⇄ IV(119-4)   119-15

[naphthoquinone with OH, OH, SO₃⁻] → [naphthoquinone with OH, O⁻, SO₃⁻] (X) + H⁺   119-16

X(119-16) → [naphthalene-1,2,5,8-tetraone with SO₃⁻] + H⁺ + 2e   119-17

II (119-2) + H⁺ ⇄ IX(119-12)   119-18
VI (119-6) + H⁺ ⇄ III(119-3)   119-19

I(119-2) + V(119-5) ⇌ [XI structure: two quinone rings linked by O······H—O hydrogen bonds top and bottom] (XI)   119-20

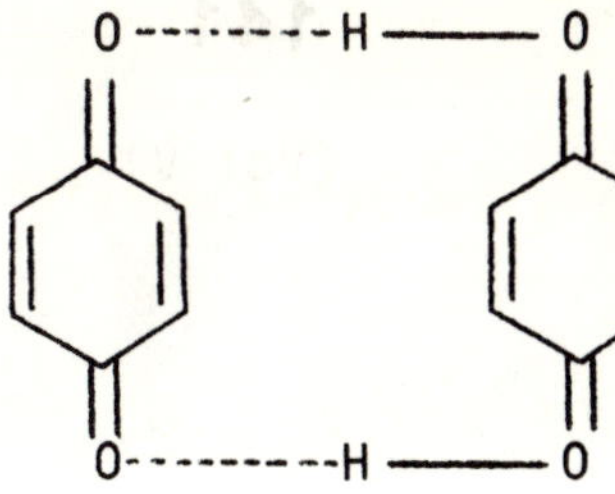

XI(119-20) + e → III(119-3) + IV(119-4)   119-21

123
(Vol V)

Mechanism:
123e: 123-1 to -5 (pH < pK) (Vol V)

123i: 123-1 to -3 (4.6 < pH < 7.0)

Proposed for Compound:
FA42

FA42

HX—⟨C₆H₄⟩—NO₂

HX = HO

124
(Vol V)

Mechanism:
124h: 124-8, -10 and -11 (strongly basic)

124j: 124i, followed by 103g

Proposed for Compound:
FA29, FB05, FB06

FA29, FB05, FB06

ArNO₂

Ar = C_6Cl_5-, $(C_2H_5)_2P(S)O$—⟨C₆H₄⟩—, or $(C_2H_5)_2P(O)O$—⟨C₆H₄⟩—

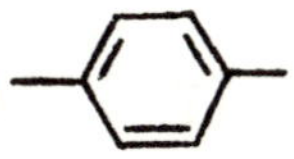

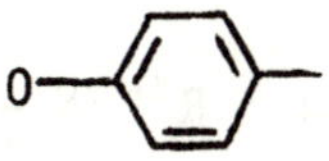

141

(Vol V)

Mechanism:
141e: 141-1 to -3

Proposed for Compound:
FA71

$$RSH = [(CH_3)_2CH]_2NCH_2CH_2SH$$

147

Mechanism:

147j: 147-20, -1, -18, -19, -16, -17 and -21 (acidic and neutral MeCN)

Proposed for Compound:
FB37, FB43, FB47, FB48, FB96, FB97, FC30, FC44, FE68, FE76, FF04, FF06, FF16

147k: 147-1, -22 and -23 (strongly basic MeCN)

FB37, FB43, FB47, FB48, FB96, FB97, FC30, FE68, FE76, FF04, FF06, FF16

147l: 147-1, and -22 to -28 (weakly basic MeCN)

FB37, FB43, FB47, FB48, FB96, FB97, FC30, FC44, FE68, FE76, FF04, FF06, FF16

147m: 147-29 and -30 (strongly basic MeCN)

FC44

147n: 147-1, -31 and -32 (acidic MeCN)

FB28, FB30, FB58, FC00, FC09, FC29, FC44, FC55, FC56, FC89, FC94, FD36, FE14

147p: 147-1, -31, -32, -25, -27 and -28 (basic MeCN)

FB35

147q: 147-33, -34 and -23 (basic MeCN)

FD41

$$C_6H_5NR^1R^2 \text{ (I)} \rightleftarrows [C_6H_5NR^1R^2]^{\dot{+}} \text{ (II)} + e \qquad 147\text{-}1$$

$$\text{II}(147\text{-}1) + \text{I}(147\text{-}1) \xrightarrow{k_2} [R^1R^2NC_6H_4C_6H_4NR^1R^2]^{\dot{+}} \text{ (III)} \; (+H_2?) \qquad 147\text{-}2$$

$$2 \, \text{II}(147\text{-}1) \xrightarrow{k_3} R^1R^2N-\underset{}{\bigcirc}-\underset{}{\bigcirc}-NR^1R^2 \text{ (IV)} + 2H^+ \qquad 147\text{-}3$$

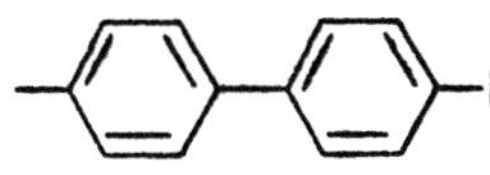

$$\text{IV}(147\text{-}3) \rightleftarrows R^1R^2\overset{+}{N}=\underset{}{\bigcirc}=\underset{}{\bigcirc}=\overset{+}{N}R^1R^2 \text{ (V)} + 2e \qquad 147\text{-}4$$

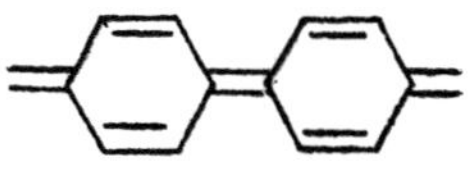

$$HR^1R^2\overset{+}{N}-\langle\text{C}_6\text{H}_4\rangle-\langle\text{C}_6\text{H}_4\rangle-\overset{+}{N}R^1R^2H \underset{}{\overset{k_5}{\rightleftarrows}} V(147\text{-}4) + 2H^+ + 2e \qquad 147\text{-}5$$

III(147-2) + I(147-1) → coupled product 147-6

In 147-7 and -8, R^2 = CH_3

$$II(147\text{-}1) + I(147\text{-}1) \rightarrow C_6H_5NR^1CH_2-\langle\text{C}_6\text{H}_4\rangle-NR^1CH_3 \text{ (VI)} + 2H^+ + e \qquad 147\text{-}7$$

$$VI(147\text{-}7) + H^+ \rightleftarrows C_6H_5\overset{+}{N}H(R^1)CH_2-\langle\text{C}_6\text{H}_4\rangle-NR^1CH_3 \text{ (VII)} \qquad 147\text{-}8$$

$$VII(147\text{-}8) \rightarrow C_6H_5NHR^1 + {}^+CH_2-\langle\text{C}_6\text{H}_4\rangle-NR^1R^2 \text{ (VIII)} \qquad 147\text{-}9$$

$$VIII(147\text{-}9) + I(147\text{-}1) \rightarrow R^1R^2N-\langle\text{C}_6\text{H}_4\rangle-CH_2-\langle\text{C}_6\text{H}_4\rangle-NR^1R^2 \text{ (IX)} + H^+ \qquad 147\text{-}10$$

$$IX(147\text{-}10) + II(147\text{-}1) \rightarrow (R^1R^2NC_6H_4)_3CH \text{ (X)} + 2H^+ + e \qquad 147\text{-}11$$

$$X(147\text{-}11) \rightleftarrows (R^1R^2NC_6H_4)_2C=\langle\text{C}_6\text{H}_4\rangle=\overset{+}{N}R^1R^2 + H^+ + e \qquad 147\text{-}12$$

In 147-13 and -14, $R^1 = R^2 = CH_3$

$$II(147\text{-}1) \rightleftarrows \text{[aryl nitrogen radical with } N(CH_3)_2 \text{]} \text{ (XI)} + H^+ \qquad 147\text{-}13$$

2XI(147-13) → IV(147-3) 147-14

2II(147-1) + IV(147-3) → 2I(147-1) + V(147-4) 147-15

$$IV(147\text{-}3) \rightarrow R^1R^2N-\langle\text{C}_6\text{H}_4\rangle-\langle\text{C}_6\text{H}_4\rangle-\overset{+}{N}R^1R^2 \text{ (XII)} + e \qquad 147\text{-}16$$

XII(147-16) → V(147-4) + e 147-17

$2\text{II}(147\text{-}1) \rightleftarrows R^1R^2\overset{+}{N}=\!\!\!\!\!\bigcirc\!\!\!\!\!\overset{H}{\underset{H}{}}\!\!\!\!\!\bigcirc\!\!\!\!\!=\overset{+}{N}R^1R^2$ (XIII)   147-18

$\text{XIII}(147\text{-}18) \xrightarrow{\text{slow}} \text{IV}(147\text{-}3) + 2H^+$ 147-19

$C_6H_5NR^1R^2H^+ \rightleftarrows \text{I}(147\text{-}1) + H^+$ 147-20

$[R^1R^2N\text{-}\!\!\bigcirc\!\!\text{-}\!\!\bigcirc\!\!\text{-}NR^1R^2]H^+ \rightleftarrows \text{IV}(147\text{-}3) + H^+$ 147-21

In 147-22 to 147-28, $R^1 = H$ and $R^2 = 4\text{-}C_6H_4X$

$\text{II}(147\text{-}1) + B \rightleftarrows C_6H_5\overset{\cdot}{N}C_6H_4X$ (XIV) $+ BH^+$ 147-22

$2\text{XIV}(147\text{-}22) \rightarrow$ [tetraaryl hydrazine structure]   147-23

$\text{II}(147\text{-}1) + \text{XIV}(147\text{-}22) \rightleftarrows C_6H_5\overset{+}{N}C_6H_4X$ (XV) $+ \text{I}(147\text{-}1)$ 147-24

$\text{XV}(147\text{-}24) + \text{I}(147\text{-}1) + B \rightarrow BH^+ +$ [structure XVI]

and/or [structure XVII]   147-25

XVI (147-25) + B → [structure: (4-X-C$_6$H$_4$)(C$_6$H$_5$)N$^+$=C$_6$H$_4$=N—C$_6$H$_4$—X] + BH$^+$ + 2e   147-26

XVII (147-25) + 2B → [phenazine structure with N-C$_6$H$_5$ groups and X substituents] (XVIII) + 2BH$^+$ + 2e   147-27

XVIII ⇌ [phenazine radical cation structure]$^{+\cdot}$ + e   147-28

In 147-29 and 30, R^1 = H and R^2 = 4-C$_6$H$_4$NHCOCH$_3$

C$_6$H$_5$NHC$_6$H$_5$NHCOCH$_3$ + 2B → C$_6$H$_5$N=⟨⟩=NCOCH$_3$ (XIX) + 2BH$^+$ + 2e   147-29

C$_6$H$_5$N$^+$—⟨⟩—NHCOCH$_3$ ⇌ XIX (147-31) + H$^+$ 147-30

In 147-31 and -32, R^1 = H and R^2 = Ar

II(147-1) → C$_6$H$_5$NArH^{2+} + e   147-31

C$_6$H$_5$NArH^{2+} ⇌ C$_6$H$_5$NAr$^+$ + H$^+$ 147-32

In 147-33 and -34, C$_6$H$_5$NR^{1}R^2 = (4-NO$_2$C$_6$H$_4$)$_2$NH

(4-NO$_2$C$_6$H$_4$)$_2$NH ⇌ (4-NO$_2$C$_6$H$_4$)$_2$N$^-$ + H$^+$ 147-33

(4-NO$_2$C$_6$H$_4$)$_2$N$^-$ → (4-NO$_2$C$_6$H$_4$)$_2$N$^\cdot$ + e   147-34

C$_6$H$_5$NR^{1}R^2 = Ar^{1}Ar^{2}NH with Ar^1 or Ar^2 = 4-XC$_6$H$_4$ and

X = Br, Cl, NO$_2$, H, OH, NH$_2$, CH$_3$, OCH$_3$, CH$_3$CO, CH$_3$CONH or SCN

156

(Vol V)

Mechanism:
156g: 156-18 to -21

156k: 156-18 and -19 as one step; -20 and -21 as one step; followed by 65c

Proposed for Compound:
FC68, FC69, FD13, FD14, FD15, FD16, FD55

FB67, FB74, FB85, FC28, FC87, FD51

$R^1CH:CHCOR^2$

R^1 = C_6H_5, 4-HOC_6H_4, 4-$CH_3C_6H_4$, 3-$CH_3OC_6H_4$, 4-$CH_3OC_6H_4$,

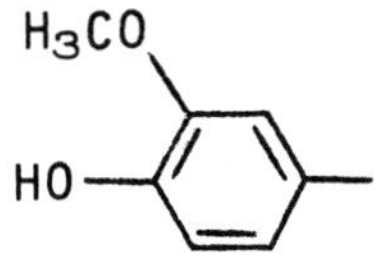

4-$C_2H_5OC_6H_4$, 4-FC_6H_4, 1-naphthyl and

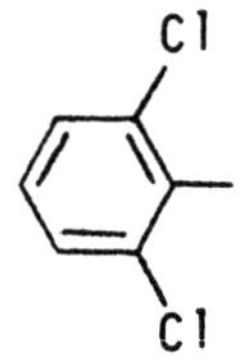

R^2 =

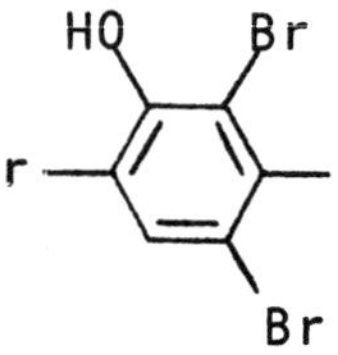

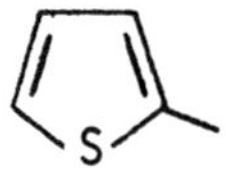

$R^1CH:CHCOR^2$ =

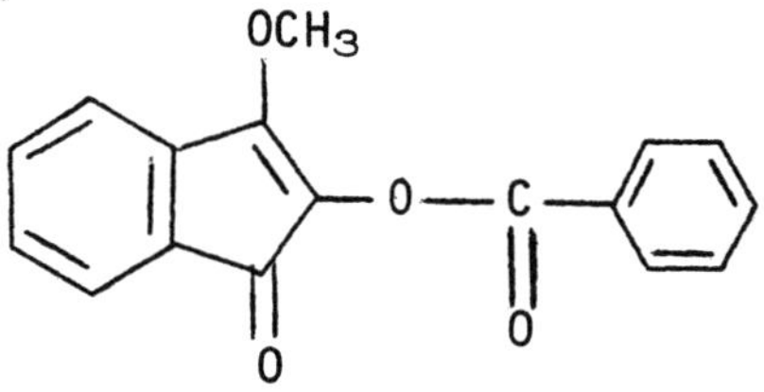

180

Mechanism:
180c: 180-1 and -3 as one step, -4 and -10

Proposed for Compound:
FC05

$R^1_2R^2R^3N^+ + e \rightarrow R^1_2R^2R^3N\cdot$ 180-1

$R^1_2R^2R^3N\cdot + H_2O \rightarrow R^1_2R^2R^3N^+ + OH^- + H\cdot$ 180-2

$R^1_2R^2R^3N\cdot \rightarrow R^1_2R^3N + (R^2)\cdot$ 180-3

$2(R^2)\cdot \rightarrow R^2R^2$ 180-4

$(R^2)\cdot + H_2O \rightarrow R^2H + OH\cdot$ 180-5

OH· $\xrightarrow{\text{deactivation}}$ H$_2$O, O$_2$, H$_2$O$_2$ 180-6

$R_2^1R^3N + H_2O \rightleftarrows R_2^1R^3NH^+ + OH^-$ 180-7

$R_2^1R^3NH^+ + e \rightarrow R_2^1R^3N + H·$ 180-8

$(R^2)· + CH_3CN \rightarrow R^2H + CH_2CN·$ 180-9

$(R^2)· + e \rightarrow (R^2)^-$ 180-10

$R^1 = R^3 = C_2H_5$

$R^2 = C_6H_5CH_2$

187

Mechanism: Proposed for Compound:
187c: 187-1, and -7 to -11   FD85

$R^1R_2^2PO + e \rightarrow R^1R_2^2PO^-$ 187-1

$R^1R_2^2PO^- + HSo \rightarrow R_2^2HPO + [R^1]· + So^-$ 187-2

In 187-3 to -6 $R^1 = XC_6H_4CO-$

$2[R^1]· \rightarrow R^1R^1$ 187-3

$R^1R^1 + e \rightarrow [R^1R^1]^{\cdot -}$ 187-4

$[R^1R^1]^{\cdot -} + e \rightarrow [R^1R^1]^{2-}$ 187-5

$[R^1R^1]^{2-} + 2H^+ \rightarrow XC_6H_4COCHOHC_6H_4X$ 187-6

In 187-7 to -11 $R^1 = XC_6H_4CH_2O$

$R^1R_2^2PO^- \xrightarrow{k} R_2^2P(O)O^- + XC_6H_4CH_2·$ 187-7

$2XC_6H_4CH_2· \rightarrow XC_6H_4CH_2CH_2C_6H_4X$ 187-8

$XC_6H_4CH_2· + HSo \rightarrow XC_6H_4CH_3 + So·$ 187-9

$XC_6H_4CH_2CH_2C_6H_4X + e \rightarrow [XC_6H_4CH_2CH_2C_6H_4X]^{\cdot -}$ 187-10

$XC_6H_4CH_3 + e \rightarrow [XC_6H_4CH_3]^{\cdot -}$ 187-11

$R^1 = 4\text{-}NO_2C_6H_4CH_2, \quad R^2 = OC_2H_5$

198

(Vol V)

Mechanism:
1981: 198-10 to -12

Proposed for Compound:
FB59, FC03, FC04, FC66,
FC67, FE32, FE33, FE34,
FE35, FE43, FE44, FE45,
FE46, FE47, FE77, FE78

$Ar^1N:NAr^2$

$Ar^1 = $ [3,5-disubstituted pyrazol-4-yl, with R^1 at positions 3 and 5, N–N in ring]

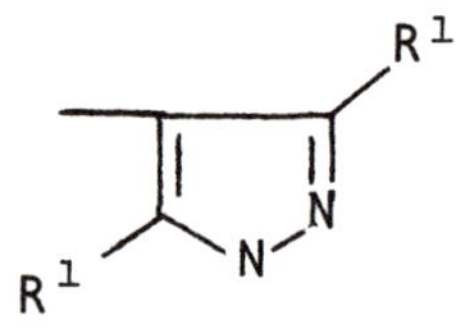

$R^1 = CH_3$ or C_6H_5

$Ar^2 = XC_6H_4$

X = H, 2-CH_3O, 4-CH_3O, 2-C_2H_5O, 4-C_2H_5O, 4-Br, 3-Cl, 4-Cl, 2-CH_3, 3-CH_3, 4-CH_3, 2,4-$(CH_3)_2$, 2,6-$(CH_3)_2$

215

(to replace old 215)
(condensed ring system)

	Mechanism:	Proposed for Compound:
215a:	215-1 to -5	BE82, BK54, CI39, CI55, CJ42, CJ54, CK16, CL02, CL54, CL57, CL58
215b:	replaced by 65e	BC85, BD15, BD16, BF50, BF51, BM09
	and by 65k	BC67, BJ72
215c:	215-1, -2, and -4 to -7	BG81
215d:	215-1, -7 followed by ece	BD00
215e:	replaced by 420b	BP78
215f:	215-1, -8 and -9	BP78
215g:	215-1, -2 and -4 (proton donor present)	BP78
215h:	215-1, -8, -12 and -13 for redn 120c for oxidn	BP78

215i: 215-1, -8, -9, -2, -4 and -5	CD37, C150, C152, CJ20, CK05, CK07, CK17, CL07, CL82, CL85, CM37, CM56
215j: replaced by 419a	CG89, C150, CK07, CK17, CL05, CL06, CL80, CL81, CL82, CM10, CM39, CM56, FC10
215k: replaced by 419b	CK17
215m: 215-1, -10, -4 and -5 (low pH)	ED80
215n: 215-1, -2, -4 and -5 (high pH)	ED80

$$Ar + e \rightleftarrows Ar^{\cdot -} \quad\quad 215\text{-}1$$

$$Ar^{\cdot -} + HSo \xrightarrow{k} ArH^{\cdot} + So^{-} \quad\quad 215\text{-}2$$

$$ArH^{\cdot} + Ar^{\cdot -} \xrightarrow{fast} ArH^{-} + Ar \quad\quad 215\text{-}3$$

$$ArH^{\cdot} + e \rightleftarrows ArH^{-} \quad\quad 215\text{-}4$$

$$ArH^{-} + HSo \xrightarrow{k_5} ArH_2 + So^{-} \quad\quad 215\text{-}5$$

$$ArH_2 + ne \rightarrow product \quad\quad 215\text{-}6$$

$$Ar^{\cdot -} + Ar \rightleftarrows ArH^{\cdot} + Ar^{1-} \text{ (proton removed)} \quad\quad 215\text{-}7$$

$$Ar^{\cdot -} + e \rightleftarrows Ar^{2-} \quad\quad 215\text{-}8$$

$$Ar^{2-} + 2HSo \rightleftarrows ArH_2 + 2So^{-} \quad\quad 215\text{-}9$$

$$ArH^{\cdot} \rightleftarrows Ar^{\cdot -} + H^{+} \quad\quad 215\text{-}10$$

$$ArH_2 \rightleftarrows ArH^{-} + H^{+} \quad\quad 215\text{-}11$$

$$ArH^{-} \rightleftarrows Ar^{2-} + H^{+} \quad\quad 215\text{-}12$$

$$ArH^{-} \rightleftarrows ArH + e \quad\quad 215\text{-}13$$

HSo = HQ for C139, C155, CJ42, CJ54, CK16, CL02, CL54, CL57, CL58

240

Mechanism:	Proposed for Compound:
240c: 240-1, and -3 to -5	FA13, FA51

$$Ar_3MX + e^{-} \rightarrow Ar_3M^{\cdot} + Cl^{-} \quad\quad 240\text{-}1$$

$Ar_3M^\cdot + HSo \rightarrow Ar_3MH + So^\cdot$ 240-2

$2Ar_3M^\cdot \rightarrow Ar_3MMAr_3$ 240-3

$Ar_3M^\cdot + e \rightarrow Ar_3M^-$ 240-4

$Ar_3M^- + HSo \rightarrow Ar_3MH + So^-$ 240-5

M = Sn

Ar = CH_3 or C_2H_5

258

Mechanism:
258b: 258-8 to -11

Proposed for Compound:
FA95, FB02

$R^1C\equiv CR^2 + e \rightarrow R^1\overline{C}=\dot{C}R^2$ 258-1

$R^1\overline{C}=\dot{C}R^2 + HSo \rightarrow R^1CH=\dot{C}R^2 + So^-$ 258-2

$R^1CH=\dot{C}R^2 + e \rightarrow R^1CH=\overline{C}R^2$ 258-3

$R^1CH=\overline{C}R^2 + HSo \rightarrow R^1CH=CHR^2 + So^-$ 258-4

$R^1CH=CHR^2 + e \rightarrow R\dot{C}H\dot{C}HR^2$ 258-5

$R^1\overline{C}H\dot{C}HR^2 + e \rightarrow R^1\overline{C}H\overline{C}HR^2$ 258-6

$R^1\overline{C}H\overline{C}HR^2 + 2HSo \rightarrow R^1CH_2CH_2R^2$ 258-7

In 258-8 to -11, $R^1 = C_6H_5\underset{\underset{OCH_3}{|}}{CH}$, $R^2 = H$

$C_6H_5\underset{\underset{OCH_3}{|}}{\overset{\overset{H}{|}}{C}}-C\equiv CH \rightleftarrows C_6H_5\underset{\underset{OCH_3}{|}}{C}=C=CH_2$ (I)   258-8

$I(258-7) + 2e + 2BH \rightarrow C_6H_5\underset{\underset{OCH_3}{|}}{\overline{C}}-CH=CH_2$ (II) $+ B^-$ 258-9

$II(258-8) + BH \rightarrow C_6H_5\underset{\underset{OCH_3}{|}}{C}=CH-CH_3$ (III)   258-10

$III(258-9) + 2e + 2BH \rightarrow C_6H_5\underset{\underset{OCH_3}{|}}{CH}CH_2CH_3$ 258-11

277

Mechanism:	Proposed for Compound:
277f: 277-7, -8, -10, -11 and -15 (DMSO,MeCN)	FA44
277g: 277-7, -16 and -4 (DMSO with proton donor)	FA44

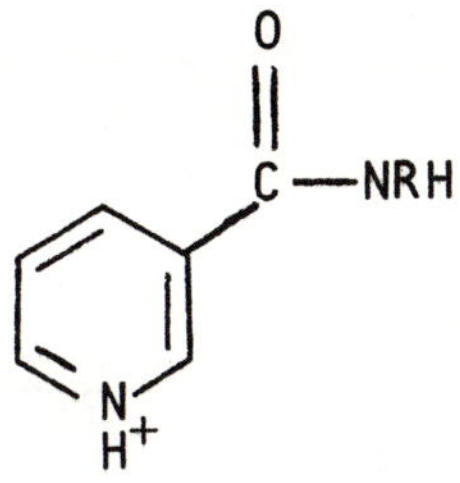

(I) ⇌ (II) + H⁺   277-1

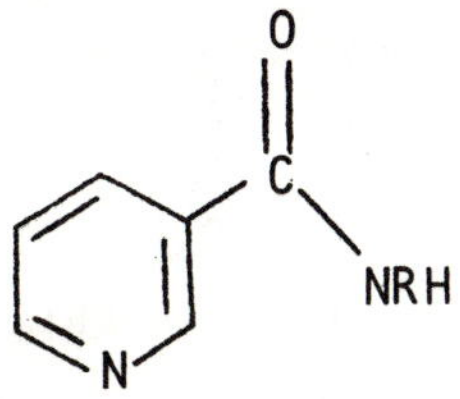

I(277-1) + e ⇌ (III)   277-2

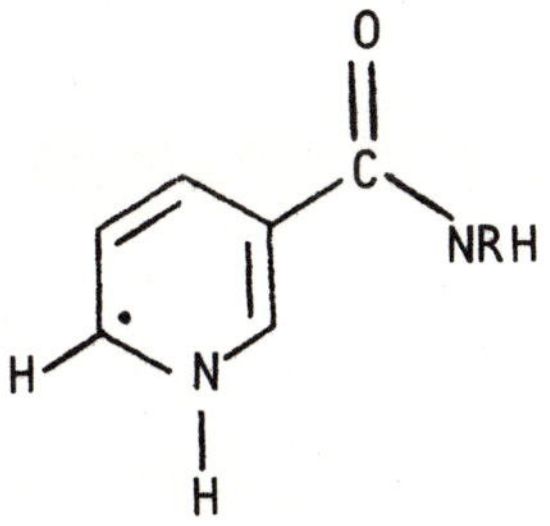

2 III(277-2) → 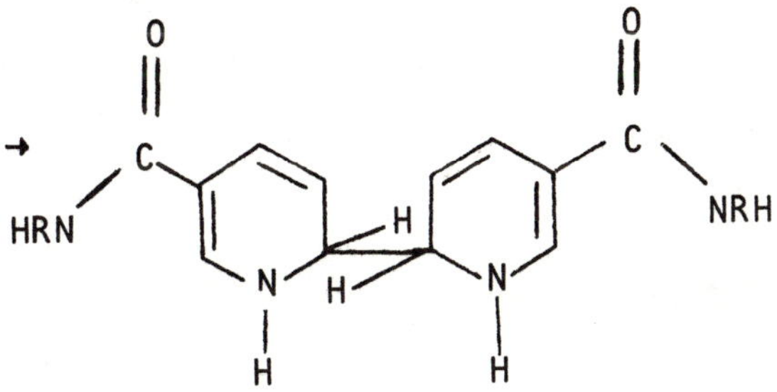 (IV)   277-3

III(277-2) + H⁺ + e → 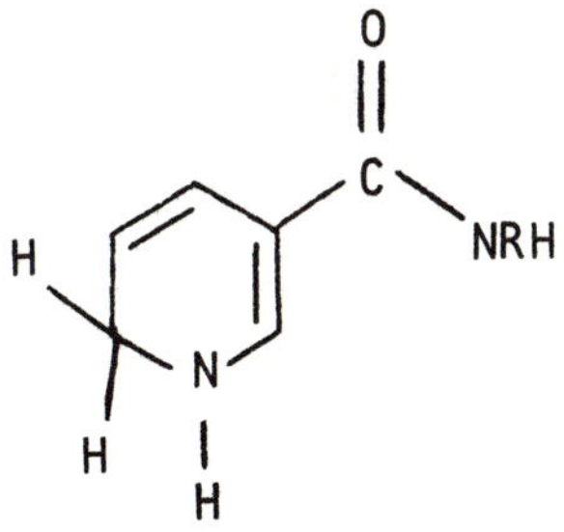 (V)   277-4

II(277-1) + H⁺ + e ⇌ III(277-2)   277-5

IV(277-3) → II(277-1) + 2H⁺ + 2e   277-6

II(277-1) + e ⇌ [structure: 1,2-dihydropyridine with C(=O)NRH at 5-position, H at 2-position] (VI)   277-7

2 VI(277-7) $\xrightarrow{k_8}$ [structure: dimer HRN-C(=O)-pyridyl-CH-CH-pyridyl-C(=O)-NRH] (VII)   277-8

VII(277-8) + 2H$^+$ → IV(277-3)   277-9

VI(277-7) + e → [structure: pyridine-3-carboxamide dianion]$^{2-}$ (VIII)   277-10

VIII(277-10) + 2H$^+$ → V(277-4)   277-11

V(277-4) → II(277-1) + 2H$^+$ + 2e   277-12

II(277-1) + 2H$^+$ + e → [structure: 1,4-dihydropyridinium with C(=O)NH$_2$ at 3-position, NH$_2^+$] (IX)   277-13

In 277-14, R = H

IX (277-13) + H$^+$ + e → [structure: 4H-nicotinamide with N$^+$H$_2$] 277-14

VII (277-8) → 2 II (277-1) + 2e 277-15

VI (277-7) + H$^+$ → III (277-2) 277-16

Code No. FA44, $k_8 = 3.5E4$ dm^{3}mol^{-1}s^{-1}

283

Mechanism: Proposed for Compound:
283a: 283-1, and -3 followed by 65c (acidic) FA62

283c: 283-2 and -4 to -7 followed by 65k (basic) FA62

283e: 283-9, and -4 to -7 followed by 65c or k FA21

$R^1C(:OH^+)CHR^2 \rightleftarrows R^1COCHR^2 + H^+$ 283-1
    | |
    X   X

$R^1COCHR^2 + B \rightleftarrows R^1CO\bar{C}R^2 + BH^+$ 283-2
    | |
    X   X

$R^1C(:OH^+)CHR^2 + 2e → R^1C=CHR^2 + X^-$ 283-3
    | |
    X   OH

$R^1COCHR^2 + 2e → R^1CO\bar{C}HR^2 + X^-$ 283-4
    |
    X

$R^1COCH_2R^2 \rightleftarrows R^1CO\bar{C}HR^2 + H^+$ 283-5

$R^1C(O^-)=CHR^2 \leftrightarrow R^1CO\bar{C}HR^2$ 283-6

$R^1C(OH):CHR^2 \rightleftarrows R^1C(O^-)=CHR^2 + H^+$ 283-7

In 283-8, X = NH$_2$

$$R^1COCHR^2 + 2H^+ + 2e \rightarrow R^1COCH_2R^2 + NH_3 \qquad 283\text{-}8$$
$$\underset{NH_2}{|}$$

$$R^1C(:OH)_2CHR^2 \rightleftarrows R^1COCHR^2 + H_2O \qquad 283\text{-}9$$
$$\underset{X}{|} \qquad\qquad \underset{X}{|}$$

$R^1 = COOC_2H_5$

$R^2 = H$

$X = Br, (CH_3)_2S^+$

328

Mechanism: Proposed for Compound:

328b: 328-1, -2 and -11   FF53

328c: 328-1 to -3   FF55

$$MP + e \rightleftarrows MP^{\cdot -} \qquad\qquad 328\text{-}1$$

$$MP^{\cdot -} + e \rightleftarrows MP^{2-} \qquad\qquad 328\text{-}2$$

$$MP^{2-} + aH^+ + ne \rightarrow MPH_a^{(2+n-a)-} \qquad\qquad 328\text{-}3$$

$$MPH_a^{(2+n-a)-} + bH^+ + n_2e \rightarrow MPH_{a+b}^{(2+n+n_2-a-b)-} \qquad\qquad 328\text{-}4$$

$$MP^{2-} + H^+ \rightleftarrows MPH^- \qquad\qquad 328\text{-}5$$

$$MPH^- \rightarrow MP + H^+ + 2e \qquad\qquad 328\text{-}6$$

$$MPH^- + e \rightleftarrows MPH^{2-} \qquad\qquad 328\text{-}7$$

$$MPH^{2-} + cH^+ + e \rightarrow MPH_{c+1}^{(3-c)-} \qquad\qquad 328\text{-}8$$

$$MPH_{c+1}^{(3-c)-} \rightarrow MPH_{c+1-d}^{(3-c+d-n_3)} + dH^+ + n_3e \qquad\qquad 328\text{-}9$$

$$MPH_{c+1-d}^{(3-c+d-n_3)-} \rightarrow MPH^- + (c-d)H^+ + (2-n_3)e \qquad\qquad 328\text{-}10$$

$$MP^{2-} + e \rightleftarrows MP^{3-} \qquad\qquad 328\text{-}11$$

$M = Mg^{2+}$

P = protoporphine(IX) dimethylester or mesoporphine(IX) dimethyl ester

$n = 3$ in 328-3 for Code No. FF55

329

Mechanism:
3291: 329-23

329m: 329-24

Proposed for Compound:
FF44, FF58, FF69, FF71,

FF66, FF77

$$\begin{bmatrix} X \\ | \\ M(III)P \\ | \\ S \end{bmatrix}^{o} + e \rightleftarrows \begin{bmatrix} X \\ | \\ M(II)P \\ | \\ S \end{bmatrix}^{-} \qquad 329\text{-}1$$

$$\begin{bmatrix} S \\ | \\ M(III)P \\ | \\ S \end{bmatrix}^{+} + e \rightleftarrows \begin{bmatrix} S \\ | \\ M(II)P \\ | \\ S \end{bmatrix}^{o} \qquad 329\text{-}2$$

$$\begin{bmatrix} S \\ | \\ M(III)P \\ | \\ S \end{bmatrix}^{+} + 2L + e \rightleftarrows \begin{bmatrix} L \\ | \\ M(II)P \\ | \\ L \end{bmatrix}^{o} + 2S \qquad 329\text{-}3$$

$$\begin{bmatrix} L \\ | \\ M(III)P \\ | \\ L \end{bmatrix}^{+} + e \rightleftarrows \begin{bmatrix} L \\ | \\ M(II)P \\ | \\ L \end{bmatrix}^{o} \qquad 329\text{-}4$$

$$\begin{bmatrix} L \\ | \\ M(II)P \\ | \\ L \end{bmatrix}^{+} + e + 2Z \rightleftarrows \begin{bmatrix} Z \\ | \\ M(II)P \\ | \\ Z \end{bmatrix}^{-} + 2L \qquad 329\text{-}5$$

$$\begin{bmatrix} Z \\ | \\ M(II)P \\ | \\ Z \end{bmatrix}^{-} + e \rightleftarrows \begin{bmatrix} Z \\ | \\ M(II)P \\ | \\ Z \end{bmatrix}^{2-} \qquad 329\text{-}6$$

$$\begin{bmatrix} X \\ | \\ M(II)P \\ | \\ S \end{bmatrix}^{-} + S \rightleftarrows \begin{bmatrix} S \\ | \\ M(II)P \\ | \\ S \end{bmatrix}^{o} + X^{-} \qquad 329\text{-}7$$

$$\begin{bmatrix} S \\ | \\ M(III)P \\ | \\ S \end{bmatrix}^+ + X^- \rightarrow \begin{bmatrix} X \\ | \\ M(III)P \\ | \\ S \end{bmatrix}^o \qquad 329\text{-}8$$

$$\begin{bmatrix} X \\ | \\ M(II)P \\ | \\ S \end{bmatrix}^- + 2L \rightleftarrows \begin{bmatrix} L \\ | \\ M(II)P \\ | \\ L \end{bmatrix}^o + S + X^- \qquad 329\text{-}9$$

$$\begin{bmatrix} S \\ | \\ M(II)P \\ | \\ S \end{bmatrix}^o + 2L \rightleftarrows \begin{bmatrix} L \\ | \\ M(II)P \\ | \\ L \end{bmatrix}^o + 2S \qquad 329\text{-}10$$

$$\begin{bmatrix} S \\ | \\ M(III)P \\ | \\ S \end{bmatrix}^+ + 2L \rightleftarrows \begin{bmatrix} L \\ | \\ M(III)P \\ | \\ L \end{bmatrix}^+ + 2S \qquad 329\text{-}11$$

$$\begin{bmatrix} X \\ | \\ M(III)P \\ | \\ S \end{bmatrix}^o + 2L \rightleftarrows \begin{bmatrix} L \\ | \\ M(III)P \\ | \\ L \end{bmatrix}^+ \qquad 329\text{-}12$$

$$\begin{bmatrix} L \\ | \\ M(III)P \\ | \\ L \end{bmatrix}^+ + 2S + e \rightleftarrows \begin{bmatrix} L \\ | \\ M(II)P \\ | \\ L \end{bmatrix}^o + 2S \qquad 329\text{-}13$$

$$\begin{bmatrix} S \\ | \\ M(II)P \\ | \\ S \end{bmatrix}^o + e + 2Z \rightleftarrows \begin{bmatrix} Z \\ | \\ M(I)P \\ | \\ Z \end{bmatrix}^- + 2S \qquad 329\text{-}14$$

$$\begin{bmatrix} L \\ | \\ M(III)P \\ | \\ L \end{bmatrix}^+ + S + e \rightleftarrows \begin{bmatrix} S \\ | \\ M(II)P \\ | \\ L \end{bmatrix}^o + L \qquad 329\text{-}15$$

TABLE III. Courses and Mechanisms of Half-Reactions

$$\begin{bmatrix} S \\ | \\ M(II)P \\ | \\ L \end{bmatrix}^{0} + e + 2Z \rightleftarrows \begin{bmatrix} Z \\ | \\ M(I)P \\ | \\ Z \end{bmatrix}^{-} + L + S \qquad 329\text{-}16$$

$$\begin{bmatrix} Z \\ | \\ M(III)P \\ | \\ Z \end{bmatrix}^{+} + e \rightleftarrows \begin{bmatrix} Z \\ | \\ M(II)P \\ | \\ Z \end{bmatrix}^{0} \qquad 329\text{-}17$$

$$\begin{bmatrix} Z \\ | \\ M(II)P \\ | \\ Z \end{bmatrix}^{0} + e \rightleftarrows \begin{bmatrix} Z \\ | \\ M(I)P \\ | \\ Z \end{bmatrix}^{-} \qquad 329\text{-}18$$

$$\begin{bmatrix} Z \\ | \\ M(I)P \\ | \\ Z \end{bmatrix}^{-} + e \rightleftarrows \begin{bmatrix} Z \\ | \\ M(I)P \\ | \\ Z \end{bmatrix}^{2-} \qquad 329\text{-}19$$

$$\begin{bmatrix} X \\ | \\ M(III)P \\ | \\ S \end{bmatrix}^{0} + S + e \rightleftarrows \begin{bmatrix} S \\ | \\ M(II)P \\ | \\ S \end{bmatrix}^{0} + X^{-} \qquad 329\text{-}20$$

$$\begin{bmatrix} Z \\ | \\ M(III)P \\ | \\ Z \end{bmatrix}^{+} \rightleftarrows \begin{bmatrix} Z \\ | \\ M(III)P \\ | \\ Z \end{bmatrix}^{2+} + e \qquad 329\text{-}21$$

$$\begin{bmatrix} Z \\ | \\ M(III)P \\ | \\ Z \end{bmatrix}^{2+} \rightleftarrows \begin{bmatrix} Z \\ | \\ M(III)P \\ | \\ Z \end{bmatrix}^{3+} + e \qquad 329\text{-}22$$

$$\begin{bmatrix} L^{1} \\ | \\ M(II)P \\ | \\ L^{2} \end{bmatrix} \rightarrow \begin{bmatrix} L^{1} \\ | \\ M(II)P \\ | \\ L^{2} \end{bmatrix}^{+} + e \qquad 329\text{-}23$$

$$\begin{bmatrix} L^{1} \\ | \\ M(II)P \\ | \\ L^{2} \end{bmatrix} \rightarrow \begin{bmatrix} L^{1} \\ | \\ M(III)P \\ | \\ L^{2} \end{bmatrix}^{+} + e \qquad 329\text{-}24$$

M = Ru

P = etioporphryn, octaethylporphine or tetraphenylporphine

L^1 = CO, pyridine or MeCN

L^2 = CO, pyridine, or MeCN

343

Mechanism:
343c: 343-1

Proposed for Compound:
FF45

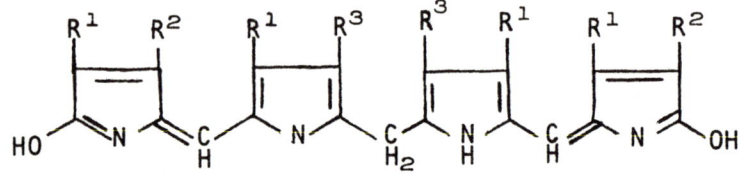

R^1 = CH_3

R^2 = CH_2:CH

R^3 = $HOCCH_2CH_2$
$\|$
O

351

Mechanism:
351f: 351-2, -24 to -27 and -12

351g: 351-1, -2, -13 and -28 to -30

Proposed for Compound:
FF85, FF86, FF87, FF89

FF81, FF83, FF84, FF88

$O \rightarrow O^{\cdot +} + e$ 351-1

$R + e \rightarrow R^{\cdot -}$ 351-2

$O^{\cdot +} + R^{\cdot -} \rightarrow O + {}^3R*$ 351-3

$O^{\cdot +} + {}^3R* \rightarrow O^{\cdot +} + R$ 351-4

$R^{\cdot -} + {}^3R* \rightarrow R^{\cdot -} + R$ 351-5

$2\,{}^3R* \rightarrow {}^1R* + R$ 351-6

$O^{\cdot +} + {}^1R* \rightarrow O^{\cdot +} + R$ 351-7

TABLE III. Courses and Mechanisms of Half-Reactions

$R^{-\cdot} + {}^1R^* \rightarrow R^{-\cdot} + R$ 351-8

${}^3R^* \rightarrow R$ 351-9

${}^1R^* \rightarrow R$ 351-10

${}^1R^* \rightarrow {}^3R^*$ 351-11

${}^1R^* \rightarrow R + h\nu$ 351-12

$O^{+\cdot} + R^{-\cdot} \rightarrow O + {}^1R^*$ 351-13

$O^{+\cdot} + R^{-\cdot} \rightarrow {}^1O^* + R$ 351-14

${}^1O^* \rightarrow O + h\nu$ 351-15

$O^{+\cdot} + R^{-\cdot} \xrightarrow{?} O + R + h\nu$ 351-16

$2\,{}^3R^* \xrightarrow{?} 2R + h\nu$ 351-17

${}^1O^* \rightarrow {}^3O^*$ 351-18

${}^3O^* + R \rightarrow O + {}^3R^*$ 351-19

$O^{+\cdot} + R^{-\cdot} \rightarrow {}^3O^* + R$ 351-20

$2\,{}^3O^* \rightarrow O + {}^1O^*$ 351-21

${}^1O^* + R \rightarrow O + {}^1R^*$ 351-22

${}^3R^* + O \rightarrow R + {}^3O^*$ 351-23

$3X^- = X_3^- + 2e$ 351-24

$R^{-\cdot} + X_3^- \rightarrow RX^{\cdot} + 2X^-$ 351-25

$RX^{\cdot} + R^{-\cdot} \rightarrow RX^{-*} + R$ 351-26

$RX^{-*} \rightarrow {}^1R^* + X^-$ 351-27

$Q \rightarrow [{}^2Q^*]^{+\cdot} + e$ 351-28

${}^1R^* + Q \rightarrow$ quenching   351-29

${}^1R^* + [{}^2Q^*]^{-\cdot} \rightarrow$ quenching   351-30

Code No.	X	O	R	Q	k_{29}	k_{30}
					\multicolumn{2}{c}{$10^{-10} dm^3 mol^{-1} s^{-1}$}	
FF81		naphthalene	naphthalene	Wurster's Blue	—	3.0±0.3
FF83		fluoranthene	fluoranthene	Wurster's Blue	1.99±0.05	2.60±0.06
FF84		pyrene	pyrene	Wurster's Blue	—	2.15±0.05
FF85	Br⁻	—	9,10-diphenyl-anthracene	—		
FF86	I⁻	—	9,10-diphenyl-anthracene			
FF87	I_3^-	—	9,10-diphenyl-anthracene			
FF88		1,2-benz-anthracene	1,2-benz-anthracene	Wurster's Blue	—	2.11±0.15
FF89	Cl⁻	—	9,10-diphenyl-anthracene	—		

356

(Vol IV)

Mechanism:

356a: 356-1, -3 and -4 (pH ≤ 7)

356b: 356-1, -2, -5 and -6 (pH ≥ 9)

356c: 356-1, -2, and -5 to -7 (pH ≥ 9)

Proposed for Compound:
FD96, FD97

FD96

FD97

358

(to replace old 358)

Mechanism:

358a: 358-1, and -5 to -7 (pH < 4)

358b: 358-2 to -4, and -8 (4 < pH < 8)

358c: 358-2 to -4, and -9 (pH > 8)

Proposed for Compound:
FD12

DE78, FD12

FD12

![structure I with OH+, H, R, OH+] (**I**) ⇌ ![structure II diketone with H, R] (II) + 2H+	358-1
II (358-1) ⇌ ![structure III anion with R] (III) + H+	358-2
III (358-2) ⟷ ![structure IV enolate with R] (IV)	358-3
![structure V enol with OH, R] (V) ⇌ IV (358-3) + H+	358-4
I + 2e → ![structure VI diol diradical with H, R] (VI)	358-5
VI (358-5) + 2e → ![structure VII dianion diol with H, R] (VII)	358-6

[Structure of VIII: indane-1,3-diol with R at position 2] (VIII) ⇌ VII(358-6) + 2H⁺   358-7

II(358-1) + 4H⁺ + 4e → VIII(358-7)   358-8

III(358-2) + 4H⁺ + 4e → [structure: indane with O⁻, OH, and R]   358-9

Code No.	R	pK_2
FD12	$O-\underset{\underset{O}{\|\|}}{C}-C_6H_5$	5.15

371

Mechanism: Proposed for Compound:
371d: 371-12, -19, -20 and -7 FB25

371e: 371-12, 21 and -22 FB25

[Structure: phenothiazine-type compound with S and X bridges] (I) + M ⇌ [same structure]$^{+}$ M$^{\cdot}$ (II) + e   371-1

TABLE III. Courses and Mechanisms of Half-Reactions

II(371-1) + H$_2$O $\rightleftarrows$ [structure with OH on S, X bridge] M·(III) + H$^+$ 371-2

III(371-2) $\rightarrow$ [structure with S=O, X bridge] (IV) + H$^+$ + M + e   371-3

2 II(371-1) $\xrightarrow{slow}$ [structure with S, X bridge]$^{2+}$ (V) + I(371-1) + 2M   371-4

V(371-4) + H$_2$O $\rightleftarrows$ IV(371-3) + 2H$^+$ 371-5

II(371-1) + H$_2$O $\rightarrow$ [structure with S=O and OH$^+$, X bridge] (VI) + M + H$^+$ + e   371-6

VI(371-6) $\rightleftarrows$ IV(371-3) + H$^+$ 371-7

II(371-1) $\rightarrow$ V(371-4) + M + e   371-8

Cl$^-$ + Pt $\rightarrow$ ClPt + e   371-9

ClPt + I(371-1) $\rightleftarrows$ [structure with S, X bridge] Cl$^+$ (VII) + Pt + e   371-10

VII(371-10) + H$_2$O $\rightarrow$ VI(371-6) + H$^+$ + Cl$^-$ 371-11

$I(371-1) \rightarrow$ [structure: phenothiazine-type with $^+S^\bullet$ and X bridge] $(VIII) + e$ 371-12

$VIII(371-12) \underset{}{\overset{slow}{\rightleftharpoons}}$ [structure with S, X, and radical on ring with H's] $(IX) + H^+$ 371-13

$VIII(371-12) + IX(371-13) \xrightarrow{fast}$ [dimeric structure with S$^+$ linkage] (X) 371-14

$VIII(371-12) \rightarrow$ [structure with carbocation on ring] $(XI) + H^+ + e$ 371-15

$I(371-1) + XI(371-15) \rightarrow X(371-14)$ 371-16

$VIII(371-12) + I(371-1) \rightarrow$ [structure with S$^\bullet$ and cation bridge] (XII) 371-17

$XII(371-17) \rightarrow X(371-14) + H^+ + e$ 371-18

VIII (371-12) + H₂O → [structure: thianthrene with OH on S] (XIII) + H⁺   371-19

VIII (371-12) + XIII (371-19) → I(371-1) + VI(371-6)   371-20

VIII (371-12) → [structure: thianthrene dication] (XIV) + e   371-21

2 VIII (371-12) ⇌ I(371-1) + XIV (371-21)   371-22

X = S

FB25	Solvent	$K_{22}(10^{12})$
	MeCN	2300
	EtCN	16000
	2-PrCN	76000
	C_6H_5CN	4900
	$MeNO_3$	1000
	$C_6H_5NO_2$	2300
	CH_2Cl_2	3300
	CH_2Cl_2 88 / F_3CCOOH 2 / $(F_3CCO)_2O$ 10	68
	F_3CCOOH 90 / $(F_3CCO)_2O$ 10	2

374

Mechanism:
374b: 374-2, -6, -7 and -12

374c: 374-1 to -5, and -8 to -12

Proposed for Compound:
FF82

FF28, FF29, FF50, FF51, FF90, FF91, FF92, FF93

$RuL_3^{2+} \rightleftarrows RuL_3^{3+} + e$ 374-1

$RuL_3^{2+} + e \rightleftarrows RuL_3^{+}$ 374-2

$RuL_3^{+} + e \rightleftarrows RuL_3$ 374-3

$RuL_3 + e \rightleftarrows RuL_3^{-}$ 374-4

$RuL_3^{+} + RuL_3^{3+} \rightarrow RuL_3^{2+*} + RuL_3^{2+}$ 374-5

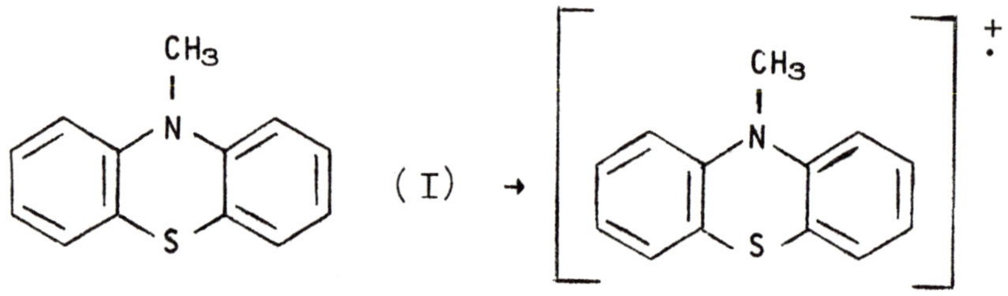

(I) → (II) + e   374-6

(N-methylphenothiazine → N-methylphenothiazine radical cation)

$II(374\text{-}6) + RuL_3^{+} \rightarrow RuL_3^{2+*} + I(374\text{-}6)$ 374-7

$RuL_3^{-} + e \rightarrow RuL_3^{2-}$ 374-8

$RuL_3^{2-} \rightarrow Ru^{2-} + 3L$ 374-9

$RuL_3 + RuL_3^{3+} \rightarrow RuL_3^{+} + RuL_3^{2+*}$ 374-10

$RuL_3^{-} + RuL_3^{3+} \rightarrow RuL_3 + RuL_3^{2+*}$ 374-11

$RuL_3^{2+*} \rightarrow RuL_3^{2+} + h\nu$ 374-12

note: RuL_3^{+} can react to produce one ecl species

RuL_3 can react twice to produce two ecl species

RuL_3^{-} can react three times to produce three ecl species

L = 2,2'-bipyridine, 1,10-phenanthroline, 2,2',2''-terpyridine or 2,4,6-tripyridyl-s-triazine

381

(to replace old 381)

Mechanism:		Proposed for Compound:
381a:	381-1, -2, -9, and -30 (pH < 1)	FA72
381b:	381-2, -3, -4, -10, and -11, followed by 381e or f (pH=1-8.3)	EC92, FA72
381c:	381-4 to -8, and -12, followed by 381g (pH=8.3-11)	EC92, FA72
381d:	381-4, -7, -8, and -13, followed by 381g (pH > 11)	EC92, FA72
381e:	381-11, -17, and -22 to -25, -27, and -30 (pH < 5)	product of 381b (FA72), FA73
381f:	381-11, -14 to -16, -19 to -24, -28, and -31 (pH=5-8)	product of 381b (FA72), FA73
381g:	381-11, -14 to -16, -18 to -21, -26, -29, and -31 (pH > 9)	product of 381c or d (FA72), FA73

[Structure: indane-1,2,3-trione with OH⁺ groups on C1 and C2] (I) ⇌ [indane-1,2,3-trione with H⁺O on C1] (II) + H⁺   381-1

II(381-1) ⇌ [indane-1,2,3-trione] (III) + H⁺   381-2

[2,2-dihydroxyindane-1,3-dione with H⁺ on C1 carbonyl] (IV) ⇌ II(381-1) + H₂O   381-3

[2,2-dihydroxyindane-1,3-dione] (V) ⇌ III(381-2) + H₂O   381-4 [structure: 2,2-dihydroxy(OH, OH2+)-indane-1,3-dione] (VI) ⇌ V(381-4) + H⁺   381-5

VI(381-5) ⇌ III(381-2) + H₃O⁺   381-6

V(381-4) ⇌ [structure: 2-(O⁻)(OH)-indane-1,3-dione] (VII) + H⁺   381-7

VII(381-7) ⇌ III(381-2) + OH⁻   381-8

I(381-1) + 2H⁺ + 4e → [structure: 2,3-dihydroxy-indan-1-one] (VIII)   381-9

II + H⁺ + 2e → [structure: 3-hydroxy-indane-1,2-dione] (IX)   381-10

IX(381-10) + H₂O ⇌ [structure: 2,2,3-trihydroxy-indan-1-one] (X)   381-11

III(381-2) + 2H⁺ + 2e → X(381-11)   381-12

VII(381-7) + 2H⁺ + 2e → (XI)   381-13

IX(381-10) ⇌ (XII) + H⁺   381-14

XII(381-14) ⟷ (XIII)   381-15

(XIV) ⇌ XIII(381-15) + H⁺   381-16

(XV) ⇌ XIV(381-16) + H⁺   381-17

XIII(381-15) ⇌ (XVI) + H⁺   381-18

XIV(381-16) ⇌ [structure: indane-1,3-dione with O⁻ at position 1, OH at position 2, =O at position 3] (XVII) + H⁺   381-19

XVII(381-19) ⟷ [structure: 2-hydroxy-indane-1,3-dione anion at C2] (XVIII)   381-20

[structure: 2-hydroxy-indane-1,3-dione with H at C2] (XIX) ⇌ XVIII(381-20) + H⁺   381-21

[structure: 1-hydroxy-2-(OH⁺)-indan-3-one with H at C1] (XX) ⇌ IX(381-10) + H⁺   381-22

XX(381-22) ⟷ [structure: 1-hydroxy-H⁺-2-hydroxy-indan-3-one] (XXI)   381-23

XXI(381-23) ⇌ XIV(381-16) + H⁺   381-24

[structure: 1-(HO⁺)-2-hydroxy-indane-1,3-dione with H at C2] (XXII) ⇌ XIX(381-21) + H⁺   381-25

XI(381-13) ⇌ [structure] (XXIII)   381-26

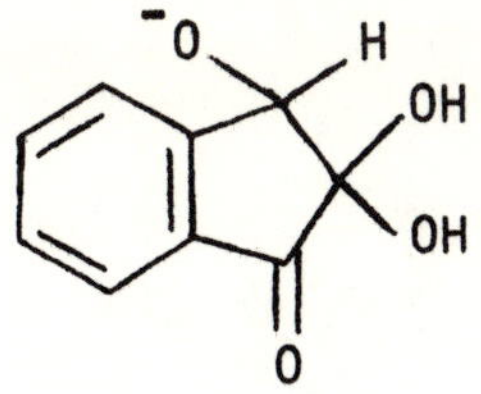

XV(381-17) or XXII(381-25) + H$^+$ + 2e → [structure] (XXIV)

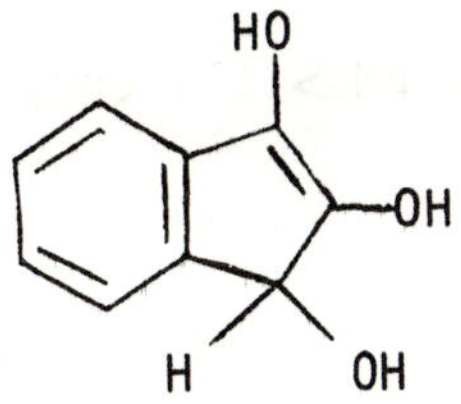

or VIII(381-9)   381-27

XIV(381-16) or XIX(381-21) + 2H$^+$ + 2e → XXIV(381-27) or

VIII(381-9)   381-28

XII(381-14) or XXIII(381-26) + 2H$^+$ + 2e → [structure] (XXV)

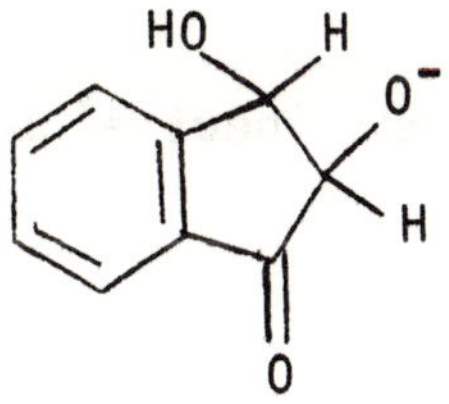

or [structure] (XXVI)   381-29

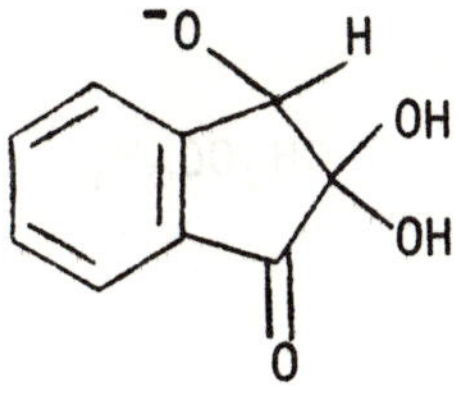

XXIV(381-27) or VIII(381-9) + 2H$^+$ + 2e → [structure]   381-30

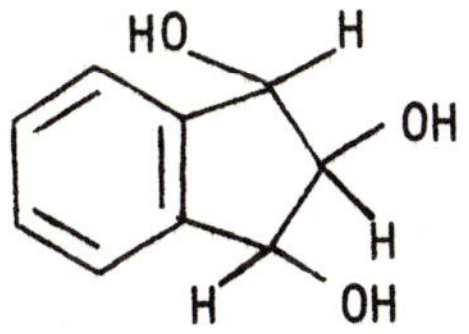

XXV(381-29) + 2H⁺ + 2e → [structure: indane-like bicyclic with HO, H, O⁻, H, OH, H substituents]   381-31

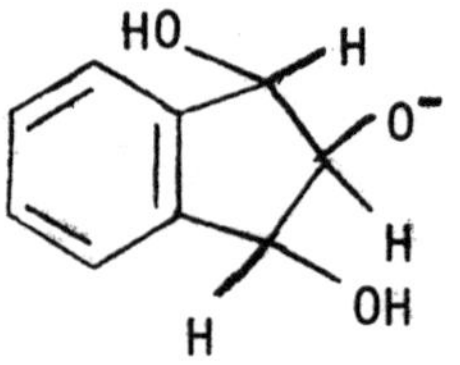

In solutions with pH > 11, competitive cleavage of ninhydrin can occur.

pK_5 8 Ap

pK_7 10

$k_{-11} = 3E-5\ s^{-1}$ at pH = 7

$pK_{16} = 5.4$

$pK_{16}' = 6.2$

$pK_{18} = 11.8$

386

(Vol V)

Mechanism:
386a: 386-1 to -12 (neutral and basic)

Proposed for Compound:
FB38, FB51, FB82, FB99, FC92

$Ar^1NHN:NAr^2$

$Ar^1 = C_6H_5$

$Ar^2 = 4\text{-}ClC_6H_4,\ C_6H_5,\ 4\text{-}NCC_6H_4,\ 4\text{-}CH_3OC_6H_4$, or $4\text{-}C_2H_5OCOC_6H_4$

397

(Vol V)

Mechanism:
397a: 397-1 to -4

Proposed for Compound:
FE86, FE87, FF08, FF49

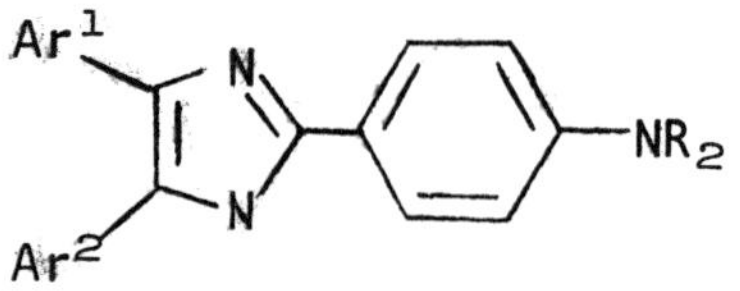

[structure: imidazole with Ar^1, Ar^2 substituents, connected to $C_6H_4\text{-}NR_2$]

$Ar^1 = Ar^2 = 4\text{-}CH_3C_6H_5,\ 4\text{-}CH_3OC_6H_4,\ 4\text{-}(CH_3)_2NC_6H_4,\ 4\text{-}C_6H_5C_6H_4$

$R = CH_3$

402

Mechanism:
402a: 402-4 to -7

402b: 402-1 to -3

Proposed for Compound:
EA33

FA09

$CH_3COCH_3 \rightleftarrows (CH_3COCH_3)_{ads}$ 402-1

$(CH_3COCH_3)_{ads} + H^+ + e \rightarrow [CH_3\dot{C}(OH)CH_3]_{ads}$ 402-2

$[CH_3\dot{C}(OH)CH_3]_{ads} + H^+ + e \xrightarrow{slow} CH_3CHOHCH_3$ 402-3

$CH_3COCH_3 + OH^- \underset{}{\overset{k_3}{\rightleftarrows}} CH_3COCH_2^- + H_2O$ 402-4

$Hg \rightarrow Hg^{2+} + 2e$ 402-5

$CH_3COCH_2^- + Hg^{2+} \rightleftarrows CH_3COCH_2Hg^+$ 402-6

$CH_3COCH_2Hg^+ + OH^- \rightleftarrows CH_3COCH_2HgOH$ 402-7

418

Mechanism:
418a: 418-1 to -4

Proposed for Compound:
FC10

$Ar + e \xrightarrow{k_1} Ar^{\cdot -}$ 418-1

$Ar^{\cdot -} + RX \xrightarrow{k_2} RAr\cdot + X^-$ 418-2

$RAr\cdot + e \rightarrow RAr^-$ 418-3

$RAr^- + RX \xrightarrow{k_4} RArR + X^-$ 418-4

$RAr^- + H_2O \xrightarrow{k_5} RArH + OH^-$ 418-5

Ar = anthracene

RX	k_2, dm^{3}mol^{-1}s^{-1}
CH$_2$:CHCH$_2$Cl	5.8E3
C$_6$H$_5$CH$_2$Cl	2.2E4
C$_6$H$_5$Br	4.5
CH$_3$(CH$_2$)$_3$Br	1.5E3
CH$_3$(CH$_2$)$_3$Cl	2.5E-1
C$_6$H$_5$OH	4.5E3
CH$_2$:CHCN	7.6E2
H$_2$O	3E-2
CH$_2$:C(CH$_3$)CN	35

419

Mechanism:		Proposed for Compound:
419a:	419-1	CG89, Cl50, CK07, CK17, CL05, CL06, CL80, CL81, CL82, CM10, CM39, CM56
419b:	419-1, -3 and -4	CK17
419c:	419-7 to -9 (HClO$_4$ added)	FE74
419d:	419-1, -5 and -6 (unbuffered)	FE74, FF05, FF41
419e:	419-1, -2 and -5	FC54, FF14
419f:	419-8, -1, and -3	FE24
419g:	419-1 and -2	FF65

$$Ar \rightarrow Ar^{+\cdot} + e \qquad 419\text{-}1$$

$$Ar^{+\cdot} \rightarrow Ar^{2+} + e \qquad 419\text{-}2$$

$$2Ar^{+\cdot} \rightleftarrows \text{product(s)} + nH^+ \text{ (exothermic)} \qquad 419\text{-}3$$

$$Ar + Ar^{+\cdot} \rightleftarrows \text{product(s)} + nH^+ \text{ (exothermic)} \qquad 419\text{-}4$$

$$2Ar^{+\cdot} \underset{}{\overset{K_5}{\rightleftarrows}} Ar^{2+} + Ar \qquad 419\text{-}5$$

$$2Ar^{+\cdot} \rightarrow Ar\text{-}Ar^{2+} \qquad 419\text{-}6$$

$$ArH_2^{2+} \rightleftarrows ArH^+ + H^+ \qquad 419\text{-}7$$

$$ArH^+ \rightleftarrows Ar + H^+ \qquad 419\text{-}8$$

$ArH^+ \rightleftarrows Ar^{2+} + H^+ + 2e$ 419-9

Code No	FC54	FE74				FF05	FF14
constant	$K_5(10^7)$	$K_5(10^2)$	$K_6, dm^3 mol^{-1}$	pK_8	pK_9	$K_6(10^2)$	K_5
Solvent							
MeCN	270	6.7	2.85E3(T=-30°C)	5.5	7.5	3.6	190
			7.35E2(T=-20°C)				
EtCN	270						600
2-PrCN	180						1900
C_6H_5CN	56						130
$MeNO_3$	120						410
$C_6H_5NO_2$	17						39
CH_2Cl_2	5.4						8.2
CH_2Cl_2 88 / F_3CCOOH 2 / $(F_3CCO)_2O$ 10	1.1						3.8
F_3CCOOH 90 / $(F_3CCO)_2O$ 10	1.1						2.5

420

Mechanism:
420a: 420-1, -2, -4 and -7
420b: 420-1, -3, -5 and -6
420c: 420-1, -2, and -4

Proposed for Compound:
FE94
BP78, FE94
BP78

$I \rightarrow II + e \quad\quad 420\text{-}1$

$II(420\text{-}1) \rightarrow III + e \quad\quad 420\text{-}2$

$II(420\text{-}1) + Nu \rightarrow IV \quad\quad 420\text{-}3$

$III(420\text{-}2) + Nu \rightarrow V \quad\quad 420\text{-}4$

$II(420\text{-}1) + IV(420\text{-}3) \rightarrow V(420\text{-}4) + I(420\text{-}1) \quad\quad 420\text{-}5$

$V(420\text{-}4) + Nu \rightarrow VI \quad\quad 420\text{-}6$

$V(420\text{-}4) + 2e \rightarrow I(420\text{-}1) + Nu \quad\quad 420\text{-}7$

Ar = C_6H_5

Nu = pyridine, OH^- or CF_3COO^-

421

Mechanism:
421a: 421-1, -5, -7 to -9, and -12 to -16
 ($pH < pK_1$)

421b: 421-1, -2, -5 to -11, -17 and -18
 ($pK_1 < pH < pK_2$)

421c: 421-2 to -6, -10, -11, and -19 to
 -21 ($pK_2 < pH < pK_3$)

Proposed for Compound:

FA06, FA26, FA99

FA06, FA26, FA99

$XCH_2C(:OH^+)COOH \rightleftarrows XCH_2COCOOH + H^+$	421-1
$XCH_2COCOOH \rightleftarrows XCH_2COCOO^- + H^+$	421-2
$XCH_2COCOO^- \rightleftarrows X\overline{C}HCOCOO^- + H^+$	421-3
$X\overline{C}HCOCOO^- \longleftrightarrow XCH:C(O^-)COO^-$	421-4
$XCH_2COCOOH + H_2O \rightleftarrows XCH_2C(OH)_2COOH$	421-5
$XCH_2COCOO^- + H_2O \rightleftarrows XCH_2C(OH)_2COO^-$	421-6
$XCH_2CCOOH \rightleftarrows XCH_2C(OH)_2COOH + H^+$ (with HO, OH_2^+)	421-7
$XCH_2CCOOH \rightleftarrows XCH_2C(:OH^+)COOH + H^+$ (with HO, OH_2^+)	421-8
$XCH_2C(:OH^+)COOH \longleftrightarrow X\overset{+}{C}H_2:C(OH)COOH$	421-9
$XCH_2C(OH)_2COOH \rightleftarrows XCH_2CCOOH + H^+$ (with HO, O^-)	421-10
$XCH_2CCOOH \rightleftarrows XCH_2COCOOH + OH^-$ (with HO, O^-)	421-11
$XCH_2C(:OH^+)COOH + 2e \rightarrow XCH_2\overline{C}(OH)COOH$	421-12
$XCH_2\overline{C}(OH)COOH \rightarrow CH_2:C(OH)COOH + X^-$	421-13
$CH_2:C(OH)COOH \rightleftarrows CH_3COCOOH$	421-14
$XCH_2C(:OH^+)COOH + 2e \rightarrow \overline{C}H_2C(:OH^+)COOH + X^-$	421-15
$\overline{C}H_2C(:OH^+)COOH \rightleftarrows CH_3COCOOH$	421-16
$XCH_2COCOOH + 2e \rightarrow \overline{C}H_2COCOOH + X^-$	421-17
$CH_3COCOOH \rightleftarrows \overline{C}H_2COCOOH + H^+$	421-18
$XCH_2COCOO^- + 2e \rightarrow \overline{C}H_2COCOO^- + X^-$	421-19
$CH_2COCOO^- \rightleftarrows \overline{C}H_2COCOO^- + H^+$	421-20

$\overline{C}H_2COCOO^- \longleftrightarrow CH_2{:}C(O^-)COO^-$ 421-21

Code No.	X	pK_2	pK_3	pK_{10}
FA06	Br	2.3	—	
FA26	$(CH_3)_2S^+$	0.7	8.3	
FA99	$4\text{-}CH_3C_6H_4S$	0.6	12.3	1.9

422

Mechanism:
422a: 422-1 to -7

Proposed for Compound:
FA00, FA02

$RSnCl_3 \rightleftarrows (RSnCl_3)_{ads}$	422-1
$(RSnCl_3)_{ads} \rightleftarrows RCl + SnCl_2$	422-2
$SnCl_2 + 2e \rightarrow Sn + 2Cl^-$	422-3
$RSnCl_3 + 2e \rightarrow RSnCl + 2Cl^-$	422-4
$RSnCl + e \rightarrow RSn + Cl^-$	422-5
$2RSnCl \rightarrow R_2SnCl_2 + Sn$	422-6
$R_2SnCl_2 + 2e \rightarrow R_2Sn + 2Cl^-$	422-7

423

Mechanism:
423a: 423-1 to -5, -8 and -9 (H_o = -2 to pH ≈ 2.5)

Proposed for Compound:
FA34, FA40

423b: 423-6, -7, and -10 to -12 (pH > 3)

FA34, FA40

TABLE III. Courses and Mechanisms of Half-Reactions

(I) ⇌ (II) + H₂O   423-1

II(423-1) ⟷ (III)   423-2

III(423-2) ⇌ (IV) + H⁺   423-3

IV(423-3) ⇌ (V) + H⁺   423-4

(VI) ⇌ V(423-4) + H₂O   423-5

V(423-4) ⇌ (VII) + H⁺   423-6

VII(423-6) ⇌ [structure: cyclohexene ring with O⁻, =O, =O, =O, =O, and HO substituents] (VIII) + H₂O   423-7

I(423-1) + 2H⁺ + 4e → [hexahydroxybenzene structure] (IX) + 2H₂O   423-8

II(423-1) + 2H⁺ + 4e → IX(423-8) + H₂O   423-9

V(423-4) + 2H⁺ + 2e → [structure with HO, =O, =O, OH, HO, HO, H, OH] (X)   423-10

X(423-10) + 2H⁺ + 2e → [structure with =O, H, HO, OH, OH, HO, HO, OH, H] (XI)   423-11

XI(423-11) $\underset{\text{steps}}{\overset{\text{several}}{\rightleftharpoons}}$ IX(423-8) + H₂O   423-12

424

(MeCN)

Mechanism:
424a: 424-3 to -5

424b: 424-2, -3, -6 and -7

424c: 424-1, -2, -3, -8 and -6

Proposed for Compound:
FA70

FA84, FA86, FC42

FB29, FB44

$$R^1R^2NOH + B \rightleftarrows R^1R^2NO^- + BH^+ \qquad 424\text{-}1$$

$$R^1R^2NO^- \rightleftarrows R^1R^2NO\cdot + e \qquad 424\text{-}2$$

$$R^1R^2NO\cdot \rightleftarrows R^1R^2NO^+ + e \qquad 424\text{-}3$$

$$R^1R^2NO^+ + R^1R^2NO\cdot \rightarrow (R^1R^2NO)_2^{+\cdot} \qquad 424\text{-}4$$

$$(R^1R^2NO)_2^{+\cdot} + R^1R^2NO\cdot \rightarrow \text{products} \qquad 424\text{-}5$$

$$R^1R^2NOH \rightleftarrows R^1R^2NO^- + H^+ \qquad 424\text{-}6$$

$$R^1R^2NOH + 2H^+ + 2e \rightarrow R^1R^2NH + H_2O \qquad 424\text{-}7$$

$$2R^1R^2NO\cdot \rightleftarrows R^1R^2NO^- + R^1R^2NO^+ \qquad 424\text{-}8$$

$R^1 = R^2 = (CH_3)_3C,\ 4\text{-}CH_3OC_6H_4$

$R^1 = 4\text{-}\cdot ONC_6H_4$

$R^2 = 4\text{-}ClC_6H_4,\ C_6H_5$

or $R^1R^2NO\cdot$ =

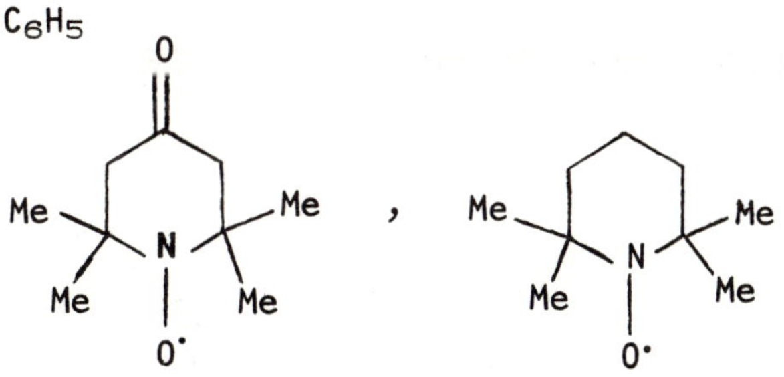

[structure: 2,2,6,6-tetramethyl-4-oxopiperidine-1-oxyl], [structure: 2,2,6,6-tetramethylpiperidine-1-oxyl]

425

(MeCN)

Mechanism:
425a: 425-1 to -6 (in presence of base)

425b: 425-2 to -6 (neutral)

Proposed for Compound:
FB86

FB86

$$\underset{\underset{OH}{|}}{Ar^1CONAr^2} + B \rightleftarrows \underset{\underset{O\cdot}{|}}{Ar^1CONAr^2} + BH^+ + e \qquad 425\text{-}1$$

$$Ar^1CONAr^2 \rightarrow Ar^1\overset{+}{CON}Ar^2 + e \qquad\qquad 425\text{-}2$$
$$ | |$$
$$ OH OH$$

(eq. 425-2) Ar¹CONAr² with OH → Ar¹CON⁺Ar² with OH + e

(eq. 425-3) Ar¹CON⁺Ar² with OH ⇌ Ar¹CONAr² with O• + H⁺

(eq. 425-4) Ar¹CONAr² with OH → Ar¹CON⁺Ar² with =O + H⁺ + 2e

(eq. 425-5) Ar¹CON⁺Ar² with =O → Ar¹CO⁺ + Ar²NO

(eq. 425-6) Ar¹CO⁺ + H₂O → ArCOOH + H⁺

$Ar^1 = Ar^2 = C_6H_5$

426

(DMF)

Mechanism:	Proposed for Compound:
426a: 426-1 and -2	FF11
426b: 426-1, -3 and -4 (proton donor present)	FF11

(I) + e ⇌ (II) 426-1

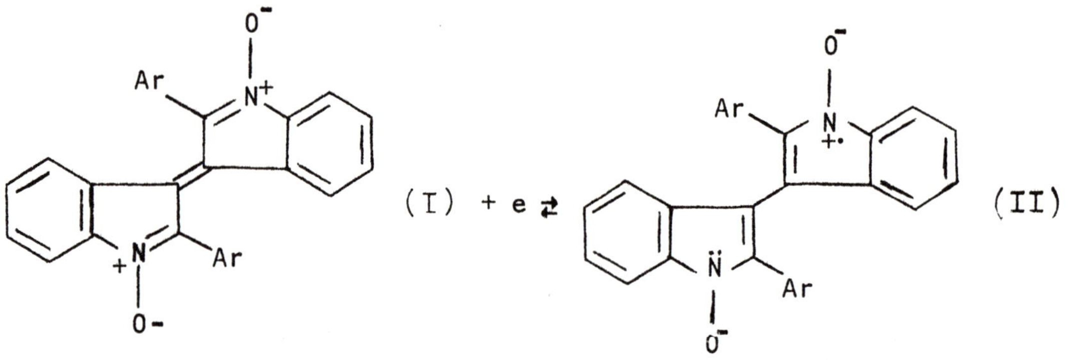

II(426-1) + e ⇌ (III) 426-2

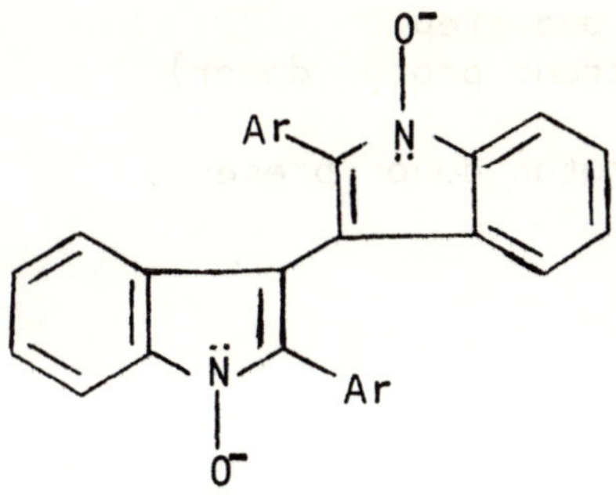

II(426-1) + HA ⇌ (IV) + A⁻ 426-3

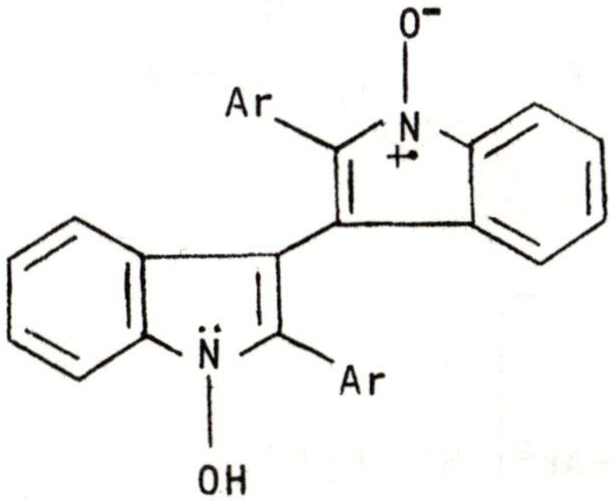

2 IV(426-3) ⇌ I(426-1) + 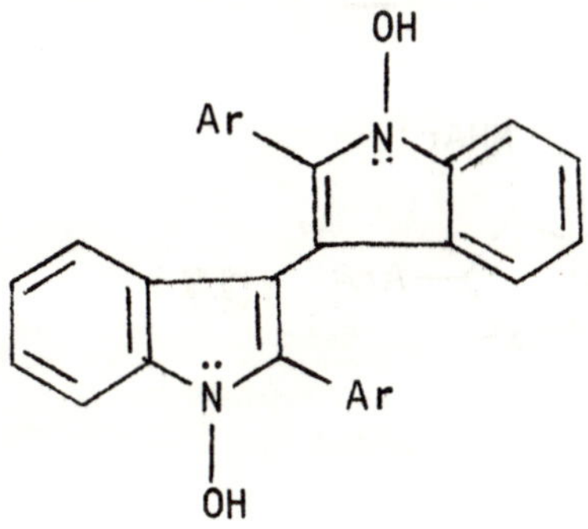 426-4

Ar = C₆H₅

427

Mechanism:	Proposed for Compound:
427a: 427-1, -2, and -3 and -4 as one step (strong proton donor present)	FD81
427b: 427-5 and -6 (no proton donor)	FE02, FE04, FE07, FE20, FE21, FE37
427c: 427-5 to -7 (weak proton donor or solvent)	FC12, FD80, FE01, FE03, FE06, FE18, FE19, FE36
427d: 427-5, -8, and -3 (weak proton donor)	FC12, FD80, FE01, FE03, FE06, FE18, FE19, FE36

427e: 427-1, -2, -3 and -4 as one step,　　　　　　FC12, FE01, FE03, FE06,
　　　　-5, -8, -3, and -4 (strong proton donor)　　FE18, FE19, FE36

427f: 427-5, -8, -3 and -6 (proton donor present)　FE02, FE04, FE07, FE20,
　　　　　　　　　　　　　　　　　　　　　　　　　　　　FE21, FE37

$[\text{Ar}^2\text{-indoleNAr}^1]\text{H}^+$ (I) ⇌ (II) + H$^+$　　　　427-1

I(427-1) + e ⇌ [III] H· 　　　　427-2

III(427-2) + e ⇌ (IV)　　　　427-3

(V) ⇌ IV(427-3) + H$^+$　　　　427-4

II(427-1) + e ⇌ (VI)·⁻　　　　427-5

VI(427-5) + e ⇌ (VII)　　　　427-6

IV(427-3) ⇌ VII(427-6) + H$^+$　　　　427-7

III(427-2) ⇌ VI(427-5) + H$^+$　　　　427-8

Proton source may be traces of H_2O, solvent, or a weak or strong proton donor

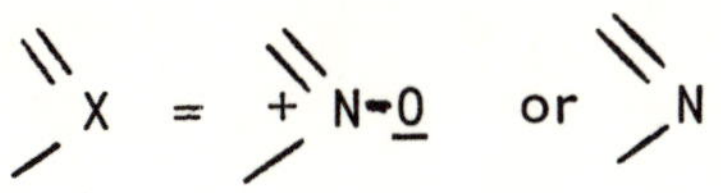

$$\diagup\!\!\!\!\!\diagdown\!\!=\!\!X \ = \ \diagup\!\!\!\!\!\diagdown\!\!=\!\!\overset{+}{N}\text{-}\underline{O} \quad \text{or} \quad \diagup\!\!\!\!\!\diagdown\!\!=\!\!N$$

Ar^1 = H, C_6H_5, 4-$CH_3OC_6H_4$, 4-$CH_3C_6H_4$, 4-$(CH_3)_2NC_6H_4$, 4-ClC_6H_4, 4-BrC_6H_4,

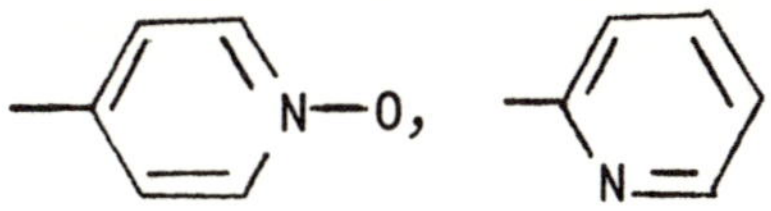

[4-methylpyridine N-oxide], [2-methylpyridine],

Ar^2 = C_6H_5

428

Mechanism:
428a: 428-1 and -2 (no nucleophile added)

428b: 428-1 to -7 (nucleophile added)

Proposed for Compound:
FE63, FE69, FE99, FF00, FF19, FF20

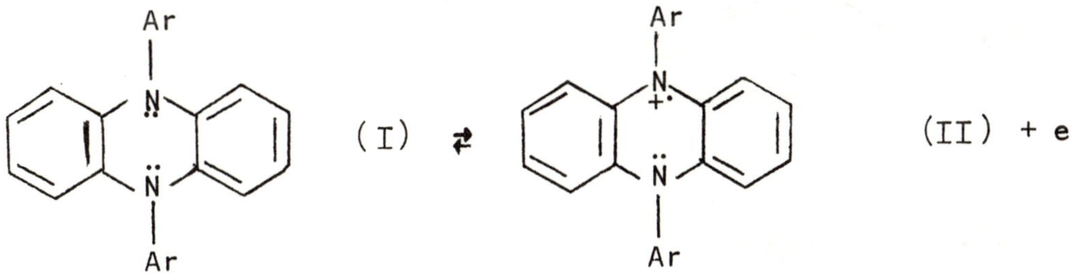

FE63, FE69, FE99, FF00, FF19, FF20

(I) ⇌ (II) + e   428-1

II(428-1) ⇌ (III) + e   428-2

III(428-2) ⟷ (IV)   428-3

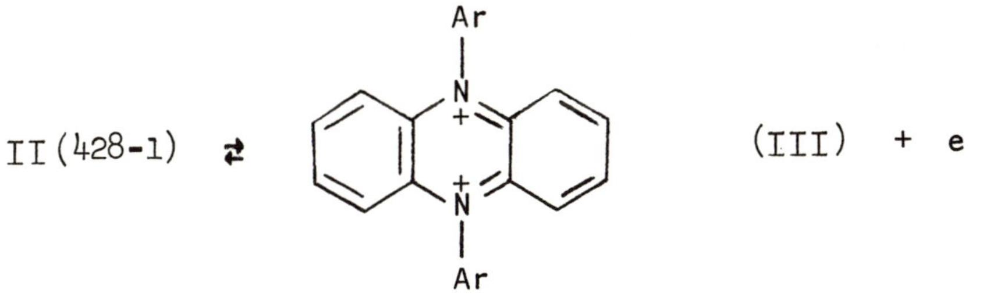

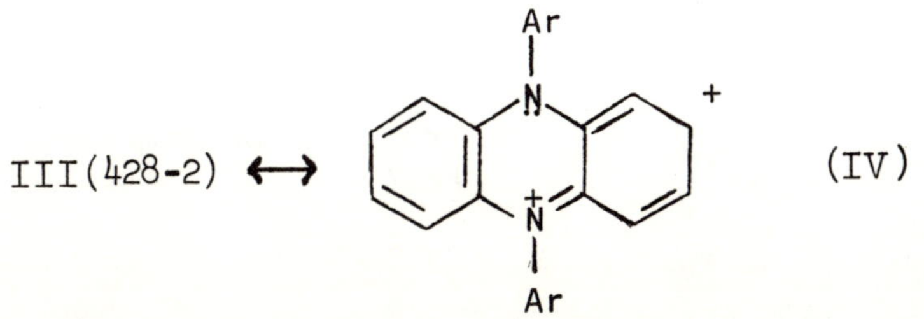

IV(428-3) + 2X⁻ → [structure] (V) + HX   428-4

V(428-4) ⇌ [structure] (VI) + 2e   428-5

VI(428-5) + 2X⁻ → [structure] (VII) + HX   428-6

VII(428-6) ⇌ [structure] + e   428-7

Ar = $4\text{-ClC}_6\text{H}_4$, C_6H_5, or $4\text{-CH}_3C_6H_4$

X⁻ = Cl or CH_3COO^-

In addition CH_3, CH_3O, or Cl can be substituted in the 2, 7 positions on the 5,10-dihydrophenazine

Mechanism:	Proposed for Compound:
429a: 429-1 to -3 (H_2SO_4 = 0.5-2)	FB57
429b: 429-1 (pH = 1-5)	FB57

TABLE III. Courses and Mechanisms of Half-Reactions

$$\text{(I)} \rightleftarrows \text{(II)} + 2H^+ + 2e \qquad 429\text{-}1$$

(I): 1,2,4-triaminobenzene with NHAr substituent
(II): quinone diimine with NAr and NH₂

$$\text{II}(429\text{-}1) + H_2O \rightarrow \text{(III)} + ArNH_2 \qquad 429\text{-}2$$

(III): 2-amino-4-iminocyclohexa-2,5-dien-1-one

$$\text{III}(429\text{-}2) + \text{I}(429\text{-}1) \rightarrow \text{[2,4-diamino-phenol]} + \text{II}(429\text{-}1) \qquad 429\text{-}3$$

430

Mechanism:	Proposed for Compound:
430a: 430-1 and -2 as one step, -3 and -4	FB50, FB98, FC57
430b: 430-1 and -5 (absence of H₂O)	FC58, FC59
430c: 430-1, -5 and -2 (presence of H₂O)	FC58, FC59

$$Ar-N{=}N-\!\!\!\left\langle\!\!\!\bigcirc\!\!\!\right\rangle\!\!\!-N(CH_3)_2 \rightarrow [Ar-N{=}N-\!\!\!\left\langle\!\!\!\bigcirc\!\!\!\right\rangle\!\!\!-N(CH_3)_2]^{2+} + 2e \qquad 430\text{-}1$$

$$[Ar-N{=}N-\!\!\!\left\langle\!\!\!\bigcirc\!\!\!\right\rangle\!\!\!-N(CH_3)_2]^{2+} + H_2O \rightarrow Ar-N{=}N-\!\!\!\left\langle\!\!\!\bigcirc\!\!\!\right\rangle\!\!\!-NHCH_3 + \text{products} \qquad 430\text{-}2$$

$$Ar-N{=}N-\!\!\!\left\langle\!\!\!\bigcirc\!\!\!\right\rangle\!\!\!-NHCH_3 + H_2O \rightarrow Ar-N{=}N-\!\!\!\left\langle\!\!\!\bigcirc\!\!\!\right\rangle\!\!\!-NH_2 + 2e + \text{products} \qquad 430\text{-}3$$

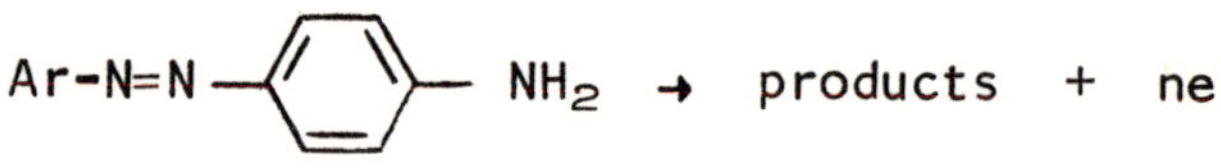

$$Ar-N=N-C_6H_4-NH_2 \rightarrow \text{products} + ne \qquad 430\text{-}4$$

In 430-5, Ar = 2-HOC$_6$H$_5$ or 4-HOC$_6$H$_5$ (structures are shown for 4-HOC$_6$H$_5$ only)

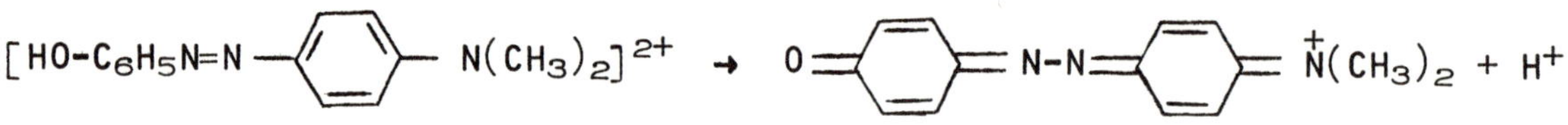

$$[HO\text{-}C_6H_5N=N\text{-}C_6H_4\text{-}N(CH_3)_2]^{2+} \rightarrow O=C_6H_4=N\text{-}N=C_6H_4=\overset{+}{N}(CH_3)_2 + H^+ \qquad 430\text{-}5$$

Ar = C$_6$H$_5$, 2-HOC$_6$H$_4$, 4-HOC$_6$H$_4$

431

Mechanism:
431a: 431-1 and -2

Proposed for Compound:
FE80

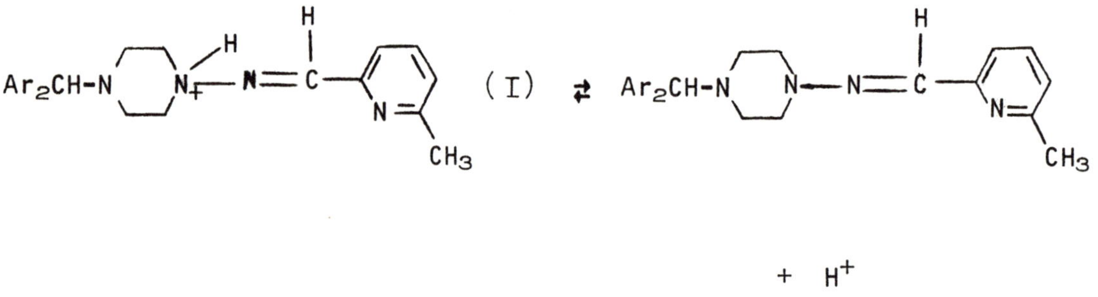

Ar$_2$CH-N(piperazine)N$^+$H—N=CH-(6-methylpyridin-2-yl) (I) ⇌ Ar$_2$CH-N(piperazine)N—N=CH-(6-methylpyridin-2-yl)

$+ H^+$ \qquad 431-1

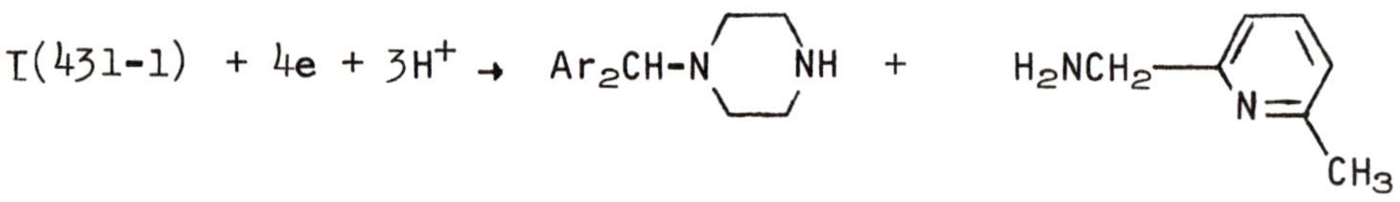

I(431-1) $+ 4e + 3H^+ \rightarrow$ Ar$_2$CH-N(piperazine)NH $+$ H$_2$NCH$_2$-(6-methylpyridin-2-yl) \qquad 431-2

432

Mechanism:		Proposed for Compound:
432a:	432-1 to -4 (pH < 8)	FB66, FC47
432b:	432-5 and -6 (pH > 10)	FC47
432c:	432-5, -7 and -8 (pH > 10)	FB66
432d:	432-3 and -4 (pH < 8)	FB62, FB64, FB65, FC45, FD73
432e:	432-6 (pH > 10)	FC45
432f:	432-7 and -8 (pH > 10)	FB64
432g:	432-9 and -10 (pH < 8)	FB63, FC43
432h:	432-7 (pH > 8)	FD73

$$\underset{\underset{OH}{|}}{\overset{\overset{O}{|}}{R-\overset{+}{N}=N-R}} \text{ (I)} \rightleftarrows \underset{\underset{O}{|}}{\overset{\overset{O}{|}}{R-N=N-R}} \text{ (II)} + H^+ \qquad 432\text{-}1$$

$$I(432\text{-}1) + 2e + H^+ \rightarrow \underset{\text{(cis + trans isomers)}}{\overset{\overset{O}{|}}{R-N=N-R}} \text{ (III)} + H_2O \qquad 432\text{-}2$$

$$\underset{\overset{+}{}}{\overset{\overset{OH}{|}}{R-N=N-R}} \text{ (IV)} \rightleftarrows III(432\text{-}2) + H^+ \qquad 432\text{-}3$$

$$IV(432\text{-}3) + 4e + 3H^+ \rightarrow RNHNHR + H_2O \qquad 432\text{-}4$$

$$II(432\text{-}1) + 2e + 2H^+ \rightarrow III(432\text{-}2) + H_2O \qquad 432\text{-}5$$

$$III(432\text{-}2) + 4e + 4H^+ \rightarrow RNHNHR + H_2O \qquad 432\text{-}6$$

In 432-7, R = cyclo hexyl and only the cis isomer is electroactive

$$\text{cis-}III(432\text{-}2) + 2e + 2H^+ \rightarrow RN=NR + H_2O \qquad 432\text{-}7$$

$$RN=NR + 2e + 2H^+ \rightarrow RNHNHR \qquad 432\text{-}8$$

$$RN=NRH^+ \rightleftarrows RN=NR + H^+ \qquad 432\text{-}9$$

$$RN=NRH^+ + 2e + H^+ \rightarrow RNHNHR \qquad 432\text{-}10$$

433

Mechanism:
433a: 433-1

Proposed for Compound:
FC06

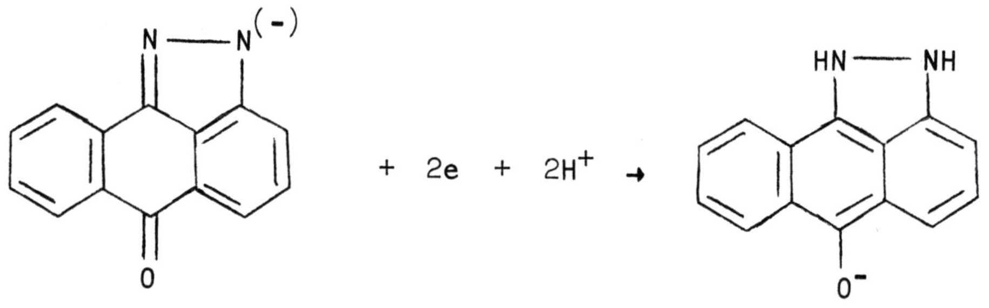

[Structure: dibenzo-fused pyrazole ketone with N=N–N(−)] + 2e + 2H$^+$ → [reduced structure with HN–NH and O$^-$]   433-1

434

(nonaqueous)

Mechanism:
434a: 434-1 to -4

Proposed for Compound:
FB54, FC01, FD24, FD25, FE12, product of 435 (FE79)

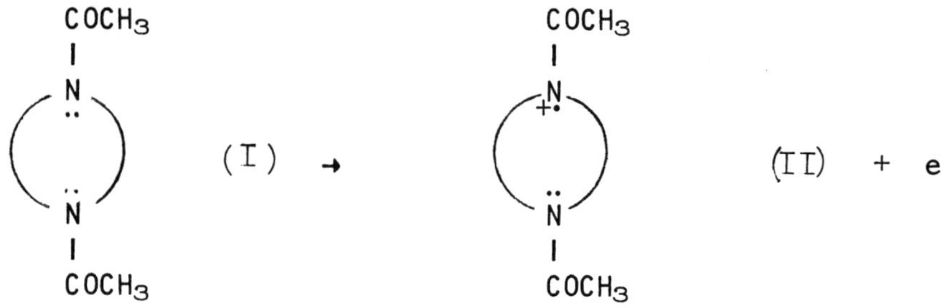

(I) [cyclic diamine with two N–COCH$_3$ groups] → (II) [radical cation] + e   434-1

II(434-1) → (III) [cyclic amine with one N–COCH$_3$] + $^+$COCH$_3$ 434-2

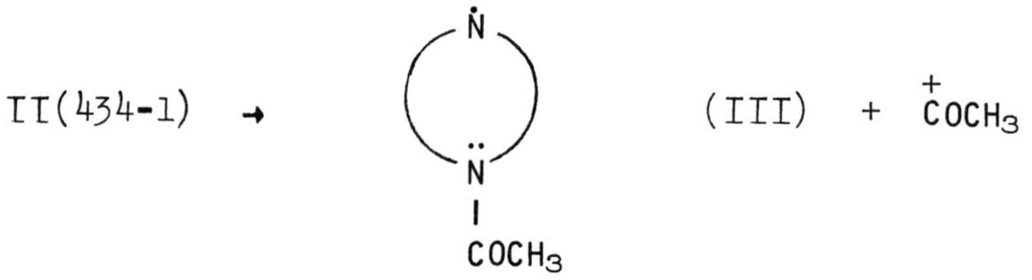

III (434-2) → [ring with N, +•N-COCH3] (IV) + e   434-3

IV (434-3) ⇌ [ring with N, N] (V) + $\overset{+}{C}OCH_3$ 434-4   434-4

I represents a diacetyl derivative of a heterocyclic species containing 2-ring nitrogens in adjacent positions or separated by 2 carbon atoms in one or more six-membered aromatic rings. The final product, V, has one more double bond than I. In I the heterocyclic rings are

[quinoxaline], [cinnoline], [1,10-phenanthroline], [phenazine],

[2,3-diphenylpyrazine], [3,6-diphenylpyridazine],

435

Mechanism:
435a: 435-1 and 2 followed by 434a

Proposed for Compound:
FE79

[Ar-C=C-Ar with N(COCH3)2 groups] → [Ar-C-C-Ar with NCOCH3 groups]   (I) + 2e + 2CH$_3$CO$^+$ 435-1

I(435-1) ⟷ [structure II: Ar-substituted diazine with two N—COCH₃ groups] (II) 435-2

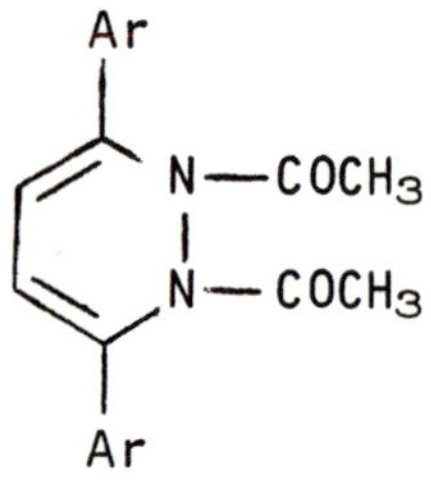

436

Mechanism:
436a: 436-1 and -2 (acidic)
436b: 436-1, -3 and -4 (neutral)

Proposed for Compound:
FF25
FF25, FF27

[pentaphenylcyclopentadienyl–OH⁺] (I) ⇌ [pentaphenylcyclopentadienone] (II) + H⁺   436-1

I(436-1) + 2e → [pentaphenylcyclopentadienyl anion with OH]   436-2

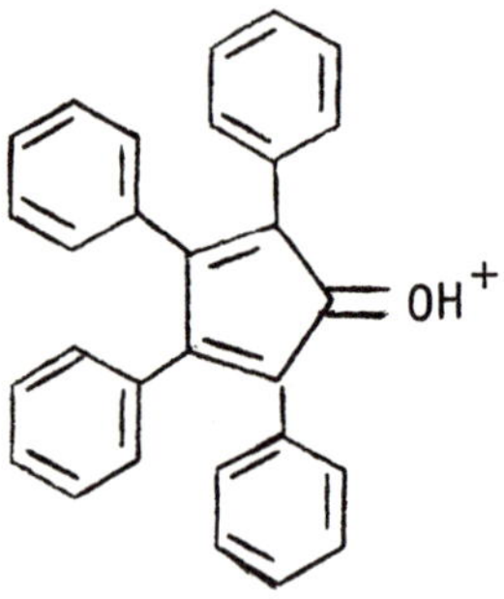

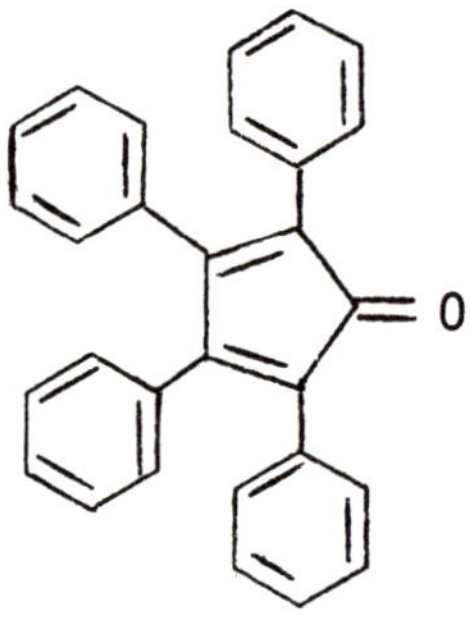

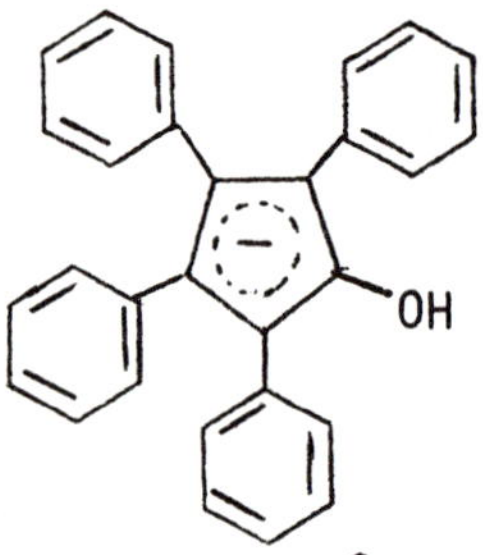

II + 2e + 2H⁺ → [tetraphenylcyclopentanone-type structure] (III)   436-3

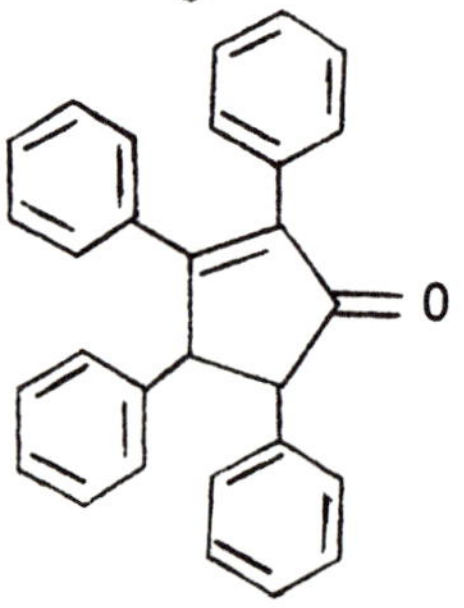

2III(436-3) + 2e + 2H⁺ → dimer   436-4

437

Mechanism:
437a: 437-1 to -8 (reactions -6 to -8 favored by presence of CF_3COOH)
437b: 437-7 and -8
437c: 437-3 and -4

Proposed for Compound:
FC98, FD40, FD60, FD74

FD58

FF48

$Ar^1(CH_2)_nAr^2 \rightarrow [Ar^1(CH_2)_nAr^2]^{+\cdot} + e$ 437-1

$2[Ar^1(CH_2)_nAr^2]^{+\cdot} \rightarrow$ [biphenyl with $(CH_2)_n Ar^2$ substituents] (I) $+ 2H^+$ 437-2

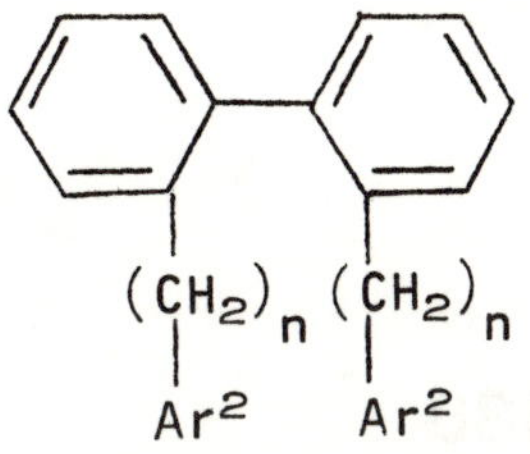

$I(437-2) \rightleftarrows$ [radical cation structure]$^{+\cdot}$ (II) $+ e$ 437-3

$II(437-3) \rightleftarrows$ [dication structure]$^{2+}$ $+ e$ 437-4

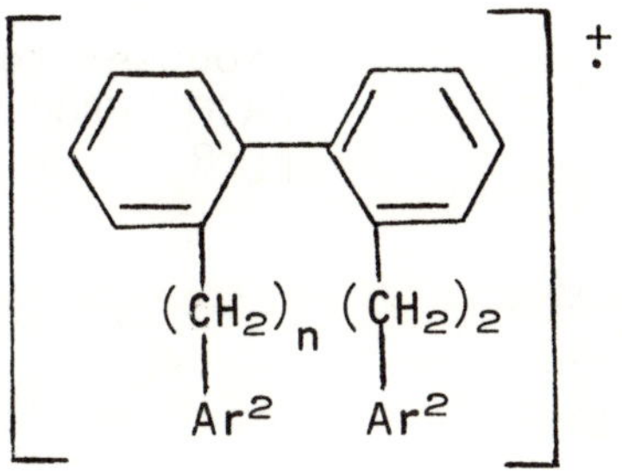

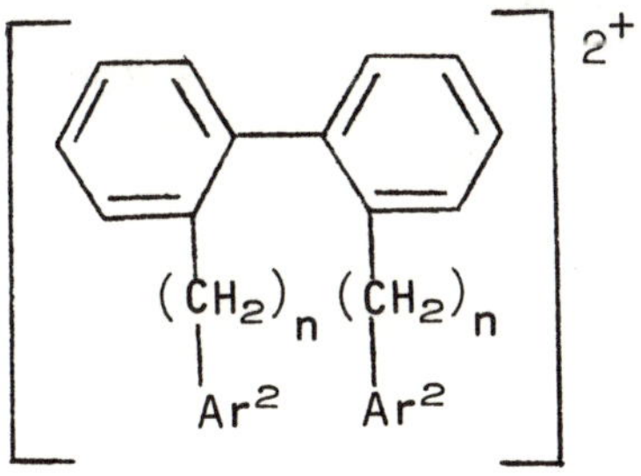

$Ar^1(CH_2)_nAr^2 \rightarrow [Ar^1(CH_2)_nAr^2]^{2+} + 2e$ 437-5

$[Ar^1(CH_2)_nAr^2]^{2+} \rightarrow$ [cyclic $(CH_2)_n$-bridged biphenyl] (III) $+ 2H^+$ 437-6

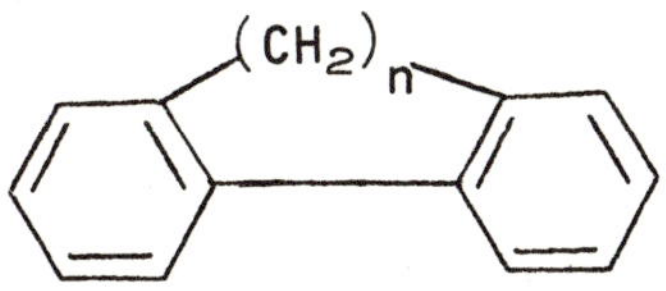

$III(437-6) \rightleftarrows$ [cyclic structure]$^{+\cdot}$ (IV) $+ e$ 437-7

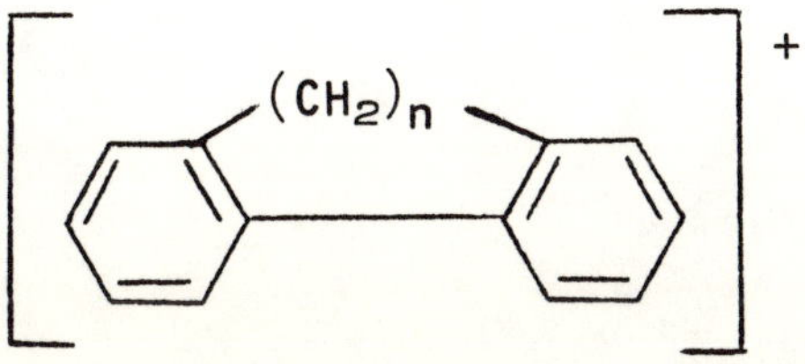

IV(437-7) ⇄ 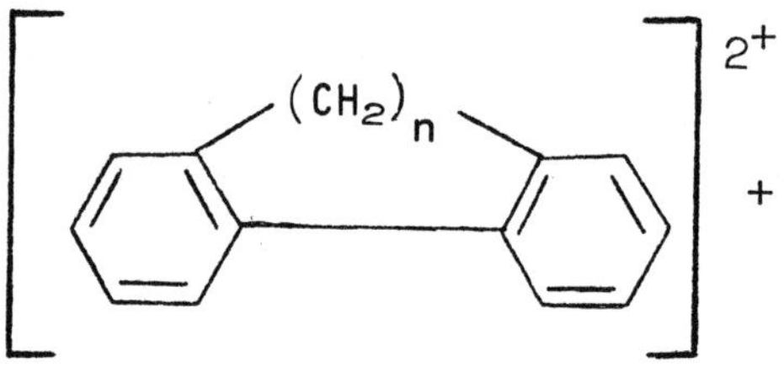 + e 437-8

Ar^1, Ar^2 = 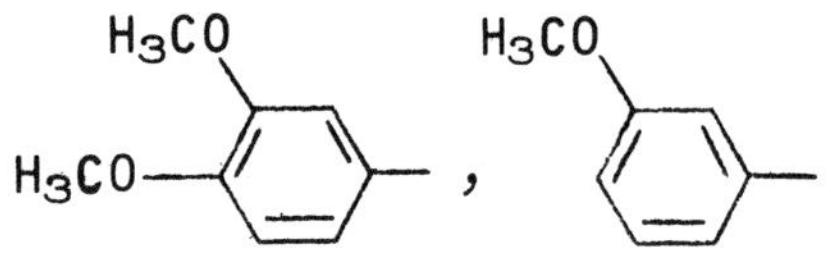

n = 1 or 2

438

Mechanism:
438a: 438-1 to -3

Proposed for Compound:
FA90, FB00, FB60, FD43, FE08

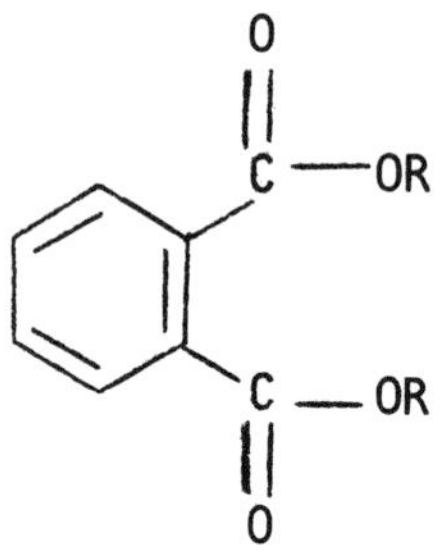 + 2e + $2H^+$ → 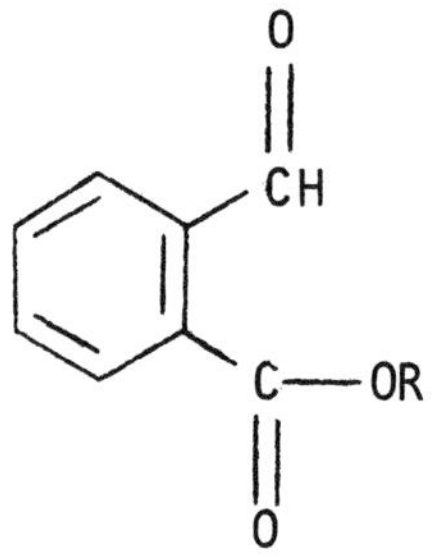 (I) + ROH 438-1

I(438-1) + OH^- ⇄ 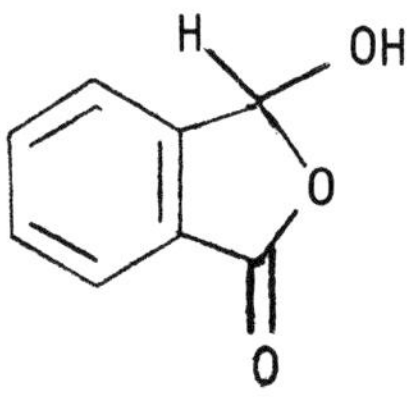 (II) + OR^- 438-2

II(438-2) + 2e + 2H$^+$ → [phthalide structure] + H$_2$O   438-3

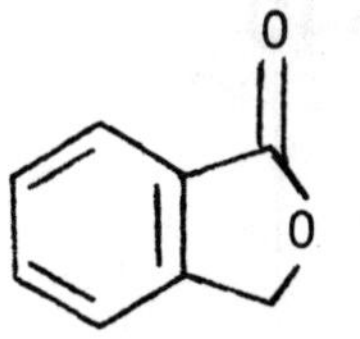

R = CH$_3$, C$_2$H$_5$, CH$_3$(CH$_2$)$_3$-, C$_6$H$_5$

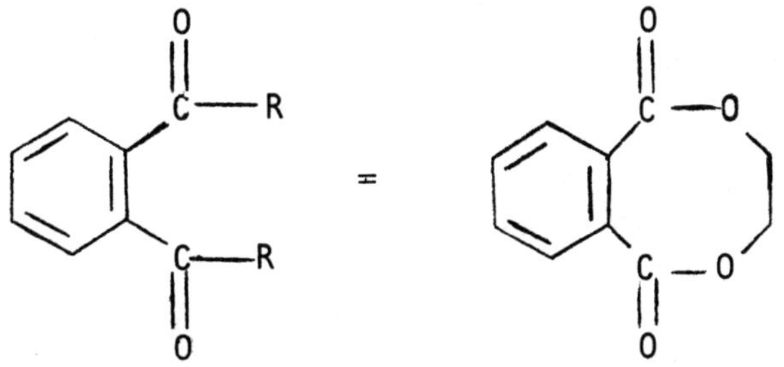

439

Mechanism:
439a: 439-1 to -4 (at Pb)

Proposed for Compound:
FA53

$$ArCOOH + H^+ \rightleftarrows ArCOOH_2^+ \quad\quad 439\text{-}1$$

$$ArCOOH_2^+ + 2e + H^+ \rightarrow ArC(OH)_2H \quad\quad 439\text{-}2$$

$$ArC(OH)_2H + H_3O^+ \rightleftarrows Ar\underset{H}{\overset{OH}{C}}-OH_2^+ + H_2O \quad\quad 439\text{-}3$$

$$Ar\underset{H}{\overset{OH}{C}}-OH_2^+ \rightleftarrows ArCHO + H_3O^+ \quad\quad 439\text{-}4$$

Ar = C$_6$H$_5$

440

Mechanism:
440a: 440-1 to -5 (nonaqueous)

Proposed for Compound:
FD57

$$ArCO(CH_2)_3COAr \; (I) \; + \; e \; \rightleftarrows \; ArCO(CH_2)_3\overset{\cdot}{\underset{\underset{O^-}{|}}{C}}Ar \; (II) \qquad 440\text{-}1$$

$$2II(440\text{-}1) \; \rightleftarrows \; I(440\text{-}1) \; + \; Ar\overset{\cdot}{\underset{\underset{O^-}{|}}{C}}(CH_2)_3\overset{\cdot}{\underset{\underset{O^-}{|}}{C}}Ar \; (III) \qquad 440\text{-}2$$

III(440-2) ⇌ [cyclopentane ring with Ar, Ar, O⁻, O⁻ on adjacent carbons] (IV)   440-3

IV(440-3) + H₂O → [cyclopentane ring with Ar, Ar, HO, O⁻] (V) + OH⁻   440-4

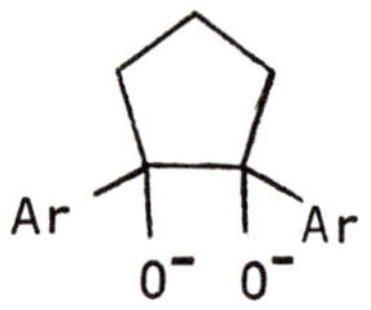

V(440-4) + H₂O → [cyclopentane ring with Ar, Ar, OH, OH] + OH⁻   440-5

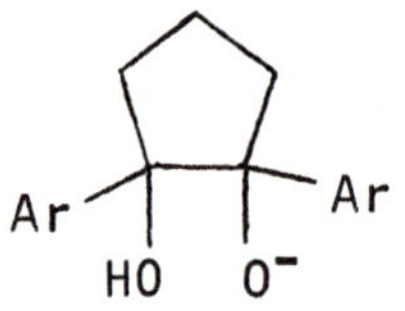

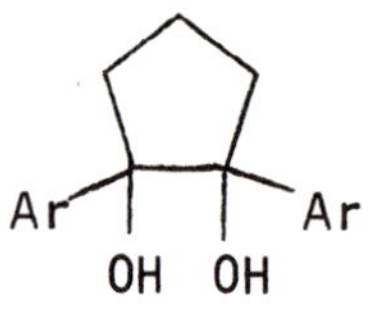

441

(DMF)

Mechanism:
441a: 441-1 to -3

Proposed for Compound:
FD26

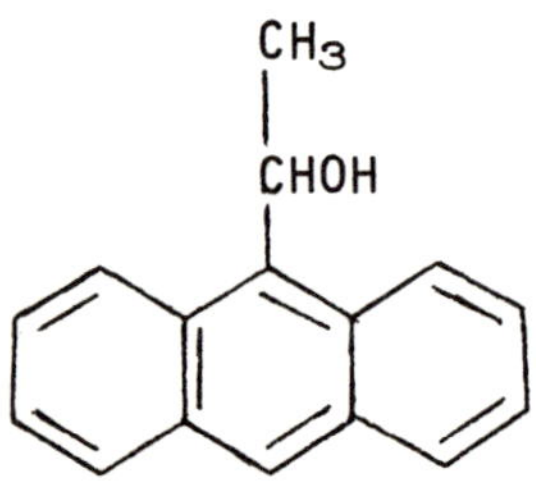

[9-(1-hydroxyethyl)anthracene] + 2e + BH → [9-(1-hydroxyethyl)-9,10-dihydroanthracene] (I) + 2B⁻   441-1

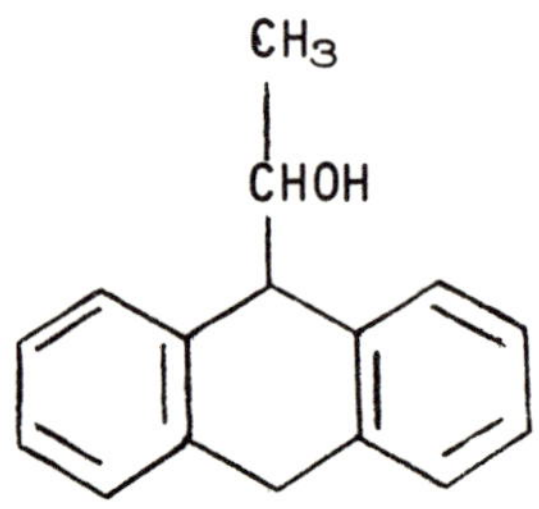

I (441-1) + B⁻ ⇌ [9-(1-hydroxyethyl)-9,10-dihydroanthracene anion structure with CH₃/CHO⁻ substituent] (II) + BH   441-2

II → [9,10-dihydroanthracen-9-ide anion]   + CH_3CHO 441-3

442

Mechanism: Proposed for Compound:
442a: 442-1 to -5 (MeCN)   FA10

$CH_2{:}CHCH_2OH \rightarrow [CH_2{:}CHCH_2OH]^{+\cdot} + e$ 442-1

$[CH_2{:}CHCH_2OH]^{+\cdot} \rightarrow$ polymer   442-2

$[CH_2{:}CHCH_2OH]^{+\cdot} \rightarrow CH_2{:}CHCHO + e + 2H^+$ 442-3

$CH_2{:}CHCHO \rightarrow$ products $+ ne$ 442-4

$CH_2{:}CHCHO \rightarrow$ polymer   442-5

443

Mechanism: Proposed for Compound:
443a: 443-1 to -4   FB52, FC64, FD42, FD75,
                                                         FE72

$Ar_2Cr \rightleftarrows Ar_2Cr^+ + e$ 443-1

$Ar_2Cr^+ \rightarrow Ar_2Cr^{2+} + e$ 443-2

$Ar_2Cr^{2+} \rightarrow 2Ar + Cr^{2+}$ 443-3

$Cr^{2+} \rightarrow Cr^{3+} + e$ 443-4

Ar = 1,3,5-CH$_3$C$_6$H$_3$, CH$_3$C$_6$H$_4$CH$_3$, C$_6$H$_5$CH$_3$, C$_6$H$_6$, (C$_6$H$_5$)$_2$

444

Mechanism:
444a: 444-1 to -6

Proposed for Compound:
FD83, FD84

$$[C_6H_6CrC_6H_5COC_6H_5]^+ + e \rightleftarrows C_6H_6CrC_6H_5COC_6H_5 \qquad 444\text{-}1$$

$$C_6H_6CrC_6H_5COC_6H_5 + e \rightleftarrows C_6H_6CrC_6H_5\dot{C}(O^-)C_6H_5 \qquad 444\text{-}2$$

$$C_6H_6CrC_6H_5\dot{C}(O^-)C_6H_5 + e \rightleftarrows C_6H_6CrC_6H_5\overline{C}(O^-)C_6H_5 \qquad 444\text{-}3$$

$$C_6H_6CrC_6H_5CH(O^-)C_6H_5 \rightleftarrows C_6H_6CrC_6H_5\overline{C}(O^-)C_6H_5 + H^+ \qquad 444\text{-}4$$

$$C_6H_6CrC_6H_5CHOHC_6H_5 \rightleftarrows C_6H_6CrC_6H_5CH(O^-)C_6H_5 + H^+ \qquad 444\text{-}5$$

$$C_6H_5CrC_6H_5CH(O^-)C_6H_5 \rightarrow (C_6H_6)_2Cr + 1/2\,C_6H_5COCH(O^-)C_6H_5 \qquad 444\text{-}6$$

TABLE IV.
COMPOUNDS INCLUDED IN TABLE I

This index is a list of the compounds that appear in Table I. It contains the names that are used in Table I and also, for most of the compounds, one or more synonyms as well as some common trivial names. All these names are arranged in alphabetical order, and each is followed by the code number assigned to the compound in Table I.

(Acetato-O)(4-aminophenyl)mercury, FA68

Acetone, FA09

(Acetonitrile)[2,3,7,8,12,13,17,18-octaethyl-21H, 23H-porphinato(2-)-N^{21}, N^{22}, N^{23}, N^{24}](pyridine)-ruthenium(III) ion, FF66

Acetonitrilopyridinooctaethylporphyrinruthenium(III) ion, FF66

Acetophenone, FA67

17α-Acetoxy-6α-chloropregna-1,4-diene-3,20-dione, FE53

17α-Acetoxy-6-chloropregna-4,6-diene-3,20-dione, FE54

17α-Acetoxy-6α-chloropregn-4-ene-3,20-dione, FE57

17α-Acetoxy-6β-chloropregn-4-ene-3,20-dione, FE58

17α-Acetoxy-6α-fluoropregn-4-ene-3,20-dione, FE59

17α-Acetoxy-6β-fluoropregn-4-ene-3,20-dione, FE60

17α-Acetoxy-6α-methylpregna-1,4-diene-3,20-dione, FE82

17α-Acetoxy-6-methylpregna-4,6-diene-3,20-dione, FE83

17α-Acetoxy-6α-methylpregn-4-ene-3,20-dione, FE84

17α-Acetoxypregna-1,4-diene-3,20-dione, FE55

17α-Acetoxypregna-4,6-diene-3,20-dione, FE56

17α-Acetoxypregn-4-ene-3,20-dione, FE61

17α-Acetoxyprogesterone, FE61

4-Acetylamino-N-(4-chlorophenyl)-benzenamine, FC29

4-Acetylaminodiphenylamine, FC44

4-Acetylamino-4'-methoxydiphenylamine, FC94

4-Acetylamino-N-phenylbenzenamine, FC44

(4-Acetylaminophenyl)(4-chlorophenyl)amine, FC29

(4-Acetylaminophenyl)(4-methoxyphenyl)amine, FC94

4-Acetyldiphenylamine, FC30

4-Acetyl-4'-methoxydiphenylamine, FC89

4-Acetyl-N-(4-methoxyphenyl)benzenamine, FC89

4-Acetyl-N-phenylbenzenamine, FC30

Acetylphenylethylene, FA96

(4-Acetylphenyl)(4-methoxyphenyl)amine, FC89

N-(4-Acetylphenyl)phenylamine, FC30

β-Acetylstyrene, FA96

Acriflavine, FC37

Acrylonitrile, FA07

Alizarine Yellow R, FB76

Allyl alcohol, FA10

2-Aminoanisole, FA58

3-Aminoanisole, FA59

4-Aminoanisole, FA60

4-Aminoazobenzene, FB50

4-Aminodiphenylamine, FB53

Aminoethane, FA05

N-{p-{[(2-Amino-4-hydroxy-6-pteridinyl)methyl]amino}benzoyl}glutamic acid, FD90

N-{p-[(2-Amino-4-hydroxypyrimidino[4,5-b]pyrazin-6-yl)methylamino]-benzoyl}glutamic acid, FD90

4-Amino-4'-methoxydiphenylamine, FC00

N-(4-Aminophenyl)aniline, FB53

N-(4-Aminophenyl)-1,4-benzenediamine, FB58

(4-Aminophenyl)ethanoatomercury(II), FA68

4-Aminophenylmercurio acetate, FA68

4-Aminophenylmercury acetate, FA68

(4-Aminophenyl)(4-methoxyphenyl)amine, FC00

1-(4-Aminophenyl)-2-phenyldiazene, FB50

2-Amino-N-phenyl-p-phenylenediamine, FB57

N-(4-Aminophenyl)-p-phenylenediamine, FB58

4-Amino-N-phenyl-o-phenylenediamine, FB57

n-Amyl alcohol, FA28

tert-Amyl alcohol, FA27

4'-Anilinoacetophenone, FC30

Anilinoazobenzene, FB51

2-Anisidine, FA58
3-Anisidine, FA59
4-Anisidine, FA60
4-(4-Anisylamino)acetophenone, FC89
Anthracene, FC10
9,10-Anthracenedione, FC07
9,10-Anthraquinone, FC07
1-(9-Anthryl)ethanol, FD26
1-Azanaphthalene, FA76
2-Azanaphthalene, FA75
Azobenzene, FB40
4,4'-Azobisphenol, FB42
trans-Azocyclohexane, FB63
trans-Azodioxycyclohexane, FB66
trans-α,α'-Azodioxytoluene, FC47
4,4'-Azodiphenol, FB42
trans-α,α'-Azotoluene, FC43
trans-Azoxy-2-chlorocyclohexane, FB62
cis-Azoxycyclohexane, FB64
trans-Azoxycyclohexane, FB65
trans-Azoxydi(1-chloro-1-phenyl-2-propyl), FD73
trans-α,α'-Azoxytoluene, FC45

Benzalacetone, FA96
1,2-Benzanthracene, FF88
Benzene, FA43
Benzeneazoaniline, FB51
Benzene chloride, FA41
1,2-Benzenedicarboxylic acid monoethenyl ester, FA90
1,4-Benzenediol, FA45
Benzenemethanol, FA56
Benzhydrol, FB95
Benzhydryl ether, FF01
Benzhydryl methyl ether, FC50
1-Benzhydryl-4-nitrosopiperazine, FD59
Benzo[c]cinnoline, FB16
1,4-Benzodiazine, FA64
2-Benzofuranyl phenyl ketone, FC70
Benzohydrol, FB95

Benzoic acid, FA53
Benzo[a]naphthalene, FC11
Benzophenone, FB83
Benzophenonebenzenechromium(0), FD84
Benzophenonebenzenechromium(I) iodide, FD83
Benzo[a]pyrazine, FA64
Benzo[c]pyridine, FA75
p-Benzoquinone, FA39
1,4-Benzoquinone, FA39
2-(Benzoyl)acetanilide, FC78
2-Benzoylbenzofuran, FC70
α-Benzoyl-α-hydroxybenzenamine, FB86
2-Benzoyloxy-1-hydroxy-3-indanone, FD21
2-Benzoyloxy-1,3-indandione, FD12
2-(Benzoyloxy)-1H-indene-1,3(2H)-dione, FD12
2-Benzoyloxy-1-methoxy-3-indenone, FD55
2-(Benzoyloxy)-3-methoxy-1H-inden-1-one, FD55
N-Benzoyl-N-phenylhydroxylamine, FB86
Benzyl alcohol, FA56
4-Benzylbiphenyl, FD82
Benzyl bromide, FA54
Benzyl ether, FC51
Benzylideneacetone, FA96
Benzylidenediphenylmethane, FE09
2-Benzylidene-1,3-indandione, FD10
Benzylpenicillenic acid, FD39
Benzyl(triethyl)ammonium bromide, FC05
Bi(4-anisyl), FC54
(9,9'-Bianthracene)-9,9'(10H,10'H)-diol, FF13
3,3'-Bianthra[1,9-c,d]pyrazole-6,6'-(2H,2'H)dione, FF09
2,2'-Biazine, FA89
(9,9'-Bi-9H-fluoren)-9,9'-diol, FE96
9,9'-Bi-9H-fluorene, FE93
9,9'-Bifluorenyl, FE93
9,9'-Bi(9-hydroxy-9,10-dihydroanthryl), FF13
Bi(2-hydroxy-1,3-dioxoinden-2-yl), FD63

TABLE IV. Compounds Included in Table I

9,9'-Bi(9-hydroxyfluorenyl), FE96

Bilirubin, FF45

3,3'-Bi(6H, 11H-6-oxopyrazolo[3,4,5-d,e]anthracene) dianion, FF09

Biphenyl, FB36

3,3'-{[1,1'-Biphenyl]-4,4'-diylbis(azo)}bis[4-aminonaphthalene] sulfonic acid, FF36

1,1'-[1,1'-Biphenyl]-4,4'-diylbis[4,5-dihydro-3,5-diphenyl-1H-pyrazole, FF65

3,3'[4,4'-Biphenylenebis(azo)]bis-(4-amino-1-naphthalenesulfonic acid), disodium salt, FF36

2,2'-Biphenylenemethane, FB77

α-(4-Biphenylyl)benzenemethanol, FD86

(4-Biphenylyl)benzyl alcohol, FD86

1-(4-Biphenylyl)ethanol, FC52

Biphenylylphenylcarbinol, FD86

3,3'-Bi[pyrazolo(3,4,5-de)anthracene-12-one](2-) ion, FF09

2,2'-Bipyridine, FA89

2,7-Bis(acetylamino)-5,10-bis(4-acetylaminophenyl)-5,10-dihydrophenazine, FF39

2,7-Bis(acetylamino)-5,10-dihydro-5,10-diphenylphenazine, FF17

4,4'-Bis(acetylamino)diphenylamine, FD36

N,N'-{[2,7-Bis(acetylamino)-5,10-phenazinediyl]di-4,1-phenylene}-bisacetamide, FF39

5,10-Bis(4-acetylaminophenyl)-2,7-dimethoxy-5,10-dihydrophenazine, FF33

N,N'-Bis(4-acetylphenyl)-N-phenyl-1,4-benzenediamine, FF16

Bis(η-benzene)chromium, FB52

Bis(η^6-biphenyl)chromium(0), FE72

Bis(η^6-biphenyl)chromium(I) iodide, FE73

Bis(π-biphenyl)chromium(I) iodide, FE73

2,3-Bis(4-biphenylyl)-2,3-butandiol, FF21

2,3-Bis[(1,1'-biphenyl)-3-yl]-2,3-butanediol, FF21

1,2-Bis(4-biphenylyl)-1,2-diphenyl-ethanediol, FF59

1,2-Bis([1,1'-biphenyl]-4-yl)-1,2-diphenyl-1,2-ethanediol, FF59

Bis(1,2-bisdiphenylphosphinoethane)-cobalt(I) perchlorate, FF74

Bis-[1,2-bis(diphenylphosphino)-ethane]iridium(I) chloride, FF75

Bis[1,2-bis(diphenylphosphino)ethane]-rhodium(I) chloride, FF76

Bis[cis-1,2-bis(diphenylphosphino)-ethylene]rhodium(I) chloride, FF72

N,N'-Bis(4-bromophenyl)-N-phenyl-1,4-benzenediamine, FE68

trans-Bis(2-chlorocyclohexyl)diazene 1-oxide, FB62

trans-Bis(2-chloro-1-methyl-2-phenyl-ethyl)diazene 1-oxide, FD73

N,N'-[5,10-Bis(4-chlorophenyl)-5,10-dihydro-2,7-phenazinediyl]bis-acetamide, FF12

trans-1,2-Bis(1-chloro-1-phenyl-2-propyl)diazene 1-oxide, FD73

trans-Bis(cyclohexyl)diazene, FB63

Biscyclopentadienyliron(II), FA93

1,4-Bis(dicyanomethylene)-2,5-cyclohexadiene, FB15

Bis(2,4-dimethoxyphenyl)amine, FD41

1,4-Bis(dimethylamino)benzene, FB07

3,7-Bis(dimethylamino)phenothiazin-5-ium chloride, FD38

3,7-Bis(dimethylamino)phenothiazin-5-ium perchlorate, FD37

3,6-Bis(dimethylamino)-9-thiazonia-10-azaanthracene perchlorate, FD37

2,7-Bis(1,1-dimethylethyl)-5,10-bis-4-(1,1-dimethylethyl)phenyl)-5,10-dihydrophenazine, FF64

2,5-Bis(1,1-dimethylethyl)-1,4-dihydro-1,1,4,4-tetraphenyl-1,4-diphosphorinium dibromide, FF54

Bis(1,1-dimethylethyl)nitroxide, FA70

Bis[4-(1,1-dimethylethyl)phenyl]-ammoniumyl, FE14

Bis(2,4-dinitrophenyl)methane, FB70

Bis(diphenylmethyl) ether, FF01

4,4'-Bis[3,5-diphenyl-Δ^2-pyrazolin-1-yl]biphenyl, FF65

Bis[2,6-di(3-pyridyl)pyridine]-ruthenium(2+) bisperchlorate, FF28, FF90

Bis[1,2-ethanediylbis(diphenylphosphine)P,P']cobalt(1+) perchlorate, FF74

Bis[1,2-ethanediylbis(diphenylphosphine)P,P']iridium(1+) chloride, FF75

Bis[1,2-ethanediylbis(diphenylphosphine)-P,P']rhodium(1+) perchlorate, FF76

Bis[1,2-ethenediylbis(diphenylphosphine)P,P']rhodium(1+) perchlorate, FF72

2,5-Bis(2-isopropoxy-2-methylpropyl)-1,4-benzenediol, FE17

2,5-Bis(2-isopropoxy-2-methylpropyl)-1,4-hydroquinone, FE17

Bis(mesitylene)chromium(0), FD75

2,5-Bis(2-methoxy-2-methylpropyl)-1,4-benzenediol, FD47

2,5-Bis(2-methoxy-2-methylpropyl)-1,4-hydroquinone, FD47

Bis(4-methoxyphenyl)amine, FC56

9,10-Bis(4-methoxyphenyl)anthracene, FF14

N,N'-Bis(4-methoxyphenyl)benzidine, FF05

N,N'-Bis(4-methoxyphenyl)[1,1'-biphenyl]-4,4'-diamine, FF05

5,5'-Bis(4-methoxyphenyl)-3,3'-[2,4-bis(4-methylphenyl)-3,4-dithia-1,6-hexamethylene]dithiolium perchlorate, FF60

1,2-Bis(4-methoxyphenyl)cyclobuteno-[3,4-b]-3,8-naphthohydroquinone dipotassium salt, FE95

4,5-Bis(4-methoxyphenyl)-2-(4-dimethylaminophenyl)imidazole, sodium salt, FE87

N,N'-Bis(4-methoxyphenyl)-N,N'-diphenylhydrazine, FF07

2,2'-Bis(3-methoxyphenylethyl)-4,4',-5,5'-tetramethoxybiphenyl, FF48

Bis(3-methoxyphenyl)methane, FC98

2,3-Bis(4-methoxyphenyl)-1,4-naphthoquinone, FE70

Bis(4-methoxyphenyl)nitroxide, FC42

N,N'-Bis(4-methoxyphenyl)-N-phenyl-1,4-benzenediamine, FF06

N,N'-Bis(4-methoxyphenyl)-N-phenyl-phenylenediamine, FF06

2,5-Bis(2-methyl-2-ethoxypropyl)-1,4-benzenediol, FD77

2,6-Bis(2-methyl-2-ethoxypropyl)-1,4-benzenediol, FD78

2,5-Bis(2-methyl-2-ethoxypropyl)hydroquinone, FD77

2,6-Bis(2-methyl-2-ethoxypropyl)hydroquinone, FD78

2-[Bis(1-methylethyl)amino]ethanethiol hydrochloride, FA71

1,2-Bis(4-methylphenyl)-1,2-diphenyldiazane, FF03

N,N'-Bis(4-methylphenyl)-N-phenyl-1,4-benzenediamine, FF04

N,N'-Bis(4-methylphenyl)-N-phenyl-1,4-phenylenediamine, FF04

2,5-Bis(4-methylphenyl)-3,3a,4-trithiapentalene, FD88

1,2-Bis(4-nitrophenyl)ethane, FC24

1,2-Bis(4-nitrophenyl)ethene, FC13

1,2-Bis(4-nitrophenyl)ethylene, FC13

Bis(4-nitrophenyl)methane, FB81

4,4'-Bis(phenylamino)biphenyl, FE74

Bis(phenylmethyl)diazene, FC43

z-Bis(phenylmethyl)diazene 1-oxide, FC45

1,2-Bis(phenylmethyl)hydrazine, FC65

2,2'-Bispyrazolanthrone, FF09

Bis(pyridine)[5,10,15,20-tetraphenyl-21H, 23H-porphinato(2-)-N^{21}, N^{22},-N^{23}n N^{24}]ruthenium, FF77

Bis(pyridino)-$\alpha,\beta,\gamma,\delta$-tetraphenylporphinylruthenium, FF77

1,3-Bis(3-pyridyl)-2-propen-1-one oxime, FB88

1,3-Bis(3-pyridyl)-2-propen-1-ylideneazenol, FB88

Bis(2,2',6',2''-terpyridine-N,N',N'')-ruthenium(2+) diperchlorate, FF28

Bis(2,2',2''-terpyridine)ruthenium(II) perchlorate, FF28

N,N'-Bis(2,3,5,6-tetramethylphenyl)-benzidine, FF41

4,4'-Bis(thiocyanato)diphenylamine, FC09

Bis(η^6-toluene)chromium(0), FC64

Bis(4-tolyl)amine, FC55

Bis[1-tolyl-2-anisyl-2(4,5-dianisyl-1,2-dithiolium-3-yl)ethyl] disulfide perchlorate, FF80

6,6'-Bis(3,3',4-trimethoxybibenzyl), FF48

Bis(2,4,6-tri-2-pyridinyl-1,3,5-triazine, N^1,N^2,N^6)ruthenium(3+) triperchlorate, FF51

Bis(2,4,6-tri-2-pyridyl-s-triazine)-ruthenium(III) perchlorate, FF51, FF93

Bis(η^6-xylene)chromium(0), FD42

2-(3-Bromobenzoyl)acetanilide, FC72

2-(4-Bromobenzoyl)acetanilide, FC73

3-Bromo-4'-dimethylaminoazobenzene, FC33

4-Bromo-4'-dimethylaminoazobenzene, FC34

4-Bromodiphenylamine, FB37

(Bromomethyl)benzene, FA54

3-Bromo-2-oxopropanoic acid, FA06

4-[(3-Bromophenyl)azo]-N,N-dimethylaniline, FC33

4-[(4-bromophenyl)azo]-N,N-dimethylaniline, FC34

4-(4-Bromophenylazo)-3,5-diphenyl-1H-pyrazole-1-carbothioamide, FE32

4-(4-Bromophenylazo)-3,5-diphenyl-1-thiocarbamoylpyrazole, FE32

4-[(4-Bromophenyl)azo]phenol, FB26

4-Bromo-N-phenylbenzenamine, FB37

3-(4-Bromophenyimino)-2-phenyl-3H-indole, FE01

3-(4-Bromophenylimino)-2-phenyl-3H-indole 1-oxide, FE02

3-(3-Bromophenyl)-3-oxo-N-phenylpropanamide, FC72

3-(4-Bromophenyl)-3-oxo-N-phenylpropanamide, FC73

β-Bromopyruvic acid, FA06

α-Bromotoluene, FA54

1,1'-(1,4-Butanediyl)bis[4-nitrobenzene], FD33

1-Butanol, FA18

2-Butanol, FA19

trans-2-Butenedinitrile, FA14

2-(2-tert-Butoxy-2-methylpropyl)-5-tert-butyl-1,4-benzenediol, FD76

2-(2-tert-Butoxy-2-methylpropyl)-5-tert-butylhydroquinone, FD76

Butyl alcohol, FA18

sec-Butyl alcohol, FA19

2-tert-Butyl-5-(2-isopropoxy-2-methylpropyl)hydroquinone, FD61

2-tert-Butyl-5-(2-methoxy-2-methyl-1-propyl)hydroquinone, FD05

tert-Butyl phenyl ketone, FB11

Capri Blue, FD37

3-(4-Carbamoylphenyl)-1-(2-thienyl)-2-propen-1-one, FC17

3-Carbamoylpyridine, FA44

3-Carbamoyl-1-β-D-ribofuranosylpyridinium hydroxide, 5'→ 5'-ester with adenosine 2'-(dihydrogenphosphate) 5'-(trihydrogen pyrophosphate), inner salt, FE27

3-Carbamoyl-1-β-D-ribofuranosylpyridinium hydroxide, 5'→ 5'-ester with adenosine 5'-(trihydrogen phosphate), inner salt, FE28

3-Carbamoyl-1-β-D-ribofuranosylpyridinium hydroxide, 5'→ 5'-ester with inosine 5'-(trihydrogen pyrophosphate), inner salt, FE26

3-Carbamoyl-1-β-D-ribofuranosylpyridinium hydroxide 5'-phosphate, inner salt, FB13

4-Carbethoxydiazoaminobenzene, FC92

Carbonylchlorobis(triphenylphosphine)-rhodium, FF57

Carbonyletioporphyrin(I)ruthenium(II), FF44

Carbonylhydridotris(triphenylphosphine)-iridium, FF78

Carbonylhydrotris(triphenylphosphine)-rhodium, FF79

Carbonyl[2,3,7,8,12,13,17,18-octaethylporphinato(2-)-N^{21}, N^{22}, N^{23}, N^{24}]-ruthenium, FF58

Carbonyl(octaethylporphyrinyl)ruthenium, FF58

Carbonyl(pyridine)[5,10,15,20-tetraphenyl-21H, 23H-porphinato(2-)-N^{21}, N^{22}, N^{23}, N^{24}]ruthenium, FF71

Carbonyl(pyridinotetraphenylporphyrinyl)ruthenium(II), FF71

Carbonyl[2,7,12,17-tetraethyl-3,8,13,18-tetramethyl-21H, 23H-porphinato-(2-)-N^{21}, N^{22}, N^{23}, N^{24}]ruthenium, FF44

Carbonyl[5,10,15,20-tetraphenyl-21H,-23H-porphinato(2-)-N^{21}, N^{22}, N^{23}, N^{24}]ruthenium, FF69

Carbonyl-($\alpha,\beta,\gamma,\delta$-tetraphenyl)-porphyrinylruthenium, FF69

Carbonyl(tetraphenylporphyrinyl)-ruthenium, FF69

3-Carboxy-4'-dimethylaminoazobenzene, FC90

4'-Carboxy-4-hydroxyazobenzene, FB80

3-Carboxy-4-hydroxy-4'-sulfoazo-benzene, disodium salt, FB68

(2-Carboxy-2-oxoethyl)dimethyl-sulfonium bromide, FA26

Chalcone, FC76

Chalcone phenylhydrazone, FE23

Chlorimipramine, FD91

6α-Chloro-17α-acetoxyprogesterone, FE57

6β-Chloro-17α-acetoxyprogesterone, FE58

4-Chloro-4'-acetylaminodiphenylamine, FC29

Chlorobenzene, FA41

2-(3-Chlorobenzoyl)acetanilide, FC74

2-(4-Chlorobenzoyl)acetanilide, FC75

2-(3-Chlorobenzylidene)-1,3-indan-dione, FD08

2-(4-Chlorobenzylidene)-1,3-indan-dione, FD09

<u>trans</u>-Chlorocarbonylbis(triphenyl-phosphine)rhodium(I), FF57

4-Chloro-N-(4-chlorophenyl)benzen-amine, FB30

6-Chloro-Δ^6-dehydro-17α-acetoxy-progesterone, FE54

6α-Chloro-Δ^1-dehydro-17α-acetoxy-progesterone, FE53

3-Chloro-10,11-dihydro-N,N-dimethyl-5H-dibenz[<u>b</u>,<u>f</u>]azepine-5-propanamine, FD91

7-Chloro-2,3-dihydro-1-methyl-5-phenyl-1H-1,4-benzodiazepine, FD27

7-Chloro-1,3-dihydro-1-methyl-5-phenyl-2H-1,4-benzodiazepin-2-one, FD22

3-Chloro-4'-dimethylaminoazobenzene, FC35

4-Chloro-4'-dimethylaminoazobenzene, FC36

3-Chloro-5-(3-dimethylaminopropyl)-dibenzo[<u>b</u>,<u>f</u>]azepine, FD91

4-Chloro-4'-hydroxyazobenzene, FB27

2-Chloro-4'-hydroxy-3'-carboxyazo-benzene, FB72

4-Chloro-4'-hydroxy-3'-carboxyazo-benzene, FB73

4-Chloro-N-hydroxy-N-(4-nitroso-phenyl)benzenamine, FB29

7-Chloro-1-methyl-1,2,3-trihydro-5-phenyl-2H-1,4-benzodiazepine, FD27

4-Chloro-4'-nitrodiphenylamine, FB28

4-Chloro-N-(4-nitrophenyl)benzen-amine, FB28

N-[4-(4-Chlorophenyl)amino]phenyl-acetamide, FC29

4-[(3-Chlorophenyl)azo]-N,N-dimethyl-aniline, FC35

4-[(4-Chlorophenyl)azo]-N,N-dimethyl-aniline, FC36

4-(3-Chlorophenylazo)-3,5-diphenyl-1H-pyrazole-1-carbothioamide, FE33

4-(4-Chlorophenylazo)-3,5-diphenyl-1H-pyrazole-1-carbothioamide, FE34

4-(3-Chlorophenylazo)-3,5-diphenyl-1-thiocarbamoylpyrazole, FE33

4-(4-Chlorophenylazo)-3,5-diphenyl-5-thiocarbamoylpyrazole, FE34

5-[(2-Chlorophenyl)azo]-2-hydroxy-benzoic acid, FB72

5-(2-Chlorophenylazo)-2-hydroxybenzoic acid, FB72

5-[(4-Chlorophenyl)azo]-2-hydroxy-benzoic acid, FB73

4-[(4-Chlorophenyl)azo]phenol, FB27

5-(2-Chlorophenylazo)salicylic acid, FB72

N-(2-Chlorophenyl)-N'-(3-carboxy-4-hydroxyphenyl)diazene, FB72

N-(4-Chlorophenyl)-N'-(3-carboxy-4-hydroxyphenyl)diazene, FB73

(4-Chlorophenyl)diazoaminobenzene, FB38

3-[5-(4-Chlorophenyl)-2-furyl]-1-(2-furyl)-2-propen-1-one, FD49

N-(4-Chlorophenyl)-N-hydroxy-4-nitroso-aniline, FB29

3-(4-Chlorophenylimino)-2-phenyl-3H-indole, FE03

3-(4-Chlorophenylimino)-2-phenyl-3H-indole 1-oxide, FE04

2-[(3-Chlorophenyl)methylene]-1H-indene-1,3(2H)-dione, FD08

2-[(4-Chlorophenyl)methylene]-1H-indene-1,3(2H)-dione, FD09

N-(4-Chlorophenyl)-4-nitroaniline, FB28

(4-Chlorophenyl)(4-nitrophenyl)amine, FB28

N-(4-Chlorophenyl)-N-(4-nitrosophenyl)hydroxylamine, FB29

3-(3-Chlorophenyl)-3-oxo-N-phenylpropanamide, FC74

3-(4-Chlorophenyl)-3-oxo-N-phenylpropanamide, FC75

1-(4-Chlorophenyl)-3-phenyltriazene, FB38

7-Chloro-1,3,4,5-tetrahydro-1-methyl-5-phenyl-3H-1,4-benzodiazepin-2-one, FD28

Chloro(triethyl)stannane, FA51

Chlorotriethyltin(IV), FA51

Chloro(trimethyl)stannane, FA13

Chlorotrimethyltin, FA13

Cibachron Blue C, FF24

Cinnamic acid nitrile, FA74

Cinnamonitrile, FA74

Cinnamyl alcohol, FA80

Cinnamyl methyl ether, FB03

Congo Red, FF36

Croconic acid, FA20

2-(4-Cyanobenzoyl)acetanilide, FD19

4-Cyanodiazoaminobenzene, FB82

Cyanoethylene, FA07

4-Cyano-β-oxo-N-phenylbenzenepropanamide, FD19

3-(4-Cyanophenyl)-3-oxo-N-phenylpropanamide, FD19

3-Cyanoquinoline, FA88

2,5-cyclohexadiene-1,4-dione, FA39

2,2'-(2,5-Cyclohexadiene-1,4-diylidene)bispropanedinitrile, FB15

Cyclohexatriene, FA43

Cyclooctatetraene, FA66

1,3,5,7-Cyclooctatetraene, FA66

Decafluorobiphenyl, FB14

Δ^1-Dehydro-17α-acetoxyprogesterone, FE55

Δ^6-Dehydro-17α-acetoxyprogesterone, FE56

Δ^1-Dehydrotestosterone, FD92

Di(4-acetylaminophenyl)amine, FD36

2,7-Di(acetylamino)-5,10-bis(4-chlorophenyl)-5,10-dihydrophenazine, FF12

5,10-Di(4-acetylaminophenyl)-2,7-di(acetylamino)-5,10-dihydrophenazine, FF39

9,10-Diacetyl-9,10-dihydro-9,10-diazaanthracene, FD25

N,N'-Diacetyl-1,4-dihydro-1,4-diazanaphthalene, FB54

9,10-Diacetyl-9,10-dihydro-9,10-diazaphenanthrene, FD24

1,4-Diacetyl-1,4-dihydro-2,3-diphenylpyrazine, FE12

1,2-Diacetyl-1,2-dihydro-4-methylcinnoline, FC01

5,6-Diacetyl-5,6-dihydro-5,6-phenanthroline, FD24

5,10-Diacetyl-5,10-dihydrophenazine, FD25

1,4-Diacetyl-1,4-dihydroquinoxaline, FB54

5,10-Di(4-acetylphenyl)-2,7-dimethoxy-5,10-dihydrophenazine, FF31

N,N'-Di(4-acetylphenyl)-N-phenylphenylenediamine, FF16

2,4-Diaminodiphenylamine, FB57

4,4'-Diaminodiphenylamine, FB58

2,8-Diamino-10-methylacridinium chloride (often used, but incorrect. See 3,6-Diamino-10-methylacridinium chloride, FC37)

3,6-Diamino-10-methylacridinium chloride, FC37

2,4-Diaminophenol, FA46

3,7-Diaminophenothiazin-5-ium chloride, FB39

3,7-Diaminophenothiazonium chloride, FB39

Di(4-aminophenyl)amine, FB58

3,7-Diamino-5-phenylphenazinium chloride, FD68

N-(2,4-Diaminophenyl)phenylamine, FB57

3,6-Diamino-10H-9-thia-10-azaanthracene chloride, FB39

9,10-Di-p-anisylanthracene, FF14

1,2-Dianisylcyclobuteno[3,4-b]-3,8-naphthohydroquinone, dipotassium salt, FE95

2,3-Dianisyl-1,4-naphthoquinone, FE70

9,10-Diazaanthracene, FB18

2,2'-Diazabiphenyl, FA89

1,5-Diazanaphthalene, FA63

4,5-Diazaphenanthrene, FB17

9,10-Diazaphenanthrene, FB16

Diazepam, FD22

Diazoaminobenzene, FB51

Dibenz[de,kl]anthracene, FD99

trans(?)-1,2-Dibenzoylethene, FD20

trans(?)-1,2-Dibenzoylethylene, FD20

1,3-Dibenzoylpropane, FD57

trans-1,2-Dibenzyldiazane, FC65

trans-1,2-Dibenzyldiazene, FC43

trans-1,2-Dibenzyldiazene 1,2-dioxide, FC47

trans-1,2-Dibenzyldiazene 1-oxide, FC45

1,2-Dibenzylidenecyclobutano[3,4-b]-3,8-naphthohydroquinone, dipotassium salt, FE91

4,5-Di(biphenylyl)-2-(dimethylaminophenyl)imidazole, sodium salt, FF49

α,α'-Dibromobibenzyl, FC22

meso-α,α'-Dibromobibenzyl, FC23

dl-1,2-Dibromo-1,2-diphenylethane, FC22

meso-1,2-Dibromo-1,2-diphenylethane, FC23

1,1'-(1,2-Dibromo-1,2-ethanediyl)-bisbenzene, FC22

N,N'-Di(4-bromophenyl)-N-phenyl-phenylenediamine, FE68

Dibutyl 1,2-benzenedicarboxylate, FD43

2,7-Di-tert-butyl-5,10-bis(4-tert-butylphenyl)-5,10-dihydrophenazine, FF64

4,4'-Di(t-butyl)diphenylamine, FE14

Di-tert-butylnitroxide, FA70

5,10-Di(4-tert-butylphenyl)-2,7-di-tert-butyl-5,10-dihydrophenazine, FF64

Dibutyl phthalate, FD43

2,5-Di-tert-butyl-1,1,4,4-tetraphenyldiphosphoniacyclohexa-2,5-diene dibromide, FF54

1,2-Dicarbethoxybenzene, FB60

1,2-Dicarbonylbutoxybenzene, FD43

1,2-Dicarbonylphenoxybenzene, FE08

1,2-Dicarbutoxybenzene, FD43

trans-Di(2-chlorocyclohexyl)diazene oxide, FB62

2,7-Dichloro-5,10-di(4-chlorophenyl)-5,10-dihydrophenazine, FE63

Dichloro(diethyl)stannane, FA17

Dichlorodiethyltin, FA17

2,7-Dichloro-5,10-di(4-nitrophenyl)-5,10-dihydrophenazine, FE62

4,4'-Dichlorodiphenylamine, FB30

Di(4-chlorophenyl)amine, FB30

5,10-Di(4-chlorophenyl)-2,7-di-(acetylamino)-5,10-dihydrophenazine, FF12

5,10-Di(4-chlorophenyl)-2,7-dichloro-5,10-dihydrophenazine, FE63

3-(2,6-Dichlorophenyl-1-(2-thienyl-2-propenone, FB67

2,6-Dichlorostyryl 2-thienyl ketone, FB67

2,3-Dicyano-2-butenedinitrile, FA31

trans-1,2-Dicyanoethene, FA14

9-Dicyanomethylene-2,4,7-trinitrofluorene, FD07

1,1-Dicyano-2-(4-methyl)phenylethene, FB08

trans-Dicyclohexyldiazene 1,2-dioxide, FB66

cis-Dicyclohexyldiazene 1-oxide, FB64

trans-Dicyclohexyldiazene 1-oxide, FB65

Dicyclopentadienyliron(II), FA93

Di[4-(1,1-dimethylethyl)phenyl]amine, FE14

Diethyl 1,2-benzenedicarboxylate, FB60

Diethyl trans-2-butenedioate, FA69

2,5-Diethyl-1,4-dihydro-1,1,4,4-tetraphenyl-1,4-diphosphorinium dibromide, FF40

Diethyl fumarate, FA69

2,17-Diethyl-1,10,19,22,23,24-hexahydro-3,7,13,18-tetramethyl-1,19-dioxo-21H-biline-8,12-dipropanoic acid, FF46

1,2-Diethylidenacenaphthene, FD23

Diethyl-p-nitrophenyl monothiophosphate, FB05

Diethyl 4-nitrophenyl phosphate, FB06

O,O-Diethyl O-(4-nitrophenyl)phosphorothioate, FB05

O,O-Diethyl O-p-nitrophenyl thiophosphate, FB05

Diethyl phthalate, FB60

Diethylstannic dichloride, FA17

2,5-Diethyl-1,1,4,4-tetraphenyldiphosphoniacyclohexa-2,5-diene dibromide, FF40

Diethyltin dichloride, FA17

1,1'-Diheptyl-4,4'-bipyridylium bis[tetrafluoroborate(1-)], FE85

N,N'-Diheptyl-4,4'-bipyridylium tetrafluoroborate, FE85

9,10-Dihydro-9-anthracenol, FC25

5,10-Dihydro-5,10-bis(4-thiocyanatophenyl)-2,7-phenazinediyl ester of thiocyanic acid, FF10

1,2-Dihydro-3,8-dihydroxy-1,2-dimethylcyclobuta[b]naphthalene, FC53

1,2-Dihydro-1,2-dimethylcyclobuta[b]naphthalene-3,8-diol, FC53

1,4-Dihydro-2,5-dimethyl-1,1,4,4-tetraphenyl-1,4-diphosphorinium dibromide, FF32

5,10-Dihydro-5,10-diphenylphenazine, FE69

N,N'-(5,10-Dihydro-5,10-diphenyl-2,7-phenazinediyl)bisacetamide, FF17

5,10-Dihydro-2,7-dithiocyanato-5,10-bis(4-thiocyanatophenyl)phenazine, FF10

1,4-Dihydro-1,1,2,4,4,5-hexaphenyl-1,4-diphosphorinium dibromide, FF62

1,4-Dihydro-1,1,2,4,4,5-hexaphenyl-1,4-diphosphorinium dichloride, FF63

3,4-Dihydro-1(2H)-naphthalenone, FA97

Dihydroninhydrin, FA73

4,5-Dihydro-1-(4-nitrophenyl)-3,5-diphenyl-1H-pyrazole, FE22

1,4-Dihydro-1,1,4,4-tetraphenyl-p-diphosphorinium dibromide, FF15

4,5-Dihydro-1,3,5-triphenyl-1H-pyrazole, FE24

4,4'-Dihydroxyazobenzene, FB42

1,4-Dihydroxybenzene, FA45

2,2'-Dihydroxy-[2,2'-bi-1H-indene]-1,1',3,3'(2H,2'H)-tetrone, FD63

1,2-Dihydroxy-1-cyclohexen-3,4,5,6-tetraone, FA34

2,2-Dihydroxy-1,3-indanedione, FA72

2,2-Dihydroxy-1H-indene-2,3(2H)-dione, FA72

2,3-Dihydroxyindenone, FA73

2,3-Dihydroxy-1H-inden-1-one, FA73

3',4-Dihydroxy-3-methoxy-2',4',6'-tribromochalcone, FD16

2,5-Dihydroxynaphthalene-1-azo-[2'-hydroxy-5'-(sodium sulfonato)benzene], FD17

3-[(2,5-Dihydroxynaphth-1-yl)azo]-4-hydroxybenzenesulfonic acid, sodium salt, FD17

N,N'-Di(4-hydroxyphenyl)diazene, FB42

17α,21-Dihydroxypregn-4-ene-3,20-dione, FE30

3',6'-Dihydroxyspiro[isobenzofuran-1(3H),9'-(9H)xanthen]-3-one, disodium salt, FD98

3',6'-Dihydroxy-2',4',5',7'-tetraiodospiro[isobenzofuran-1(3H),9'-(9H)xanthen]-3-one, disodium salt, FD97

3',4-Dihydroxy-2',4',6'-tribromochalcone, FC69

Diisopropylaminoethanethiol hydrochloride, FA71

3,3'-Dimethoxybibenzyl, FD40

4,4'-Dimethoxybiphenyl, FC54

2,7-Dimethoxy-5,10-bis(4-methoxyphenyl)-5,10-dihydrophenazine, EF20

2,4-Dimethoxy-N-(2,4-dimethoxyphenyl)benzenamine, FD41

1,2-Dimethoxy-4-[2-(3,4-dimethoxyphenyl)ethyl]benzene, FD74

4,4'-Dimethoxydiphenylamine, FC56

2,7-Dimethoxy-5,10-diphenyl-5,10-dihydrophenazine, FF00

N,N'-[2,7-Dimethoxy-5,10-phenazinediyl)di-4,1-phenylene]bisacetamide, FF33

1,1'-[(2,7-Dimethoxy-5,10-phenazinediyl)di-1,4-phenylene]bisethanone, FF31

5,10-Di(4-methoxyphenyl)-2,7-dimethoxy-5,10-dihydrophenazine, FF20

1,2-Di(4-methoxyphenyl)-1,2-diphenyldiazane, FF07

1,2-Di(3-methoxyphenyl)ethane, FD40

1-(3,4-Dimethoxyphenyl)-2-(3-methoxyphenyl)ethane, FD60

1-(3,4-Dimethoxyphenyl)propene, FB12

1,2-Dimethoxy-4-propenylbenzene, FB12

3,4-Dimethoxypropenylbenzene, FB12

4-(Dimethylamino)azobenzene, FC57

4'-Dimethylaminoazobenzene-3-sulfonic acid, FC62

4-Dimethylaminoazobenzene-4-sulfonic acid, FC63

4-Dimethylamino-3'-bromoazobenzene, FC33

4-Dimethylamino-4'-bromoazobenzene, FC34

4-Dimethylamino-3'-carboxyazobenzene, FC90

4-Dimethylamino-4'-carboxyazobenzene, FC91

4-Dimethylamino-3'-chloroazobenzene, FC35

4-Dimethylamino-4'-chloroazobenzene, FC36

4-Dimethylaminodiphenyldiazene, FC57

4-Dimethylamino-3'-fluoroazobenzene, FC38

4-Dimethylamino-4'-fluoroazobenzene, FC39

4-Dimethylamino-2'-hydroxyazobenzene, FC58

4-Dimethylamino-4'-hydroxyazobenzene, FC59

4-Dimethylamino-3'-iodoazobenzene, FC40

4-Dimethylamino-4'-iodoazobenzene, FC41

4-Dimethylamino-3'-methoxyazobenzene, FD02

4-Dimethylamino-4'-methoxyazobenzene, FD03

4-Dimethylamino-3'-methylazobenzene, FD00

4-Dimethylamino-4'-methylazobenzene, FD01

4-Dimethylamino-3'-nitroazobenzene, FC48

4-Dimethylamino-4'-nitroazobenzene, FC49

4-Dimethylaminophenylazobenzene, FC57

3-(4-Dimethylaminophenylazo)benzenesulfonic acid, FC62

4-(4-Dimethylaminophenylazo)benzenesulfonic acid, FC63

3-(4-Dimethylaminophenylazo)benzoic acid, FC90

4-(4-Dimethylaminophenylazo)benzoic acid, FC91

2-(4-Dimethylaminophenylazo)phenol, FC58

4-(4-Dimethylaminophenylazo)phenol, FC59

4-(4-Dimethylaminophenylazo)toluene, FD01

2-(4-Dimethylaminophenyl)-4,5-bis(4-methylphenyl)imidazole, sodium salt, FE86

2-(4'-Dimethylaminophenyl)-4,5-di(4''-biphenyl)imidazole, sodium salt, FF49

2-(4'-Dimethylaminophenyl)-4,5-di(4-methoxyphenyl)imidazole, sodium salt, FE87

2-(4-Dimethylaminophenyl)-4,5-di(4-methylphenyl)imidazole, sodium salt, FE86

3-[[p-(Dimethylamino)phenyl]imino]-2-phenyl-3H-indole, FE36

3-(4-Dimethylaminophenylimino)-2-phenylindole, FE36

3-{[p-(Dimethylamino)phenyl]imino}-2-phenyl-3H-indole 1-oxide, FE37

3-(4-Dimethylaminophenylimino)-2-phenylindole 1-oxide, FE37

1-(4-Dimethylaminophenyl)-2-phenyldiazene, FC57

1-(4-Dimethylaminophenyl)-2-(4-sulfophenyl)diazene, FC63

3-[4-Dimethylaminophenyl]-1-(2-thienyl)-2-propen-1-one, FC88

4-Dimethylamino-ω-thenoylstyrene, FC88

Dimethyl 1,2-benzenedioate, FB00

2,7-Dimethyl-5,10-bis(4-methylphenyl)-5,10-dihydrophenazine, FF19

[Dimethyl 7,12-diethenyl-3,8,13,17-tetramethyl-21H, 23H-porphine-2,18-dipropanoato(2-)-N^{21}, N^{22},-N^{23}, N^{24}]magnesium, FF53

[Dimethyl 7,12-diethyl-3,8,13,17-tetramethyl-21H,23H-porphine-2,18-dipropanoato(2-)-N^{21},N^{22},-N^{23},N^{24}]magnesium, FF55

4,4'-Dimethyldiphenylamine, FC55

2,7-Dimethyl-5,10-diphenyl-5,10-dihydrophenazine, FE99

3,5-Dimethyl-4-(2-ethoxyphenylazo)-1-thiocarbamoylpyrazole, FC66

3,5-Dimethyl-4-(4-ethoxyphenylazo)-1-thiocarbamoylpyrazole, FC67

4-(1,1-Dimethylethyl)-N-[4-(1,1-dimethylethyl)phenyl]benzenamine, FE14

2-(1,1-Dimethylethyl)-5-(2-ethoxy-2-methylpropyl)-1,4-benzenediol, FD44

2-(1,1-Dimethylethyl)-6-(2-ethoxy-2-methylpropyl)-1,4-benzenediol, FD45

2-(1,1-Dimethylethyl)-5-(2-ethoxy-2-methylpropyl)-1,4-hydroquinone, FD44

2-(1,1-Dimethylethyl)-6-(2-ethoxy-2-methylpropyl)-1,4-hydroquinone, FD45

2-(1,1-Dimethylethyl)-5-(2-methyl-2-methoxypropyl)-1,4-hydroquinone, FD05

2-(1,1-Dimethylethyl)-5-[2-methyl-2-(1-methylethoxy)propyl]-1,4-benzenediol, FD61

2-(1,1-Dimethylethyl)-5-[2-methyl-2-(1-methylethoxy)propyl]-1,4-hydroquinone, FD61

Di(1-methylethyl)(thioethyl)amine hydrochloride, FA71

Dimethylglycol phthalate, FA90

Dimethyl ketone, FA09

Dimethylmethanone, FB83

N,N-Dimethyl-4-(3-methoxyphenylazo)-benzenamine, FD02

3,5-Dimethyl-4-(2-methoxyphenylazo)-1-thiocarbamoylpyrazole, FC03

3,5-Dimethyl-4-(4-methoxyphenylazo)-1-thiocarbamoylpyrazole, FC04

N,N-Dimethyl-4[(3-methylphenylazo]-benzenamine, FD00

N,N-Dimethyl-4-[(4-methylphenyl)azo]-benzenamine, FD01

N,N-Dimethyl-4-[(3-nitrophenyl)azo]-aniline, FC48

N,N-Dimethyl-4-[(4-nitrophenyl)azo]-aniline, FC49

Di(4-methylphenyl)amine, FC55

4-(2,4-Dimethylphenylazo)-3,5-diphenyl-1H-pyrazole-1-carbothioamide, FE77

4-(2,6-Dimethylphenylazo)-3,5-diphenyl-1H-pyrazole-1-carbothioamide, FE78

4-(2,4-Dimethylphenylazo)-3,5-diphenyl-1-thiocarbamoylpyrazole, FE77

4-(2,6-Dimethylphenylazo)-3,5-diphenyl-1-thiocarbamoylpyrazole, FE78

3,5-Dimethyl-4-(phenylazo)-1-thiocarbamoylpyrazole, FB59

5,10-Di(4-methylphenyl)-2,7-dimethyl-5,10-dihydrophenazine, FF19

N,N'-Di(4-methylphenyl)-N,N'-diphenylhydrazine, FF03

2,2-Dimethyl-1-phenyl-1-propanone, FB11

Dimethyl phthalate, FB00

2-(1,1-Dimethylpropyl)-5-(2-ethoxy-2-methylpropyl)-1,4-hydroquinone, FD62

2-(1,1-Dimethylpropyl)-5-(2-methoxy-2-methylpropyl)-1,4-benzenediol, FD46

2-(1,1-Dimethylpropyl)-5-(2-methoxy-2-methylpropyl)hydroquinone, FD46

3-Dimethylsulfonio-2-oxopropanoic acid bromide, FA26

β-Dimethylsulfoniumpyruvic acid bromide, FA26

2,5-Dimethyl-1,1,4,4-tetraphenyl-diphosphoniacyclohexa-2,5-diene dibromide, FF32

peri-Dinaphthalene, FD99

1,2-Dinitrobenzene, FA36

1,3-Dinitrobenzene, FA37

1,4-Dinitrobenzene, FA38

4,4'-Dinitrobibenzyl, FC24

2,2'-Dinitrobiphenyl, FB19

3,3'-Dinitrobiphenyl, FB20

4,4'-Dinitrobiphenyl, FB21

4,4'-Dinitrodiphenylamine, FB35

4,4'-Dinitro-1,4-diphenylbutane, FD33

4,4'-Dinitrodiphenylmethane, FB81

4,4'-Dinitrodiphenylpropane, FC83

1,5-Di(5-nitro-2-furyl)-1,4-pentadien-3-one amidinohydrazone hydrochloride, FB89

1,5-Di(5-nitro-2-furyl)-1,4-pentadien-3-one aminohydrazonium chloride, FB89

2,4-Dinitro-4'-hydroxy-3'-carboxyazobenzene, FB69

Di(4-nitrophenyl)amine, FB35

5-[(2,4-Dinitrophenyl)azo]-2-hydroxybenzoic acid, FB69

5-(2,4-Dinitrophenylazo)salicylic acid, FB69

1,4-Di(4-nitrophenyl)butane, FD33

N-(2,4-Dinitrophenyl)-N'-(3-carboxy-4-hydroxyphenyl)diazene, FB69

5,10-Di(4-nitrophenyl)-2,7-dichloro-5,10-dihydrophenazine, FE62

1,3-Di(4-nitrophenyl)propane, FC83

4,4'-Dinitrostilbene, FC13

6,11-Dioxido-2,4-diphenyldibenzo[b,h]biphenylene, dipotassium salt, FF35

9,10-Dioxoanthracene, FC07

1,1'-Dioxy-2,21-diphenyl-$\Delta^{3,3'}$-bi-3H-indole, FF11

1,2-Diphenoxybenzene, FE08

Diphenyl, FB36

α,α-Diphenylallyl alcohol, FC84

3,3-Diphenylallyl alcohol, FC85

Diphenylamine, FB47

9,10-Diphenylanthracene, FE94, FF85-FF87, FF89

Diphenyl 1,2-benzenedicarboxylate, FE08

α,α-Diphenylbenzenemethanol, FD87

N,N'-Diphenylbenzidine, FE74

N,N'-Diphenyl-[1,1'-biphenyl]-4,4'-diamine, FE74

5,5'-Diphenyl-3,3'-[2,5-bis(4-methoxyphenyl)-3,4-dithia-1,6-hexamethyleno]dithiolium perchlorate, FF52

N,N'-(1,4-Diphenyl-1,3-butadiene-1,4-diyl)bis[N-acetylacetamide], FE79

trans(?)-1,4-Diphenyl-2-buten-1,4-dione, FD20

Diphenylcarbinol, FB95

1,2-Diphenylcyclobuta[b]naphthalene-3,8-diol, FE67

1,2-Diphenylcyclobuta[b]naphthalene-3,8-diol dipotassium salt, FE64

1,2-Diphenylcyclobuteno[3,4-b]-3,8-naphthohydroquinone, FE67

2,3-Diphenyl-1,4-diacetyl-1,4-diazabenzene, FE12

2,3-Diphenyl-1,4-diacetyl-1,4-H-pyrazine, FE12

Diphenyldiazene, FB40

5,10-Diphenyl-2,7-dimethoxy-5,10-dihydrophenazine, FF00

5,10-diphenyl-2,7-dimethyl-5,10-dihydrophenazine, FE99

trans(?)-1,4-Diphenyl-1,4-dioxo-2-butene, FD20

5,5'-Diphenyl-3,3'-(2,5-diphenyl-3,4-dithia-1,6-hexamethylene)dithiolium diperchlorate, FF47

Diphenyl diselenide, FB46

2,5-Diphenyl-[1,2]dithiolo[1,5,b]-[1,2]dithiole-7-S^{IV}, FD56

Diphenylene disulfide, FB25

Diphenylenemethane, FB77

1,1-Diphenylethene, FC19

trans-1,2-Diphenylethene, FC21

unsym-Diphenylethylene, FC19

1,1-Diphenylethylene, FC19

trans-1,2-Diphenylethylene, FC21

Diphenyl ketone, FB83

Diphenylmethanol, FB95

Diphenylmethoxymethane, FC50

3,5-Diphenyl-4-(2-methoxyphenyl-1-thiocarbamoylpyrazole, FE43

3,5-Diphenyl-4-(4-methoxyphenyl)-1-thiocarbamoylpyrazole, FE44

1,2-Diphenyl-1-methylethene, FC82

Diphenylmethyl methyl ether, FC50

1-Diphenylmethyl-4-[(6-methyl-2-pyridyl)methyleneamino]piperazine, FE80

4-(Diphenylmethyl)-N-[(6-methyl-2-pyridinyl)methylene]-1-piperazinamine, FE80

1-(Diphenylmethyl)-4-nitrosopiperazine, FD59

3,5-Diphenyl-4-(2-methylphenylazo)-1-thiocarbamoylpyrazole, FE45

3,5-Diphenyl-4-(3-methylphenylazo)-thiocarbamoylpyrazole, FE46

3,5-Diphenyl-4-(4-methylphenylazo)-1-thiocarbamoylpyrazole, FE47

2,3-Diphenyl-1,4-naphthoquinone, FE31

Diphenyl 4-nitrobenzyl phosphate, FD85

1,5-Diphenyl-1,5-pentandione, FD57

3,5-Diphenyl-4-(phenylazo)-1H-pyrazole-1-carbothioamide, FE35

3,5-Diphenyl-4-phenylazo-4-thiocarbamoylpyrazole, FE35

Diphenyl phthalate, FE08

1,1-Diphenylpropargyl alcohol, FC77

1,2-Diphenylpropene, FC82

1,1-Diphenylprop-2-en-1-ol, FC84

3,3-Diphenylprop-2-en-1-ol, FC85

1,3-Diphenylpropenone, FC76

trans-1,3-Diphenyl-2-propen-1-one, FC76

1,3-Diphenyl-2-propen-1-one phenylhydrazone, FE23

1,1-Diphenylprop-2-yn-1-ol, FC77

2,5-Diphenyl-1,1,4,4-tetraethyl-1,4-diphosphoniacyclohexa-2,5-diene dichloride, FE84

2,5-Diphenyl-1,1,4,4-tetra(4-fluorophenyl)diphosphoniacyclohexa-2,5-diene dichloride, FF61

1,3-Diphenyl-1-triazene, FB51

2,5-Diphenyl-3,3a,4-trithiapentalene, FD56

Dipotassium 1,2-diphenylbuteno-[3,4-b]-3,8-naphthohydroquinone, FE64

Dipyridinotetraphenylporphyrinylruthenium, FF77

2,2'-Dipyridyl, FA89

Disodium 2',4',5',7'-tetraiodofluorescein, FD97

α,α'-Dithiobisformamidine, FA03

β,β'-Dithiobis[3-(4-methylstyryl)-5-anisyl-1,2-dithiolium] perchlorate, FF60

β,β'-Dithiobis[3-(4-methylsytryl)-4,5-dianisyl-1,2-dithiolium] perchlorate, FF73

β,β'-Dithiobis[3-(4'-methylsytryl)-5-phenyl-1,2-dithiolium] perchlorate, FF52

β,β'-Dithiobis(3-styryl-4,5-diphenyl-1,2-dithiolium) perchlorate, FF70

β,β'-Dithiobis(3-stryl-5-phenyl-1,2-dithiolium) perchlorate, FF47

4,4'-Dithiocyanatodiphenylamine, FC09

Di(4-thiocyanatophenyl)amine, FC09

5,10-Di(4-thiocyanatophenyl)-2,7-dithiocyanato-5,10-dihydrophenazine, FF10

2,17-divinyl-1,10,19,22,23,24-hexahydro-3,7,13,18-tetramethyl-1,19-dioxo-21H-biline-8,12-dipropionic acid, FF45

Erythrosine, FD97

1,1'-(1,2-Ethandiyl)bis[3-methoxybenzene], FD40

Ethanol, FA04

1,1',1'', 1'''-(1,2-Ethenediylidene)-tetrakis[4-nitrobenzene], FE92

Ethenetetracarbonitrile, FA31

1,1'-Ethenylidenebisbenzene, FC19

1-Ethenyl-4-methoxybenzene, FA79

4-Ethoxycarbonyldiazoaminobenzene, FC92

3-(4-Ethoxycarbonylphenyl)-1-phenyltriazene, FC92

(3-Ethoxy-2,3-dioxopropyl)dimethylsulfonium bromide, FA62

2-(2-Ethoxy-2-methylpropyl)-6-(2,2-dimethylethyl) 1,4-benzenediol, FD45

2-(2-Ethoxy-2-methylpropyl)-6-(2,2-dimethylethyl)hydroquinone, FD45

2-(2-Ethoxy-2-methylpropyl)-5-(1,1-dimethylethyl)-1,4-hydroquinone, FD44

4-(2-Ethoxyphenylazo)-3,5-dimethylpyrazole-1-carbothioamide, FC66

4-(4-Ethoxyphenylazo)-3,5-dimethyl-1H-pyrazole-1-carbothioamide, FC67

3-(4-Ethoxyphenyl)-1-(2-thienyl)-2-propen-1-one, FC87

4-Ethoxy-ω-thenoylstyrene, FC87

Ethyl alcohol, FA04

Ethylamine, FA05

(N-Ethyl-1-azonia-2-naphthyl)(N-ethyl-1-aza-1,4-dihydronaphth-4-ylidene)methane iodide, FE50

1-(N-Ethyl-1-azonia-2-naphthyl)-3-(N-ethyl-1-aza-1,2-dihydronaphth-2-ylidene)-1-propene bromide, FE88

(N-Ethyl-1-azonia-2-naphthyl)(N-ethyl-1-azonia-1,2-dihydro-2-ylidene)methane iodide, FE49

4-Ethylbiphenyl, FC32

Ethyl 3-bromo-2-oxopropanoate, FA21

Ethyl β-bromopyruvate, FA21

Ethyl trans-2-butenoate, FA49

Ethyl cinnamate, FB10

Ethyl crotonate, FA49

Ethyl α-cyanocinnamate, FB49

Ethyl 2-cyano-3-phenyl-2-propenoate, FB49

Ethyl 3-dimethylsulfonio-2-oxopropanoate bromide, FA62

Ethyl dimethylsulfoniopyruvate bromide, FA62

Ethyl β-dimethylsulfonium-α-oxopropanoate bromide, FA62

Ethyl 3-dimethylsulfonium-2-oxopropanoate bromide, FA62

Ethyleneglycol phthalate, FA90

3-Ethyl-2-[7-(3-ethyl-2(3H)-benzothiazolylidene)-1,3,5-heptatrienyl]benzothiazolium perchlorate, FE89

3-Ethyl-2-[(3-ethyl-2(3H)-benzothiazolylidene)methyl]benzothiazolium bromide, FD89

3-Ethyl-2-[2-[(3-ethyl-2(3H)-benzothiazolylidene)methyl]-1-butenyl]-benzothiazolium bromide, FE52

3-Ethyl-2-[(3-ethyl-2(3H)-benzothiazolylidene)-2-methyl-1-propenyl]benzothiazolium bromide, FE38

3-Ethyl-2-[5-(3-ethyl-2(3H)-benzothiazolylidene)-1,3-pentadienylbenzothiazolium iodide, FE51

3-Ethyl-2-[3-(3-ethyl-2(3H)-benzothiazolylidene)-2-propenyl]benzothiazolium bromide, FE25

1-Ethyl-2[3-(3-ethyl-2(3H)-benzothiazolylidene)-2-propenyl]-quinolinium bromide, FE48

Ethyl S-ethyl-β-mercaptopyruvate, FA61

1-Ethyl-2-[(1-ethyl-2(1H)-quinolinylidene)methyl]quinolinium iodide, FE49

1-Ethyl-2-[(1-ethyl-4(1H)-quinolinylidene)methyl]quinolinium iodide, FE50

1-Ethyl-2-[3-(1-ethyl-2(1H)-quinolinylidene)-1-propenyl]quinolinium bromide, FE88

2-Ethyl-1(N-ethyl-1H-1-thia-3-azonia-2-indenyl)-3-(N-ethyl-1-thia-3-azaindan-2-ylidene)-1-propane bromide, FE52

Ethyl 3-ethylthio-2-oxopropanoate, FA61

Ethyl ethylthiopyruvate, FA61

N-Ethyl-N-hydroxy-β-methoxy-3-(trifluoromethyl)benzeneethanamine, FB61

N-Ethyl-N-hydroxy-β-methoxy-β-(3-trifluoromethylphenyl)ethylamine, FB61

S-Ethyl-β-mercaptopyruvic acid, FA25

N-Ethyl-2-methoxy-2-(3-trifluoromethylphenyl)ethylhydroxylamine, FB61

Ethyl 2-oxopropanoate, FA24

Ethyl β-phenylacrylate, FB10

Ethyl phenyl ketone, FA81

Ethyl 3-phenylpropenoate, FB10

Ethyl 4-(3-phenyl-1-triazenyl)-benzoate, FC92

Ethyl phthalate, FB60

Ethyl pyruvate, FA24

Ethylstannic trichloride, FA02

1-(N-Ethyl-1H-1-thia-3-azonia-2-indenyl)-3-(N-ethyl-1-aza-1,2-dihydronaphth-2-ylidene-1-propene bromide, FE48

1-(N-Ethyl-1H-1-thia-3-azonia-2-indenyl)-5-(N-ethyl)-thia-3-azaindan-2-ylidene)-1,3-pentadiene iodide, FE51

1-(N-Ethyl-1H-1-thia-3-azonia-2-indenyl)-3-(N-ethyl-1-thia-3-azaindan-2-ylidene)-1-propene bromide, FE25

1-(N-Ethyl-1H-1-thia-3-azonia-2-indenyl)-7-(N-ethyl-1-thia-3-azoniaindan-2-ylidene)-1,3,5-heptatriene perchlorate, FE89

3-(Ethylthio)-2-oxopropanoic acid, FA25

β-Ethylthiopyruvic acid, FA25

Ethyltin(IV) trichloride, FA02

α-Ethynylbenzenemethanol, FA78

α-Ethynyl-α-methylbenzenemethanol, FA94

α-Ethynyl-α-phenylbenzenemethanol, FC77

Etioporphyrin I, FF42

Ferricinium picrate, FD18

Ferricinium 2,4,6-trinitrophenolate, FD18

Ferrocene, FA93

Fluoranthene, FF83

Fluorene, FB77

9H-Fluorene, FB77

9H-Fluoren-9-ol, FB84

9-Fluorenol, FB84

9-Fluorenone, FB71

Fluorescein, disodium salt, FD98

6α-Fluoro-17α-acetoxyprogesterone, FE59

6β-Fluoro-17α-acetoxyprogesterone, FE60

6α-Fluoro-17,21-dihydroxy-16α-methylpregn-4-ene-3,20-dione, FE39

6β-Fluoro-17,21-dihydroxy-16α-methylpregn-4-ene-3,20-dione, FE40

3-Fluoro-4'-dimethylaminoazobenzene, FC38

4-Fluoro-4'-dimethylaminoazobenzene, FC39

6α-Fluoro-17β-hydroxyandrost-4-en-3-one, FD93

6β-Fluoro-17β-hydroxyandrost-4-en-3-one, FD94

6α-Fluoro-16α-methyl-Δ^4-pregnene-17α,21-diol-3,20-dione, FE39

6β-Fluoro-16α-methyl-Δ^4-pregnene-17α,21-diol-3,20-dione, FE40

4-[(3-Fluorophenyl)azo]-N,N-dimethylaniline, FC38

4-[(4-Fluorophenyl)azo]-N,N-dimethyl-1-aniline, FC39

3-(4-Fluorophenyl)-1-(2-thienyl)-2-propen-1-one, FB74

4-Fluorostyryl 2-thienyl ketone, FB74

6α-Fluorotestosterone, FD93

6β-Fluorotestosterone, FD94

Folic acid, FD90

Formamidine disulfide, FA03

Fumaronitrile, FA14

1-(2-Furyl)-3-[5-(4-chlorophenyl)-2-furyl]-2-propen-1-one, FD49

1-(2-Furyl)-2-methyl-3-(5-phenyl-2-furyl)-2-propen-1-one, FD66

1-(2-Furyl)-3[5-(4-nitrophenyl)-2-furyl]-2-propen-1-one, FD50

1-(2-Furyl)-3-(5-phenyl-2-furyl)-2-propen-1-one, FD53

Hexafluorobenzene, FA30

1-Hexanol, FA50

1,1,2,4,4,5-Hexaphenyldiphosphonia-cyclohexa-2,5-diene dibromide, FF62

1,1,2,4,4,5-Hexaphenyldiphosphonia-cyclohexa-2,5-diene dichloride, FF63

n-Hexyl alcohol, FA50

trans-α,α'-Hydrazotoluene, FC65

Hydridomonocarbonyltris(triphenylphosphine)iridium(I), FF78

Hydridomonocarbonyltris(triphenylphosphine)rhodium(I), FF79

Hydroquinol, FA45

Hydroquinone, FA45

3-Hydroxyamino-1-(2-hydroxy-4-methoxyphenyl)-3-(3-pyridyl)-1-propanone oxime, FD04

3-Hydroxylamino-1-(2-hydroxy-4-methoxyphenyl)-3-(3-pyridyl)-propylideneazenol, FD04

3-(Hydroxyamino)-1-(2-hydroxyphenyl)-3-(2-pyridyl)-1-propanone oxime, FC60

3-Hydroxyamino-1-(2-hydroxyphenyl)-3-(3-pyridyl)-1-propanone oxime, FC61

3-Hydroxyamino-1-(2-hydroxyphenyl)-3-(3-pyridyl)propylideneazenol, FC61

17β-Hydroxyandrosta-1,4-dien-3-one, FD92

17β-Hydroxy-Δ^4-androsten-3-one, FD95

17β-Hydroxy-4-androsten-3-one, FD95

8-Hydroxy-1-azanaphthalene oxime, FA77

4-Hydroxyazobenzene, FB41

4-Hydroxyazobenzene-4'-sulfonic acid, FB45

8-Hydroxy-1-benzazine, FA77

1-(2-Hydroxybenzene)-3-(2-pyridine)-2-propen-1-one, FC14

1-(2-Hydroxybenzene)-3-(3-pyridine)-2-propen-1-one, FC15

1-(2-Hydroxybenzene)-3-(4-pyridine)-2-propen-1-one, FC16

8-Hydroxybenzo[b]pyrazine, FA77

2-(3-Hydroxybenzylidene)-1,3-indandione, FD11

4-Hydroxy-4'-bromoazobenzene, FB26

1-Hydroxybutane, FA18

2-Hydroxybutane, FA19

4-Hydroxy-3-carboxyazobenzene, FB79

4-Hydroxy-3-carboxy-2'-chloroazobenzene, FB72

4-Hydroxy-3-carboxy-4'-chloroazobenzene, FB73

4-Hydroxy-3-carboxy-2',4'-dinitroazobenzene, FB69

4-Hydroxy-3-carboxy-4'-nitroazobenzene, FB76

4-[(4-Hydroxy-3-carboxyphenyl)azo]-benzenesulfonic acid, disodium salt, FB68

4-Hydroxy-1,3-diaminobenzene, FA46

9-Hydroxy-9,10-dihydroanthracene, FC25

2-Hydroxy-4'-dimethylaminoazobenzene, FC58

4-Hydroxy-4'-dimethylaminoazobenzene, FC59

2-Hydroxy-5-(2,4-dinitrophenylazo)-benzoic acid, FB69

4-Hydroxydiphenylamine, FB48

4-Hydroxydiphenyldiazene, FB41

3-Hydroxy-3,3-diphenyl-1-propene, FC84

3-Hydroxy-3,3-diphenyl-1-propyne, FC77

Hydroxyethane, FA04

9-(1-Hydroxyethyl)anthracene, FD26

4-(1-Hydroxyethyl)biphenyl, FC52

1-(1-Hydroxyethyl)naphthalene, FB55

2-(1-Hydroxyethyl)naphthalene, FB56

9-Hydroxyfluorene, FB84

1-Hydroxyhexane, FA50

3-Hydroxy-1-(2-hydroxy-4-methoxyphenyl)-3-(2-pyridyl)-1-propanone oxime, FC95

3-Hydroxy-1-(2-hydroxy-4-methoxyphenyl)-3-(3-pyridyl)-1-propanone oxime, FC96

3-Hydroxy-1-(2-hydroxy-4-methoxyphenyl)-3-(4-pyridyl)-1-propanone, oxime, FC97

3-Hydroxy-1-(2-hydroxy-4-methoxyphenyl)-3-(2-pyridyl)propylideneazenol, FC95

3-Hydroxy-1-(2-hydroxy-4-methoxyphenyl)-3-(3-pyridyl)propylideneazenol, FC96

3-Hydroxy-1-(2-hydroxy-4-methoxyphenyl)-3-(4-pyridyl)propylideneazenol, FC97

5-Hydroxy-4[(2-hydroxy-1-naphthyl)-azo]-3-methyl-1-phenylpyrazole, FE10

3-Hydroxy-4-[(2-hydroxy-1-naphthyl)-azo]-7-nitronaphthalenesulfonic acid, sodium salt, FE00

4-Hydroxy-ω-(3-hydroxy-2,4,6-tribromo)styrene, FC69

4-Hydroxy-4'-iodoazobenzene, FB31

Hydroxymethane, FA01

1-(2-Hydroxy-4-methoxyphenyl)-3-hydroxy-3-(2-pyridyl)-1-propanone oxime, FC95

1-(2-Hydroxy-4-methoxyphenyl)-3-(4-pyridyl)-1,3-propanedione dioxime, FC93

1-(2-Hydroxy-4-methoxyphenyl)-3-(2-pyridyl)-2-propen-1-one, FC79

1-(2-Hydroxy-4-methoxyphenyl)-3-(3-pyridyl)-2-propen-1-one, FC80

1-(2-Hydroxy-4-methoxyphenyl)-3-(4-pyridyl)-2-propen-1-one, FC81

3-(4-Hydroxy-3-methoxyphenyl)-1-(2,4,6-tribromo-3-hydroxyphenyl)-2-propen-1-one, FD16

3'-Hydroxy-3-methoxy-2',4',6'-tribromochalcone, FD14

3'-Hydroxy-4-methoxy-2',4',6'-tribromochalcone, FD15

N-Hydroxy-β-methoxy-3-(trifluoromethyl)benzeneethanamine, FB01

N-Hydroxy-β-methoxy-β-(3'-trifluoromethylphenyl)ethylamine, FB01

17β-Hydroxy-6α-methylandrosta-4-en-3-one, FE15

17β-Hydroxy-6β-methylandrost-4-en-3-one, FE16

4-Hydroxy-3'-methylazobenzene, FB90

4-Hydroxy-4'-methylazobenzene, FB91

2-Hydroxy-2-methylbutane, FA27

3-Hydroxy-3-methyl-1-butyne, FA22

9-Hydroxy-9-methylfluorene, FC26

5-Hydroxy-3-methyl-1-phenylpyrazole-4-azo-1'-(2'-hydroxynaphthalene), FE10

3'-Hydroxy-4-methyl-2',4',6'-tribromochalcone, FD13

4-Hydroxy-2'-nitroazobenzene, FB32

4-Hydroxy-3'-nitroazobenzene, FB33

4-Hydroxy-4'-nitroazobenzene, FB34

4-Hydroxynitrobenzene, FA42

2-Hydroxy-5-(2-nitrophenylazo)-benzoic acid, FB75

2-Hydroxy-5-[(4-nitrophenyl)azo]-benzoic acid, FB76

N-Hydroxy-4-nitroso-N-phenylbenzamine, FB44

2-Hydroxy-6-nitro-4-sulfonaphthalene-1-azo-(1'-hydroxy-2'-naphthalene), sodium salt, FE00

1-Hydroxy-3-oxoindan-2-yl benzoate, FD21

1-Hydroxypentane, FA28

4-Hydroxyphenylazobenzene, FB41

4-[(4-Hydroxyphenyl)azo]benzenesulfonic acid, FB45

2-Hydroxy-5-phenylazobenzoic acid, FB79

4-(4-Hydroxyphenylazo)benzoic acid, FB80

N-Hydroxy-N-phenylbenzamide, FB86

N-(4-Hydroxyphenyl)-N'-(4-bromophenyl)diazene, FB26

3-Hydroxy-3-phenyl-1-butyne, FA94

N-(4-Hydroxyphenyl)-N'-(4-chlorophenyl)diazene, FB27

p-Hydroxy-m-phenylenediamine, FA46

1-(2-Hydroxyphenyl)-3-hydroxylamino-3-(2-pyridyl)-1-propanone oxime, FC60

1-(2-Hydroxyphenyl)-3-hydroxylamino-3-(3-pyridyl)-1-propanone oxime, FC61

1-(2-Hydroxyphenyl)-3-hydroxy-3-(2-pyridyl)-2-propen-1-one, FC18

N-(4-Hydroxyphenyl)-N'-(4-iodophenyl)-diazene, FB31

N-(4-Hydroxyphenyl)-N'-(4-methoxyphenyl)diazene, FB80

2-[(2-Hydroxyphenyl)methylene]-1H-indene-1,3(2H)-dione, FD11

N-(4-Hydroxyphenyl)-N'-(2-nitrophenyl)diazene, FB32

N-(4-Hydroxyphenyl)-N'-(3-nitrophenyl)diazene, FB33

N-(4-Hydroxyphenyl)-N'-(4-nitrophenyl)diazene, FB34

1-(4-Hydroxyphenyl)-2-phenyldiazene, FB41

3-Hydroxy-3-phenyl-1-propyne, FA78

1-(2-Hydroxyphenyl)-3-(4-pyridyl)-1,3-propanedione dioxime, FC31

1-(2-Hydroxyphenyl)-3-(2-pyridyl)-2-propen-1-one, FC14

1-(2-Hydroxyphenyl)-3-(3-pyridyl)-2-propen-1-one, FC15

1-(2-Hydroxyphenyl)-3-(4-pyridyl)-2-propen-1-one, FC16

1-(4-Hydroxyphenyl)-2-(4-sulfophenyl)-diazene, FB45

3-(2-Hydroxyphenyl)1-(2,4,6-tribromo-3-hydroxyphenyl)-2-propen-1-one, FC69

1-Hydroxypropane, FA11

2-Hydroxypropane, FA12

3-Hydroxy-1-propene, FA10

3-Hydroxypropyne, FA08

2'-Hydroxy-3-(2-pyridyl)acrylophenone, FC14

2'-Hydroxy-3-(3-pyridyl)acrylophenone, FC15

2'-Hydroxy-3-(4-pyridyl)acrylophenone, FC16

8-Hydroxyquinoline, FA77

2-Hydroxy-5-[(4-sulfophenyl)azo]-benzoic acid, disodium salt, FB68

2-{[3-Hydroxy-5-(3-sulfophenyl)]-S-triazinylamino}benzenesulfonic acid, FF24

3'-Hydroxy-2',4',6'-tribromochalcone, FC68

4-Hydroxy-ω-(2,4,6-tribromo-3-hydroxy)-styrene, FC69

ω-(3-Hydroxy-2,4,6-tribromo)styrene, FC68

Hydroxytriphenylarsonium perchlorate, FD72

N,N'-(Iminodi-1,4-phenylene)bis-acetamide, FD36

3-Imino-2-phenyl-3H-indole, FC12

2H-Indan-1,3-dion-2-yl benzoate, FD12

1,2,3-Indantrione (2-monohydrate), FA72

3-Iodo-4'-dimethylaminoazobenzene, FC40

4-Iodo-4'-dimethylaminoazobenzene, FC41

4-[(3-Iodophenyl)azo]-N,N-dimethylaniline, FC40

4-[(4-Iodophenyl)azo]-N,N-dimethylaniline, FC41

4-[(4-Iodophenyl)azo]phenol, FB31

Isobutyrophenone, FB04

Isophorone, FA83

Isopropenyl methyl ketone, FA23

Isopropyl alcohol, FA12

Isopropylidene acetone, FA48

Isoquinoline, FA75

Leuco Crystal Violet, FE90

Magnesium(II)mesoporphyrin IX dimethyl ester, FF55

Magnesium(II)protoporphyrin IX dimethyl ester, FF53

3-Mercapto-N-[(2-benzyl-5-oxo-2-oxazolin-4-ylidene)methyl]-3-mercaptovaline, FD39

Mesitylene, FA82

Mesityl oxide, FA48

Mesobilirubin(IX)α, FF46

Mesoporphin IX dimethyl ester complex with Mg(II), FF55

Methanol, FA01

4-Methoxy-4'-aminodiphenylamine, FC00

2-Methoxyaniline, FA58

3-Methoxyaniline, FA59

4-Methoxyaniline, FA60

2-Methoxybenzenamine, FA58

3-Methoxybenzenamine, FA59

4-Methoxybenzenamine, FA60

2-(3-Methoxybenzoyl)acetanilide, FD31

2-(4-Methoxybenzoyl)acetanilide, FD32

2-(4-Methoxybenzylidene)-1,3-indandione, FD54

4-Methoxydiazoaminobenzene, FB99

4-Methoxydiphenylamine, FB97

3-(3-Methoxy-4-hydroxyphenyl)-1-(2,4,6-tribromo-3-hydroxyphenyl)-2-propen-1-one, FD16

9-Methoxy-9-methylfluorene, FC86

3-Methoxy-β-oxo-N-phenylbenzenepropanamide, FD31

4-Methoxy-β-oxo-N-phenylbenzenepropanamide, FD32

4-(4-Methoxyphenylamino)acetophenone, FC89

N-[4-[(4-Methoxyphenyl)amino]phenyl]-acetamide, FC94

1-[4-[(4-Methoxyphenyl)amino]phenyl]-ethanone, FC89

4-[(4-Methoxyphenyl)azo]-N,N-dimethylbenzenamine, FD03

4-[(2-Methoxyphenyl)azo]-3,5-dimethyl-1H-pyrazole-1-carbothioamide, FC03

4-[(4-Methoxyphenyl)azo]-3,5-dimethyl-1H-pyrazole-1-carbothioamide, FC04

4-[(2-Methoxyphenyl)azo]-3,5-diphenyl-1H-pyrazole-1-carbothioamide, FE43

4-(4-Methoxyphenylazo)-3,5-diphenyl-1H-pyrazole-1-carbothioamide, FE44

4-Methoxy-N-phenylbenzenamine, FB97

(4-Methoxyphenyl)ethene, FA79

3-(4-Methoxyphenylimino)-2-phenyl-3H-indole, FE19

3-(4-Methoxyphenylimino)-2-phenyl-3H-indole 1-oxide, FE21

4-Methoxy-N-(2-phenyl)-3H-indolylidene)benzamine, FE19

2-[(4-Methoxyphenyl)methylene]-1H-indene-1,3(2H)-dione, FD54

3-(3-Methoxyphenyl)-3-oxo-N-phenyl-propanamide, FD31

TABLE IV. Compounds Included in Table I

3-(4-Methoxyphenyl)-3-oxo-N-phenyl-propanamide, FD32

N-(4-Methoxyphenyl)-4-phenylenediamine, FC00

(4-Methoxyphenyl)phenylmethyl methyl ether, FC99

1-(4-Methoxyphenyl)-3-phenyl-1-triazene, FB99

1-Methoxy-1-phenylpropene, FB02

3-Methoxy-1-phenylpropene, FB03

3-Methoxy-3-phenyl-1-propyne, FA95

3-(3-Methoxyphenyl)-1-(2,4,6-tribromo-3-hydroxyphenyl)-2-propen-1-one, FD14

3-(4-Methoxyphenyl)-1-(2,4,6-tribromo-3-hydroxyphenyl)-2-propen-1-one, FD15

(1-Methoxy-1-propen-1-yl)benzene, FB02

(3-Methoxy-1-propenyl)benzene, FB03

(1-Methoxy-2-propyn-1-yl)benzene, FA95

p-Methoxystyrene, FA79

α-Methoxy-3-(trifluoromethyl)benzeneacetaldehyde oxime, FA92

2-Methoxy-2-(3-trifluoromethyl)ethylazanol, FB01

α-Methoxy-α-(3'-trifluoromethylphenyl)acetadoxime, FA92

β-Methoxy-β-(3-trifluoromethylphenyl)ethylhydroxylamine, FB01

2-Methoxy-2(3-trifluoromethylphenyl)-1-(hydroxyazylene)ethane, FA92

2-Methoxy-2(3-trifluoromethylphenyl)-1-(hydroxyimino)ethane, FA92

Methoxytriphenylmethane, FE13

6α-Methyl-17α-acetoxyprogesterone, FE84

4-Methylaminoazobenzene, FB98

1-(4-Methylaminophenyl)-2-phenyldiazene, FB98

6α-Methyl-Δ^4-androsten-17β-ol-3-one, FE15

6β-Methyl-Δ^4-androsten-17β-ol-3-one, FE16

α-Methyl-9-anthracenemethanol, FD26

3-Methyl-2-azanaphthalene, FA91

Methylbenzene, FA55

2-(3-Methylbenzoyl)acetanilide, FD29

2-(4-Methylbenzoyl)acetanilide, FD30

2-(4-Methylbenzylidene)-1,3-indandione, FD52

4-(Methylbenzylidenemalonitrile, FB08

2-Methyl-1,2-bis(3-pyridyl)-1-propanone, FC46

2-Methyl-2-butanol, FA27

3-Methyl-3-buten-2-one, FA23

2-Methyl-3-butyn-2-ol, FA22

1-Methyl-3-carbamoylpyridinium chloride, FA57

Methyl cinnamate, FA98

6-Methyl-Δ^6-dehydro-17α-acetoxyprogesterone, FE83

6α-Methyl-Δ^1-dehydro-17α-acetoxyprogesterone, FE82

4-Methyl-1,2-diacetyl-1,2-dihydro-1,2-diazanaphthalene, FC01

4-Methyl-4'-dimethylaminoazobenzene, FD01

2-[2-Methyl-2(1,1-dimethylethoxy)-propyl]-5-(1,1-dimethylethyl)-1,4-hydroquinone, FD76

4-Methyldiphenylamine, FB96

2-Methyl-1,2-di(3-pyridyl)-1-propanone, FC46

2,2'-Methylenebiphenyl, FB77

1,1'-Methylenebis[3-methoxybenzene], FC98

Methylene Blue, FD38

2-(3,4-Methylenedioxybenzylidene)-1,3-indandione, FD48

6-(2-Methyl-2-ethoxypropyl-2-1,1-dimethylethyl)-1,4-hydroquinone, FD45

2-(2-Methyl-2-ethoxypropyl)-5-(1,1-dimethylpropyl)-1,4-hydroquinone, FD62

2-Methyl-1-(N-ethyl-1H-1-thia-3-azonia-2-indenyl)-3-(N-ethyl-1-thia-3-azaindan-2-ylidene)-1-propene bromide, FE38

9-Methyl-9H-fluoren-9-ol, FC26

3-Methyl-4'-hydroxyazobenzene, FB90

4-Methyl-4'-hydroxyazobenzene, FB91

4,4',4''-Methylidynetris[N,N-dimethylaniline], FE90

3-Methylisoquinoline, FA91

Methyl 3-mercaptopropanoate, FA16

Methyl 3-mercaptopropionate, FA16

S-Methyl-β-mercaptopyruvic acid, FA15

2-(2-Methyl-2-methoxypropyl)-5-(1,1-dimethylpropyl)-1,4-hydroquinone, FD46

2-[2-Methyl-2-(1-methylethoxy)propyl]-5-(1,1-dimethylethyl)-1,4-hydroquinone, FD61

4-Methyl-N-(4-methylphenyl)benzenamine, FC55

2-Methylnaphthalene, FB09

α-Methyl-1-naphthalenemethanol, FB55

α-Methyl-2-naphthalenemethanol, FB56

1-Methylnicotinamide chloride, FA57

3-Methyl-4-oxo-2-pentene, FA47

3-Methyl-β-oxo-N-phenylbenzenepropanamide, FD29

4-Methyl-β-oxo-N-phenylbenzenepropanamide, FD30

Methylpentafluorobenzene, FA52

3-Methyl-3-penten-2-one, FA47

4-Methyl-3-penten-2-one, FA48

10-Methylphenothiazine, FB87, FF82

Methyl β-phenylacrylate, FA98

N-Methyl-4-phenylazobenzenamine, FB98

4-(2-Methylphenylazo)-3,5-diphenyl-pyrazole-1-carbothioamide, FE45

4-(3-Methylphenylazo)-3,5-diphenyl-1H-pyrazole-1-carbothioamide, FE46

4-(4-Methylphenylazo)-3,5-diphenyl-1H-pyrazole-1-carbothioamide, FE47

4-[(3-Methylphenyl)azo]phenol, FB90

4-[(4-Methylphenyl)azo]phenol, FB91

4-Methyl-N-phenylbenzenamine, FB96

1-(3-Methylphenyl)-2-(4-hydroxyphenyl)diazene, FB90

1-(4-Methylphenyl)-2-(4-hydroxyphenyl)diazene, FB91

Methyl phenyl ketone, FA67

2-[(4-Methylphenyl)methylene]-1H-indene-1,3(2H)-dione, FD52

[(4-Methylphenyl)methylene]propanedinitrile, FB08

3-(3-Methylphenyl)-3-oxo-N-phenyl-propanamide, FD29

3-(4-Methylphenyl)-3-oxo-N-phenyl-propanamide, FD30

2-(4-Methylphenyl)-5-phenyl [1,2]-dithiolo[1,5-b][1,2]dithiole-7-S^N, FD67

2-Methyl-1-phenyl-3-(5-phenyl-2-furyl)-2-propen-1-one, FE11

2-(4-Methylphenyl)-5-phenyl-3,3a,4-trithiapentalene, FD67

2-Methyl-1-phenylpropanone, FB04

Methyl 3-phenylpropenoate, FA98

Methyl 3-phenyl-2-propen-1-yl ether, FB03

Methyl 1-phenyl-2-propyn-1-yl ether, FA95

3-(2-Methylphenyl)-1-(2-thienyl)-2-propen-1-one, FC27

3-(4-Methylphenyl)-1-(2-thienyl)-2-propen-1-one, FC28

3-[(4-Methylphenyl)thio]-2-oxopropanoic acid, FA99

β-(4-Methylphenylthio)pyruvic acid, FA99

3-(4-Methylphenyl)-1-(2,4,6-tribromo-3-hydroxyphenyl)-2-propen-1-one, FD13

Methyl propargylic ether, FA95

2-Methylpropenyl methyl ketone, FA48

1-Methylpyridinium-3-carbamoyl chloride, FA57

Methylstannic trichloride, FA00

6α-Methyltestosterone, FE15

6β-Methyltestosterone, FE16

2-Methyl-3-(all trans-3,7,11,15-tetramethyl-2,6,10,14-hexadecatetraenyl)-1,4-naphthoquinone, FF34

Methylthionine chloride, FD38

3-Methylthio-2-oxopropanoic acid, FA15

Methyl 3-thiopropanoate, FA16

3-Methylthiopyruvic acid, FA15

Methyltin(IV) trichloride, FA00

Methyl triphenylmethyl ether, FE13

Metyrapone, FC46

Mordant Orange I, FB76

C.I. Mordant Yellow 10Y, FB68

Naphthalene, FF81

3-(1-Naphthyl)-1-(2-thienyl)-2-propen-1-one, FD51

1,5-Naphthyridine, FA63

Niacinamide, FA44

TABLE IV. Compounds Included in Table I

Nicotinamide, FA44

Nicotinamide adenine dinucleotide, FE28

Nicotinamide mononucleotide, FB13

Nicotinic acid amide, FA44

Ninhydrin, FA72

4-Nitrobenzyl diphenyl phosphate, FD85

3-Nitro-4'-dimethylaminoazobenzene, FC48

4-Nitro-4'-dimethylaminoazobenzene, FC49

4-Nitrodiphenylamine, FB43

2-Nitro-4'-hydroxy-3'-carboxyazobenzene, FB75

4-Nitro-4'-hydroxy-3'-carboxyazobenzene, FB76

4'-Nitro-N-(4-nitrophenyl)benzenamine, FB35

Nitropentachlorobenzene, FA29

4-Nitrophenol, FA42

4-[(2-nitrophenyl)azo]phenol, FB32

4-[(3-Nitrophenyl)azo]phenol, FB33

4-[(4-Nitrophenyl)azo]phenol, FB34

5-(p-Nitrophenylazo)salicylic acid, FB76

4-Nitro-N-phenylbenzenamine, FB43

N-(2-Nitrophenyl)-N'-(3-carboxy-4-hydroxyphenyl)diazene, FB75

N-(4-Nitrophenyl)-N'-(3-carboxy-4-hydroxyphenyl)diazene, FB76

1-(4-Nitrophenyl)-3,5-diphenyl-Δ^2-pyrazoline, FE22

3-[5-(4-Nitrophenyl)-2-furanyl-1-(2-furanyl)-2-propen-1-one, FD50

(4-Nitrophenyl)methyl diphenyl phosphate, FD85

4-Nitrosodiphenylhydroxylamine, FB44

N-(4-Nitrosophenyl)-N-phenylhydroxylamine, FB44

Nitrovin, FB89

NMN, FB13

Octadecyldimethylbenzyl ammonium chloride, FF89

2,3,7,8,12,13,17,18-Octaethylporphine, FF56

Octafluoronaphthalene, FA87

β-Oxo-N-phenylbenzenepropanamide, FC78

1-Oxo-1,2,3,4-tetrahydronaphthalene, FA97

4-[3-Oxo-3-(2-thienyl)-1-propenyl]-benzamide, FC17

Oxotriphenylphosphorane, FD69

Oxotriphenylphosphorus, FD69

1,1'-[Oxybis(methylene)]bisbenzene, FC51

1,1',1'',1'''-(Oxydimethylidyne)-tetrakisbenzene, FF01

Paraoxon, FB06

Parathion, FB05

Pentachloronitrobenzene, FA29

Pentafluorobenzene, FA32

Pentafluoromethylbenzene, FA52

2,3,4,5,6-Pentafluorotoluene, FA52

1-Pentanol, FA28

Pentaphenylphosphorane, FF30

n-Pentyl alcohol, FA28

Perfluorobenzene, FA30

Perfluorobiphenyl, FB14

Perfluoronaphthalene, FA87

Perylene, FD99

9,10-Phenanthraquinone, FC08

Phenanthrene, FC11

Phenanthrenequinone, FC08

o-Phenanthroline, FB17

1,10-Phenanthroline, FB17

4,5-Phenanthroline, FB17

Phenazine, FB18

Phenosafranine, FD68

4-(Phenylamino)acetophenone, FC30

N-Phenyl-4-aminophenol, FB48

4-(Phenylamino)phenol, FB48

N-[4-(Phenylamino)phenyl]acetamide, FC44

1-[4-(Phenylamino)phenyl]ethanone, FC30

9-Phenylanthracene, FE05

4-(Phenylazo)aniline, FB50

4-(Phenylazo)benzenamine, FB50

4-Phenylazophenol, FB41

5-Phenylazosalicylic acid, FB79

N-Phenylbenzenamine, FB47

Phenylbenzene, FB36

N-Phenyl-1,2-benzenediamine, FB53

α-Phenylbenzenemethanol, FB95

N'-Phenyl-1,2,4-benzenetriamine, FB57

trans-2-Phenyl-1-benzoylethene, FC76

α-Phenyl-[1,1'-biphenyl]-4-methanol, FD86

2-Phenyl-3-(4-bromophenylimino)-indolenine, FE01

2-Phenyl-3-[(4-bromophenyl)imino]-indolenine 1-oxide, FE02

4-Phenyl-3-buten-2-one, FA96

2-Phenyl-3-butyn-2-ol, FA94

ω-Phenylcarbamoylacetophenone, FC78

ω-Phenylcarbamoyl-3-bromoacetophenone, FC72

ω-Phenylcarbamoyl-4-bromoacetophenone, FC73

ω-Phenylcarbamoyl-3-chloroacetophenone, FC74

ω-Phenylcarbamoyl-4-chloroacetophenone, FC75

ω-Phenylcarbamoyl-4-cyanoacetophenone, FD19

ω-Phenylcarbamoyl-3-methoxyacetophenone, FD31

ω-Phenylcarbamoyl-4-methoxyacetophenone, FD32

ω-Phenylcarbamoyl-3-methylacetophenone, FD29

ω-Phenylcarbamoyl-4-methylacetophenone, FD30

cis-1-Phenyl-2-carbethoxy-2-cyanoethylene, FB49

Phenyl carbinol, FA56

N-Phenyl-N'-(3-carboxy-4-hydroxyphenyl)diazene, FB79

2-Phenyl-3-(4-chlorophenylimino)-indolenine, FE03

2-Phenyl-3-(4-chlorophenylimino)-indolenine 1-oxide, FE04

β-Phenylcinnamyl alcohol, FC85

1-Phenyl-3-(4-cyanophenyl)-1-triazene, FB82

Phenylcyclooctatetraene, FC20

1-Phenyl-1,3,5,7-cyclooctatetraene, FC20

2-Phenyl-3-{[4-(dimethylamino)phenyl]-imino}indolenine, FE36

2-Phenyl-3-{[4-(dimethylamino)phenyl]-imino}indolenine 1-oxide, FE37

4-Phenyldiphenylmethane, FD82

1-Phenyl-2-(1,3-diphenylpropen-3-ylidene)diazane, FE23

Phenyl diselenide, FB46

(Phenyldiseleno)benzene, FB46

1-Phenylethanone, FA67

1-Phenyl-1-hydroxymethane, FA56

2-Phenyl-3-iminoindolenine, FC12

N-(2-Phenyl-3H-indole-3-ylidene)-4-pyridinamine N,N'-dioxide, FD81

Phenylmethanol, FA56

2-Phenyl-3-[(4-methoxyphenyl)imino]-indolenine, FE19

2-Phenyl-3-[(4-methoxyphenyl)imino]-indolenine 1-oxide, FE21

4-(Phenylmethyl)biphenyl, FD82

2-(Phenylmethyl)-4-[(carboxy-2-methyl-2-mercapto-1-propylamino)-methylidene]-4-oxo-Δ^2-1,3-oxazetine, FD39

2-(Phenylmethylene)-1H-indene-1,3(2H)-dione, FD10

2-Phenyl-3-[(4-methylphenyl)imino]-indolenine, FE18

2-Phenyl-3-[(4-methylphenyl)imino]-indolenine 1-oxide, FE20

1-Phenyl-2-methylpropanone, FB04

N-Phenyl-4-nitrobenzenamine, FB43

2-Phenyl-3-(1'-oxide-4'-pyridyl)-iminoindolenine 1-oxide, FD81

5-Phenylphenazinium chloride, FD65

3-Phenyl-2-[3-(3-phenyl-2(3H)-benzothiazolylidene)-1-propenyl]benzothiazolium iodide, FF26

N-Phenyl-p-phenylenediamine, FB53

2-Phenyl-3-(phenylimino)-3H-indole, FE06

2-Phenyl-3-(phenylimino)indolenine, FE06

2-Phenyl-3-(phenylimino)indolenine 1-oxide, FE07

2-Phenyl-3-phenylimino-3H-indole 1-oxide, FE07

2-Phenyl-3-(2-phenyl-3H-indol-3-ylidene)-3H-indole 1,1'-dioxide, FF11

N-Phenyl-3-phenyl-3-oxopropanamide, FC78

1-Phenyl-1-propanone, FA81

1-Phenylpropargyl alcohol, FA78

3-Phenyl-2-propenoic acid nitrile, FA74

3-Phenylprop-2-en-1-ol, FA80

3-Phenyl-2-propen-1-ol, FA80

1-Phenylpropen-3-yl methyl ether, FB03

1-Phenyl-2-propyn-1-ol, FA78

2-Phenyl-3-(4-pyridylimino)-3H-indole 1,1-dioxide, FD81

2-Phenyl-3-(2-pyridylimino)-3H-indole 1-oxide, FD80

2-Phenyl-3-(2-pyridylimino)indolenine 1-oxide, FD80

1-(N-Phenyl-1H-1-thia-3-azonia-2-indenyl)-3-(N-phenyl-1-thia-3-azaindan-2-ylidene)-1-propene iodide, FF26

3-Phenyl-1-(2-thienyl)-2-propen-1-one, FB85

2-Phenyl-3-(p-tolylimino)-3H-indole, FE18

2-Phenyl-3-(p-tolylimino)-3H-indole 1-oxide, FE20

4-(3-Phenyl-1-triazenyl)benzoic acid, ethyl ester, FC92

4-(3-Phenyl-1-triazenyl)benzonitrile, FB82

3-Phenyl-1(2,4,6-tribromo-3-hydroxyphenyl)-2-propen-1-one, FC68

2-Piperonylidene-1,3-indandione, FD48

Pivalophenone, FB11

Δ^4-Pregnene-17α,21-diol-3,20-dione, FE30

Δ^4-Pregnene-3,20-dione, FE29

4-Pregnene-3,20-dione, FE29

Progesterone, FE29

1-Propanol, FA11

2-Propanol, FA12

2-Propanone, FA09

Propargyl alcohol, FA08

2-Propenenitrile, FA07

2-Propen-1-ol, FA10

4-Propenyl-1,2-dimethoxybenzene, FB12

Propiophenone, FA81

Propyl alcohol, FA11

2-Propyn-1-ol, FA08

Protoporphine IX dimethyl ester complex with Mg(II), FF53

Pteroylglutamic acid, FD90

Pynacryptol Green, FD65

Pyrazolanthrone, FC06

Pyrazolo[a,b,c-m,n]-9(10H)anthrone anion, FC06

Pyrene, FF84

3-Pyridinecarboxamide, FA44

2-Pyridoyl-4-pyridoylmethane bis-(oxime), FB92

2-Pyridoyl-3-pyridoylmethane bis-(oxime), FB93

3-Pyridoyl-4-pyridoylmethane bis-(oxime), FB94

1-(2-Pyridyl)-3-hydroxyamino-3-(3-pyridyl)-1-propanone oxime, FC02

1-(2-Pyridyl)-3-(2-hydroxyphenyl)-1-propen-3-one, FC14

1-(3-Pyridyl)-3-(2-hydroxyphenyl)-1-propen-3-one, FC15

1-(4-Pyridyl)-3-(2-hydroxyphenyl)-1-propen-3-one, FC16

1-(2-Pyridyl)-3-(4-pyridyl)-1,3-propanedione dioxime, FB92

1-(3-Pyridyl)-3-(2-pyridyl)-1,3-propanedione dioxime, FB93

1-(3-Pyridyl)-3-(4-pyridyl)-1,3-propanedione dioxime, FB94

1-(2-Pyridyl-3-(3-pyridyl)prop-1,3-diylidene diazenol, FB93

1-(2-Pyridyl)-3-(4-pyridyl)prop-1,3-diylidene diazenol, FB92

1-(3-Pyridyl)-3-(4-pyridyl)prop-1,3-diylidene diazenol, FB94

1-(2-Pyridyl)-3-(4-pyridyl)-2-propen-1-one, FB78

Quinoline, FA76

3-Quinolinecarbonitrile, FA88

8-Quinolinol, FA77

Quinone, FA39

Quinoxaline, FA64

Reactive Blue 2, FF24

Rhodizonic acid, FA34

Rubrene, FF94, FF95

Ruthenium(II) carbonyletioporphyrin-(I), FF44

Sodium deamino nicotinamide adenine dinucleotide, FE26

Sodium 2-(4-dimethylaminophenyl)-4,5-di(4-biphenylyl)imidazol-1-ide, FF49

Sodium 2-(4-dimethylaminophenyl)-4,5-di(4-methoxyphenyl)imidazol-1-ide, FE87

Sodium 2-(4-dimethylaminophenyl)-4,5-di(4-methylphenyl)imidazol-1-ide, FE86

Sodium diphenyldiazobis-α-naphthylamine sulfonate, FF36

Sodium nicotinamide adenine dinucleotide phosphate, FE27

Sodium 2,4,5-tris(4-dimethylaminophenyl)imidazol-1-ide, FF08

Solochrome Black PVS, FD17

Solochrome Black WDFA, FE00

Solochrome Red ERS, FE10

Solochrome Yellow 2GS, FB68

Stesolid, FD22

trans-Stilbene, FC21

Styryl 2-thienyl ketone, FB85

2,2',2''-Terpyridine, FC71

2,2':6'2''-Terpyridine, FC71

Testosterone, FD95

N,N,N',N'-Tetraacetyl-1,4-diamino-1,4-diphenyl-1,3-butadiene, FE79

2',4',5',7'-Tetrabromo-3',6'-dihydroxyspiro[isobenzofuran-1(3H),-9-[9H]-xanthen]-3-one, disodium salt, FD96

2,4,5,7-Tetrabromo-R-fluoroscein, FD96

2',4',5',7'-Tetrabromofluorescein, disodium salt, FD96

Tetrabutylammonium acetate, FD79

Tetrabutylammonium bromide, FF85

Tetrabutylammonium iodide, FF86

Tetrabutylammonium triiodide, FF87

Tetracyanoethylene, FA31

7,7,8,8-Tetracyanoquinodimethane, FB15

Tetracyclone, FF25

1,1,4,4-Tetraethyl-1,4-dihydro-2,5-diphenyl-1,4-diphosphorinium dichloride, FE81

2,4,6,8-Tetraethyl-1,3,5,7-tetramethylporphin, FF42

2,7,12,17-Tetraethyl-3,8,13,18-tetramethyl-21H, 23H-porphine, FF42

1,2,3,5-Tetrafluorobenzene, FA33

1,2,2a,8a-Tetrahydro-1,2-bis(phenylmethylenecyclobuta[b]naphthalene-3,8-dione, dipotassium salt, FE91

1,2,3,4-Tetrahydro-1,1,4,4-tetraphenyl-p-diphosphorinium dibromide, FF18

2,2,4,5-Tetrahydroxy-4-cyclopentene-1,3-dione, FA20

2',4',5',7'-Tetraiodofluorescein, disodium salt, FD97

Tetrakis(4-chlorophenyl)diazane, FE65

Tetrakis(4-chlorophenyl)hydrazine, FE65

1,1,4,4-Tetrakis(4-fluorophenyl)-1,4-dihydro-2,5-diphenyl-1,4-diphosphorinium dichloride, FF61

4,4',5,5'-Tetrakis(4-methoxyphenyl)-3,3'-[2,5-bis(4-methylphenyl)-1,6-bis(4-methoxyphenyl)-3,4-dithia-1,6-hexylene]dithiolium perchlorate, FF80

4,4',5,5'-Tetrakis(4-methoxyphenyl)-3,3'-[2,5-bis(4-methylphenyl)-3,4-dithia-1,6-hexylene]dithiolium perchlorate, FF73

2,3,4,5-Tetrakis(4-methoxyphenyl)-[1,2]dithiolo[1,5-b][1,2]dithiole-7-S^{IV}, FF43

2,3,4,5-Tetrakis(4-methoxyphenyltrithiapentalene, FF43

Tetrakis(4-methylphenyl)diazane, FF23

Tetrakis(4-methylphenyl)hydrazine, FF23

Tetrakis(4-nitrophenyl)diazane, FE66

Tetrakis(4-nitrophenyl)ethene, FE92

Tetrakis(4-nitrophenyl)ethylene, FE92

Tetrakis(4-nitrophenyl)hydrazine, FE66

1-Tetralone, FA97

3,3',4,4'-Tetramethoxybibenzyl, FD74

2,2',4,4'-Tetramethoxydiphenylamine, FD41

1,3,6,7-Tetramethyl-4,5-dicarboxyethyl-2,8-divinyl(b-13)-dihydrobilenone, FF45

2,2,6,6-Tetramethyl-4-oxopiperidine nitroxide, FA84

2,2,6,6-Tetramethyl-1-oxopiperidinium cation, FA85

2,2,6,6-Tetramethyl-4-oxo-1-piperidinyloxy, FA84

N,N,N',N'-Tetramethyl-4-phenylenediamine, FB07

2,2,6,6-Tetramethylpiperidine nitroxide, FA86

2,2,6,6-Tetramethylpiperidine nitroxide cation, FA85

2,2,6,6-Tetramethyl-1-piperidinyloxy, FA86

2,2,6,6-Tetramethyl-4-piperidone-1-oxyl, FA84

2,2',4,4'-Tetranitrodiphenylmethane, FB70

2,4,5,7-Tetranitrofluorene-Δ^9,α-malonitrile, FD06

(2,4,5,7-Tetranitro-9H-fluoren-9-ylidene)propanedinitrile, FD06

1,2,4,7-Tetraphenylcyclooctatetraene, FF37

1,3,5,7-Tetraphenylcyclooctatetraene, FF38

Tetraphenylcyclopentadienone, FF25

2,3,4,5-Tetraphenyl-2,4-cyclopentadien-1-one, FF25

2,3,4,5-Tetraphenyl-2-cyclopenten-1-one, FF27

1,1,2,2-Tetraphenyldiazane, FE75

4,4',5,5'-Tetraphenyl-3,3'-(2,5-diphenyl-3,4-dithia-1,6-hexamethylene)-dithiolium perchlorate, FF70

1,1,4,4-Tetraphenyldiphosphoniacyclohexa-2,5-diene dibromide, FF15

1,1,4,4-Tetraphenyldiphosphoniacyclohexa-2-ene dibromide, FF18

1,1,4,4-Tetraphenyldiphosphoniacyclohexane dibromide, FF22

1,1,4,4-Tetraphenyl-p-diphosphorinanium dibromide, FF22

1,1,2,2-Tetraphenyl-1,2-ethandiylium, FE97

Tetraphenylethene, FE97

Tetraphenylethylene, FE97

1,1,2,2-Tetraphenylhydrazine, FE75

Tetraphenylphosphonium perchlorate, FE71

$\alpha,\beta,\gamma,\delta$-Tetraphenylporphine, FF67

5,10,15,20-Tetraphenylporphine, FF67

Thenoyl-4-carbamoylstyrene, FC17

ω-Thenoyl-4-carboxamidostyrene, FC17

Thenoyl-2,6-dichlorostyrene, FB67

ω-Thenoyl-4-dimethylaminosytrene, FC88

ω-Thenoyl-4-ethoxystyrene, FC87

Thenoyl-4-fluorostyrene, FB74

4-Thenoyl-2-methylstyrene, FC27

ω-Thenoyl-4-methylstyrene, FC28

ω-Thenoylstyrene, FB85

ω-Thenoyl-2,3,4-trimethoxystyrene, FD34

ω-Thenoyl-3,4,5-trimethoxystyrene, FD35

Thianthracene, FB25

Thianthrene, FB25

Thianthrene 5,5-dioxide, FB23

Thianthrene 5,10-dioxide, FB24

Thianthrene monoxode, FB22

Thianthrene 5-oxide, FB22

Thiathrene sulfone, FB23

1-[β-(2-Thienoyl)vinyl]-2,6-dichlorobenzene, FB67

1-(2-Thienyl)-3-(4-carboxamidophenyl)-2-propen-1-one, FC17

1-(2-Thienyl)-3-(4-ethoxyphenyl)-2-propenone, FC87

1-(2-Thienyl)-3-(4-fluorophenyl)-2-propenone, FB74

1-(2-Thienyl)-3-(2-methylphenyl)-2-propenone, FC27

1-(2-Thienyl)-3-(4-methylphenyl)-2-propenone, FC28

1-(2-Thienyl)-3-(2,3,4-trimethoxyphenyl)-2-propen-1-one, FD34

1-(2-Thienyl)-3-(3,4,5-trimethoxyphenyl)-2-propen-1-one, FD35

1-Thiocarbamoyl-3,5-dimethyl-4-(2'-ethoxyphenyl)azopyrazole, FC66

1-Thiocarbamoyl-3,5-dimethyl-4-(4'-ethoxyphenylazo)pyrazole, FC67

1-(Thiocarbamoyl-3,5-dimethyl-4-(2-methoxyphenylazo)pyrazole, FC03

1-Thiocarbamoyl-3,5-dimethyl-4-(4-methoxyphenylazo)pyrazole, FC04

1-Thiocarbamoyl-3,5-dimethyl-4-(phenylazo)pyrazole, FB59

1-Thiocarbamoyl-3,5-diphenyl-4-(4-bromophenylazo)pyrazole, FE32

1-Thiocarbamoyl-3,5-diphenyl-4-(3-chlorophenylazo)pyrazole, FE33

1-Thiocarbamoyl-3,5-diphenyl-4-(4-chlorophenylazo)pyrazole, FE34

1-Thiocarbamoyl-3,5-diphenyl-4-(2,4-dimethylphenylazo)pyrazole, FE77

1-Thiocarbamoyl-3,5-diphenyl-4-(2,6-dimethylphenylazo)pyrazole, FE78

1-Thiocarbamoyl-3,5-diphenyl-4-(2-methoxyphenylazo)pyrazole, FE43

1-Thiocarbamoyl-3,5-diphenyl-4-(4-methoxyphenylazo)pyrazole, FE44

1-Thiocarbamoyl-3,5-diphenyl-4-(2-methylphenylazo)pyrazole, FE45

1-Thiocarbamoyl-3,5-diphenyl-4-(3-methylphenylazo)pyrazole, FE46

1-Thiocarbamoyl-3,5-diphenyl-4-(4-methylphenylazo)pyrazole, FE47

1-Thiocarbamoyl-3,5-diphenyl-4-(phenylazo)pyrazole, FE35

3-Thiocyanato-N-(4-thiocyanatophenyl)-benzenamine, FC09

Thiocyanic acid, iminodi-4,1-phenylene ester, FC09

Thionine, FB39

Thioperoxydicarbonimidic diamide, FA03

Thiophos 3422, FB05

Toluene, FA55

1-(4-Tolyl)-2-dicyanoethylene, FB08

β-(p-Tolylthio)pyruvic acid, FA99

1-(2,4,6-Tribromo-3-hydroxyphenyl)-3-(4-hydroxyphenyl)-2-propen-1-one, FC69

1-(2,4,6-Tribromo-3-hydroxyphenyl)-3-phenyl-2-propen-1-one, FC68

ω-(2,4,6-Tribromo-3-hydroxy)styrene, FC68

N,N,N-Tributyl-1-butaneaminium acetate, FD79

Trichloro(ethyl)stannane, FA02

Trichloro(ethyl)tin, FA02

Trichloro(methyl)stannane, FA00

Trichloro(methyl)tin, FA00

2,4,5-Tri(4-dimethylaminophenyl)-imidazole, sodium salt, FF08

N,N,N-Triethylbenzenemethanaminium bromide, FC05

Triethyl(benzyl)ammonium bromide, FC05

Triethylstannic chloride, FA51

Triethyltin(IV) chloride, FA51

1,3,5-Trifluorobenzene, FA35

3,3',4-Trimethoxybibenzyl, FD60

2,3,7-Trimethoxy-9,10-dihydrophenanthrene, FD58

2,3,4-Trimethoxy-ω-thenoylstyrene, FD34

3,4,5-Trimethoxy-ω-thenoylstyrene, FD35

α,α,α-Trimethylacetophenone, FB11

1,3,5-Trimethylbenzene, FA82

3,5,5-Trimethyl-2-cyclohexen-1-one, FA83

Trimethylstannic chloride, FA13

Trimethyltin chloride, FA13

(2,4,7-Trinitro-9H-fluoren-9-ylidene)-malonitrile, FD07

(2,4,7-Trinitro-9H-fluoren-9-ylidene)-propanedinitrile, FD07

N,N'-Triphenyl-1,4-benzenediamine, FE76

Triphenylcarbinol, FD87

N,N,N'-Triphenyl-1,4-diaminobenzene, FE76

2,3,5-Triphenyl[1,2]dithiolo[1,5-b]-[1,2]dithiole-7-S IV, FE41

Triphenylethene, FE09

Triphenylethylene, FE09

Triphenylmethanol, FD87

N,N,N'-Triphenylphenylenediamine, FE76

Triphenyl phosphate, FD70

Triphenylphosphine, FD71

Triphenylphosphine oxide, FD69

1,3,5-Triphenyl-Δ^2-pyrazoline, FE24

2,4,6-Triphenylpyrylium perchlorate, FE42, FF95

2,5,6-Triphenyltrithiapentalene, FE41

2,3,5-Triphenyl-1,6,6a-trithiopentalene, FE41

2,4,6-Tri(2-pyridyl)-s-triazine, FD64

Trisbipyridineruthenium(II) diperchlorate, FF29

Tris(2,2'-bipyridine-N,N')ruthenium-(2+) diperchlorate, FF29, FF82, FF91

Tris(4-dimethylaminophenyl)methane, FE90

2,3,5-Tris(4-methoxyphenyl)[1,2]-dithiolo[1,5-b][1,2]dithiole-7-S^{IV}, FE98

2,4,6-Tris(4-methoxyphenyl)pyrylium perchlorate, FF02

2,3,5-Tris(4-methoxyphenyl)trithiapentalene, FE98

Tris(1,10-phenanthroline)ruthenium-(II) perchlorate, FF50, FF92

Trityl alcohol, FD87

Trityl methyl ether, FE13

Valium, FD22

4-Vinylanisole, FA79

Vinyl cyanide, FA07

Vitamin K$_2$, FF34

Vival, FD22

Wurster's Blue, FB07

TABLE V.
FUNCTIONAL-GROUP INDEX

This index divides the 580 compounds that appear in Table I into 107 groups and subgroups of chemically related compounds. It is preceded by a list of these groups which begins on the following page.

So that none of the final categories would contain more than about 50 compounds, many main groups have been divided and some have been subdivided. Frequently the division was made, first according to the molecular frame (e.g., alicyclic, aliphatic, aromatic or benzenoid, and heterocyclic compounds) and then according to the number of substituents present, but for halogenated compounds the first division was according to the kind of halogen atom, and organometallic ones were divided according to the metallic atom.

In each category there appear, in alphabetical order, the names of the compounds, together with the code numbers assigned to them in Table I, in which the functional group named is, or may be, electroactive. Compounds in which a given functional group is present but is believed to be electroinactive are not listed under the name of that group. Compounds for which the products of the half-reaction are unknown are listed under each of the groups they contain.

ACETYLENES, see HYDROCARBONS and UNSATURATED COMPOUNDS

ACIDS, CARBOXYLIC (see also HYDROGEN ION REDUCTIONS, KETONES, and UNSATURATED COMPOUNDS)
　Aliphatic
　Aromatic

ACID SALTS, see ACIDS, CARBOXYLIC

ACIDS, SULFONIC, see SULFUR COMPOUNDS

ALCOHOLS, see HYDROXY COMPOUNDS

AMIDES
　Aliphatic, not Aryl-substituted
　Aliphatic, Aryl-substituted
　Aliphatic, Heterocyclic-substituted
　Aromatic
　Heterocyclic
　Thioamides

AMINES (see also NITROSO COMPOUNDS)
　Aliphatic
　Anilines
　　N-Unsubstituted
　　N-Monosubstituted
　　N,N-Disubstituted
　Other Aromatic
　Cyclic
　Heterocyclic
　Quarternary

AMINE SALTS, see AMINES

AMINO ACIDS, see ACIDS, CARBOXYLIC

ARSENIC, see PHOSPHORUS

AZINES, see IMINES

AZO AND AZOXY COMPOUNDS
　Azobenzenes
　Others

AZOXY, see AZO

BENZOQUINONES, see QUINONES

CARBAMATES, see ESTERS, NONCARBOXYLIC

CARBOHYDRATES AND SUGARS

CATALYTIC HYDROGEN-ION REDUCTION

DYES

EPOXIDES, see ETHERS

ESTERS, CARBOXYLIC

ESTERS, NONCARBOXYLIC
　Phosphates

ETHERS AND EPOXIDES

HALOGEN COMPOUNDS
　Aliphatic
　　Bromo
　　Chloro
　　Fluoro
　Alicyclic (see also Steroids)
　Aromatic
　　Monobromo
　　Polybromo
　　Monoiodo
　　Monochloro
　　Polychloro
　　Fluoro
　Heterocyclic
　　Bromo
　　Chloro
　　Iodo
　Steroids
　　Chloro
　　Fluoro

HETEROCYCLIC COMPOUNDS
　Nitrogen Heterocyclic
　　Pyridines
　　One Ring-Nitrogen Atoms, Monocyclic (except Pyridines)
　　One Ring-Nitrogen Atom, Polycyclic
　　Two Ring-Nitrogen Atoms, Monocyclic
　　Two Ring-Nitrogen Atoms, Polycyclic
　　Three Ring-Nitrogen Atoms
　　Four Ring-Nitrogen Atoms (except Purines and Porphines)
　Oxygen Heterocyclic (see also ANHYDRIDES, CARBOXYLIC; CARBOHYDRATES AND SUGARS; and LACTONES
　Sulfur Heterocyclic (see also SULFUR COMPOUNDS)

HYDRAZIDES, see HYDRAZINES

HYDRAZINES, HYDRAZIDES, AND HYDRAZONES (see also CARBAZIDES)

HYDRAZONES, see HYDRAZINES

HYDROCARBONS

HYDROXY COMPOUNDS (see also QUINONES)
　Aliphatic
　Alicyclic
　Aromatic
　　Phenols, Unsubstituted and Monosubstituted
　　Phenols, Polysubstituted
　　Two or More Rings
　Heterocyclic
　Hydroxylamines

HYDROXY SALTS, see HYDROXY COMPOUNDS

IMINES AND OTHER AZOMETHINES (see also CARBAZIDES, HYDRAZINES, and OXIMES)
　Aliphatic
　Cyclic Imines

ISOTHIOCYANATES AND THIOCYANATES

KETONES (see also HETEROCYCLIC COMPOUNDS, IMIDES, and UNSATURATED COMPOUNDS)
　Dialkyl Ketones
　Alkyl-Aryl Ketones
　Alkyl-Heterocyclic Ring Ketones
　Other Alkyl Ketones
　Diaryl Ketones
　Aryl-Heterocyclic Ring Ketones
　Ketosteroids
　Other Carbocyclic Ketones
　Heterocyclic Ketones
　Ketoacids and Derivatives

LACTAMS, see LACTONES

LACTONES AND LACTAMS (see also HETEROCYCLIC COMPOUNDS)

METHOXIMES, see OXIMES

NITRATES, see ESTERS, NONCARBOXYLIC

NITRILES

NITRO COMPOUNDS
　Nitrobenzenes
　Nitroaromatics with Condensed Rings
　Nitro Derivatives of O-Heterocyclics

NITROSO AND NITROSYL COMPOUNDS

ORGANOMETALLICS
　Co
　Cr
　Fe
　Hg
　Ir
　Mg
　Rh
　Ru
　Sn

N-OXIDES
　Pyridine N-Oxides
　Other Heterocyclic N-Oxides
　Others

OXIMES AND ALKOXIMES
　Aldoximes
　Ketoximes

PHENOLS, see HYDROXY COMPOUNDS

PHOSPHORUS AND ARSENIC (see also ESTERS, NONCARBOXYLIC; ORGANOMETALLICS, and SULFUR COMPOUNDS)

QUARTERNARY COMPOUNDS, see AMINES

QUINONES AND HYDROQUINONES (see also HYDROXY COMPOUNDS)
　Benzoquinones
　Naphthoquinones
　Other Polycyclic Quinones

QUINONIMINES

SELENIUM AND TELLURIUM COMPOUNDS

SUGARS, see CARBOHYDRATES

SULFATES, see ESTERS, NONCARBOXYLIC

SULFIDES, see SULFUR COMPOUNDS

SULFONAMIDES, see SULFUR COMPOUNDS

SULFONES, see SULFUR COMPOUNDS

SULFONIC ACIDS, see SULFUR COMPOUNDS

SULFUR COMPOUNDS (see also ESTERS, NONCARBOXYLIC; HETEROCYCLIC COMPOUNDS; ISOTHIOCYANATES; and ORGANOMETALLICS)
　Thiols
　Sulfides, Thioethers, and Sulfonium Ions
　Disulfides
　Thiones, Thioamides, Thiobarbiturates, Thiopyrimidines and Thiopurines
　Sulfoxides, Sulfones, and S-Dioxides
　Sulfuric and Thiosulfuric Acid Derivatives, Sulfonates, Sulfinates, and Sulfonamides
　Phosphoric and Arsenic Acid Derivatives (see also ORGANOMETALLICS)

TELLURIUM, see SELENIUM

THIOCYANATES, see ISOTHIOCYANATES

THIOLS, see SULFUR COMPOUNDS

THIOUREAS, see UREAS

UNSATURATED COMPOUNDS
　Ethylenic
　　Double Bond Isolated from Non-hydrocarbon Functional Group
　　α,β-Unsaturated Ketones
　　α,β-Unsaturated Acids, Thioacids, Esters, and Amides
　　α,β-Unsaturated Amines, Halogens, Alcohols, Ethers, Imines, and Nitrocompounds

α,β-Unsaturated Nitriles
α,β-Unsaturated Phosphines
Alicyclic
α,β-Unsaturated Ketosteroids
Heterocyclic With Side Chain
Unsaturated
Acetylenic

UREAS AND THIOUREAS (see also
HETEROCYCLIC COMPOUNDS; and KETONES,
Heterocyclic)

ACETYLENES, see HYDROCARBONS and UNSATURATED COMPOUNDS

ACIDS, CARBOXYLIC (see also HYDROGEN ION REDUCTIONS, KETONES, and UNSATURATED COMPOUNDS)

Aliphatic

benzylpenicillenic acid, FD39

bilirubin, FF45

3-bromo-2-oxopropanoic acid, FA06

2,17-diethyl-1,10,19,22,23,24-hexahydro-3,7,13,18-tetramethyl-1,19-dioxo-21H-biline-8,12-dipropanoic acid, FF46

3-dimethylsulfonio-2-oxopropanoic acid bromide, FA26

3-(ethylthio)-2-oxopropanoic acid, FA25

folic acid, FD90

3-[(4-methylphenyl)thio]-2-oxopropanoic acid, FA99

3-methylthio-2-oxopropanoic acid, FA15

phenylglyoxalic acid, FA65

Aromatic

benzoic acid, FA53

3-carboxy-4-hydroxy-4'-sulfoazobenzene disodium salt, FB68

5-(2-chlorophenylazo)-2-hydroxybenzoic acid, FB72

5-(4-chlorophenylazo)-2-hydroxybenzoic acid, FB73

3-(4-dimethylaminophenylazo)-benzoic acid, FC90

4-(4-dimethylaminophenylazo)-benzoic acid, FC91

eosin, FD96

erythrosine, FD97

fluorescein, disodium salt, FD98

2-hydroxy-5-(2,4-dinitrophenylazo)benzoic acid, FB69

2-hydroxy-5-(2-nitrophenylazo)-benzoic acid, FB75

2-hydroxy-5-(4-nitrophenylazo)-benzoic acid, FB76

2-hydroxy-5-phenylazobenzoic acid, FB79

4-(4-hydroxyphenylazo)benzoic acid, FB80

ACID SALTS, see ACIDS CARBOXYLIC

ACIDS, SULFONIC, see SULFUR COMPOUNDS

ALCOHOLS, see HYDROXY COMPOUNDS

AMIDES

Aliphatic, not Aryl-substituted

N,N,N'N'-tetraacetyl-1,4-diamino-1,4-diphenyl-1,3-butadiene, FE79

Aliphatic, Aryl-substituted

4-acetylaminodiphenylamine, FC44

4-acetylamino-4'-methoxydiphenylamine, FC94

2,7-bis(acetylamino)-5,10-bis-(4-acetylaminophenyl)-5,10-dihydrophenazine, FF39

2,7-bis(acetylamino)-5,10-di(4-chlorophenyl)-5,10-dihydrophenazine, FF12

2,7-bis(acetylamino)-5,10-dihydro-5,10-diphenylphenazine, FF17

4,4'-bis(acetylamino)diphenylamine, FD36

5,10-bis(4-acetylaminophenyl)-2,7-dimethoxy-5,10-dihydrophenazine, FF33

3-(3-bromophenyl)-3-oxo-N-phenylpropanamide, FC72

3-(4-bromophenyl)-3-oxo-N-phenylpropanamide, FC73

4-chloro-4'-acetylaminodiphenylamine, FC29

3-(3-chlorophenyl)-3-oxo-N-phenylpropanamide, FC74

3-(4-chlorophenyl)-3-oxo-N-phenylpropanamide, FC75

3-(4-cyanophenyl)-3-oxo-N-phenylpropanamide, FD19

3-(3-methoxyphenyl)-3-oxo-N-phenylpropanamide, FD31

3-(4-methoxyphenyl)-3-oxo-N-phenylpropanamide, FD32

3-(3-methylphenyl)-3-oxo-N-phenylpropanamide, FD29

AMIDES (cont.)

Aliphatic, Aryl-substituted (cont.)

3-(4-methylphenyl)-3-oxo-N-phenyl-propanamide, FD30

β-oxo-N-phenylbenzenepropanamide, FC78

Aliphatic, Heterocyclic-substituted

1,4-diacetyl-1,4-dihydro-2,3-diphenylpyrazine, FE12

1,2-diacetyl-1,2-dihydro-4-methylcinnoline, FC01

5,6-diacetyl-5,6-dihydro-5,6-phenanthroline, FD24

5,10-diacetyl-5,10-dihydro-phenazine, FD25

1,4-diacetyl-1,4-dihydroquinoxaline, FB54

Aromatic

N-benzoyl-N-phenylhydroxylamine, FB86

3-(4-carbamoylphenyl)-1-(2-thienyl)-2-propenone, FC17

folic acid, FD90

Heterocyclic

3-carbamoylpyridine, FA44

7-chloro-1,3,4,5-tetrahydro-1-methyl-5-phenyl-3H-1,4-benzodiazepin-2-one, FD28

Diazepam, FD22

1-methylpyridinium-3-carbamoyl chloride, FA57

nicotinamide adenine dinucleotide, FE28

nicotinamide mononucleotide, FB13

sodium deamino nicotinamide adenine dinucleotide, FE26

sodium nicotinamide adenine dinucleotide phosphate, FE27

Thioamides

4-(4-bromophenylazo)-3,5-diphenyl-1-thiocarbamoylpyrazole, FE32

4-(3-chlorophenylazo)-3,5-diphenyl-1-thiocarbamoylpyrazole, FE33

4-(4-chlorophenylazo)-3,5-diphenyl-5-thiocarbamoylpyrazole, FE34

3,5-dimethyl-4-(2-ethoxyphenylazo)-1-thiocarbamoylpyrazole, FC66

3,5-dimethyl-4-(4-ethoxyphenylazo)-1-thiocarbamoylpyrazole, FC67

3,5-dimethyl-4-phenylazo-1-thiocarbamoylpyrazole, FB59

3,5-dimethyl-4-(2-methoxyphenylazo)-1-thiocarbamoylpyrazole, FC03

3,5-dimethyl-4-(2-methoxyphenylazo)-1-thiocarbamoylpyrazole, FC03

4-(2,4-dimethylphenylazo)-3,5-diphenyl-1-thiocarbamoylpyrazole, FE77

4-(2,6-dimethylphenylazo)-3,5-diphenyl-1-thiocarbamoylpyrazole, FE78

3,5-diphenyl-4-(2-methoxyphenyl-1-thiocarbamoylpyrazole, FE43

3,5-diphenyl-4-(4-methoxyphenyl)-1-thiocarbamoylpyrazole, FE44

3,5-diphenyl-4-(2-methylphenylazo)-1-thiocarbamoylpyrazole, FE45

3,5-diphenyl-4-(3-methylphenylazo)-thiocarbamoylpyrazole, FE46

3,5-diphenyl-4-(4-methylphenylazo)-1-thiocarbamoylpyrazole, FE47

3,5-diphenyl-4-phenylazo-4-thiocarbamoylpyrazole, FE35

formamidine disulfide, FA03

AMINES (see also NITROSO COMPOUNDS)

Aliphatic

benzylpenicillenic acid, FD39

2-[bis(1-methylethyl)amino]ethanethiol hydrochloride, FA71

chlorimipramine, FD91

1,5-di(5-nitro-2-furyl)-1,4-pentadien-3-one amidinohydrazone hydrochloride, FB89

ethylamine, FA05

N-ethyl-N-hydroxy-β-methoxy-β-(3-trifluoromethylphenyl)-ethylamine, FB61

3-(hydroxyamino)-1-(2-hydroxyphenyl)-3-(2-pyridyl)-1-propanone oxime, FC60

AMINES (see also NITROSO COMPOUNDS) (cont.)
  Aliphatic (cont.)

    3-hydroxyamino-1-(2-hydroxyphenyl)-3-(3-pyridyl)-1-propanone oxime, FC61

    β-methoxy-β-(3-trifluoromethylphenyl)ethylhydroxylamine, FB01

  Anilines
    N-Unsubstituted

      4-aminoazobenzene, FB50

      4-aminodiphenylamine, FB53

      4-amino-4'-methoxydiphenylamine, FC00

      4-aminophenylmercury acetate, FA68

      2-anisidine, FA58

      3-anisidine, FA59

      4-anisidine, FA60

      2,4-diaminodiphenylamine, FB57

      4,4'-diaminodiphenylamine, FB58

      2,4-diaminophenol, FA46

    N-Monosubstituted

      4-acetylaminodiphenylamine, FC44

      4-acetylamino-4'-methoxydiphenylamine, FC94

      4-acetyldiphenylamine, FC30

      4-acetyl-4'-methoxydiphenylamine, FC89

      4-aminodiphenylamine, FB53

      4-amino-4'-methoxydiphenylamine, FC00

      5-[(4-amino-3-sulfoanthraquinon-1-yl)amino]-2-{[3-hydroxy-5-(3-sulfophenyl)]-s-triazinylamino}benzenesulfonic acid, FF24

      N-benzoyl-N-phenylhydroxylamine, FB86

      4,4'-bis(acetylamino)diphenylamine, FD36

      N,N'-bis(4-bromophenyl)-N-phenylphenylenediamine, FE68

      N,N'-bis(4-methoxyphenyl)benzidine, FF05

      N,N'-bis(4-methoxyphenyl)-N-phenylphenylenediamine, FF06

      N,N'-bis(4-methylphenyl)-N-phenyl-1,4-phenylenediamine, FF04

      N,N'-bis(2,3,5,6-tetramethylphenyl)benzidine, FF41

      4-bromodiphenylamine, FB37

      4-chloro-4'-acetylaminodiphenylamine, FC29

      4-chloro-4'-nitrodiphenylamine, FB28

      N,N'-di(4-acetylphenyl)-N-phenylphenylenediamine, FF16

      2,4-diaminodiphenylamine, FB57

      4,4'-diaminodiphenylamine, FB58

      4,4'-di(t-butyl)diphenylamine, FE14

      4,4'-dichlorodiphenylamine, FB30

      4,4'-dimethoxydiphenylamine, FC56

      4,4'-dimethyldiphenylamine, FC55

      4,4'-dinitrodiphenylamine, FB35

      diphenylamine, FB47

      N,N'-diphenylbenzidine, FE74

      4,4'-dithiocyanatodiphenylamine, FC09

      folic acid, FD90

      4-methoxydiphenylamine, FB97

      4-methylaminoazobenzene, FB98

      4-methyldiphenylamine, FB96

      4-nitrodiphenylamine, FB43

      4-nitrosodiphenylhydroxylamine, FB44

      4-(phenylamino)phenol, FB48

      2,2',4,4'-tetramethoxydiphenylamine, FD41

      N,N,N'-triphenylphenylenediamine, FE76

    N,N-Disubstituted

      N,N'-bis(4-bromophenyl)-N-phenylphenylenediamine, FE68

      4,5-bis(4-methoxyphenyl)-2-(4-dimethylaminophenyl)imidazole sodium, FE87

      N,N'-bis(4-methoxyphenyl)-N-phenylphenylenediamine, FF06

AMINES (see also NITROSO COMPOUNDS) (cont.)
  Anilines
    N,N-Disubstituted (cont.)

    N,N'-bis(4-methylphenyl)-N-phenyl-1,4-phenylenediamine, FF04

    3-bromo-4'-dimethylamineazobenzene, FC33

    4-bromo-4'-dimethylamineazobenzene, FC34

    3-chloro-4'-dimethylaminoazobenzene, FC35

    4-chloro-4'-dimethylaminoazobenzene, FC36

    N-(4-chlorophenyl)-N-(4-nitrosophenyl)hydroxylamine, FB29

    N,N'-di(4-acetylphenyl)-N-phenyl-phenylenediamine, FF16

    4,5-di(biphenylyl)-2-(dimethylaminophenyl)imidazole, sodium salt, FF49

    4-(dimethylamino)azobenzene, FC57

    4'-dimethylaminoazobenzene-3-sulfonic acid, FC62

    4'-dimethylaminoazobenzene-4-sulfonic acid, FC63

    4-dimethylamino-3'-fluoroazobenzene, FC38

    4-dimethylamino-4'-fluoroazobenzene, FC39

    4-dimethylamino-2'-hydroxyazobenzene, FC58

    4-dimethylamino-4'-hydroxyazobenzene, FC59

    4-dimethylamino-3'-iodoazobenzene, FC40

    4-dimethylamino-4'-iodoazobenzene, FC41

    4-dimethylamino-3'-methoxyazobenzene, FD02

    4-dimethylamino-4'-methoxyazobenzene, FD03

    4-dimethylamino-3'-methylazobenzene, FD00

    4-dimethylamino-4'-methylazobenzene, FD01

    4-dimethylamino-3'-nitroazobenzene, FC48

    4-dimethylamino-4'-nitroazobenzene, FC49

    3-(4-dimethylaminophenylazo)-benzoic acid, FC90

    4-(4-dimethylaminophenylazo)-benzoic acid, FC91

    2-(4-dimethylaminophenyl)-4,5-bis(4-methylphenyl)imidazole sodium, FE86

    3-(4-dimethylaminophenylimino)-2-phenylindole, FE36

    3-(4-dimethylaminophenylimino)-2-phenylindole 1-oxide, FE37

    3-(4-dimethylaminophenyl)-1-(2-thienyl)-2-propen-1-one, FC88

    N,N,N'-triphenylphenylenediamine, FE76

    2,4,5-tris(4-dimethylaminophenyl)-imidazole sodium, FF08

    tris(4-dimethylaminophenyl)methane, FE90

    Wurster's Blue, FB07

  Other Aromatic

    5[(4-amino-3-sulfoanthraquinon-1-yl)amino]-2-{[3-hydroxy-5-(3-sulfophenyl)]-s-triazinylamino}benzenesulfonic acid, FF24

    Congo Red, FF36

  Cyclic

    chlorimipramine, FD91

    7-chloro-2,3-dihydro-1-methyl-5-phenyl-1H-1,4-benzodiazepine, FD27

    7-chloro-1,3,4,5-tetrahydro-1-methyl-5-phenyl-3H-1,4-benzodiazepin-2-one, FD28

    1,4-diacetyl-1,4-dihydroquinoxaline, FB54

  Heterocyclic

    Capri Blue, FD37

    3,6-diamino-10-methylacridinium chloride, FC37

    3,7-diaminophenothiazin-5-ium chloride, FB39

    3,7-diamino-5-phenylphenazinium chloride, FD68

AMINES (see also NITROSO COMPOUNDS) (cont.)
  Heterocyclic (cont.)
    folic acid, FD90
    Methylene Blue, FD38

  Quaternary
    tetrabutylammonium acetate, FD79
    triethyl(benzyl)ammonium bromide, FC05

AMINE SALTS, see AMINES

AMINO ACIDS, see ACIDS, CARBOXYLIC

ARSENIC, see PHOSPHORUS

AZINES, see IMINES

AZO AND AZOXY COMPOUNDS

  Azobenzenes

    4-aminoazobenzene, FB50
    azobenzene, FB40
    4,4'-azobisphenol, FB42
    trans-α,α'-azotoluene, FC43
    3-bromo-4'-dimethylamineazobenzene, FC33
    4-bromo-4'-dimethylamineazobenzene, FC34
    4-(4-bromophenylazo)phenol, FB26
    3-carboxy-4-hydroxy-4'-sulfoazobenzene disodium salt, FB68
    3-chloro-4'-dimethylaminoazobenzene, FC35
    4-chloro-4'-dimethylaminoazobenzene, FC36
    5-(2-chlorophenylazo)-2-hydroxybenzoic acid, FB72
    5-(4-chlorophenylazo)-2-hydroxybenzoic acid, FB73
    4-(4-chlorophenylazo)phenol, FB27
    4-(dimethylamino)azobenzene, FC57
    4'-dimethylaminoazobenzene-3-sulfonic acid, FC62
    4'-dimethylaminoazobenzene-4-sulfonic acid, FC63
    4-dimethylamino-3'-fluoroazobenzene, FC38
    4-dimethylamino-4'-fluoroazobenzene, FC39
    4-dimethylamino-2'-hydroxyazobenzene, FC58
    4-dimethylamino-4'-hydroxyazobenzene, FC59
    4-dimethylamino-3'-iodoazobenzene, FC40
    4-dimethylamino-4'-iodoazobenzene, FC41
    4-dimethylamino-3'-methoxyazobenzene, FD02
    4-dimethylamino-4'-methoxyazobenzene, FD03
    4-dimethylamino-3'-methylazobenzene, FD00
    4-dimethylamino-4'-methylazobenzene, FD01
    4-dimethylamino-3'-nitroazobenzene, FC48
    4-dimethylamino-4'-nitroazobenzene, FC49
    3-(4-dimethylaminophenylazo)benzoic acid, FC90
    4-(4-dimethylaminophenylazo)benzoic acid, FC91
    2-hydroxy-5-(2,4-dinitrophenylazo)benzoic acid, FB69
    2-hydroxy-5-(2-nitrophenylazo)benzoic acid, FB75
    2-hydroxy-5-(4-nitrophenylazo)benzoic acid, FB76
    4-(4-hydroxyphenylazo)benzenesulfonic acid, FB45
    2-hydroxy-5-phenylazobenzoic acid, FB79
    4-(4-hydroxyphenylazo)benzoic acid, FB80
    4-(4-iodophenylazo)phenol, FB31
    4-methylaminoazobenzene, FB98
    4-(3-methylphenylazo)phenol, FB90
    4-(4-methylphenylazo)phenol, FB91

AZO AND AZOXY COMPOUNDS (cont.)

Azobenzenes (cont.)

4-(2-nitrophenylazo)phenol, FB32

4-(3-nitrophenylazo)phenol, FB33

4-(4-nitrophenylazo)phenol, FB34

4-phenylazophenol, FB41

Others

trans-azocyclohexane, FB63

trans-azodioxycyclohexane, FB66

trans-α,α'-azodioxytoluene, FC47

trans-azoxy-2-chlorocyclohexane, FB62

cis-azoxycyclohexane, FB64

trans-azoxycyclohexane, FB65

trans-azoxydi(1-chloro-1-phenyl-2-propyl), FD73

trans-α,α'-azoxytoluene, FC45

4-(4-bromophenylazo)-3,5-diphenyl-1-thiocarbamoylpyrazole, FE32

4-(3-chlorophenylazo)-3,5-diphenyl-1-thiocarbamoylpyrazole, FE33

4-(4-chlorophenylazo)-3,5-diphenyl-5-thiocarbamoylpyrazole, FE34

1-(4-chlorophenyl)-3-phenyltriazene, FB38

Congo Red, FF36

4-cyanodiazoaminobenzene, FB82

diazoaminobenzene, FB51

3-[(2,5-dihydroxynaphth-1-yl)azo]-4-hydroxybenzenesulfonic acid, sodium salt, FD17

3,5-dimethyl-4-(2-ethoxyphenyl)azo)-1-thiocarbamoylpyrazole, FC66

3,5-dimethyl-4-(4-ethoxyphenylazo)-1-thiocarbamoylpyrazole, FC67

3,5-dimethyl-4-(2-methoxyphenylazo)-1-thiocarbamoylpyrazole, FC03

3,5-dimethyl-4-(4-methoxyphenylazo)-1-thiocarbamoylpyrazole, FC04

4-(2,4-dimethylphenylazo)-3,5-diphenyl-1-thiocarbamoylpyrazole, FE77

4-(2,6-dimethylphenylazo)-3,5-diphenyl-1-thiocarbamoylpyrazole, FE78

3,5-dimethyl-4-(phenylazo)-1-thiocarbamoylpyrazole, FB59

3,5-diphenyl-4-(2-methoxyphenyl-1-thiocarbamoylpyrazole, FE43

3,5-diphenyl-4-(4-methoxyphenyl)-1-thiocarbamoylpyrazole, FE44

3,5-diphenyl-4-(2-methylphenylazo)-1-thiocarbamoylpyrazole, FE45

3,5-diphenyl-4-(3-methylphenylazo)thiocarbamoylpyrazole, FE46

3,5-diphenyl-4-(4-methylphenylazo)-1-thiocarbamoylpyrazole, FE47

3,5-diphenyl-4-phenylazo-4-thiocarbamoylpyrazole, FE35

3-(4-ethoxycarbonylphenyl)-1-phenyltriazene, FC92

4-methoxydiazoaminobenzene, FB99

Solochrome Black WDFA, FE00

Solochrome Red ERS, FE10

AZOXY, see AZO

BENZOQUINONES, see QUINONES

CARBAMATES, see ESTERS, NONCARBOXYLIC

CARBOHYDRATES AND SUGARS

nicotinamide adenine dinucleotide, FE28

nicotinamide mononucleotide, FB13

sodium deamino nicotinamide adenine dinucleotide, FE26

sodium nicotinamide adenine dinucleotide phosphate, FE27

CATALYTIC HYDROGEN-ION REDUCTION

4-(4-bromophenylazo)phenol, FB26

4-(4-chlorophenylazo)phenol, FB27

3,5-dimethyl-4-(2-ethoxyphenylazo)-1-thiocarbamoylpyrazole, FC66

3,5-dimethyl-4-(4-ethoxyphenylazo)-1-thiocarbamoylpyrazole, FC67

CATALYTIC HYDROGEN-ION REDUCTION (cont.)

4-(4-hydroxyphenylazo)benzenesulfonic acid, FB45

2-hydroxy-5-(phenylazo)benzoic acid, FB79

4-(4-hydroxyphenylazo)benzoic acid, FB80

4-(4-iodophenylazo)phenol, FB31

4-(3-nitrophenylazo)phenol, FB33

4-phenylazophenol, FB41

1-(2-pyridyl)-3-(4-pyridyl)-2-propene-1-one, FB78

DYES

Capri Blue, FD37

3-carboxy-4-hydroxy-4'-sulfoazobenzene disodium salt, FB68

Cibachron Blue C, FF24

Congo Red, FF34

eosin, FD96

erythrosine, FD97

fluoroescein, disodium salt, FD98

2-hydroxy-5-(4-nitrophenylazo)-benzoic acid, FB76

Methylene Blue, FD38

Solochrome Black PVS, FD17

Solochrome Black WDFA, FE00

Solochrome Red ERS, FE10

tris(4-dimethylaminophenyl)-methane, FE90

Wurster's Blue, FB07

EPOXIDES, see ETHERS

ESTERS, CARBOXYLIC

17α-acetoxy-6α-chloropregna-1,4-diene-3,20-dione, FE53

17α-acetoxy-6-chloropregna-4,6-diene-3,20-dione, FE54

17α-acetoxy-6α-chloropregn-4-ene-3,20-dione, FE57

17α-acetoxy-6β-chloropregn-4-ene-3,20-dione, FE58

17α-acetoxy-6α-fluoropregn-4-ene-3,20-dione, FE59

17α-acetoxy-6β-fluoropregn-4-ene-3,20-dione, FE60

17α-acetoxy-6α-methylpregna-1,4-diene-3,20-dione, FE82

17α-acetoxy-6-methylpregna-4,6-diene-3,20-dione, FE83

17α-acetoxy-6α-methylpregn-4-ene-3,20-dione, FE84

17α-acetoxypregna-1,4-diene-3,20-dione, FE55

17α-acetoxypregna-4,6-diene-3,20-dione, FE56

17α-acetoxypregn-4-ene-3,20-dione, FE61

2-benzoyloxy-1-hydroxy-3-indanone, FD21

2-benzoyloxy-1,3-indandione, FD12

2-benzoyloxy-1-methoxy-3-indenone, FD55

dibutyl phthalate, FD43

diethyl fumarate, FA69

diethyl phthalate, FB60

dimethyl phthalate, FB00

diphenyl phthalate, FE08

3-(4-ethoxycarbonylphenyl)-1-phenyltriazene, FC92

ethyl-3-bromo-2-oxopropanoate, FA21

ethyl trans-2-butenoate, FA49

ethyl cinnamate, FB10

ethyl α-cyanocinnamate, FB49

ethyl 3-dimethylsulfonio-2-oxopropanoate bromide, FA62

ethyleneglycol phthalate, FA90

ethyl 3-ethylthio-2-oxopropanoate, FA61

ethyl 2-oxopropanoate, FA24

magnesium(II)mesoporphyrin IX dimethyl ester, FF55

magnesium(II)protoporphyrin IX dimethyl ester, FF53

methyl cinnamate, FA98

methyl 3-mercaptopropanoate, FA16

ESTERS, NONCARBOXYLIC

Phosphates

    diethyl 4-nitrophenyl phosphate, FB06

    O,O-diethyl O-(4-nitrophenyl)-phosphorothioate, FB05

    diphenyl 4-nitrobenzyl phosphate, FD85

    nicotinamide adenine dinucleotide, FE28

    nicotinamide mononucleotide, FB13

    sodium deamino nicotinamide adenine dinucleotide, FE26

    sodium nicotinamide adenine dinucleotide phosphate, FE27

    triphenyl phosphate, FD70

ETHERS AND EPOXIDES

    4-acetylamino-4'-methoxydiphenyl-amine, FC94

    4-acetyl-4'-methoxydiphenylamine, FC89

    4-amino-4'-methoxydiphenylamine, FC00

    2-anisidine, FA58

    3-anisidine, FA59

    4-anisidine, FA60

    benzhydryl methyl ether, FC50

    2-benzoyloxy-1-methoxy-3-indenone, FD55

    benzyl ether, FC51

    5,10-bis(4-acetylaminophenyl)-2,7-dimethoxy-5,10-dihydrophenazine, FF33

    bis(diphenylmethyl) ether, FF01

    2,5-bis(2-isopropoxy-2-methyl-propyl)hydroquinone, FE17

    2,5-bis(2-methoxy-2-methylpropyl)-1,4-hydroquinone, FD47

    N,N'-bis(4-methoxyphenyl)benzidine, FF05

    5,5'-bis(4-methoxyphenyl)-3,3'-[2,4-bis(4-methylphenyl)-3,4-dithia-1,6-hexamethylene]dithiolium perchlorate, FF60

    1,2-bis(4-methoxyphenyl)cyclobuteno[3,4-b]-3,8-naphthohydroquinone, dipotassium, FE95

    4,5-bis(4-methoxyphenyl)-2-(4-dimethylaminophenyl)imidazole, sodium, FE87

    N,N'-bis(4-methoxyphenyl)-N,N'-diphenylhydrazine, FF07

    2,2'-bis(3-methoxyphenylethyl)-4,4',5,5'-tetramethoxybiphenyl, FF48

    bis(3-methoxyphenyl)methane, FC98

    2,3-bis(4-methoxyphenyl)-1,4-naphthoquinone, FE70

    bis(4-methoxyphenyl)nitroxide, FC42

    N,N'-bis(4-methoxyphenyl)-N-phenylphenylenediamine, FF06

    2,5-bis(2-methyl-2-ethoxypropyl)-hydroquinone, FD77

    2,6-bis(2-methyl-2-ethoxypropyl)-hydroquinone, FD78

    2-(2-tert-butoxy-2-methylpropyl)-5-tert-butylhydroquinone, FD76

    2-tert-butyl-5-(2-methoxy-2-methyl-1-propyl)hydroquinone, FD05

    5,10-di(4-acetylphenyl)-2,7-dimethoxy-5,10-dihydrophenazine, FF31

    N,N'-di(4-acetylphenyl)-N-phenylphenylenediamine, FF16

    9,10-di-p-anisylanthracene, FF14

    4,4'-dimethoxybiphenyl, FC54

    2,7-dimethoxy-5,10-bis(4-methoxyphenyl)-5,10-dihydrophenazine, FF20

    4,4'-dimethoxydiphenylamine, FC56

    2,7-dimethoxy-5,10-diphenyl-5,10-phenyl-5,10-dihydrophenazine, FF00

    1,2-di(3-methoxyphenyl)ethane, FD40

    1,2-dimethoxy-4-propenylbenzene, FB12

    4-dimethylamino-3'-methoxyazobenzene, FD02

    4-dimethylamino-4'-methoxyazobenzene, FD03

    3,5-dimethyl-4-(2-ethoxyphenylazo)-1-thiocarbamoylpyrazole, FC66

    3,5-dimethyl-4-(4-ethoxyphenylazo)-1-thiocarbamoylpyrazole, FC67

ETHERS AND EPOXIDES (cont.)

2-(1,1-dimethylethyl)-5-(2-ethoxy-2-methylpropyl)-1,4-hydroquinone, FD44

2-(1,1-dimethylethyl)-6-(2-ethoxy-2-methylpropyl)-1,4-hydroquinone, FD45

2-(1,1-dimethylethyl)-5-[2-methyl-2-(1-methylethoxy)propyl]-1,4-hydroquinone, FD61

3,5-dimethyl-4-(2-methoxyphenylazo)-1-thiocarbamoylpyrazole, FC03

3,5-dimethyl-4-(4-methoxyphenylazo)-1-thiocarbamoylpyrazole, FC04

2-(1,1-dimethylpropyl)-5-(2-ethoxy-2-methylpropyl)hydroquinone, FD62

2-(1,1-dimethylpropyl)-5-(2-methoxy-2-methylpropyl)hydroquinone, FD46

3,5-diphenyl-4-(2-methoxyphenyl)-1-thiocarbamoylpyrazole, FE43

3,5-diphenyl-4-(4-methoxyphenyl)-1-thiocarbamoylpyrazole, FE44

2,5-diphenyl-1,1,4,4-tetrakis(4-methoxyphenyl)diphosphoniacyclohexa-2,5-diene dibromide, FF68

3-(4-ethoxyphenyl)-1-(2-thienyl)-2-propen-1-one, FC87

N-ethyl-N-hydroxy-β-methoxy-β-(3-trifluoromethylphenyl)-ethylamine, FB61

3-hydroxyamino-1-(2-hydroxy-4-methoxyphenyl)-3-(3-pyridyl)-1-propanone oxime, FD04

3-hydroxy-1-(2-hydroxy-4-methoxyphenyl)-3-(2-pyridyl)-1-propanone oxime, FC95

3-hydroxy-1-(2-hydroxy-4-methoxyphenyl)-3-(3-pyridyl)-1-propanone oxime, FC96

3-hydroxy-1-(2-hydroxy-4-methoxyphenyl)-3-(4-pyridyl)-1-propanone oxime, FC97

1-(2-hydroxy-4-methoxyphenyl)-3-(4-pyridyl)-1,3-propanedione dioxime, FC93

1-(2-hydroxy-4'-methoxyphenyl)-3-(2-pyridyl)-2-propen-1-one, FC79

1-(2-hydroxy-4-methoxyphenyl)-3-(3-pyridyl)-2-propen-1-one, FC80

1-(2-hydroxy-4-methoxyphenyl-3-(4-pyridyl)-2-propen-1-one, FC81

3-(4-hydroxy-3-methoxyphenyl)-1-(2,4,6-tribromo-3-hydroxyphenyl)-2-propen-1-one, FD16

2-(4-methoxybenzylidene)-1,3-indandione, FD54

4-methoxydiazoaminobenzene, FB99

4-methoxydiphenylamine, FB97

9-methoxy-9-methylfluorene, FC86

3-(4-methoxyphenylimino)-2-phenyl-3H-indole, FE19

3-(4-methoxyphenylimino)-2-phenyl-3H-indole 1-oxide, FE21

3-(3-methoxyphenyl)-3-oxo-N-phenylpropanamide, FD31

3-(4-methoxyphenyl)-3-oxo-N-phenylpropanamide, FD32

(4-methoxyphenyl)phenylmethyl methyl ether, FC99

1-methoxy-1-phenylpropene, FB02

3-methoxy-1-phenylpropene, FB03

3-methoxy-3-phenyl-1-propyne, FA95

3-(3-methoxyphenyl)-1-(2,4,6-tribromo-3-hydroxyphenyl)-2-propen-1-one, FD14

3-(4-methoxyphenyl)-1-(2,4,6-tribromo-3-hydroxyphenyl)-2-propen-1-one, FD15

p-methoxystyrene, FA79

α-methoxy-α-(3'-trifluoromethylphenyl)acetaldoxime, FA92

β-methoxy-β-(3-trifluoromethylphenyl)ethyl hydroxylamine, FB01

methyl triphenylmethyl ether, FE13

4,4',5,5'-tetrakis(4-methoxyphenyl)-3,3'-[2,5-bis(4-methylphenyl)-1,6-bis(4-methoxyphenyl)-3,4-dithia-1,6-hexylene]dithiolium perchlorate, FF80

4,4',5,5'-tetrakis(4-methoxyphenyl)-3,3'-[2,5-bis(4-methylphenyl)-3,4-dithia-1,6-hexylene]dithiolium perchlorate, FF73

2,3,4,5-tetrakis(4-methoxyphenyltrithiapentalene, FF43

3,3',4,4'-tetramethoxybibenzyl, FD74

ETHERS AND EPOXIDES (cont.)

2,2',4,4'-tetramethoxydiphenyl-amine, FD41

1-(2-thienyl)-3-(2,3,4-tri-methoxyphenyl)-2-propen-1-one, FD34

1-(2-thienyl)-3-(3,4,5-tri-methoxyphenyl)-2-propen-1-one, FD35

3,3'4-trimethoxybibenzyl, FD60

2,3,7-trimethoxy-9,10-dihydro-phenanthrene, FD58

2,4,6-tris(4-methoxyphenyl)-pyrylium perchlorate, FF02

2,3,5-tris(4-methoxyphenyl)tri-thiapentalene, FE98

HALOGEN COMPOUNDS

Aliphatic

Bromo

benzyl bromide, FA54

3-bromo-2-oxopropanoic acid, FA06

dl-1,2-dibromo-1,2-diphenyl-ethane, FC22

meso-1,2-dibromo-1,2-diphenylethane, FC23

ethyl 3-bromo-2-oxopropanoate, FA21

Chloro

trans-azoxydi(1-chloro-1-phenyl-2-propyl), FD73

Fluoro

N-ethyl-N-hydroxy-β-methoxy-β-(3-trifluoromethylphenyl)-ethylamine, FB61

α-methoxy-α-(3'-trifluoromethyl-phenyl)acetaldoxime, FA92

β-methoxy-β-(3-trifluoromethyl-phenyl)ethylhydroxylamine, FB01

Alicyclic (see also Steroids)

trans-azoxy-2-chlorocyclohexane, FB62

Aromatic

Monobromo

3-bromo-4'-dimethylamineazo-benzene, FC33

4-bromo-4'-dimethylaminoazo-benzene, FC34

4-bromodiphenylamine, FB37

4-(4-bromophenylazo)-3,5-diphen-yl-1-thiocarbamoylpyrazole, FE32

4-(4-bromophenylazo)phenol, FB26

3-(4-bromophenylimino)-2-phenyl-3H-indole, FE01

3-(4-bromophenylimino)-2-phenyl-3H-indole 1-oxide, FE02

3-(3-bromophenyl)-3-oxo-N-phenyl-propanamide, FC72

3-(4-bromophenyl)-3-oxo-N-phenyl-propanamide, FC73

Polybromo

N,N'-bis(4-bromophenyl)-N-phenyl-phenylenediamine, FE68

3-(4-hydroxy-3-methoxyphenyl)-1-(2,4,6-tribromo-3-hydroxy-phenyl)-2-propen-1-one, FD16

3-(3-methoxyphenyl)-1-(2,4,6-tri-bromo-3-hydroxyphenyl)-2-propen-1-one, FD14

3-(4-methoxyphenyl)-1-(2,4,6-tri-bromo-3-hydroxyphenyl)-2-propen-1-one, FD15

3-(4-methylphenyl)-1-(2,4,6-tri-bromo-3-hydroxyphenyl)-2-propen-1-one, FD13

1-(2,4,6-tribromo-3-hydroxy-phenyl)-3-(4-hydroxyphenyl)-2-propen-1-one, FC69

1-(2,4,6-tribromo-3-hydroxy-phenyl)-3-phenyl-2-propen-1-one, FC68

Monoiodo

4-dimethylamino-3'-iodoazobenzene, FC40

4-dimethylamino-4'-iodoazobenzene, FC41

4-[(4-iodophenyl)azo]phenol, FB31

HALOGEN COMPOUNDS (cont.)

Aromatic

Monochloro

4-chloro-4'-acetylaminodiphenylamine, FC29

chlorobenzene, FA41

2-(3-chlorobenzylidene)-1,3-indandione, FD08

2-(4-chlorobenzylidene)-1,3-indandione, FD09

3-chloro-4'-dimethylaminoazobenzene, FC35

4-chloro-4'-dimethylaminoazobenzene, FC36

4-chloro-4'-nitrodiphenylamine, FB28

4-(3-chlorophenylazo)-3,5-diphenyl-1-thiocarbamoylpyrazole, FE33

4-(4-chlorophenylazo)-3,5-diphenyl-5-thiocarbamoylpyrazole, FE34

5-(2-chlorophenylazo)-2-hydroxybenzoic acid, FB72

5-(4-chlorophenylazo)-2-hydroxybenzoic acid, FB73

4-[(4-chlorophenyl)azo]phenol, FB27

3-[5-(4-chlorophenyl)-2-furyl]-1-(2-furyl)-2-propen-1-one, FD49

3-(4-chlorophenylimino)-2-phenyl-3H-indole, FE03

3-(4-chlorophenylimino)-2-phenyl-3H-indole 1-oxide, FE04

N-(4-chlorophenyl)-N-(4-nitrosophenyl)hydroxylamine, FB29

3-(3-chlorophenyl)-3-oxo-N-phenylpropanamide, FC74

3-(4-chlorophenyl)-3-oxo-N-phenylpropanamide, FC75

1-(4-chlorophenyl)-3-phenyltriazene, FB38

Polychloro

2,7-di(acetylamino)-5,10-di(4-chlorophenyl)-5,10-dihydrophenazine, FF12

2,7-dichloro-5,10-bis(4-chlorophenyl)-5,10-dihydrophenazine, FE63

4,4'-dichlorodiphenylamine, FB30

3-(2,6-dichlorophenyl)-1-(2-thienyl-2-propenone, FB67

pentachloronitrobenzene, FA29

tetrakis(4-chlorophenyl)hydrazine, FE65

Fluoro

decafluorobiphenyl, FB14

4-dimethylamino-3'-fluoroazobenzene, FC38

4-dimethylamino-4'-fluoroazobenzene, FC39

2,5-diphenyl-1,1,4,4-tetra(4-fluorophenyl)diphosphoniacyclohexa-2,5-diene dichloride, FF61

3-(4-fluorophenyl)-1-(2-thienyl)-2-propen-1-one, FB74

hexafluorobenzene, FA30

methylpentafluorobenzene, FA52

octafluoronaphthalene, FA87

pentafluorobenzene, FA32

1,2,3,5-tetrafluorobenzene, FA33

1,3,5-trifluorobenzene, FA35

Heterocyclic

Bromo

eosin, FD96

Chloro

chlorimipramine, FD91

7-chloro-2,3-dihydro-1-methyl-5-phenyl-1H-1,4-benzodiazepine, FD27

7-chloro-1,3,4,5-tetrahydro-1-methyl-5-phenyl-3H-1,4-benzodiazepin-2-one, FD28

Diazepam, FD22

2,7-dichloro-5,10-bis(4-chlorophenyl)-5,10-dihydrophenazine, FE63

HALOGEN COMPOUNDS (cont.)

Heterocyclic

Chloro (cont.)

2,7-dichloro-5,10-di(4-nitrophenyl)-5,10-dihydrophenazine, FE62

Iodo

erythrosine, FD97

Steroids

Chloro

17α-acetoxy-6α-chloropregna-1,4-diene-3,20-dione, FE53

17α-acetoxy-6-chloropregna-4,6-diene-3,20-dione, FE54

17α-acetoxy-6α-chloropregn-4-ene-3,20-dione, FE57

17α-acetoxy-6β-chloropregn-4-ene-3,20-dione, FE58

Fluoro

17α-acetoxy-6α-fluoropregn-4-ene-3,20-dione, FE59

17α-acetoxy-6β-fluoropregn-4-ene-3,20-dione, FE60

6α-fluoro-17,21-dihydroxy-16α-methylpregn-4-ene-3,20-dione, FE39

6β-fluoro-17,21-dihydroxy-16α-methylpregn-4-ene-3,20-dione, FE40

6α-fluorotestosterone, FD93

6β-fluorotestosterone, FD94

HETEROCYCLIC COMPOUNDS

Nitrogen Heterocyclic

Pyridines

2,2'-bipyridine, FA89

bis[2,6-di(3-pyridyl)pyridine]ruthenium(2+) bisperchlorate, FF28

1,3-bis(3-pyridyl)-2-propen-1-one oxime, FB88

bis(2,4,6-tri-2-pyridyl-s-triazine)ruthenium(III) perchlorate, FF51

3-carbamoylpyridine, FA44

N,N'-diheptyl-4,4'-bipyridyliumtetrafluoroborate, FE85

1-diphenylmethyl-4-[(6-methyl-2-pyridyl)methyleneamino]piperazine, FE80

3-hydroxyamino-1-(2-hydroxy-4-methoxyphenyl)-3-(3-pyridyl)-1-propanone oxime, FD04

3-(hydroxyamino)-1-(2-hydroxyphenyl)-3-(2-pyridyl)-1-propanone oxime, FC60

3-hydroxyamino-1-(2-hydroxyphenyl)-3-(3-pyridyl)-1-propanone oxime, FC61

3-hydroxy-1-(2-hydroxy-4-methoxyphenyl)-3-(2-pyridyl)-1-propanone oxime, FC95

3-hydroxy-1-(2-hydroxy-4-methoxyphenyl)-3-(3-pyridyl)-1-propanone oxime, FC96

3-hydroxy-1-(2-hydroxy-4-methoxyphenyl)-3-(4-pyridyl)-1-propanone oxime, FC97

1-(2-hydroxy-4-methoxyphenyl)-3-(4-pyridyl)-1,3-propanedione dioxime, FC93

1-(2-hydroxy-4'-methoxyphenyl)-3-(2'-pyridyl)-2-propen-1-one, FC79

1-(2-hydroxy-4-methoxyphenyl)-3-(3-pyridyl)-2-propen-1-one, FC80

1-(2-hydroxy-4-methoxyphenyl)-3-(4-pyridyl)-2-propen-1-one, FC81

1-(2-hydroxyphenyl)-3-hydroxy-3-(2-pyridyl)-2-propen-1-one, FC18

1-(2-hydroxyphenyl)-3-(4-pyridyl)-1,3-propanedione dioxime, FC31

1-(2-hydroxyphenyl)-3-(2-pyridyl)-2-propen-1-one, FC14

1-(2-hydroxyphenyl)-3-(3-pyridyl)-2-propen-1-one, FC15

1-(2-hydroxyphenyl)-3-(4-pyridyl)-2-propen-1-one, FC16

2-methyl-1,2-di(3-pyridyl)-1-propanone, FC46

1-methylpyridinium-3-carbamoyl chloride, FA57

nicotinamide adenine dinucleotide, FE28

HETEROCYCLIC COMPOUNDS (cont.)

Nitrogen Heterocyclic

Pyridines (cont.)

nicotinamide mononucleotide, FB13

2-phenyl-3-(4-pyridylimino)-3H-indole 1,1-dioxide, FD81

2-phenyl-3-(2-pyridylimino)-3H-indole 1-oxide, FD80

1-(2-pyridyl)-3-hydroxyamino-3-(3-pyridyl)-1-propanone oxime, FC02

1-(2-pyridyl)-3-(4-pyridyl)-1,3-propanedione dioxime, FB92

1-(3-pyridyl)-3-(2-pyridyl)-1,3-propanedione dioxime, FB93

1-(3-pyridyl)-3-(4-pyridyl)-1,3-propanedione dioxime, FB94

1-(2-pyridyl)-3-(4-pyridyl)-2-propen-1-one, FB78

sodium deamino nicotinamide adenine dinucleotide, FE26

sodium nicotinamide adenine dinucleotide phosphate, FE27

2,2',2''-terpyridine, FC71

2,4,6-tri(2-pyridyl)-s-triazine, FD64

tris(2,2'-bipyridine-N,N')-ruthenium(2+) diperchlorate, FF29

One Ring-Nitrogen Atom, Monocyclic (Except Pyridines)

benzylpenicillenic acid, FD39

bilirubin, FF45

2,17-diethyl-1,10,19,22,23,24-hexahydro-3,7,13,18-tetramethyl-1,19-dioxo-21H-biline-8,12-dipropanoic acid, FF46

2,2,6,6-tetramethyl-4-oxo-1-piperidinyloxy, FA84

2,2,6,6-tetramethylpiperidine-nitroxide cation, FA85

2,2,6,6-tetramethyl-1-piperidinyloxy, FA86

One Ring-Nitrogen Atom, Polycyclic

3-(4-bromophenyimino)-2-phenyl-3H-indole, FE01

3-(4-bromophenylimino)-2-phenyl-3H-indole 1-oxide, FE02

Capri Blue, FD37

chlorimipramine, FD91

3-(4-chlorophenylimino)-2-phenyl-3H-indole, FE03

3-(4-chlorophenylimino)-2-phenyl-3H-indole 1-oxide, FE04

3-cyanoquinoline, FA88

3,6-diamino-10-methylacridinium chloride, FC37

3,7-diaminophenothiazin-5-ium chloride, FB39

3-(4-dimethylaminophenylimino)-2-phenylindole, FE36

3-(4-dimethylaminophenylimino)-2-phenylindole 1-oxide, FE37

3-ethyl-2-[7-(3-ethyl-2(3H)-benzothiazolylidene)-1,3,5-heptatrienyl]benzothiazolium perchlorate, FE89

3-ethyl-2-[(3-ethyl-2(3H)-benzothiazoylidene)methyl]benzothiazolium bromide, FD89

3-ethyl-2-[2-[(3-ethyl-2(3H)-benzothiazolylidene)methyl]-1-butenyl]-benzothiazolium bromide, FE52

3-ethyl-2-[(3-ethyl-2(3H)-benzothiazolylidene)-2-methyl-1-propenyl]benzothiazolium bromide, FE38

3-ethyl-2-[5-(3-ethyl-2(3H)-benzothiazolylidene)-1,3-pentadienyl-benzothiazolium iodide, FE51

3-ethyl-2-[3-(3-ethyl-2(3H)-benzothiazolylidene)-1-propenyl]-benzothiazolium bromide, FE25

1-ethyl-2[3-(3-ethyl-2(3H)-benzothiazolylidene)-1-propenyl]-quinolinium bromide, FE48

1-ethyl-2-[(1-ethyl-2(1H)-quinolinylidene)methyl]quinolinium iodide, FE49

1-ethyl-2-[(1-ethyl-4(1H)-quinolinylidene)methyl]quinolinium iodide, FE50

HETEROCYCLIC COMPOUNDS (cont.)

Nitrogen Heterocyclic

One Ring-Nitrogen Atom, Polycyclic (cont.)

1-ethyl-2-[3-(1-ethyl-2(1H)-quinolinylidene)-1-propenyl]-quinolinium bromide, FE88

8-hydroxyquinoline, FA77

3-imino-2-phenyl-3H-indole, FC12

isoquinoline, FA75

3-(4-methoxyphenylimino)-2-phenyl-3H-indole, FE19

3-(4-methoxyphenylimino)-2-phenyl-3H-indole 1-oxide, FE21

Methylene Blue, FD38

3-methylisoquinoline, FA91

10-methylphenothiazine, FB87

3-phenyl-2-[3-(3-phenyl-2(3H)-benzothiazolylidene)-1-propenyl]benzothiazolium iodide, FF26

2-phenyl-3-(phenylimino)-3H-indole, FE06

2-phenyl-3-phenylimino-3H-indole 1-oxide, FE07

2-phenyl-3-(2-phenyl-3H-indol-3-ylidene)-3H-indole 1,1'-dioxide, FF11

2-phenyl-3-(4-pyridylimino)-3H-indole 1,1-dioxide, FD81

2-phenyl-3-(2-pyridylimino)-3H-indole 1-oxide, FD80

2-phenyl-3-(p-tolylimino)-3H-indole, FE18

2-phenyl-3-(p-tolylimino)-3H-indole 1-oxide, FE20

quinoline, FA76

Two Ring-Nitrogen Atoms, Monocyclic

1-benzhydryl-4-nitrosopiperazine, FD59

4,4'-bis[3,5-diphenyl-Δ^2-pyrazolin-1-yl]biphenyl, FF65

4,5-bis(4-methoxyphenyl)-2-(4-dimethylaminophenyl)imidazole sodium, FE87

4-(4-bromophenylazo)-3,5-diphenyl-1-thiocarbamoylpyrazole, FE32

4-(3-chlorophenylazo)-3,5-diphenyl-1-thiocarbamoylpyrazole, FE33

4-(4-chlorophenylazo)-3,5-diphenyl-5-thiocarbamoylpyrazole, FE34

1,4-diacetyl-1,4-dihydro-2,3-diphenylpyrazine, FE12

4,5-di(biphenylyl)-2-(dimethylaminophenyl)imidazole sodium salt, FF49

2-(4-dimethylaminophenyl)-4,5-bis-(4-methylphenyl)imidazole sodium, FE86

3,5-dimethyl-4-(2-ethoxyphenylazo)-1-thiocarbamoylpyrazole, FC66

3,5-dimethyl-4-(4-ethoxyphenylazo)-1-thiocarbamoylpyrazole, FC67

3,5-dimethyl-4-(2-methoxyphenylazo)-1-thiocarbamoylpyrazole, FC03

3,5-dimethyl-4-(4-methoxyphenylazo)-1-thiocarbamoylpyrazole, FC04

4-(2,4-dimethylphenylazo)-3,5-diphenyl-1-thiocarbamoylpyrazole, FE77

4-(2,6-dimethylphenylazo)-3,5-diphenyl-1-thiocarbamoylpyrazole, FE78

3,5-dimethyl-4-(phenylazo)-1-thiocarbamoylpyrazole, FB59

3,5-diphenyl-4-(2-methoxyphenyl-1-thiocarbamoylpyrazole, FE43

3,5-diphenyl-4-(4-methoxyphenyl)-1-thiocarbamoylpyrazole, FE44

1-diphenylmethyl-4-[(6-methyl-2-pyridyl)methyleneamino]-piperazine, FE80

3,5-diphenyl-4-(2-methylphenylazo)-1-thiocarbamoylpyrazole, FE45

3,5-diphenyl-4-(3-methylphenylazo)thiocarbamoylpyrazole, FE46

3,5-diphenyl-4-(4-methylphenylazo)-1-thiocarbamoylpyrazole, FE47

3,5-diphenyl-4-phenylazo-4-thiocarbamoylpyrazole, FE35

1-(4-nitrophenyl)-3,5-diphenyl-Δ^2-pyrazoline, FE22

Solochrome Red ERS, FE10

2,4,5-tri(4-dimethylaminophenyl)-imidazole sodium, FF08

1,3,5-triphenyl-Δ^2-pyrazoline, FE24

HETEROCYCLIC COMPOUNDS (cont.)

Nitrogen Heterocyclic (cont.)

Two Ring-Nitrogen Atoms, Polycyclic

3,3'-bi-6H, 11H-6-oxopyrazolo-[3,4,5-d,e]anthracene dianion, FF09

2,7-bis(acetylamino)-5,10-bis-(4-acetylaminophenyl)-5,10-dihydrophenazine, FF39

2,7-bis(acetylamino)-5,10-di(4-chlorophenyl)-5,10-dihydrophenazine, FF12

2,7-bis(acetylamino)-5,10-dihydro-5,10-diphenylphenazine, FF17

5,10-bis(4-acetylaminophenyl)-2,7-dimethoxy-5,10-dihydrophenazine, FF33

7-chloro-2,3-dihydro-1-methyl-5-phenyl-1H-1,4-benzodiazepine, FD27

7-chloro-1,3,4,5-tetrahydro-1-methyl-5-phenyl-3H-1,4-benzodiazepin-2-one, FD28

1,2-diacetyl-1,2-dihydro-4-methylcinnoline, FC01

5,6-diacetyl-5,6-dihydro-5,6-phenanthroline, FD24

5,10-diacetyl-5,10-dihydrophenazine, FD25

1,4-diacetyl-1,4-dihydroquinoxaline, FB54

5,10-di(4-acetylphenyl)-2,7-dimethoxy-5,10-dihydrophenazine, FF31

3,7-diamino-5-phenylphenazinium chloride, FD68

9,10-diazaphenanthrene, FB16

Diazepam, FD22

2,7-dichloro-5,10-bis(4-chlorophenyl)-5,10-dihydrophenazine, FE63

2,7-dichloro-5,10-di(4-nitrophenyl)-5,10-dihydrophenazine, FE62

2,7-dicyanato-5,10-(4-cyanatophenyl)-5,10-dihydrophenazine, FF10

5,10-dihydro-5,10-diphenylphenazine, FE69

2,7-dimethoxy-5,10-bis(4-methoxyphenyl)-5,10-dihydrophenazine, FF20

2,7-dimethoxy-5,10-diphenyl-5,10-dihydrophenazine, FF00

2,7-dimethyl-5,10-bis(4-methylphenyl)-5,10-dihydrophenazine, FF19

2,7-dimethyl-5,10-diphenyl-5,10-phenazine, FE99

1,5-naphthyridine, FA63

1,10-phenanthroline, FB17

phenazine, FB18

5-phenylphenazinium chloride, FD65

pyrazolo[a,b,c-m,n]-9(10H)anthrone anion, FC06

quinoxaline, FA64

tris(1,10-phenanthroline ruthenium(II) perchlorate, FF50

Three Ring-Nitrogen Atoms

5-[(4-amino-3-sulfoanthraquinon-1-yl)amino]-2-{[3-hydroxy-5-(3-sulfophenyl)]-s-triazinylamino}benzenesulfonic acid, FF24

2,4,6-tri(2-pyridyl)-s-triazine, FD64

Four Ring-Nitrogen Atoms (Except Purines and Porphines)

folic acid, FD90

nicotinamide adenine dinucleotide, FE28

sodium deamino nicotinamide adenine dinucleotide, FE26

sodium nicotinamide adenine dinucleotide phosphate, FE27

Porphines

acetonitrilopyridinooctaethylporphyrin ruthenium(III) ion, FF66

bis(pyridino)-α,β,γ-δ-tetraphenylporphinylruthenium, FF77

carbonyl(octaethylporphyrinyl)ruthenium, FF58

carbonyl(pyridinotetraphenylporphyrinyl)ruthenium(II), FF71

carbonyl(tetraphenylporphyrinyl)ruthenium, FF69

HETEROCYCLIC COMPOUNDS (cont.)

Nitrogen Heterocyclic

Porphines (cont.)

   carbonyletioporphyrin(I), ruthenium(II), FF44

   etioporphyrin I, FF42

   magnesium(II)mesoporphyrin IX dimethylester, FF55

   magnesium(II)protoporphyrin IX dimethylester, FF53

   2,3,7,8,12,13,17,18-octaethylporphine, FF56

   5,10,15,20-tetraphenylporphine, FF67

Oxygen Heterocyclic (see also ANHYDRIDES, CARBOXYLIC; CARBOHYDRATES AND SUGARS; and LACTONES)

   2-benzoylbenzofuran, FC70

   benzylpenicillenic acid, FD39

   3-[5-(4-chlorophenyl)-2-furyl]-1-(2-furyl)-2-propen-1-one, FD49

   1,5-di(5-nitro-2-furyl)-1,4-pentadien-3-one amidinohydrazone hydrochloride, FB89

   1-(2-furyl)-2-methyl-3-(5-phenyl-2-furyl)-2-propen-1-one, FD66

   1-(2-furyl)-3-[5-(4-nitrophenyl-2-furyl]-2-propen-1-one, FD50

   3-(5-phenyl-2-furyl)-1-(2-furyl)-2-propen-1-one, FD53

   2-(3,4-methylenedioxybenzylidene)-1,3-indandione, FD48

   2-methyl-1-phenyl-3-(5-phenyl-2-furyl)-2-propen-1-one, FE11

   nicotinamide adenine dinucleotide, FE28

   sodium deamino nicotinamide adenine dinucleotide, FE26

   sodium nicotinamide adenine dinucleotide phosphate, FE27

   tetraphenylcyclopentadienone, FF25

   2,4,6-triphenylpyrylium perchlorate, FE42

   2,4,6-tris(4-methoxyphenyl)-pyrylium perchlorate, FF02

Sulfur Heterocyclic (see also SULFUR COMPOUNDS)

   5,5'-bis(4-methoxyphenyl)-3,3'-[2,4-bis(4-methylphenyl)-3,4-dithia-1,6-hexamethylene]dithiolium perchlorate, FF60

   2,5-bis(4-methylphenyl)-3,3a,4-trithiapentalene, FD88

   Capri Blue, FD37

   3-(4-carbamoylphenyl)-1-(2-thienyl)-2-propenone, FC17

   3,7-diaminophenothiazin-5-ium chloride, FB39

   3-(2,6-dichlorophenyl)-1-(2-thienyl)-2-propenone, FB67

   3-(4-dimethylaminophenyl)-1-(2-thienyl)-2-propen-1-one, FC88

   5,5'-diphenyl-3,3'-[2,5-bis(4-methoxyphenyl)-3,4-dithia-1,6-hexamethyleno]dithiolium perchlorate, FF52

   5,5'-diphenyl-3,3'-(2,5-diphenyl-3,4-dithia-1,6-hexamethylene)-dithiolium diperchlorate, FF47

   2,5-diphenyl-3,3a,4-trithiapentalene, FD56

   3-(4-ethoxyphenyl)-1-(2-thienyl)-2-propen-1-one, FC87

   3-ethyl-2-[7-(3-ethyl-2(3H)-benzothiazolylidene)-1,3,5-heptatrienyl]-benzothiazolium perchlorate, FE89

   3-ethyl-2-[(3-ethyl-2(3H)-benzothiazoylidene)methyl]benzothiazolium bromide, FD89

   3-ethyl-2-[2-(3-ethyl-2(3H)-benzothiazolylidene)methyl]-1-butenyl]-benzothiazolium bromide, FE52

   3-ethyl-2-[(3-ethyl-2(3H)-benzothiazolylidene)-2-methyl-1-propenyl]benzothiazolium bromide, FE38

   3-ethyl-2-[5-(3-ethyl-2(3H)-benzothiazolylidene)-1,3-pentadienylbenzothiazolium iodide, FE51

HETEROCYCLIC COMPOUNDS (cont.)

Sulfur Heterocyclic (see also SULFUR COMPOUNDS) (cont.)

   3-ethyl-2-[3-(3-ethyl-2(3H)-benzothiazolylidene)-1-propenyl]benzothiazolium bromide, FE25

   1-ethyl-2-[3-(3-ethyl-2(3H)-benzothiazolylidene)-1-propenyl]quinolinium bromide, FE48

   3-(4-fluorophenyl)-1-(2-thienyl)-2-propen-1-one, FB74

   Methylene Blue, FD38

   10-methylphenothiazine, FB87

   2-(4-methylphenyl)-5-phenyl-3,3a,4-trithiapentalene, FD67

   3-(2-methylphenyl)-1-(2-thienyl)-2-propen-1-one, FC27

   3-(4-methylphenyl)-1-(2-thienyl)-2-propen-1-one, FC28

   3-(1-naphthyl)-1-(2-thienyl)-2-propen-1-one, FD51

   3-phenyl-2-[3-(3-phenyl-2(3H)-benzothiazolylidene)-1-propenyl]benzothiazolium iodide, FF26

   3-phenyl-1-(2-thienyl)-2-propen-1-one, FB85

   4,4',5,5'-tetrakis(4-methoxyphenyl)-3,3'-[2,5-bis(4-methylphenyl)-1,6-bis(4-methoxyphenyl)-3,4-dithia-1,6-hexylene]dithiolium perchlorate, FF80

   4,4',5,5'-tetrakis(4-methoxyphenyl)-3,3'-[2,5-bis(4-methylphenyl)-3,4-dithia-1,6-hexylene]dithiolium perchlorate, FF73

   2,3,4,5-tetrakis(4-methoxyphenyl-trithiapentalene, FF43

   4,4',5,5'-tetraphenyl-3,3'-(2,5-diphenyl-3,4-dithia-1,6-hexamethylene)dithiolium perchlorate, FF70

   thianthrene, FB25

   thianthrene 5,5-dioxide, FB23

   thianthrene 5,10-dioxide, FB24

   thianthrene 5-oxide, FB22

   1-(2-thienyl)-3-(2,3,4-trimethoxyphenyl)-2-propen-1-one, FD34

   1-(2-thienyl)-3-(3,4,5-trimethoxyphenyl)-2-propen-1-one, FD35

   2,3,5-triphenyl-1,6,6a-trithiapentalene, FE41

   2,3,5-tris(4-methoxyphenyl)trithiapentalene, FE98

HYDRAZIDES, see HYDRAZINES

HYDRAZINES, HYDRAZIDES, AND HYDRAZONES (see also CARBAZIDES)

   N,N'-bis(4-methoxyphenyl)-N,N'-diphenylhydrazine, FF07

   1-(4-chlorophenyl)-3-phenyltriazene, FB38

   4-cyanodiazoaminobenzene, FB82

   diazoaminobenzene, FB51

   N,N'-di(4-methylphenyl)-N,N'-diphenylhydrazine, FF03

   1,5-di(5-nitro-2-furyl)-1,4-pentadien-3-one amidinohydrazone hydrochloride, FB89

   1,3-diphenyl-2-propen-1-one phenylhydrazone, FE23

   3-(4-ethoxycarbonylphenyl)-1-phenyltriazene, FC92

   α,α'-hydrazotoluene, FC65

   tetrakis(4-chlorophenyl)hydrazine, FE65

   tetrakis(4-methylphenyl)hydrazine, FF23

   tetrakis(4-nitrophenyl)hydrazine, FE66

   1,1,2,2-tetraphenylhydrazine, FE75

HYDRAZONES, see HYDRAZINES

HYDROCARBONS

   anthracene, FC10

   benzene, FA43

   4-benzylbiphenyl, FD82

   9,9'-bifluorenyl, FE93

   9,9'-bi(9-hydroxy-9,10-dihydroanthryl), FF13

HYDROCARBONS (cont.)

9,9'-bi(9-hydroxyfluorenyl), FE96
biphenyl, FB36
1,2-bis(4-biphenylyl)-1,2-diphenylethanediol, FF59
1,3,5,7-cyclooctatetraene, FA66
9,10-di-p-anisylanthracene, FF14
1,2-diethylidenacenaphthene, FD23
9,10-diphenylanthracene, FE94
4-ethylbiphenyl, FC32
fluorene, FB77
9-hydroxy-9,10-dihydroanthracene, FC25
9-hydroxy-9-methylfluorene, FC26
9-methoxy-9-methylfluorene, FC86
2-methylnaphthalene, FB09
perylene, FD99
phenanthrene, FC11
9-phenylanthracene, FE05
phenylcyclooctatetraene, FC20
1,2,4,7-tetraphenylcyclooctatetraene, FF37
1,3,5,7-tetraphenylcyclooctatetraene, FF38
toluene, FA55
1,3,5-trimethylbenzene, FA82

HYDROXY COMPOUNDS (see also QUINONES)

Aliphatic

benzhydrol, FB95
benzyl alcohol, FA56
(4-biphenylyl)benzyl alcohol, FD86
1-(4-biphenylyl)ethanol, FC52
2,3-bis(4-biphenylyl)-2,3-butandiol, FF21
1,2-bis(4-biphenylyl)-1,2-diphenylethanediol, FF59
1-butanol, FA18
2-butanol, FA19
1,1-diphenylprop-2-en-1-ol, FC84
3,3'-diphenylprop-2-en-1-ol, FC85
1,1-diphenylprop-2-yn-1-ol, FC77
ethanol, FA04
6α-fluoro-17,21-dihydroxy-16α-methylpregn-4-ene-3,20-dione, FE39
6β-fluoro-17,21-dihydroxy-16α-methylpregn-4-ene-3,20-dione, FE40
1-hexanol, FA50
9-(1-hydroxyethyl)anthracene, FD26
1-(1-hydroxyethyl)naphthalene, FB55
2-(1-hydroxyethyl)naphthalene, FB56
3-hydroxy-1-(2-hydroxy-4-methoxyphenyl)-3-(2-pyridyl)-1-propanone oxime, FC95
3-hydroxy-1-(2-hydroxy-4-methoxyphenyl)-3-(3-pyridyl)-1-propanone oxime, FC96
3-hydroxy-1-(2-hydroxy-4-methoxyphenyl)-3-(4-pyridyl)-1-propanone oxime, FC97
1-(2-hydroxyphenyl)-3-hydroxy-3-(2-pyridyl)-2-propane-1-one, FC18
methanol, FA01
2-methyl-2-butanol, FA27
2-methyl-3-butyn-2-ol, FA22
1-pentanol, FA28
2-phenyl-3-butyn-2-ol, FA94
3-phenyl-2-propen-1-ol, FA80
1-phenyl-2-propyn-1-ol, FA78
Δ^4-pregnene-17α,21-diol-3,20-dione, FE30
1-propanol, FA11
2-propanol, FA12
2-propen-1-ol, FA10
2-propyn-1-ol, FA08
triphenylmethanol, FD87

Alicyclic

2-benzoyloxy-1-hydroxy-3-indanone, FD21
bi(2-hydroxy-1,3-dioxoinden-2-yl), FD63
9,9'-bi(9-hydroxyfluorenyl), FE96
2,3-dihydroxy-1H-inden-1-one, FA73
9-fluorenol, FB84

HYDROXY COMPOUNDS (see also QUINONES) (cont.)

Alicyclic (cont.)

6α-fluoro-17,21-dihydroxy-16α-methylpregn-4-ene-3,20-dione, FE39

6β-fluoro-17,21-dihydroxy-16α-methylpregn-4-ene-3,20-dione, FE40

6α-fluorotestosterone, FD93

6β-fluorotestosterone, FD94

17β-hydroxyandrosta-1,4-dien-3-one, FD92

9-hydroxy-9,10-dihydroanthracene, FC25

9-hydroxy-9-methylfluorene, FC26

6α-methyltestosterone, FE15

6β-methyltestosterone, FE16

testosterone, FD95

2,2,4,5-tetrahydroxy-4-cyclopentene-1,3-dione, FA20

Aromatic

Phenols, Unsubstituted and Monosubstituted

4,4'-azobisphenol, FB42

4-[(4-bromophenyl)azo]phenol, FB26

4-(4-chlorophenyl)azo]phenol, FB27

3-[(2,5-dihydroxynaphth-1-yl)azo]-4-hydroxybenzenesulfonic acid, sodium salt, FD17

4-dimethylamino-2'-hydroxyazobenzene, FC58

4-dimethylamino-4'-hydroxyazobenzene, FC59

hydroquinone, FA45

3-(hydroxyamino)-1-(2-hydroxyphenyl)-3-(2-pyridyl)-1-propanone oxime, FC60

3-hydroxyamino-1-(2-hydroxyphenyl)3-(3-pyridyl)-1-propanone oxime, FC61

2-(3-hydroxybenzylidene)-1,3-indandione, FD11

4-[(4-hydroxyphenyl)azo]benzenesulfonic acid, FB45

4-(4-hydroxyphenylazo)benzoic acid, FB80

1-(2-hydroxyphenyl)-3-hydroxy-3-(2-pyridyl)-2-propen-1-one, FC18

1-(2-hydroxyphenyl)-3-(4-pyridyl)-1,3-propanedione dioxime, FC31

1-(2-hydroxyphenyl)-3-(2-pyridyl)-2-propen-1-one, FC14

1-(2-hydroxyphenyl)-3-(3-pyridyl)-2-propen-1-one, FC15

1-(2-hydroxyphenyl)-3-(4-pyridyl)-2-propen-1-one, FC16

4-[(4-iodophenyl)azo]phenol, FB31

4-[(3-methylphenyl)azo]phenol, FB90

4-[(4-methylphenyl)azo]phenol, FB91

4-nitrophenol, FA42

4-[(2-nitrophenyl)azo]phenol, FB32

4-[(3-nitrophenyl)azo]phenol, FB33

4-[(4-nitrophenyl)azo]phenol, FB34

4-(phenylamino)phenol, FB48

4-phenylazophenol, FB41

1-(2,4,6-tribromo-3-hydroxyphenyl)-3-(4-hydroxyphenyl)-2-propen-1-one, FC69

Phenols, Polysubstituted

2,5-bis(2-isopropoxy-2-methylpropyl)hydroquinone, FE17

2,5-bis(2-methoxy-2-methylpropyl)-1,4-hydroquinone, FD47

1,6-bis(2-methyl-2-ethoxypropyl)hydroquinone, FD78

2,5-bis(2-methyl-2-ethoxypropyl)hydroquinone, FD77

2-(2-tert-butoxy-2-methylpropyl)-5-tert-butylhydroquinone, FD76

2-tert-butyl-5-(2-methoxy-2-methyl-1-propyl)hydroquinone, FD05

3-carboxy-4-hydroxy-4'-sulfoazobenzene disodium salt, FB68

5-(chlorophenylazo)-2-hydroxybenzoic acid, FB72

HYDROXY COMPOUNDS (see also QUINONES) (cont.)

Aromatic

Phenols, Polysubstituted (cont.)

5-(4-chlorophenylazo)-2-hydroxybenzoic acid, FB73

2,4-diaminophenol, FA46

1,2-dihydroxy-1-cyclohexen-3,4,5,6-tetraone, FA34

2-(1,1-dimethylethyl)-5-(2-ethoxy-2-methylpropyl)-1,4-hydroquinone, FD44

2-(1,1-dimethylethyl)-6-(2-ethoxy-2-methylpropyl)-1,4-hydroquinone, FD45

2-(1,1-dimethylethyl)-5-[2-methyl-2(1-methylethoxy)-propyl]-1,4-hydroquinone, FD61

2-(1,1-dimethylpropyl)-5-(2-ethoxy-2-methylpropyl)hydroquinone, FD62

2-(1,1-dimethylpropyl)-5-(2-methoxy-2-methylpropyl)hydroquinone, FD46

3-hydroxyamino-1-(2-hydroxy-4-methoxyphenyl)-3-(3-pyridyl)-1-propanone oxime, FD04

2-hydroxy-5-(2,4-dinitrophenylazo)benzoic acid, FB69

3-hydroxy-1-(2-hydroxy-4-methoxyphenyl)-3-(2-pyridyl)-1-propanone oxime, FC95

3-hydroxy-1-(2-hydroxy-4-methoxyphenyl)-3-(3-pyridyl)-1-propanone oxime, FC96

3-hydroxy-1-(2-hydroxy-4-methoxyphenyl)-3-(4-pyridyl)-1-propanone oxime, FC97

1-(2-hydroxy-4-methoxyphenyl)-3-(4-pyridyl)-1,3-propanedione dioxime, FC93

1-(2-hydroxy-4-methoxyphenyl)-3-(2-pyridyl)-2-propen-1-one, FC79

1-(2-hydroxy-4-methoxyphenyl)-3-(3-pyridyl)-2-propen-1-one, FC80

1-(2-hydroxy-4-methoxyphenyl)-3-(4-pyridyl)-2-propen-1-one, FC81

3-(4-hydroxy-3-methoxyphenyl)-1-(2,4,6-tribromo-3-hydroxyphenyl)-2-propen-1-one, FD16

2-hydroxy-5-(2-nitrophenylazo)-benzoic acid, FB75

2-hydroxy-5-(4-nitrophenylazo)-benzoic acid, FB76

2-hydroxy-5-phenylazobenzoic acid, FB79

3-(3-methoxyphenyl)-1-(2,4,6-tribromo-3-hydroxyphenyl)-2-propen-1-one, FD14

3-(4-methoxyphenyl)-1-(2,4,6-tribromo-3-hydroxyphenyl)-2-propen-1-one, FD15

3-(4-methylphenyl)-1-(2,4,6-tribromo-3-hydroxyphenyl)-2-propen-1-one, FD13

tetrahydroxy-1,4-benzoquinone, FA40

1-(2,4,6-tribromo-3-hydroxyphenyl)-3-phenyl-2-propen-1-one, FC68

Two or more Rings

9,9'-bi(9-hydroxy-9,10-dihydroanthryl), FF13

1,2-bis(4-methoxyphenyl)cyclobuteno[3,4-b]-3,8-naphthohydroquinone, dipotassium salt, FE95

1,2-dibenzylidenecyclobutano-[3,4-b]-3,8-naphthohydroquinone, dipotassium salt, FE91

1,2-dihydro-1,2-dimethylcyclobuta[b]naphthalene-3,8-diol, FC53

3-[(2,5-dihydroxynaphth-1-yl)-azo]-4-hydroxybenzenesulfonic acid, sodium salt, FD17

6,11-dioxido-2,4-diphenyldibenzo-[b,h]biphenylene, dipotassium salt, FF35

1,2-diphenylcyclobuta[b]naphthalene-3,8-diol, FE67

1,2-diphenylcyclobuta[b]naphthalene-3,8-diol, dipotassium salt, FE64

Solochrome Black WDFA, FE00

Solochrome Red ERS, FE10

HYDROXY COMPOUNDS (see also QUINONES) (cont.)

Heterocyclic (see also CARBOHYDRATES)

5-[(4-amino-3-sulfoanthraquinon-1-yl)amino]-2-{[3-hydroxy-5-(3-sulfophenyl]-s-triazinylamino-benzenesulfonic acid, FF24

2,17-diethyl-1,10,19,22,23,24-hexahydro-3,7,13,18-tetramethyl-1,19-dioxo- H-biline-8,12-dipropanoic acid, FF46

eosin, FD96

erythrosine, FD97

fluorescein disodium salt, FD98

folic acid, FD90

8-hydroxyquinoline, FA77

nicotinamide adenine dinucleotide, FE28

sodium deamino nicotinamide adenine dinucleotide, FE26

sodium nicotinamide adenine dinucleotide phosphate, FE27

Solochrome Red ERS, FE10

Hydroxylamines

N-benzoyl-N-phenylhydroxylamine, FB86

N-(4-chlorophenyl)-N-(4-nitrosophenyl)hydroxylamine, FB29

N-ethyl-N-hydroxy-β-methoxy-β-(3-trifluoromethylphenyl)-ethylamine, FB61

3-hydroxyamino-1-(2-hydroxy-4-methoxyphenyl)-3-(3-pyridyl)-1-propanone oxime, FD04

3-hydroxyamino-1-(2-hydroxyphenyl)-3-(2-pyridyl)-1-propanone oxime, FC60

3-hydroxyamino-1-(2-hydroxyphenyl)-3-(3-pyridyl)-1-propanone oxime, FC61

β-methoxy-β-(3-trifluoromethylphenyl)ethylhydroxylamine, FB01

4-nitrosodiphenylhydroxylamine, FB44

1-(2-pyridyl)-3-hydroxyamino-3-(3-pyridyl)-1-propanone oxime, FC02

HYDROXY SALTS, see HYDROXY COMPOUNDS

IMINES AND OTHER AZOMETHINES (see also CARBAZIDES, HYDRAZINES, and OXIMES)

Aliphatic

3-(4-bromophenylimino)-2-phenyl-3H-indole, FE01

3-(4-bromophenylimino)-2-phenyl-3H-indole 1-oxide, FE02

3-(4-chlorophenylimino)-2-phenyl-3H-indole, FE03

3-(4-chlorophenylimino)-2-phenyl-3H-indole 1-oxide, FE04

3-(4-dimethylaminophenylimino)-2-phenylindole, FE36

3-(4-dimethylaminophenylimino)-2-phenylindole 1-oxide, FE37

1,5-di(5-nitro-2-furyl)-1,4-pentadien-3-one amidinohydrazone hydrochloride, FB89

1-diphenylmethyl-4-[(6-methyl-2-pyridyl)methyleneamino]piperazine, FE80

1,3-diphenyl-2-propen-1-one phenylhydrazone, FE23

3-imino-2-phenyl-3H-indole, FC12

3-(4-methoxyphenylimino)-2-phenyl-3H-indole, FE19

3-(4-methoxyphenylimino)-2-phenyl-3H-indole 1-oxide, FE21

2-phenyl-3-(phenylimino)-3H-indole, FE06

2-phenyl-3-phenylimino-3H-indole 1-oxide, FE07

2-phenyl-3-(4-pyridylimino)-3H-indole 1,1-dioxide, FD81

2-phenyl-3-(2-pyridylimino)-3H-indole 1-oxide, FD80

2-phenyl-3-(p-tolylimino)-3H-indole, FE18

2-phenyl-3-(p-tolylimino)-3H-indole 1-oxide, FE20

Cyclic Imines

3,3'-bi(6H,11H-6-oxopyrazolo-[3,4,5-d,e]anthracene) dianion, FF09

3-(4-bromophenylimino)-2-phenyl-3H-indole, FE01

IMINES AND OTHER AZOMETHINES (see also CARBAZIDES, HYDRAZINES, AND OXIMES) (cont.)

Cyclic Imines (cont.)

3-(4-bromophenylimino)-2-phenyl-3 3H-indole 1-oxide, FE02

carbonyletioporphyrin(I), ruthenium(II), FF44

7-chloro-2,3-dihydro-1-methyl-5-phenyl-1H-1,4-benzodiazepine, FD27

3-(4-chlorophenylimino)-2-phenyl-3H-indole, FE03

3-(4-chlorophenylimino)-2-phenyl-3H-indole 1-oxide, FE04

Diazepam, FD22

3-(4-dimethylaminophenylimino)-2-phenylindole, FE36

3-(4-dimethylaminophenylimino)-2-phenylindole 1-oxide, FE37

etioporphyrin I, FF42

3-imino-2-phenyl-3H-indole, FC12

3-(4-methoxyphenylimino)-2-phenyl-3H-indole, FE19

3-(4-methoxyphenylimino)-2-phenyl-3H-indole 1-oxide, FE21

2-phenyl-3-(phenylimino)-3H-indole, FE06

2-phenyl-3-phenylimino-3H-indole 1-oxide, FE07

2-phenyl-3-(2-phenyl-3H-indol-3-ylidene)-3H-indole 1,1'-dioxide, FF11

2-phenyl-3-(4-pyridylimino)-3H-indole 1,1-dioxide, FD81

2-phenyl-3-(2-pyridylimino-3H-indole 1-oxide, FD80

2-phenyl-3-(p-tolylimino)-3H-indole, FE18

2-phenyl-3-(p-tolylimino)-3H-indole 1-oxide, FE20

ISOTHIOCYANATES AND THIOCYANATES

2,7-dicyanato-5,10-(4-cyanatophenyl)-5,10-dihydrophenazine, FF10

4,4'-dithiocyanatodiphenylamine, FC09

KETONES (see also HETEROCYCLIC COMPOUNDS, IMIDES, and UNSATURATED COMPOUNDS)

Dialkyl Ketones

acetone, FA09

3-methyl-3-buten-2-one, FA23

3-methyl-3-penten-2-one, FA47

4-methyl-3-penten-2-one, FA48

4-phenyl-3-buten-2-one, FA96

Alkyl-Aryl Ketones

acetophenone, FA67

4-acetyldiphenylamine, FC30

4-acetyl-4'-methoxydiphenylamine, FC89

benzophenonebenzenechromium(0), FD84

benzophenonebenzenechromium(I) iodide, FD83

3-(3-bromophenyl)-3-oxo-N-phenylpropanamide, FC72

3-(4-bromophenyl)-3-oxo-N-phenylpropanamide, FC73

3-(3-chlorophenyl)-3-oxo-N-phenylpropanamide, FC74

3-(4-chlorophenyl)-3-oxo-N-phenylpropanamide, FC75

5,10-di(4-acetylphenyl)-2,7-dimethoxy-5,10-dihydrophenazine, FF31

N,N'-di(4-acetylphenyl)-N-phenyl-phenylenediamine, FF16

trans(?)-1,2-dibenzoylethylene, FD20

1,3-dibenzoylpropane, FD57

2,2-dimethyl-1-phenyl-1-propanone, FB11

trans 1,3-diphenyl-2-propen-1-one, FC76

1-(2-hydroxy-4-methoxyphenyl)-3-(2-pyridyl)-2-propen-1-one, FC79

1-(2-hydroxy-4-methoxyphenyl)-3-(3-pyridyl)-2-propen-1-one, FC80

1-(2-hydroxy-4-methoxyphenyl)-3-(4-pyridyl)-2-propen-1-one, FC81

3-(4-hydroxy-3-methoxyphenyl)-1-(2,4,6-tribromo-3-hydroxyphenyl)-2-propen-1-one, FD16

1-(2-hydroxyphenyl)-3-hydroxy-3-(2-pyridyl)-2-propen-1-one, FC18

KETONES (see also HETEROCYCLIC COMPOUNDS, IMIDES, AND UNSATURATED COMPOUNDS) (cont.)

Alkyl-Aryl Ketones (cont.)

1-(2-hydroxyphenyl)-3-(2-pyridyl)-2-propen-1-one, FC14

1-(2-hydroxyphenyl)-3-(3-pyridyl)-2-propen-1-one, FC15

1-(2-hydroxyphenyl)-3-(4-pyridyl)-2-propen-1-one, FC16

3-(3-methoxyphenyl)-3-oxo-N-phenylpropanamide, FD31

3-(4-methoxyphenyl)-3-oxo-N-phenylpropanamide, FD32

3-(3-methoxyphenyl)-1-(2,4,6-tribromo-3-hydroxyphenyl)-2-propen-1-one, FD14

3-(4-methoxyphenyl)-1-(2,4,6-tribromo-3-hydroxyphenyl)-2-propen-1-one, FD15

3-(3-methylphenyl)-3-oxo-N-phenylpropanamide, FD29

3-(4-methylphenyl)-3-oxo-N-phenylpropanamide, FD30

2-methyl-1-phenyl-3-(5-phenyl-2-furyl)-2-propen-1-one, FE11

2-methyl-1-phenylpropanone, FB04

3-(4-methylphenyl)-1-(2,4,6-tribromo-3-hydroxyphenyl)-2-propen-1-one, FD13

β-oxo-N-phenylbenzenepropanamide, FC78

propiophenone, FA81

1-(2,4,6-tribromo-3-hydroxyphenyl)-3-(4-hydroxyphenyl)-2-propen-1-one, FC69

1-(2,4,6-tribromo-3-hydroxyphenyl)-3-phenyl-2-propen-1-one, FC68

5,6-diacetyl-5,6-dihydro-5,6-phenanthroline, FD24

5,10-diacetyl-5,10-dihydrophenazine, FD25

1,4-diacetyl-1,4-dihydroquinoxaline, FB54

3-(2,6-dichlorophenyl)-1-(2-thienyl)-2-propenone, FB67

3-(4-dimethylaminophenyl)-1-(2-thienyl)-2-propen-1-one, FC88

3-(4-ethoxyphenyl)-1-(2-thienyl)-2-propen-1-one, FC87

3-(4-fluorophenyl)-1-(2-thienyl)-2-propen-1-one, FB74

1-(2-furyl)-2-methyl-3-(5-phenyl-2-furyl)-2-propen-1-one, FD66

1-(2-furyl)-3[5-(4-nitrophenyl)-2-furyl]-2-propen-1-one, FD50

1-(2-furyl)-3-(5-phenyl-2-furyl)-2-propen-1-one, FD53

1-(2-hydroxyphenyl)-3-hydroxy-3-(2-pyridyl)-2-propen-1-one, FC18

2-methyl-1,2-di(3-pyridyl)-1-propanone, FC46

3-(2-methylphenyl)-1-(2-thienyl)-2-propen-1-one, FC27

3-(4-methylphenyl)-1-(2-thienyl)-2-propen-1-one, FC28

3-(1-naphthyl)-1-(2-thienyl)-2-propen-1-one, FD51

3-phenyl-1-(2-thienyl-2-propen-1-one, FB85

1-(2-pyridyl)-3-(4-pyridyl)-2-propen-1-one, FB78

1-(2-thienyl)-3-(2,3,4-trimethoxyphenyl)-2-propen-1-one, FD34

1-(2-thienyl)-3-(3,4,5-trimethoxyphenyl)-2-propen-1-one, FD35

Alkyl-Heterocyclic Ring Ketones

3-(4-carbamoylphenyl)-1-(2-thienyl)-2-propenone, FC17

3-[5-(4-chlorophenyl)-2-furyl]-1-(2-furyl)-2-propen-1-one, FD49

1,4-diacetyl-1,4-dihydro-2,3-diphenylpyrazine, FE12

1,2-diacetyl-1,2-dihydro-4-methylcinnoline, FC01

Other Alkyl Ketones

17α-acetoxy-6α-chloropregna-1,4-diene-3,20-dione, FE53

17α-acetoxy-6-chloropregna-4,6-diene-3,20-dione, FE54

17α-acetoxy-6α-chloropregn-4-ene-3,20-dione, FE57

17α-acetoxy-6β-chloropregn-4-ene-3,20-dione, FE58

KETONES, (see also HETEROCYCLIC COMPOUNDS, IMIDES, AND UNSATURATED COMPOUNDS) (cont.)

Other Alkyl Ketones (cont.)

17α-acetoxy-6α-fluoropregn-4-ene-3,20-dione, FE59

17α-acetoxy-6β-fluoropregn-4-ene-3,20-dione, FE60

17α-acetoxy-6α-methylpregna-1,4-diene-3,20-dione, FE82

17α-acetoxy-6-methylpregna-4,6-diene-3,20-dione, FE83

17α-acetoxy-6α-methylpregn-4-ene-3,20-dione, FE84

17α-acetoxypregna-1,4-diene-3,20-dione, FE55

17α-acetoxypregna-4,6-diene-3,20-dione, FE56

17α-acetoxypregn-4-ene-3,20-dione, FE61

6α-fluoro-17,21-dihydroxy-16α-methylpregn-4-ene-3,20-dione, FE39

6β-fluoro-17,21-dihydroxy-16α-methylpregn-4-ene-3,20-dione, FE40

Δ^4-pregnene-17α,21-diol-3,20-dione, FE30

progesterone, FE29

Diaryl Ketones

benzophenone, FB83

Aryl-Heterocyclic Ring Ketones

2-benzoylbenzofuran, FC70

Ketosteroids

17α-acetoxy-6α-chloropregna-1,4-diene-3,20-dione, FE53

17α-acetoxy-6-chloropregna-4,6-diene-3,20-dione, FE54

17α-acetoxy-6α-chloropregn-4-ene-3,20-dione, FE57

17α-acetoxy-6β-chloropregn-4-ene-3,20-dione, FE58

17α-acetoxy-6α-fluoropregn-4-ene-3,20-dione, FE59

17α-acetoxy-6β-fluoropregn-4-ene-3,20-dione, FE60

17α-acetoxy-6α-methylpregna-1,4-diene-3,20-dione, FE82

17α-acetoxy-6-methylpregna-4,6-diene-3,20-dione, FE83

17α-acetoxy-6α-methylpregn-4-ene-3,20-dione, FE84

17α-acetoxypregna-1,4-diene-3,20-dione, FE55

17α-acetoxypregna-4,6-diene-3,20-dione, FE56

17α-acetoxypregn-4-ene-3,20-dione, FE61

6α-fluoro-17,21-dihydroxy-16α-methylpregn-4-ene-3,20-dione, FE39

6β-fluoro-17,21-dihydroxy-16α-methylpregn-4-ene-3,20-dione, FE40

6α-fluorotestosterone, FD93

6β-fluorotestosterone, FD94

17β-hydroxyandrosta-1,4-dien-3-one, FD92

6α-methyltestosterone, FE15

6β-methyltestosterone, FE16

Δ^4-pregnene-17α,21-diol-3,20-dione, FE30

progesterone, FE29

testosterone, FD95

Other Carbocyclic Ketones

2-benzoyloxy-1-hydroxy-3-indanone, FD21

2-benzoyloxy-1,3-indandione, FD12

2-benzoyloxy-1-methoxy-3-indenone, FD55

2-benzylidene-1,3-indandione, FD10

bi(2-hydroxy-1,3-dioxoinden-2-yl), FD63

2-(3-chlorobenzylidene)-1,3-indandione, FD08

2-(4-chlorobenzylidene)-1,3-indandione, FD09

1,2-dihydroxy-1-cyclohexen-3,4,5,6-tetraone, FA34

2,3-dihydroxy-1H-inden-1-one, FA73

9-fluorenone, FB71

KETONES, (see also HETEROCYCLIC COMPOUNDS, IMIDES, AND UNSATURATED COMPOUNDS) (cont.)

Other Carbocyclic Ketones (cont.)

2-(3-hydroxybenzylidene)-1,3-indandione, FD11

2-(4-methoxybenzylidene)-1,3-indandione, FD54

2-(4-methylbenzylidene)-1,3-indandione, FD52

2-(3,4-methylenedioxybenzylidene)-1,3-indandione, FD48

ninhydrin, FA72

2,2,4,5-tetrahydroxy-4-cyclopentene-1,3-dione, FA20

1-tetralone, FA97

2,3,4,5-tetraphenyl-2-cyclopenten-1-one, FF27

3,5,5-trimethyl-2-cyclohexen-1-one, FA83

Heterocyclic Ketones

3,3'-bi(6H,11H-6-oxopyrazolo-[3,4,5-d,e]anthracene) dianion, FF09

eosin, FD96

erythrosine, FD97

fluorescein disodium salt, FD98

pyrazolo[a,b,c-m,n]-9(10H)-anthrone anion, FC06

2,2,6,6-tetramethyl-4-oxo-1-piperidinyloxy, FA84

Ketoacids and Derivatives

3-bromo-2-oxopropanoic acid, FA06

3-dimethylsulfonio-2-oxopropanoic acid bromide, FA26

ethyl 3-bromo-2-oxopropanoate, FA21

ethyl 3-dimethylsulfonio-2-oxopropanoate bromide, FA62

ethyl 3-ethylthio-2-oxopropanoate acid, FA61

ethyl 2-oxopropanoate, FA24

3-(ethylthio)-2-oxopropanoic acid, FA25

3-[(4-methylphenyl)thio]-2-oxopropanoic acid, FA99

3-methylthio-2-oxopropanoic acid, FA15

phenylglyoxalic acid, FA65

LACTAMS, see LACTONES

LACTONES AND LACTAMS (see also HETEROCYCLIC COMPOUNDS)

benzylpenicillenic acid, FD39

bilirubin, FF45

erythrosine, FD97

eosin, FD96

fluorescein disodium salt, FD98

nicotinamide adenine dinucleotide, FE28

sodium deamino nicotinamide adenine dinucleotide, FE26

sodium nicotinamide adenine dinucleotide phosphate, FE27

METHOXIMES, see OXIMES

NITRATES, see ESTERS, NONCARBOXYLIC

NITRILES

acrylonitrile, FA07

trans-2-butenedinitrile, FA14

cinnamonitrile, FA74

4-cyanodiazoaminobenzene, FB82

3-(4-cyanophenyl)-3-oxo-N-phenylpropanamide, FD19

3-cyanoquinoline, FA88

2,2'-(2,5-cyclohexadiene-1,4-diylidene)bispropanedinitrile, FB15

ethyl α-cyanocinnamate, FB49

4-methylbenzylidene malonitrile, FB08

tetracyanoethylene, FA31

(2,4,5,7-tetranitro-9H-fluoren-9-ylidene)propanedinitrile, FD06

NITRILES (cont.)

(2,4,7-trinitro-9H-fluoren-9-ylidene)propanedinitrile, FD07

NITRO COMPOUNDS

Nitrobenzenes

bis(2,4-dinitrophenyl)methane, FB70

1,2-bis(4-nitrophenyl)ethane, FC24

1,2-bis(4-nitrophenyl)ethene, FC13

bis(4-nitrophenyl)methane, FB81

4-chloro-4'-nitrodiphenylamine, FB28

2,7-dichloro-5,10-di(4-nitrophenyl)-5,10-dihydrophenazine, FE62

diethyl 4-nitrophenyl phosphate, FB06

O,O-diethyl-O-(4-nitrophenyl)-phosphorothioate, FB05

4-dimethylamino-3'-nitroazobenzene, FC48

4-dimethylamino-4'-nitroazobenzene, FC49

1,2-dinitrobenzene, FA36

1,3-dinitrobenzene, FA37

1,4-dinitrobenzene, FA38

2,2'-dinitrobiphenyl, FB19

3,3'-dinitrobiphenyl, FB20

4,4'-dinitrobiphenyl, FB21

4,4'-dinitrodiphenylamine, FB35

4,4'-dinitro-1,4-diphenylbutane, FD33

4,4'-dinitrodiphenylpropane, FC83

diphenyl 4-nitrobenzyl phosphate, FD85

1-(2-furyl)-3[5-(4-nitrophenyl)-2-furyl]-2-propen-1-one, FD50

2-hydroxy-5-(2,4-dinitrophenylazo)benzoic acid, FB69

2-hydroxy-5-(2-nitrophenylazo)-benzoic acid, FB75

2-hydroxy-5-(4-nitrophenylazo)-benzoic acid, FB76

4-nitrodiphenylamine, FB43

4-nitrophenol, FA42

4-[(2-nitrophenyl)azo]phenol, FB32

4-[(3-nitrophenyl)azo]phenol, FB33

4-[(4-nitrophenyl)azo]phenol, FB34

1-(4-nitrophenyl)-3,5-diphenyl-Δ^2-pyrazoline, FE22

pentachloronitrobenzene, FA29

tetrakis(4-nitrophenyl)ethylene, FE92

tetrakis(4-nitrophenyl)hydrazine, FE66

Nitroaromatics with Condensed Rings

Solochrome Black, WDFA, FE00

(2,4,5,7-tetranitro-9H-fluoren-9-ylidene)propanedinitrile, FD06

(2,4,7-trinitro-9H-fluoren-9-ylidene)propanedinitrile, FD07

Nitro Derivatives of O-Heterocycles

1,5-di(5-nitro-2-furyl)-1,4-pentadien-3-one amidinohydrazone hydrochloride, FB89

NITROSO AND NITROSYL COMPOUNDS

1-benzhydryl-4-nitrosopiperazine, FD59

bis(4-methoxyphenyl)nitroxide, FC42

N-(4-chlorophenyl)-N-(4-nitrosophenyl)hydroxylamine, FB29

di-*tert*-butylnitroxide, FA70

4-nitrosodiphenylhydroxylamine, FB44

ORGANOMETALLICS

Co

bis(1,2-bisdiphenylphosphinoethane)cobalt(I) perchlorate, FF74

ORGANOMETALLICS (cont.)

Cr

benzophenonebenzenechromium(0), FD84

benzophenonebenzenechromium(I), iodide, FD83

bis(η-benzene)chromium), FB52

bis(η^6-biphenyl)chromium(0), FE72

bis(η^6-biphenyl)chromium(I) iodide, FE73

bis(mesitylene)chromium(0), FD75

bis(η^6-toluene)chromium(0), FC64

bis(η^6-xylene)chromium(0), FD42

Fe

ferricinium picrate, FD18

ferrocene, FA93

Hg

4-aminophenylmercury acetate, FA68

Ir

bis-[1,2-bis(diphenylphosphino)ethane]iridium(I) chloride, FF75

hydridomonocarbonyltris(triphenylphosphine)iridium(I), FF78

Mg

magnesium(II)mesoporphyrin IX dimethyl ester, FF55

magnesium(II)protoporphyrin IX dimethyl ester, FF53

Rh

bis[1,2-bis(diphenylphosphino)ethane]rhodium(I) chloride, FF76

bis-[cis-1,2-bis(diphenylphosphino)ethylene]rhodium(I) chloride, FF72

carbonyl(chloro)bis(triphenylphosphine)rhodium, FF57

hydridomonocarbonyltris(triphenylphosphine)rhodium(I), FF79

Ru

acetonitrilopyridinooctaethylporphyrinruthenium(III) ion, FF66

bis[2,6-di(3-pyridyl)pyridine]ruthenium(2+) bisperchlorate, FF28

bis(pyridino)-$\alpha,\beta,\gamma,\delta$-tetraphenylporphinylruthenium, FF77

bis(2,4,6-tri-2-pyridyl-s-triazine)ruthenium(III) perchlorate, FF51

carbonyletioporphyrin(I)ruthenium(II), FF44

carbonyl(octaethylporphyrinyl)ruthenium, FF58

carbonyl(pyridinotetraphenylporphyrinyl)ruthenium(II), FF71

carbonyl(tetraphenylporphyrinyl)ruthenium, FF69

tris(2,2'-bipyridine-N,N')ruthenium(2+) diperchlorate, FF29

tris(1,10-phenanthrolineruthenium(II) perchlorate, FF50

Sn

chloro(triethyl)stannane, FA51

chloro(trimethyl)stannane, FA13

dichloro(diethyl)stannane, FA17

trichloro(ethyl)stannane, FA02

trichloro(methyl)stannane, FA00

N-OXIDES

Pyridine N-Oxides

2,2,6,6-tetramethyl-4-oxo-1-piperidinyloxy, FA84

2,2,6,6-tetramethylpiperidine nitroxide cation, FA85

2,2,6,6-tetramethyl-1-piperidinyloxy, FA86

N-OXIDES (cont.)

Other Heterocyclic N-Oxides

3-(4-bromophenylimino)-2-phenyl-3H-indole 1-oxide, FE02

3-(4-chlorophenylimino)-2-phenyl-3H-indole 1-oxide, FE04

3-(4-dimethylaminophenylimino)-2-phenylindole 1-oxide, FE37

3-(4-methoxyphenylimino)-2-phenyl-3H-indole 1-oxide, FE21

2-phenyl-3-phenylimino-3H-indole 1-oxide, FE07

2-phenyl-3-(2-phenyl-3H-indol-3-ylidene)-3H-indole 1,1'-dioxide, FF11

2-phenyl-3-(4-pyridylimino)-3H-indole 1,1-dioxide, FD81

2-phenyl-3-(2-pyridylimino)-3H-indole 1-oxide, FD80

2-phenyl-3-(p-tolylimino)-3H-indole 1-oxide, FE20

Others

trans-azodioxycyclohexane, FB66

trans-α,α'-azodioxytoluene, FC47

trans-azoxy-2-chlorocyclohexane, FB62

cis-azoxycyclohexane, FB64

trans-azoxycyclohexane, FB65

trans-azoxydi(1-chloro-1-phenyl-2-propyl), FD73

trans-α,α'-azoxytoluene, FC45

bis(4-methoxyphenyl)nitroxide, FC42

di-tert-butylnitroxide, FA70

OXIMES AND ALKOXIMES

Aldoximes

α-methoxy-α-(3'-trifluoromethylphenyl)acetaldoxime, FA92

Ketoximes

1,3-bis(3-pyridyl)-2-propen-1-one oxime, FB88

3-hydroxyamino-1-(2-hydroxy-4-methoxyphenyl)-3-(3-pyridyl)-1-propanone oxime, FD04

3-hydroxyamino-1-(2-hydroxyphenyl)-3-(2-pyridyl)-1-propanone oxime, FC60

3-hydroxyamino-1-(2-hydroxyphenyl)-3-(3-pyridyl)-1-propanone oxime, FC61

3-hydroxy-1-(2-hydroxy-4-methoxyphenyl)-3-(2-pyridyl)-1-propanone oxime, FC95

3-hydroxy-1-(2-hydroxy-4-methoxyphenyl)-3-(3-pyridyl)-1-propanone oxime, FC96

3-hydroxy-1-(2-hydroxy-4-methoxyphenyl)-3-(4-pyridyl)-1-propanone oxime, FC97

1-(2-hydroxy-4-methoxyphenyl)-3-(4-pyridyl)-1,3-propanedione dioxime, FC93

1-(2-hydroxyphenyl)-3-(4-pyridyl)-1,3-propanedione dioxime, FC31

1-(2-pyridyl)-3-hydroxyamino-3-(3-pyridyl)-1-propanone oxime, FC02

1-(2-pyridyl)-3-(4-pyridyl)-1,3-propanedione dioxime, FB92

1-(3-pyridyl)-3-(2-pyridyl)-1,3-propanedione dioxime, FB93

1-(3-pyridyl)-3-(4-pyridyl)-1,3-propanedione dioxime, FB94

PHENOLS, see HYDROXY COMPOUNDS

PHOSPHORUS AND ARSENIC (see also ESTERS, NONCARBOXYLIC; ORGANOMETALLICS, and SULFUR COMPOUNDS)

bis(1,2-bisdiphenylphosphinoethane)cobalt(I) perchlorate, FF74

bis-[1,2-bis(diphenylphosphino)ethane]iridium(I) chloride, FF75

bis[1,2-bis(diphenylphosphino)ethane]rhodium(I) chloride, FF76

bis-[cis-1,2-bis(diphenylphosphino)ethylene]rhodium(I) chloride, FF72

carbonyl(chloro)bis(triphenylphosphine)rhodium, FF57

PHOSPHORUS AND ARSENIC (see also ESTERS, NONCARBOXYLIC; ORGANOMETALLICS, AND SULFUR COMPOUNDS) (cont.)

  2,5-di-tert-butyl-1,1,4,4-tetraphenyldiphosphiniacyclohexa-2,5-diene dibromide, FF54

  diethyl 4-nitrophenyl phosphate, FB06

  O,O-diethyl O-(4-nitrophenyl)-phosphorothioate, FB05

  2,5-diethyl-1,1,4,4-tetraphenyldiphosphoniacyclohexa-2,5-diene dibromide, FF40

  2,5-dimethyl-1,1,4,4-tetraphenyldiphosphoniacyclohexa-2,5-diene dibromide, FF32

  diphenyl 4-nitrobenzyl phosphate, FD85

  2,5-diphenyl-1,1,4,4-tetra(4-fluorophenyl)diphosphoniacyclohexa-2,5-diene dichloride, FF61

  2,5-diphenyl-1,1,4,4-tetrakis(4-methoxyphenyl)diphosphoniacyclohexa-2,5-diene dibromide, FF68

  1,1,2,4,4,5-hexaphenyldiphosphoniacyclohexa-2,5-diene dibromide, FF62

  1,1,2,4,4,5-hexaphenyldiphosphoniacyclohexa-2,5-diene dichloride, FF63

  hydridomonocarbonyltris(triphenylphosphine)iridium(I), FF78

  hydridomonocarbonyltris(triphenylphosphine)rhodium(I), FF79

  hydroxytriphenylarsonium perchlorate, FD72

  nicotinamide adenine dinucleotide, FE28

  nicotinamide mononucleotide, FB13

  pentaphenylphosphorane, FF30

  sodium deamino nicotinamide adenine dinucleotide, FE26

  sodium nicotinamide adenine dinucleotide phosphate, FE27

  1,1,4,4-tetraethyl-1,4-dihydro-2,5-diphenyl-1,4-diphosphorinium dichloride, FE81

  1,1,4,4-tetraphenyldiphosphoniacyclohexa-2,5-diene dibromide, FF15

  1,1,4,4-tetraphenyldiphosphoniacyclohexa-2-ene dibromide, FF18

  1,1,4,4-tetraphenyldiphosphoniacyclohexane dibromide, FF22

  tetraphenylphosphonium perchlorate, FE71

  triphenyl phosphate, FD70

  triphenylphosphine, FD71

  triphenylphosphine oxide, FD69

QUARTERNARY COMPOUNDS, see AMINES

QUINONES AND HYDROQUINONES (see also HYDROXY COMPOUNDS)

Benzoquinones

  1,4-benzoquinone, FA39

  2,5-bis(2-isopropoxy)-2-methylpropyl)hydroquinone, FE17

  2,5-bis(2-methoxy-2-methylpropyl)-hydroquinone, FD47

  2,5-bis(2-methyl-2-ethoxypropyl)-hydroquinone, FD77

  2,6-bis(2-methyl-2-ethoxypropyl)-hydroquinone, FD78

  2-(2-tert-butoxy-2-methylpropyl)-5-tert-butylhydroquinone, FD76

  2-tert-butyl-5-(2-methoxy-2-methyl-1-propyl)hydroquinone, FD05

  2-(1,1-dimethylethyl)-5-(2-ethoxy-2-methylpropyl)hydroquinone, FD44

  2-(1,1-dimethylethyl)-6-(2-ethoxy-2-methylpropyl)hydroquinone, FD45

  2-(1,1-dimethylethyl)-5-[2-methyl-2(1-methylethoxy)propyl]-hydroquinone, FD61

  2-(1,1-dimethylpropyl)-5-(2-ethoxy-2-methylpropyl)hydroquinone, FD62

  2-(1,1-dimethylpropyl)-5-(2-methoxy-2-methylpropyl)hydroquinone, FD46

  hydroquinone, FA45

  tetrahydroxy-1,4-benzoquinone, FA40

QUINONES AND HYDROQUINONES (see also HYDROXY COMPOUNDS) (cont.)

Naphthoquinones

2,3-bis(4-methoxyphenyl-1,4-naphthoquinone, FE70

2,3-diphenyl-1,4-naphthoquinone, FE31

vitamin K_2, FF34

Other Polycyclic Quinones

5-[(4-amino-3-sulfoanthraquinon-1-yl)amino]-2-{[3-hydroxy-5-(3-sulfophenyl)]-s-triazinyl-amino}benzenesulfonic acid, FF24

9,10-anthraquinone, FC07

1,2-bis(4-methoxyphenyl)cyclobuteno[3,4-b]-3,8-naphthohydroquinone dipotassium salt, FE95

1,2-dibenzylidenecyclobutano-[3,4-b]-3,8-naphthohydroquinone dipotassium salt, FE91

6,11-dioxido-2,4-diphenyldibenzo-[b,h]-biphenylene dipotassium salt, FF35

1,2-diphenylcyclobuta[b]naphthalene-3,8-diol, FE67

1,2-diphenylcyclobuta[b]naphthalene-3,8-diol dipotassium salt, FE64

9,10-phenanthraquinone, FC08

QUINONIMES

3,3'-bi(6H,11H-6-oxopyrazolo-[3,4,5-d,e]anthracene) dianion, FF09

SELENIUM AND TELLURIUM COMPOUNDS

diphenyl diselenide, FB46

SUGARS, see CARBOHYDRATES

SULFATES, see ESTERS, NONCARBOXYLIC

SULFIDES, see SULFUR COMPOUNDS

SULFONAMIDE, see SULFUR COMPOUNDS

SULFONES, see SULFUR COMPOUNDS

SULFONIC ACIDS, see SULFUR COMPOUNDS

SULFUR COMPOUNDS (see also ESTERS, NONCARBOXYLIC; HETEROCYCLIC COMPOUNDS; ISOTHIOCYANATES; and ORGANOMETALLICS)

Thiols

benzylpenicillenic acid, FD39

2-[bis(1-methylethyl)amino]-ethanethiol hydrochloride, FA71

methyl 3-mercaptopropanoate, FA16

Sulfides, Thioethers, and Sulfonium Ions

3-dimethylsulfonio-2-oxopropanoic acid bromide, FA26

ethyl 3-dimethylsulfonio-2-oxopropanoate bromide, FA62

ethyl 3-ethylthio-2-oxopropanoate, FA61

3-(ethylthio)-2-oxopropanoic acid, FA25

3-[(4-methylphenyl)thio]-2-oxopropanoic acid, FA99

3-methylthio-2-oxopropanoic acid, FA15

Disulfides

5,5'-bis(4-methoxyphenyl)-3,3'-[2,4-bis(4-methylphenyl)-3,4-dithia-1,6-hexamethylene]dithiolium perchlorate, FF60

formamidine disulfide, FA03

4,4',5,5'-tetrakis(4-methoxyphenyl)-3,3'-[2,5-bis(4-methylphenyl)-1,6-bis(4-methoxyphenyl)-3,4-dithia-1,6-hexylene]-dithiolium perchlorate, FF80

4,4',5,5'-tetrakis(4-methoxyphenyl)-3,3'-[2,5-bis(4-methylphenyl)-3,4-dithia-1,6-hexylene]-dithiolium perchlorate, FF73

4,4',5,5'-tetraphenyl-3,3'-(2,5-diphenyl-3,4-dithia-1,6-hexamethylene)dithiolium perchlorate, FF70

SULFUR COMPOUNDS (see also ESTERS, NONCARBOXYLIC; HETEROCYCLIC COMPOUNDS; ISOTHIOCYANATES; and ORGANOMETALLICS) (cont.)

Thiones, Thioamides, Thiobarbiturates, Thiopyrimidines, and Thiopurines

4-(4-bromophenylazo)-3,5-diphenyl-1-thiocarbamoylpyrazole, FE32

4-(3-chlorophenylazo)-3,5-diphenyl-1-thiocarbamoylpyrazole, FE33

4-(4-chlorophenylazo)-3,5-diphenyl-5-thiocarbamoylpyrazole, FE34

3,5-dimethyl-4-(2-ethoxyphenylazo)-1-thiocarbamoylpyrazole, FC66

3,5-dimethyl-4-(4-ethoxyphenylazo)-1-thiocarbamoylpyrazole, FC67

3,5-dimethyl-4-(2-methoxyphenylazo)-1-thiocarbamoylpyrazole, FC03

3,5-dimethyl-4-(4-methoxyphenylazo)-1-thiocarbamoylpyrazole, FC04

4-(2,4-dimethylphenylazo)-3,5-diphenyl-1-thiocarbamoylpyrazole, FE77

4-(2,6-dimethylphenylazo)-3,5-diphenyl-1-thiocarbamoylpyrazole, FE78

3,5-dimethyl-4-(phenylazo)-1-thiocarbamoylpyrazole, FB59

3,5-diphenyl-4-(2-methoxyphenyl-1-thiocarbamoylpyrazole, FE43

3,5-diphenyl-4-(4-methoxyphenyl)-1-thiocarbamoylpyrazole, FE44

3,5-diphenyl-4-(2-methylphenylazo)-1-thiocarbamoylpyrazole, FE45

3,5-diphenyl-4-(3-methylphenylazo)thiocarbamoylpyrazole, FE46

3,5-diphenyl-4-(4-methylphenylazo)-1-thiocarbamoylpyrazole, FE47

3,5-diphenyl-4-(phenylazo-4-thiocarbamoylpyrazole, FE35

Sulfoxides, Sulfones, and S-dioxides

thianthrene 5,5-dioxide, FB23

thianthrene 5,10-dioxide, FB24

thianthrene 5-oxide, FB22

Sulfuric and Thiosulfuric Acid Derivatives, Sulfonates, Sulfinates, and Sulfonamides

5-[(4-amino-3-sulfoanthraquinon-1-yl)amino]-2-{[3-hydroxy-5-(3-sulfophenyl)]-s-triazinylamino}benzenesulfonic acid, FF24

3-carboxy-4-hydroxy-4'-sulfoazobenzene disodium salt, FB68

Congo Red, FF36

4'-dimethylaminoazobenzene-3-sulfonic acid, FC62

4-dimethylaminoazobenzene-4-sulfonic acid, FC63

4-[(4-hydroxyphenyl)azo]benzenesulfonic acid, FB45

Solochrome Black PVS, FD17

Solochrome Black WDFA, FE00

Phosphoric and Arsenic Acid Derivatives (see also ORGANOMETALLICS)

O,O-diethyl O-(4-nitrophenyl)-phosphorothioate, FB05

TELLURIUM, see SELENIUM

THIOCYANATES, see ISOTHIOCYANATES

THIOLS, see SULFUR COMPOUNDS

THIOUREAS, see UREAS

UNSATURATED COMPOUNDS

Ethylenic

Double Bond Isolated from Non-hydrocarbon Functional Group

1,2-bis(4-nitrophenyl)ethene, FC13

1,2-dibenzylidenecyclobutano-[3,4-b]-3,8-naphthohydroquinone dipotassium salt, FE91

1,2-diethylidenacenaphthene, FD23

1,2-dimethoxy-4-propenylbenzene, FB12

1,1-diphenylethene, FC19

1,2-diphenylpropene, FC82

UNSATURATED COMPOUNDS (cont.)

Ethylenic

Double Bond Isolated from Non-hydrocarbon Functional Group (cont.)

p-methoxystyrene, FA79

trans-stilbene, FC21

tetrakis(4-nitrophenyl)ethylene, FE92

tetraphenylethylene, FE97

triphenylethene, FE09

vitamin K_2, FF34

α,β-Unsaturated Ketones

2-benzylidene-1,3-indandione, FD10

3-(4-carbamoylphenyl)-1-(2-thienyl)-2-propenone, FC17

2-(3-chlorobenzylidene)-1,3-indandione, FD08

2-(4-chlorobenzylidene)-1,3-indandione, FD09

3-[5-(4-chlorophenyl)-2-furyl]-1-(2-furyl)-2-propen-1-one, FD49

trans(?)-1,2-dibenzoylethylene, FD20

3-(2,6-dichlorophenyl)-1-(2-thienyl-2-propenone, FB67

3-(4-dimethylaminophenyl)-4-(2-thienyl)-2-propen-1-one, FC88

trans-1,3-diphenyl-2-propen-1-one, FC76

3-(4-ethoxyphenyl)-1-(2-thienyl)-2-propen-1-one, FC87

3-(4-fluorophenyl)-1-(2-thienyl)-2-propen-1-one, FB74

1-(2-furyl)-2-methyl-3-(5-phenyl-2-furyl)-2-propen-1-one, FD66

1-(2-furyl)-3-[5-(4-nitrophenyl)-2-furyl]-2-propen-1-one, FD50

1-(2-furyl)-3-(5-phenyl-2-furyl)-2-propen-1-one, FD53

2-(3-hydroxybenzylidene)-1,3-indandione, FD11

1-(2-hydroxy-4-methoxyphenyl)-3-(2-pyridyl)-2-propen-1-one, FC79

1-(2-hydroxy-4-methoxyphenyl)-3-(3-pyridyl)-2-propen-1-one, FC80

1-(2-hydroxy-4-methoxyphenyl)-3-(4-pyridyl)-2-propen-1-one, FC81

3-(4-hydroxy-3-methoxyphenyl)-1-2,4,6-tribromo-3-hydroxyphenyl)-2-propen-1-one, FD16

2-(4-methoxybenzylidene)-1,3-indandione, FD54

3-(3-methoxyphenyl)-1-(2,4,6-tribromo-3-hydroxyphenyl)-2-propan-1-one, FD14

3-(4-methoxyphenyl)-1-(2,4,6-tribromo-3-hydroxyphenyl)-2-propen-1-one, FD15

2-(4-methylbenzylidene)-1,3-indandione, FD52

3-methyl-3-buten-2-one, FA23

2-(3,4-methylenedioxybenzylidene)-1,3-indandione, FD48

3-methyl-3-penten-2-one, FA47

4-methyl-3-penten-2-one, FA48

2-methyl-1-phenyl-3-(5-phenyl-2-furyl)-2-propen-1-one, FE11

3-(2-methylphenyl)-1-(2-thienyl)-2-propen-1-one, FC27

3-(4-methylphenyl)-1-(2-thienyl)-2-propen-1-one, FC28

3-(4-methylphenyl)-1-(2,4,6-tribromo-3-hydroxyphenyl)-2-propen-1-one, FD13

3-(1-naphthyl)-1-(2-thienyl)-2-propen-1-one, FD51

4-phenyl-3-buten-2-one, FA96

3-phenyl-1-(2-thienyl)-2-propen-1-one, FB85

1-(2-pyridyl)-3-(4-pyridyl)-2-propen-1-one, FB78

2,2,4,5-tetrahydroxy-4-cyclopentene-1,3-dione, FA20

2,3,4,5-tetraphenyl-2-cyclopenten-1-one, FF27

1-(2-thienyl)-3-(2,3,4-trimethoxyphenyl)-2-propen-1-one, FD34

1-(2-thienyl)-3-(3,4,5-trimethoxyphenyl)-2-propen-1-one, FD35

1-(2,4,6-tribromo-3-hydroxyphenyl)-3-(4-hydroxyphenyl)-2-propen-1-one, FC69

1-(2,4,6-tribromo-3-hydroxyphenyl)-3-phenyl-2-propen-1-one, FC68

3,5,5-trimethyl-2-cyclohexen-1-one, FA83

UNSATURATED COMPOUNDS (cont.)

Ethylenic (cont.)

α,β-Unsaturated Ketosteroids (cont.)

17α-acetoxy-6β-fluoropregn-4-ene-3,20-dione, FE60

17α-acetoxy-6α-methylpregna-1,4-diene-3,20-dione, FE82

17α-acetoxy-6-methylpregna-4,6-diene-3,20-dione, FE83

17α-acetoxy-6α-methylpregn-4-ene-3,20-dione, FE84

17α-acetoxypregna-1,4-diene-3,20-dione, FE55

17α-acetoxypregna-4,6-diene-3,20-dione, FE56

17α-acetoxypregn-4-ene-3,20-dione, FE61

6α-fluoro-17,21-dihydroxy-16α-methylpregn-4-ene-3,20-dione, FE39

6β-fluoro-17,21-dihydroxy-16α-methylpregn-4-ene-3,20-dione, FE40

6α-fluorotestosterone, FD93

6β-fluorotestosterone, FD94

17β-hydroxyandrosta-1,4-dien-3-one, FD92

6α-methyltestosterone, FE15

6β-methyltestosterone, FE16

Δ^4-pregnene-17α,21-diol-3,20-dione, FE30

progesterone, FE29

testosterone, FD95

Heterocyclic With Side Chain Unsaturated

bilirubin, FF45

1,3-bis(3-pyridyl)-2-propen-1-one oxime, FB88

3-(4-carbamoylphenyl)-1-(2-thienyl)-2-propen-1-one, FC17

3-[5-(4-chlorophenyl)-2-furyl]-1-(2-furyl)-2-propen-1-one, FD49

2-(2,6-dichlorophenyl)-1-(2-thienyl)-2-propen-1-one, FB67

2,17-diethyl-1,10,19,22,23,24-hexahydro-3,7,13,18-tetramethyl-1,19-dioxo-21H-biline-8,12-dipropanoic acid, FF46

3-(4-dimethylaminophenyl)-1-(2-thienyl)-2-propen-1-one, FC88

1,5-di(5-nitro-2-furyl)-1,4-pentadien-3-one amidinohydrazone hydrochloride, FB89

3-(4-ethoxyphenyl)-1-(2-thienyl)-2-propen-1-one, FC87

etioporphyrin I, FF42

3-ethyl-2-[7-(3-ethyl-2(3H)-benzothiazolylidene)-1,3,5-heptatrienyl]-benzothiazolium perchlorate, FE89

3-ethyl-2-[2-(3-ethyl-2(3H)-benzothiazolylidene)methyl]-1-butenyl]-benzothiazolium bromide, FE52

3-ethyl-2-[(3-ethyl-2(3H)-benzothiazolylidene)-2-methyl-1-propenyl]benzothiazolium bromide, FE38

3-ethyl-2-[5-(3-ethyl-2(3H)-benzothiazolylidene)-1,3-pentadienylbenzothiazolium iodide, FE51

3-ethyl-2-[3-(3-ethyl-2(3H)-benzothiazolylidene)-1-propenyl]benzothiazolium bromide, FE25

1-ethyl-2-[3-(3-ethyl-2(3H)-benzothiazolylidene)-1-propenyl]quinolinium bromide, FE48

1-ethyl-2-[(1-ethyl-2(1H)-quinolinylidene)methyl]quinolinium iodide, FE49

1-ethyl-2-[(1-ethyl-4(1H)-quinolinylidene)methyl]quinolinium iodide, FE50

1-ethyl-2-[3-(1-ethyl-2(1H)-quinolinylidene)-1-propenyl]-quinolinium bromide, FE88

3-(4-fluorophenyl)-1-(2-thienyl)-2-propen-1-one, FB74

1-(2-furyl)-2-methyl-3-(5-phenyl-2-furyl)-2-propen-1-one, FD66

1-(2-furyl-3-[5-(4-nitrophenyl)-2-furyl]-2-propen-1-one, FD50

1-(2-furyl)-3-(5-phenyl-2-furyl)-2-propen-1-one, FD53

1-(2-hydroxy-4-methoxyphenyl)-3-(3-pyridyl)-2-propen-1-one, FC80

UNSATURATED COMPOUNDS (cont.)

  Ethylenic (cont.)

   α,β-Unsaturated Acids, Thioacids, Esters, and Amides

   diethyl fumarate, FA69

   ethyl trans-2-butenoate, FA49

   ethyl cinnamate, FB10

   ethyl α-cyanocinnamate, FB49

   methyl cinnamate, FA98

   N,N,N',N'-tetraacetyl-1,4-diamino-1,4-diphenyl-1,3-butadiene, FE79

   α,β-Unsaturated Amines, Halogens, Alcohols, Ethers, Imines, and Nitrocompounds

   17α-acetoxy-6α-chloropregna-1,4-diene-3,20-dione, FE53

   17α-acetoxy-6-chloropregna-4,6-diene-3,20-dione, FE54

   17α-acetoxy-6α-chloropregn-4-ene-3,20-dione, FE57

   17α-acetoxy-6β-chloropregn-4-ene-3,20-dione, FE58

   17α-acetoxy-6α-fluoropregn-4-ene-3,20-dione, FE59

   17α-acetoxy-6β-fluoropregn-4-ene-3,20-dione, FE60

   1,2-bis(4-nitrophenyl)ethene, FC13

   1,3-bis(3-pyridyl)-2-propen-1-one oxime, FB88

   1,5-di(5-nitro-2-furyl)-1,4-pentadien-3-one amidinohydrazone hydrochloride, FB89

   1,1-diphenylprop-2-en-1-ol, FC84

   3,3-diphenylprop-2-en-1-ol, FC85

   1,3-diphenyl-2-propen-1-one phenylhydrazone, FE23

   6α-fluoro-17,21-dihydroxy-16α-methylpregn-4-ene-3,20-dione, FE39

   6β-fluoro-17,21-dihydroxy-16α-methylpregn-4-ene-3,20-dione, FE40

   6α-fluorotestosterone, FD93

   6β-fluorotestosterone, FD94

   1-methoxy-1-phenylpropene, FB02

   3-methoxy-1-phenylpropene, FB03

   3-phenyl-2-propen-1-ol, FA80

   2-propen-1-ol, FA10

   2,2,4,5-tetrahydroxy-4-cyclopentene-1,3-dione, FA20

   α,β-Unsaturated Nitriles

   acrylonitrile, FA07

   trans-2-butenedinitrile, FA14

   cinnamonitrile, FA74

   2,2'-(2,5-cyclohexadiene-1,4-diylidene)bispropanedinitrile, BA06

   ethyl α-cyanocinnamate, FB49

   2,2'-(2,5-cyclohexadiene-1,4-diylidene)bispropanedinitrile, FB15

   4-methylbenzylidene malonitrile, FB08

   tetracyanoethylene, FA31

   (2,4,5,7-tetranitro-9H-fluoren-9-ylidene)propanedinitrile, FD06

   (2,4,7-trinitro-9H-fluoren-9-ylidene)propanedinitrile, FD07

   α,β-Unsaturated Phosphines

   bis-[cis-1,2-bis(diphenylphosphino)ethylene]rhodium(I) chloride, FF72

  Alicyclic

   1,3,5,7-cyclooctatetraene, FA66

   2,2,4,5-tetrahydroxy-4-cyclopentene-1,3-dione, FA20

   3,5,5-trimethyl-2-cyclohexen-1-one, FA83

   α,β-Unsaturated Ketosteroids

   17α-acetoxy-6α-chloropregna-1,4-diene-3,20-dione, FE53

   17α-acetoxy-6-chloropregna-4,6-diene-3,20-dione, FE54

   17α-acetoxy-6α-chloropregn-4-ene-3,20-dione, FE57

   17α-acetoxy-6β-chloropregn-4-ene-3,20-dione, FE58

   17α-acetoxy-6α-fluoropregn-4-ene-3,20-dione, FE59

UNSATURATED COMPOUNDS (cont.)

Ethylenic (cont.)

Heterocyclic With Side Chain Unsaturated (cont.)

1-(2-hydroxy-4-methoxyphenyl)-3-(4-pyridyl)-2-propen-1-one, FC81

magnesium(II)mesoporphyrin IX dimethyl ester, FF55

magnesium(II)protoporphyrin IX dimethyl ester, FF53

2-(3,4-methylenedioxybenzylidene)-1,3-indandione, FD48

2-methyl-1-phenyl-3-(5-phenyl-2-furyl)-2-propen-1-one, FE11

3-(2-methylphenyl)-1-(2-thienyl)-2-propen-1-one, FC27

3-(4-methylphenyl)-1-(2-thienyl)-2-propen-1-one, FC28

3-phenyl-2-[3-(3-phenyl-2(3H)-benzothiazolylidene)-1-propenyl]benzothiazolium iodide, FF26

3-phenyl-1-(2-thienyl)-2-propen-1-one, FB85

1-(2-pyridyl)-3-(4-pyridyl)-2-propen-1-one, FB78

ruthenium(II) carbonyletioporphyrin(I), FF44

1-(2-thienyl)-3-(2,3,4-trimethoxyphenyl)-2-propen-1-one, FD34

1-(2-thienyl)-3-(3,4,5-trimethoxyphenyl)-2-propen-1-one, FD35

Acetylenic

1,1-diphenylprop-2-yn-1-ol, FC77

3-methoxy-3-phenyl-1-propyne, FA95

2-methyl-3-butyn-2-ol, FA22

2-phenyl-3-butyn-2-ol, FA95

1-phenyl-2-propyn-1-ol, FA78

2-propyn-1-ol, FA08

UREAS AND THIOUREAS (see also HETEROCYCLIC COMPOUNDS; and KETONES, Heterocyclic)

4-(4-bromophenylazo)-3,5-diphenyl-1-thiocarbamolypyrazole, FE32

4-(3-cholorphenylazo)-3,5-diphenyl-1-thiocarbamoylpyrazole, FE33

4-(4-chlorophenylazo)-3,5-diphenyl-5-thiocarbamoylpyrazole, FE34

3,5-dimethyl-4-(2-ethoxyphenylazo)-1-thiocarbamoylpyrazole, FC66

3,5-dimethyl-4-(4-ethoxyphenylazo)-1-thiocarbamoylpyrazole, FC67

3,5-dimethyl-4-(2-methoxyphenylazo)-1-thiocarbamoylpyrazole, FC03

4-(2,4-dimethylphenylazo)-3,5-diphenyl-1-thiocarbamoylpyrazole, FE77

4-(2,6-dimethylphenylazo)-3,5-diphenyl-1-thiocarbamoylpyrazole, FE78

3,5-dimethyl-4-(phenylazo)-1-thiocarbamoylpyrazole, FB59

1,5-di(5-nitro-2-furyl)-1,4-pentadien-3-one amidinohydrazone hydrochloride, FB89

3,5-diphenyl-4-(2-methoxyphenyl-1-thiocarbamoylpyrazole, FE43

3,5-diphenyl-4-(4-methoxyphenyl)-1-thiocarbamoylpyrazole, FE44

3,5-diphenyl-4-(2-methylphenylazo)-1-thiocarbamoylpyrazole, FE45

3,5-diphenyl-4-(3-methylphenylazo)-1-thiocarbamoylpyrazole, FE46

3,5-diphenyl-4-(4-methylphenylazo)-1-thiocarbamoylpyrazole, FE47

3,5-diphenyl-4-phenylazo-4-thiocarbamoylpyrazole, FE35

formamidine disulfide, FA03

TABLE VI.
CHEMICAL ABSTRACTS SERVICE REGISTRY NUMBERS

This is a list of the Chemical Abstracts Service Registry Numbers given under the names of the compounds appearing in Table I. Registry Numbers have not been assigned by the Abstracts Service to some of the compounds in Table I, and these compounds are therefore not included in this table. The Registry Numbers are listed in numerical order, and each is followed by the code number assigned to the corresponding compound in Table I.

53-59-8, FE27	91-22-5, FA76	123-31-9, FA45
53-84-9, FE28	91-57-6, FB09	127-66-2, FA94
54-36-4, FC46	92-52-4, FB36	131-11-3, FB00
55-80-1, FD00	92-82-0, FB18	136-17-4, FB57
56-38-2, FB05	92-85-3, FB25	136-35-6, FB51
57-83-0, FE29	93-55-0, FA81	141-79-7, FA48
58-22-0, FD95	95-86-3, FA46	148-24-3, FA77
58-72-0, FE09	98-86-2, FA67	150-74-3, FC39
59-30-3, FD90	98-92-0, FA44	151-68-8, FE82
60-09-3, FB50	99-65-0, FA37	151-69-6, FE53
60-11-7, FC57	100-02-7, FA42	152-58-9, FE30
61-73-4, FD38	100-22-1, FB07	198-55-0, FD99
64-17-5, FA04	100-25-4, FA38	230-17-1, FB16
65-85-0, FA53	100-39-0, FA54	254-79-5, FA63
66-71-7, FB17	100-51-6, FA56	302-22-7, FE54
67-56-1, FA01	101-54-2, FB53	302-23-8, FE61
67-63-0, FA12	101-64-4, FC00	303-49-1, FD91
67-64-1, FA09	101-70-2, FC56	304-88-1, FB86
70-23-5, FA21	102-54-5, FA93	311-45-5, FB06
71-23-8, FA11	103-26-4, FA98	313-72-4, FA87
71-36-3, FA18	103-30-0, FC21	319-89-1, FA40
71-41-0, FA28	103-33-3, FB40	332-54-7, FC38
71-43-2, FA43	103-36-6, FB10	336-79-8, FE60
71-58-9, FE84	103-50-4, FC51	363-72-4, FA32
75-04-7, FA05	104-54-1, FA80	366-18-7, FA89
75-85-4, FA27	104-94-9, FA60	367-00-0, FB74
76-84-6, FD87	106-51-4, FA39	372-38-3, FA35
78-59-1, FA83	107-13-1, FA07	378-59-6, FE39
78-92-2, FA19	107-18-6, FA10	392-56-3, FA30
81-93-6, FD68	107-19-7, FA08	425-51-4, FE56
82-68-8, FA29	108-67-8, FA82	434-90-2, FB14
84-11-7, FC08	108-88-3, FA55	439-14-5, FD22
84-62-8, FE08	108-90-7, FA41	448-71-5, FF42
84-65-1, FC07	111-27-3, FA50	479-33-4, FF25
84-66-2, FB60	115-19-5, FA22	485-47-2, FA72
84-74-2, FD43	115-86-6, FD70	486-25-9, FB71
85-01-8, FC11	119-61-9, FB83	502-02-3, FC63
86-40-8, FC37	119-65-3, FA75	514-73-8, FE51
86-73-7, FB77	120-12-7, FC10	528-29-0, FA36
90-04-0, FA58	122-37-2, FB48	529-34-0, FA97
91-01-0, FB95	122-39-4, FB47	530-48-3, FC19
91-19-0, FA64	122-57-6, FA96	531-91-9, FE74

536-90-3, FA59
537-65-5, FB58
565-62-8, FA47
574-42-5, FF01
581-64-6, FB39
595-33-5, FE83
596-31-6, FE13
602-55-1, FE05
603-48-5, FE90
605-35-0, FD71
611-63-2, FC25
611-70-1, FB04
611-73-4, FA65
613-42-3, FD82
614-47-1, FC76
617-35-6, FA24
620-84-8, FB96
620-93-9, FC55
621-90-9, FB98
623-70-1, FA49
623-91-6, FA69
629-20-9, FA66
634-21-9, FE50
635-52-0, FE75
635-65-4, FF45
637-69-4, FA79
670-54-2, FA31
736-30-1, FC24
742-01-8, FE24
764-42-1, FA14
771-56-2, FA52
791-28-6, FD69
814-78-8, CA23
833-81-8, FC82
836-30-6, FB43
846-48-0, FD92
866-55-7, FA17
917-23-7, FF67
938-16-9, FB11
951-02-0, FB24
958-96-3, FB20
959-66-0, FC78

962-01-6, FD29
962-05-0, FC75
965-01-5, FD32
977-96-8, FE49
993-16-8, FA00
994-31-0, FA51
1005-24-9, FA57
1016-09-7, FC50
1017-22-7, FC82
1033-90-5, FD56
1066-45-1, FA13
1066-57-5, FA02
1094-61-7, FB13
1113-59-3, FA06
1125-80-0, FA91
1148-79-4, FC71
1172-02-7, FD07
1207-72-3, FB87
1208-86-2, FB97
1271-54-1, FB52
1435-60-5, FB34
1484-88-4, FE42
1499-10-1, FE94
1517-72-2, FB55
1518-16-7, FB15
1528-74-1, FB21
1530-12-7, FE93
1597-68-8, FD93
1666-13-3, FB46
1689-64-1, FB84
1689-82-3, FB41
1698-25-5, FD59
1745-32-0, FE38
1786-61-7, FE00
1807-53-0, FF23
1817-74-9, FB81
1817-76-1, FB70
1821-27-8, FB35
1851-07-6, FE26
1852-58-0, FD94
2011-53-2, FB33
2050-16-0, FB42

2132-80-1, FC54
2159-74-2, FB63
2196-95-4, FF11
2243-76-7, FB76
2315-20-0, FB89
2316-28-1, FF22
2362-50-7, FB22
2367-82-0, FA33
2406-25-9, FA70
2436-96-6, FB19
2477-73-8, FE57
2491-74-9, FC49
2491-76-1, FC36
2496-15-3, FC59
2497-33-8, FB91
2497-38-3, FB80
2501-02-2, FC13
2564-83-2, FA86
2588-88-7, FF30
2619-57-0, FD28
2643-00-7, FC42
2658-74-4, FE58
2670-67-9, FE88
2683-82-1, FF56
2703-27-7, FB27
2703-28-8, FB31
2724-85-8, FB32
2826-25-7, FB08
2875-24-3, FC14
2875-25-4, FC15
2875-27-6, FC16
2896-70-0, FA84
2898-12-6, FD27
2918-83-4, FB45
2935-90-2, FA16
3009-50-5, FD03
3010-57-9, FD01
3028-95-3, FE52
3035-94-7, FB26
3073-51-6, FE96
3147-53-3, FB79
3256-06-2, FA03

3395-76-4, FC43	6344-90-7, FC79	15635-95-7, FF29
3422-75-1, FD30	6396-84-5, FC58	15746-82-4, FF28
3490-33-9, FD45	6622-61-3, FC80	15875-51-1, FD52
3490-43-5, FD44	6676-96-6, FB90	15875-54-4, FD09
3490-44-6, FD77	6962-04-5, FB30	15875-55-5, FD08
3562-73-0, FC52	7093-78-9, FD41	15924-61-5, FF15
3601-19-2, FE80	7228-47-9, FB56	16094-76-1, FE41
3665-72-3, FE69	7317-52-4, FF27	16277-67-1, FB03
3682-35-7, FD64	7364-21-8, FC99	16568-56-2, FF46
3696-00-2, FB59	7401-39-0, FC81	17095-31-7, FD89
3789-77-3, FC35	7421-76-3, FD54	17185-29-4, FF79
3805-65-0, FC34	7478-17-3, FD25	17208-03-6, FD05
3805-67-2, FC41	7512-20-1, FD26	17208-10-5, FD61
3808-71-7, FC40	7598-80-3, FD86	17208-12-7, FE17
3837-55-6, FC48	7626-68-8, FC65	17208-13-8, FD76
3923-51-1, FC84	10252-45-6, FE22	17208-16-1, FD78
3923-52-2, FC77	10368-11-3, FC83	17250-25-8, FF78
3988-77-0, FB85	10426-00-3, FF21	17389-14-9, FE25
4070-75-1, FD20	10534-59-5, FD79	17576-88-4, FC33
4187-87-5, CA78	10577-64-7, FD85	17785-89-6, FC94
4360-47-8, FA74	11077-21-7, FD18	18440-53-4, FF03
4393-30-0, FF13	12087-58-0, FC64	18542-43-3, FA15
4541-69-9, FB02	12099-17-1, FE73	18625-21-3, FC62
4603-00-3, FC20	12131-29-8, FD42	18852-45-4, FE21
4625-55-2, FB99	13018-79-6, FD37	18852-46-5, FE20
4801-14-3, FC85	13027-48-0, FC22	18852-47-6, FD80
5103-42-4, FD63	13152-79-9, FD23	19099-38-8, FF37
5165-73-1, FE07	13224-48-1, FF59	19606-98-5, FE76
5169-66-4, FE37	13251-86-0, FE15	19808-49-2, FD81
5197-95-5, FC05	13252-06-7, FE16	19838-82-5, FB38
5339-39-9, FC09	13274-97-0, FF18	19859-51-9, FE71
5381-33-9, FD10	13309-09-6, FB78	20005-35-0, FD49
5707-44-8, FC32	13440-24-9, FC23	20005-37-2, FD50
5963-51-9, FD74	13787-49-0, FD72	20266-41-5, FD98
6028-94-0, FC28	13938-94-8, FF57	20266-43-7, FD96
6054-99-5, FB68	14385-59-2, FC88	20266-45-9, FD97
6263-83-8, FD57	14533-87-0, FB49	20439-99-0, FE81
6268-49-1, FC91	14724-63-1, FF53	20691-83-2, FD02
6272-40-8, FC70	15043-47-7, FF76	20691-84-3, FC90
6283-24-5, FA68	15362-58-0, FF62	20894-65-9, FD51
6311-22-4, FC26	15390-38-2, FF75	20983-67-9, FB28
6339-33-9, FC12	15517-55-2, FD06	21461-09-6, FB75

TABLE VI. Chemical Abstracts Service Registry Numbers

21461-10-9, FB72	31054-06-5, FE19	37799-62-5, FE23
21461-12-1, FB73	31054-08-7, FE03	38443-42-4, FD67
21557-88-0, FF32	31055-09-8, FE01	38505-21-4, FD31
21557-91-5, FF40	31083-67-7, FE02	38505-24-7, FC74
21791-68-4, FF51	31438-32-1, FE66	38505-26-9, FD19
22268-66-2, FE89	31438-34-3, FE65	38674-90-7, FC44
22754-44-5, FF72	31438-41-2, FE63	38755-22-5, FE98
22873-66-1, FF50	31438-42-3, FF19	38755-23-6, FF43
23073-34-9, FE06	31438-43-4, FF20	39194-34-8, FC86
23600-83-1, FC30	31909-32-7, FF49	39927-30-5, FF68
23689-01-2, FC89	31909-33-8, FE86	39927-31-6, FF61
24279-81-0, FE36	31909-34-9, FE87	40650-95-1, FB69
24672-76-2, FF14	31909-35-0, FF08	40940-98-5, FD53
24743-47-3, FE34	32073-84-0, FF69	41480-68-6, FF54
24743-48-4, FE32	32174-84-8, FE12	41480-71-1, FF63
24743-49-5, FE43	32348-13-3, FB67	41480-75-5, FA71
24743-51-9, FE44	32367-67-2, FE14	41636-35-5, FF58
24743-54-2, FE77	33085-81-3, FE72	41751-82-0, FF71
24743-56-4, FE78	33753-12-7, FE31	41854-80-2, FD33
24749-13-1, FE35	34529-59-4, FC92	43040-07-9, FF65
24749-14-2, FE45	34529-60-7, FB82	43070-16-2, FF66
24749-15-3, FE47	34690-41-0, FF77	43145-31-9, FF44
24749-16-4, FE33	34839-21-9, FF07	45842-10-2, FA85
25299-18-7, FD11	34846-64-5, FA88	46910-91-2, FC87
25559-51-7, FC72	34881-50-0, FF74	47797-98-8, FE92
25559-52-8, FC73	35087-43-5, FF38	50874-13-0, FA95
26049-06-9, FB66	35590-50-2, FC95	50982-47-3, FC53
28073-96-3, FC68	35590-57-9, FC96	50982-49-5, FE64
28073-99-6, FC69	35590-59-1, FC97	50982-54-2, FE91
28074-00-2, FD14	35699-38-8, FA99	50982-55-3, FE70
28074-01-3, FD15	35872-22-1, FC04	51033-95-5, FA25
28074-02-4, FD16	36256-94-7, FC60	51033-96-6, FA61
28074-05-7, FD13	36256-96-9, FD04	51033-97-7, FA26
28548-57-4, FB44	36256-97-0, FC02	51033-98-8, FA62
29147-32-8, FC66	36256-98-1, FC31	51095-48-8, FC98
29147-33-9, FC67	36256-99-2, FC93	51095-50-2, FF48
29147-34-0, FC03	36257-01-9, FB93	51095-51-3, FD58
29382-68-1, FA90	36257-03-7, FB94	51210-88-9, FB88
29510-58-5, FE67	36257-04-2, FC61	51930-45-1, FE11
29874-34-8, FD48	36475-32-8, FB92	51930-46-2, FD66
31054-04-3, FE04	36530-85-5, FE85	52346-47-1, FD83
31054-05-4, FE18	36707-27-4, FD40	52445-45-1, FD84

52886-84-7, FE48	59130-88-0, FE99	61228-21-5, FF12
52886-85-8, FF26	59130-97-1, FA73	61228-23-7, FE62
53094-46-5, FC27	59131-00-9, FF05	61236-15-5, FD36
53094-47-6, FD34	59190-81-7, FB62	61236-16-6, FC29
53094-48-7, FD35	59190-83-9, FD73	61255-13-8, FF33
53094-49-8, FC17	59245-29-3, FF41	62958-59-2, FF06
53151-14-7, FD17	59410-64-9, FD12	62958-79-6, FF04
53428-42-5, FE46	59410-65-0, FD55	62958-80-9, FE68
53513-44-3, FF55	59856-51-8, FE97	63056-19-9, FA20
54349-79-0, FB12	60389-44-8, FB54	63183-44-8, FA34
54446-36-5, FB37	60389-45-9, FE79	63373-56-8, FF02
56795-80-3, FB29	60389-46-0, FC01	64756-74-7, FB01
57497-40-2, FB64	61228-17-9, FF39	64756-75-8, FB61
58379-14-9, FB65	61228-19-1, FF10	64756-76-9, FA92
59130-87-9, FF00	61228-20-4, FF31	

TABLE VII.
INDEX OF SOLVENTS EMPLOYED

This table is an index that lists the solvents (except pure water) and mixtures of solvents that appear in Table I. It enables the user to identify all of the entries in that table that contain information pertaining to any particular solvent or mixture of solvents.

The entry

Butyronitrile, FC20, FF38

signifies that data are given in Table I for the compounds designated by code numbers FC20 and FF38 in nominally pure butyronitrile as the solvent. We have considered a solvent to be nominally pure whenever there was no deliberate addition of a second solvent or whenever the stated purity of the solvent was at least 99%. This criterion does not enable the user of this table to distinguish between entries for, say, reagent-grade acetonitrile and those for acetonitrile that had been rigorously purified to remove as much as possible of the water and other proton donors.

The entry

Benzene-methanol (60:40), FA00, FA02, FA13, FA17, FA 51

provides references to data obtained in binary mixtures containing 60% of benzene and 40% of methanol by volume. Binary (and other) mixtures are listed in the alphabetical order of the names of their constituents, so that mixtures of benzene with acetronitrile precede those of benzene with methanol. Binary mixtures containing the same constituents in different proportions are listed in the order of decreasing percentages, which is always by volume unless otherwise stated, of the constituents whose name is given first. Thus ethanol-water (95:5) precedes ethanol-water (75:25). Occasionally in this series we decided to include data even though the information given in the original literature left us more or less uncertain about the proportions of the components of a mixture. These cases are represented by entries of the form

Ethanol-water (96?:4?), FF25, FF27

which denotes a mixture of ethanol, and water where there was some evidence offered that the percentages were 96 and 4, but it was not directly stated.

The citations for binary mixtures are given only once, in the entry for the constituent whose name occurs first in the alphabet. A cross-reference is always given under the name of the other constituent and in the following form:

Methanol-benzene, *see* Benzene-methanol

Ternary mixtures are treated similarly. For a mixture of trifluoroacetic anhydride-dichloromethane-trifluoroacetic acid, the citation is given in the entry for dichloromethane, and in that entry it appears as part of the subentry for mixtures of dichloromethane-trifluoroacetic acid-trifluoroacetic anhydride, so that the three constituents are named in alphabetical order:

Dichloromethane
-trifluoroacetic acid
-trifluoroacetic anhydride (88:02:10), FB25, FC54, FF14

All ternary mixtures are fully cross-referenced so that they can be located in the entries for all of their constituents.

TABLE VII. Index of Solvents Employed

Acetic acid-water (80:20), FB22, FB23, FB24, FB25
Acetone, FB83
    -water (50:50), FC33, FC35, FC38, FC40, FC48, FC57, FC62, FC90, FD00, FD02
        (10:90), FD91
Acetonitrile, FA10, FA31, FA36, FA37, FA38, FA39, FA44, FA45, FA47, FA54, FA57,
        FA64, FA66, FA70, FA75, FA76, FA79, FA84, FA85, FA86, FA89, FA91,
        FA93, FB07, FB12, FB15, FB16, FB17, FB18, FB19, FB20, FB21, FB25,
        FB28, FB29, FB30, FB35, FB37, FB43, FB44, FB47, FB48, FB49, FB53,
        FB58, FB70, FB71, FB81, FB86, FB87, FB96, FB97, FC00, FC05, FC09,
        FC10, FC20, FC24, FC29, FC30, FC42, FC44, FC54, FC55, FC56, FC71,
        FC83, FC89, FC94, FC98, FD06, FD07, FD33, FD36, FD40, FD41, FD57,
        FD58, FD60, FD64, FD71, FD72, FD74, FD79, FD85, FD99, FE05, FE14,
        FE22, FE23, FE24, FE42, FE62, FE63, FE65, FE66, FE68, FE69, FE71,
        FE74, FE75, FE76, FE85, FE86, FE87, FE90, FE92, FE94, FE99, FF00,
        FF02, FF03, FF04, FF05, FF06, FF07, FF08, FF10, FF12, FF14, FF16,
        FF17, FF19, FF20, FF23, FF28, FF29, FF31, FF33, FF37, FF38, FF39,
        FF41, FF48, FF49, FF50, FF51, FF64, FF65, FF66, FF72, FF74, FF75,
        FF76, FF81, FF82, FF83, FF84, FF85, FF86, FF87, FF88, FF89, FF90,
        FF91, FF92, FF93, FF94, FF95
    -water (99.95:00.05), FA07, FA14, FA23, FA47, FA48, FA49, FA69, FA74, FA83,
                FA96, FA98, FB08, FB10, FB49, FC76
        (99.9 :00.1), FA49, FD57
        (99.5 :00.5), FA96
        (99   :01   ), FA07, FA14, FA69, FA74, FA96, FA98, FB10, FC76
        (98   :02   ), FA98, FB10, FB49, FD57
        (97-99:03-01), FB08
        (97   :03   ), FB08
        (96   :04   ), FA69, FB49
        (95   :05   ), FA74, FA83, FA96, FA98, FB08, FB49, FC76
        (90   :10   ), FA07, FA14, FA74, FA98, FB10, FB49, FC76
        (85   :15   ), FA14
    -benzene (50:50), FF80
    -dichloromethane (50:50), FD56, FD67, FD88, FE41, FE98, FF43, FF47, FF52,
                FF60, FF70, FF73
    -toluene (70:30), FF78, FF79
        (50:50), FF57

Benzene-acetonitrile, <u>see</u> acetonitrile-benzene
    -methanol (60:40), FA00, FA02, FA13, FA17, FA51
Benzonitrile, FB25, FC54, FF14, FF96
Butyronitrile, FC20, FF38

Dichloromethane, FB25, FC54, FF14, FF42, FF44, FF56, FF58, FF67, FF69, FF71, FF77
    -acetonitrile, <u>see</u> acetonitrile-dichloromethane
    -trifluoroacetic acid (80:20), FC98, FD40, FD60, FD74, FE94
        -trifluoroacetic anhydride (88:02:10), FB25, FC54, FF14
Dimethylformamide, FA08, FA10, FA22, FA36, FA37, FA38, FA56, FA66, FA78, FA80,
            FA93, FA94, FA95, FB02, FB03, FB09, FB19, FB20, FB21, FB36,
            FB40, FB52, FB54, FB55, FB56, FB70, FB71, FB77, FB81, FB84,
            FB95, FC01, FC07, FC08, FC10, FC12, FC13, FC19, FC20, FC21,
            FC22, FC23, FC24, FC25, FC26, FC32, FC50, FC51, FC52, FC64,
            FC77, FC82, FC83, FC84, FC85, FC86, FC99, FD23, FD24, FD25,
            FD26, FD33, FD42, FD49, FD50, FD53, FD63, FD66, FD69, FD71,
            FD75, FD80, FD81, FD82, FD83, FD84, FD85, FD86, FD87, FE01,
            FE02, FE03, FE04, FE06, FE07, FE09, FE11, FE12, FE13, FE18,
            FE19, FE20, FE21, FE31, FE32, FE33, FE34, FE35, FE36, FE37,
            FE42, FE43, FE44, FE45, FE46, FE47, FE64, FE70, FE71, FE72,
            FE73, FE77, FE78, FE79, FE81, FE91, FE92, FE93, FE94, FE95,
            FE96, FE97, FF01, FF13, FF15, FF18, FF21, FF22, FF30, FF32,
            FF35, FF40, FF53, FF54, FF55, FF59, FF61, FF62, FF63, FF68,
            FF93, FF97

```
         -water (99.995:00.005), FF11
                (99.9  :00.1   ), FA07, FA23, FA48, FA83, FA98, FC76, FD20
                (99    :01     ), FA07, FA48, FA83, FD20
                (98    :02     ), FA36, FA37, FA38, FA98, FB21
                (96    :04)    ), FD20
                (95    :05     ), FA07, FA98, FB10
              (90-99.9:10-00.1), FA96
                (90    :10     ), FA07, FB10
                (80    :20     ), FA98
                (20    :80     ), FB88, FB92, FB93, FB94, FC01, FC31, FC60, FC61, FC93,
                                  FC95, FC96, FC97, CD04
                (10    :90     ), FD22, FD27, FD28
Dimethylsulfoxide, FA44, FA57, FA93, FB13, FC10, FE26, FE27, FE28, FF45
Dioxane, FC81
     -water (75:25), FB78, FC14, FC15, FC16, FC79
            (50:50), FB50, FB98, FC34, FC36, FC39, FC41, FC49, FC57, FC58, FC59,
                    FC63, FC80, FC91, FD01, FD03

Ethanol; FA93, FA96, FC76
     -water (96?:04?), FF25, FF27
            (95 :05 ), FA88
            (75 :25 ), FB66
            (50 :50 ), FB67, FB74, FB78, FB85, FC06, FC14, FC15, FC16, FC17, FC18,
                      FC27, FC28, FC53, FC79, FC80, FC81, FC87, FC88, FD05, FD13,
                      FD14, FD15, FD16, FD34, FD35, FD44, FD45, FD46, FD47, FD51,
                      FD61, FD62, FD73, FD76, FD77, FD78, FD83, FD92, FD93, FD94,
                      FD95, FE15, FE16, FE17, FE29, FE30, FE39, FE40, FE48, FE56,
                      FE57, FE58, FE59, FE60, FE61, FE67, FE82, FE83, FE84, FF09
            (48 :52 ), FC70
            (40 :60 ), FA90, FB00, FB60, FC72, FC73, FC74, FC75, FC78, FD19, FD29,
                      FD30, FD31, FD32, FD43, FE08
            (30 :70 ), FB62, FB63, FB64, FB65, FB66, FB68, FB69
            (25 :75 ), FA16, FC43, FC45, FC47, FC65
            (20 :80 ), FA06, FA21, FA24, FD08, FD09, FD10, FD11, FD48, FD52, FD54
            (10 :90 ), FB57, FC45
            (05 :95 ), FA99, FD39
            (03 :97 ), FA15, FA25, FA61

Fluorosulfuric acid, FA30, FA32, FA33, FA35, FA52, FA87, FB14

Hexamethylphosphoramide, FB40, FC10, FC19, FC21, FC82, FE09, FE97
Hydrofluoric acid, FA41, FA43, FA55, FA82, FC10, FC11, FD99

Isobutyronitrile, FB25, FC54, FF14

Methanol, FA23, FA67, FA81, FA97, FB04, FB11
     -?%, FA29, FA42, FB05, FB06
     -water (70:30), FF34
            (50:50), FB38, FB39, FB46, FB51, FB82, FB99, FC37, FC92, FD37, FD38,
                    FD65, FD68, FD70, FD89, FD96, FD97, FD98, FE25, FE38, FE49,
                    FE50, FE51, FE52, FE53, FE54, FE55, FE88, FE89, FF26, FF36
            (45:55), FD38
            (40:60), FB51, FD38, FD89, FD96, FD97, FD98, FE38, FE88, FF26
            (36:64), FA67
            (25:75), FA25
            (10:90), FA92, FB01, FB61, FD59, FE80
     -benzene, see benzene-methanol
     -triethanolamine-water (45:5:50), FB89

Nitrobenzene, FB25, FC54, FF14
Nitromethane, FB25, FC10, FC42, FC54, FC57, FD99, FE94, FF14
```

Propionitrile, FB25, FC54, FF14

Toluene-acetonitrile, see acetonitrile-toluene
Triethanolamine-methanol-water, see methanol-triethanolamine-water
Trifluoroacetic acid-dichloromethane-trifluoroacetic anhydride, see dichloromethane-trifluoroacetic acid-trifluoroacetic anhydride
    -dichloromethane, see dichloromethane-trifluoroacetic acid
    -trifluoroacetic anhydride (90:10), FC54
    -water (90:00.2:09.8), FF14
Trifluoroacetic anhydride-dichloromethane-trifluoroacetic acid, see dichloromethane-trifluoroacetic acid-trifluoroacetic anhydride

Water (see introduction to this table)
    -acetonitrile, see Acetonitrile-water
    -dimethylformamide, see Dimethylformamide-water
    -dioxane, see Dioxane-water
    -ethanol, see Ethanol-water
    -methanol, see Methanol-water

TABLE VIII.
INDEX OF TECHNIQUES EMPLOYED

This is an index that lists the techniques employed in obtaining the data that appear in Table I. It enables the user to identify all of the entries in that table that contain information obtained by any particular technique.

In the typical entry

Chronoamperometry(IR), FE22, FE23, FE24, FF74

the name is the one recommended by the International Union of Pure and Applied Chemistry (*Pure and Applied Chemistry*, **45**, 81, 1976). Cross-references are provided liberally, so that, for example, the entries for polarographic coulometry can be found by users who seek them under "dropping-electrode coulometry", "millicoulometry", or "microcoulometry", and these cross-references also serve to notify the user that the name followed by the citations is recommended in preference to one cross-referenced to it.

Following each recommended name there appears, in parentheses, the two-letter symbol by which the technique is denoted in Column 6 (or, occasionally, in Column 21 or 26) of Table I. This enables the user to tell, without having to consult the list of abbreviations repeatedly, what part of the information contained in a long entry in Table I has been obtained by the technique in which he is interested.

Finally, for every technique except polarography itself, there appear in alphanumeric order the code numbers for all of the compounds about which information obtained by that technique appears in Table I. There are many entries of the form

Alternating-current chronopotentiometry(EF)

in which the name of a technique is followed by a two-letter symbol that identifies it as a recommended name, but in which no code numbers follow the symbol. Such an entry signifies that the technique is included in our sphere of activity but has produced no information that is given in this volume. At the other extreme, the entry for polarography is of this same form but for a very different reason: as was stated previously, a list of the code numbers of the compounds for which polarographic data are given would be too long to be of any use. We estimate that it would have contained about 360 citations. If this figure is compared with the total of 415 citations given below, it becomes apparent that, during the period covered by these volumes, the contribution made by polarography to our fundamental knowledge of the electrochemical behaviors of organic, biochemical, and organometallic substances appreciably exceeded that of all the other techniques combined.

Ac polarography (PV), FA57, FD90, FF28, FF34
    higher-harmonic (PH)
        with phase-sensitive rectification, (PB)
    in-phase (PVI), FD90
    quadrature (PVQ)
Ac voltammetry (VV)
    in-phase (VVI)
Amperometry (IO)
    differential (ID)
    with two indicator electrodes (IB)

Cathode-ray polarography, see Polarography, single-sweep
Chronoamperometry (IR) FE22, FE23, FE24, FF74
    convective (CA)
    double potential-step (IU)
    dropping-electrode, with linear potential sweep, see Polarography, single-sweep
    polarographic (IV)
    rotating-disc-electrode, see Chronoamperometry, convective
    stirred-pool-electrode, see Chronoamperometry, convective
    thin-layer (TI)
    with linear potential sweep (IL), FA10, FA53, FA75, FA76, FA86, FA88, FA91,
                           FB16, FB21, FB22, FB23, FB24, FB25, FB71,
                           FB77, FB84, FB87, FD56, FD67, FD88, FD89,
                           FD96, FD97, FD98, FE22, FE23, FE38, FE41,
                           FE70, FE88, FE92, FE93, FE98, FF26, FF43,
                           FF47, FF52, FF60, FF70, FF73
    semi-integral (XT), FA36, FA37, FA38, FB07, FB19, FB20, FB21, FB70, FB71,
        FB81, FC13, FC24, FC83, FD33, FE92
    with non-linear potential sweep (IM), FD60
Chronocoulometry (QR)
    convective (CC)
    double potential-step (QU)
    potential-step, see Chronocoulometry
    rotating-disc-electrode (CC)
    stirred-pool-electrode (CC)
    thin-layer (TQ)
Chronopotentiometry (ER)
    ac (EF)
    alternating voltage (EK)
    current-cessation, see Chronopotentiometry, current-step
    current-reversal, see Chronopotentiometry, current-step
    current-step (EE)
    cyclic (EY)
    derivative (EM)
    programmed-current (EC)
    with linear current-sweep (EL)
    with superimposed ac (EV)
Coulometry
    controlled-current with a reagent precursor (QT)
        without a reagent precursor (QP), FD60
    controlled-potential (QE), FA15, FA25, FA26, FA70, FA86, FA95, FB66,
                          FB97, FC05, FC47, FC50, FC52, FC99, FD41,
                          FE23, FE24, FE71, FE94, FE96, FF13, FF21
    dropping-electrode, see Coulometry, polarographic
    polarographic (PQ)

Dropping electrode coulometry, see Coulometry, polarographic
Dynamic capacity, measurement of (NF)

Electrochemiluminescence (HN), FF85, FF86, FF87, FF88, FF89, FF90, FF91, FF92, FF93, FF94, FF95
Electrogravimetry (QW)
Electrolysis, controlled-potential (CP) FA53, FA54, FA67, FA97, FB22, FB25, FF59

Faradaic impedance, measurement of (FI)
Faradaic rectification, high-level (HL)

Linear-sweep voltammetry, see Chronoamperometry with linear potential-sweep

Microcoulometry, see Coulometry, polarographic
Millicoulometry, see Coulometry, polarographic

Non-faradaic admittance, see Dynamic capacity

Oscillopolarography (PO), FD70
Oscillovoltammetry (VO)

Polarography (PY)
    ac, see Ac polarography
    alternating-voltage (PI)
    cathode-ray, see Polarography, single-sweep
    current-scanning (PC)
    cyclic triangular-wave, see Polarography, triangular-wave, cyclic
    demodulation (FP)
    derivative (PD)
    differential (PF)
    differential pulse, see Polarography, pulse, differential
    double-tone (PX)
    higher-harmonic ac, see Ac polarography, higher harmonic
    incremental-charge ($\overline{DQ}$)
    in-phase ac, see Ac polarography, in-phase
    intermodulation (PL)
    Kalousek (PK), FD49, FD50, FD53, FD66, FD96, FE11, FE73, FF45
    modulation (MM)
    multisweep (PM)
    oscillographic, see Oscillopolarography
    pulse (PP), FB05, FB06
        derivative (DP)
        differential (DI), FA29, FA42, FA92, FB01, FB05, FB61, FB89, FD90, FD91
    rf (RP)
    single-sweep (PW), FA14, FA23, FA47, FA48, FA49, FA68, FA73, FA74, FA83, FA96, FA98, FB08, FB10, FB49, FC76, FD57
    square-wave (PG)
    staircase (PS)
    Tast (PT), FA88, FD59, FE80
    triangular-wave (PA), FA72, FA73, FD63, FF30, FF76
        cyclic (PR), FA06, FA34, FA72, FD69, FD71, FE71, FF55

Stirred-pool-electrode chronoamperometry, see Chronoamperometry, convective

Titration
    controlled-potential coulometric, see Coulometry, controlled-potential coulometric (QT), FD60

Voltammetry
    ac, see Ac voltammetry
    current-scanning (VC)
    cyclic triangular-wave, see Voltammetry, triangular-wave, cyclic
    derivative (VD) FB50, FB98, FC34, FC36, FC39, FC41, FC57, FC58, FC59, FC91, FD01, FD03
    differential (VF)
    hydrodynamic (VY), FA05, FA41, FA43, FA45, FA55, FA58, FA59, FA60,
        FA63, FA79, FA82, FA84, FA86, FB12, FB28, FB29,
        FB30, FB35, FB37, FB43, FB44, FB47, FB48, FB50,
        FB52, FB53, FB58, FB86, FB96, FB97, FB98, FC00,
        FC09, FC10, FC11, FC29, FC30, FC33, FC34, FC35,
        FC36, FC38, FC39, FC40, FC41, FC42, FC44, FC48,
        FC49, FC55, FC56, FC57, FC58, FC59, FC62, FC63,
        FC64, FC89, FC90, FC91, FC94, FD00, FD01, FD02,
        FD03, FD36, FD41, FD42, FD60, FD72, FD75, FD99,
        FE14, FE42, FE62, FD63, FE65, FE66, FE68, FE69,
        FE72, FE74, FE75, FE76, FE86, FE87, FE90, FE94,
        FE99, FF00, FF02, FF03, FF04, FF05, FF06, FF07,
        FF08, FF10, FF11, FF12, FF16, FF17, FF19, FF20,
        FF23, FF29, FF31, FF33, FF39, FF41, FF49, FF64,
        FF78, FF79
        triangular (YA)
            cyclic (YR)
    linear-sweep, see Chronoamperometry with linear potential-sweep
    multisweep (VM)
    oscillographic, see Oscillovoltammetry
    triangular-wave (VA), FA01, FA04, FA09, FA10, FA11, FA12, FA18, FA19,
        FA27, FA28, FA30, FA31, FA32, FA33, FA35, FA44,
        FA45, FA46, FA50, FA52, FA57, FA64, FA70, FA84,
        FA85, FA86, FA87, FA89, FA93, FB14, FB15, FB17,
        FB18, FB25, FB29, FB40, FB43, FB44, FB54, FB57,
        FB71, FB77, FB84, FB86, FB97, FC01, FC10, FC19,
        FC21, FC22, FC23, FC54, FC56, FC71, FC82, FD06,
        FD07, FD18, FD24, FD25, FD38, FD41, FD58, FD64,
        FD71, FD72, FD80, FD81, FD90, FE01, FE04, FE05,
        FE09, FE12, FE20, FE24, FE74, FE79, FE90, FE93,
        FE94, FE96, FF11, FF14, FF20, FF28, FF29, FF34,
        FF38, FF42, FF45, FF48, FF50, FF51, FF65, FF71
        convective (CT)
        cyclic (VR), FA39, FA66, FA86, FB13, FB96, FC42, FC45, FC98, FD40,
        FD60, FD74, FD85, FE24, FE26, FE27, FE28, FE31, FE64,
        FE91, FE95, FE97, FF35, FF44, FF45, FF56, FF58, FF66,
        FF67, FF69, FF77
    with linear current-sweep, see Voltammetry, current-scanning

TABLE IX.
INDEX OF INDICATOR ELECTRODES EMPLOYED

This is an index that lists the indicator and working electrodes employed in obtaining the data that appear in Table I. It enables the user to identify all of the entries in that table that contain information obtained with any particular indicator or working electrode.

In the typical entry

Carbon, glassy, stationary (glC: xxns) FA57

the initial portion identifies the electrode used and gives its configuration whenever possible and, in parentheses, the abbreviation used to denote the electrode in Column 11 of Table I. This enables the user to tell, without having to consult the list of abbreviations repeatedly, what part of the information contained in a long entry in Table I has been obtained with the electrode and configuration in which he is interested. Following the abbreviation, there appear the code numbers of the compounds that have been studied with the electrode. Finally, liberal cross-references are provided to ensure that no entry that might be of interest will be overlooked.

Precious though space in the original literature certainly is, it does not seem to us to be valuable enough to justify the frequency with which sentences like "The graphite indicator electrode was described previously" are used by authors meticulous in specifying the other circumstances under which their data were obtained. As it is well known that electrochemical properties depend on the configuration of the indicator electrode as well as on the material from which it is contructed, our original intention was to provide a complete index of both configurations and materials. A few cases in which the reference number appended to a sentence like the one quoted led only to an exactly identical sentence in another article convinced us, however, that this intention was unlikely to be achievable, and we have therefore confined ourselves to transcribing the infomation provided. We recognize that this complicated the use of this table, even with the cross-references provided, and urge the authors of future papers to describe their electrodes as carefully as they do the other facets of their work.

There is one null entry in this table, and that is the one for the dropping mercury electrode. This reflects the same fact that led to the omission of polarography from Table VIII. There are some compounds in Table I for which no information obtained with a dropping mercury electrode is given in this volume. The great majority of these are compounds that undergo oxidation at relatively positive potentials and that were therefore studied with platinum, graphite, carbon-paste, or other indicator electrodes. Many such compounds are not amenable to investigation with dropping mercury electrodes, but there are many others that are. Some of these can also be reduced within the range of potentials accessible with a dropping mercury electrode while some undergo oxidation within that range under certain conditions but not under others; for some there may be polarographic data in the prior literature, while for others the possibility of obtaining a polarographic response may not have been investigated. In any event, an entry documenting the use of the dropping mercury electrode in obtaining the data included in this volume would have contained 353 entries and would have been too long to be of any practical use.

TABLE IX. Index of Indicator Electrodes Employed

Aluminum (Al:nsns), FA54, FC05

Carbon; <u>see also</u> Graphite
    glassy (glC:nsns), FA63
        disc, rotating (glC:rodi), FA82, FA84, FA86, FB86, FC10, FC42, FD99, FF45
            stationary (glC:xxdi), FA10, FA84, FA86, FB86, FC10
        periodic displacement of solution (glC:pdns), FD72
        stationary (glC:xxns), FA57
    paste (Cp:nsns), FA46
        stationary (Cp:xxns), FB57

Gold
    stationary (Au:xxns), FF45

Graphite
    paste (Nujol)
        disc, stationary (g(Nujol):xxdi), FD96, FD97, FD98, FE38, FE88
        stationary (g(Nujol):xxns), FD89, FF26
    tubular, stationary, flowing solution (g:xftu), FA58, FA59, FA60
    wax-impregnated, disc
        stationary (gw:xxdi), FA31, FB15, FD06, FD07

Lead
    rotating (Pb:rons), FA53
    wire, stationary (Pb:xxwi), FA53

Mercury (Hg:nsns), FA26, FE71
    assumed (Hg?:nsns), FA95, FC50, FC99
    hanging drop (HMDE), FA09, FA36, FA37, FA38, FA44, FA57, FA63, FA66, FB07,
        FB15, FB19, FB20, FB21, FB70, FB71, FB77, FB81, FB84,
        FC10, FC13, FC22, FC23, FC24, FC83, FD33, FD38, FD80,
        FD81, FD90, FE01, FE04, FE20, FE26, FE27, FE28, FE92,
        FE93, FE96, FF11, FF34, FF38, FF45, FF71
    -plated platinum, wire, stationary (Pt(Hg):xxwi), FD85
    pool (Hg:nspo), FA15, FB66, FC47
        stirred (Hg:srpo), FA25, FC05, FE96, FF13, FF21, FF59
    sessile drop, stationary (Hg:xxsd), FB40, FC19, FC21, FC82, FE09, FE97

Nickel/nickel hydroxide
    rotating (Ni/Ni(OH)$_2$:rons), FA05

Palladium
    stationary (Pd:xxns), FA01
Platinum (Pt:nsns), FD41, FF95
    assumed (Pt?:nsns), FB97
    bead
        stationary (Pt:xxbe), FC98, FD40, FD58, FD60, FD71, FD74, FF42, FF44,
        FF48, FF56, FF58, FF66, FF67, FF71, FF77
    button
        stationary (Pt:xxbu), FA39, FA45, FA64, FA75, FA76, FA91, FA93, FB16,
        FB18, FB25, FC54, FD18, FD56, FD67, FD88, FE05,
        FE41, FE98, FF14, FF43, FF47, FF52, FF60, FF70,
        FF73, FF80

Platinum (cont.)
    coil (Pt:nsco), FF85, FF86, FF87, FF89
    disc, rotating (Pt:rodi), FA10, FA41, FA43, FA45, FA55, FA79, FA84, FA86,
                          FB12, FB28, FB29, FB30, FB35, FB37, FB43, FB44,
                          FB47, FB48, FB50, FB53, FB58, FB86, FB96, FB97,
                          FB98, FC00, FC09, FC10, FC11, FC29, FC30, FC33,
                          FC34, FC35, FC36, FC38, FC39, FC40, FC41, FC42,
                          FC44, FC48, FC49, FC55, FC56, FC57, FC58, FC59,
                          FC62, FC63, FC89, FC90, FC91, FC94, FD00, FD01,
                          FD02, FD03, FD36, FD41, FD99, FE14, FE42, FE62,
                          FE63, FE65, FE66, FE68, FE69, FE74, FE75, FE76,
                          FE86, FE87, FE94, FE99, FF00, FF02, FF03, FF04,
                          FF05, FF06, FF07, FF08, FF10, FF11, FF12, FF16,
                          FF17, FF19, FF20, FF23, FF31, FF33, FF39, FF41,
                          FF49, FF64
        stationary (Pt:xxdi), FA10, FA84, FA85, FA86, FA89, FB17, FB29, FB44,
                          FB87, FC42, FC56, FC71, FD41, FD64, FE31, FE64,
                          FE70, FE74, FE91, FE94, FE95, FF11, FF20, FF28,
                          FF29, FF35, FF50, FF51
    flag, periodic displacement of solution (Pt:pdfl), FD72
        stationary (Pt:xxfl), FD72
    gauze (Pt:nsgz), FA70, FA86, FB22, FB25
        stationary electrode, flowing solution (Pt:xfgz), FD60, FE94
    hemisphere, periodic displacement of solution (Pt:pdhb), FB52, FC64, FD42,
                                                                         FD75, FE72
    periodic displacement of solution (Pt:pdns), FF78, FF79
    ring-disc, rotating (Pt:rord), FA79, FB12, FF29, FF93
    rotating (Pt:rons), FD60
    stationary (Pt:xxns), FA01, FA04, FA11, FA12, FA18, FA19, FA27, FA28,
                    FA50, FB18, FB43, FB54, FB96, FB97, FC01, FE12,
                    FE22, FE23, FE24, FE79, FE90, FF38, FF65, FF97,
    wire
        stationary (Pt:xxwi), FA30, FA32, FA33, FA35, FA52, FA70, FA86, FA87,
                        FB14, FB22, FB23, FB24, FB25, FF82, FF90, FF91,
                        FF92, FF93

Zinc oxide
    flag, stationary (ZnO:xxfl), FF94

TABLE X.
KEY TO LITERATURE CITATIONS

This table is provided to permit decoding the literature references that appear in Column 14 of Table I and obtaining full citations from them.

The form and significance of those references are described in the introduction of Table I. To illustrate the use of this table we shall suppose that a full citation is wanted for the information given in this volume for benzyl alcohol. In Table I this compound appears under code number FA56, and the literature reference is given as EA019-0629 in Column 14 on the first of the lines dealing with this compound.

As was discussed in the introduction to Table I, the letters "EA" denote the journal. The next three digits give the number of the volume, in this case 19. As that number contains only two digits, it is preceded by a zero (if the volume number contained only one digit, it would be preceded by two zeros). The last four digits give the page number. Here this is 0629. Occasionally one, two, or three zeros are needed to give the total of four digits. Hence the reference is to volume **19**, page 629, of the journal denoted by the two-letter code "EA".

To identify the journal that is denoted by this abbreviation, this table must be inspected. The journals are arranged in the alphabetical order of the two-letter codes assigned to them, and each such code is followed in parentheses by the full title of the journal to which it corresponds. It can be found that the two-letter code "EA" denotes *Electrochimica Acta*. Hence the reference is to the *Electrochimica Acta*, volume **19**, page 629, and this table provides the further information that the paper was published in 1974.

Names of some journals published in other languages than English are given in transliteration of the original title. When data were obtained from English translations (e.g., of Russian journals) the English translation of the name of the journal is given.

Beneath the abbreviation and the name of the journal, volume numbers are given consecutively along with the corresponding years. Under each volume number and year there are references to all of the publications from which data were taken and are given in Table I. These references are given in exactly the same form as is used in Table I. and are followed by the names of the authors. The references to each volume of each journal are listed in the order of increasing page numbers.

For example, the reference "AA096-0345" is the third one which appears underneath volume **96**(1978) of the journal *Analytica Chimica Acta*. It gives the authors' names as E. Jacobsen and M. W. Bjørnsen. Consequently the full citation in the usual form is E. Jacobsen and M. W. Bjørnsen, *Analytica Chimica Acta*, **96**, 345 (1978).

Following the author's name there appears the code numbers of all of the compounds appearing in this volume for which data were taken from the reference given.

Data for more than one compound were often obtained from a single publication. In such a case the code numbers of all the compounds are given in alphanumeric order. Usually these additional code numbers denote compounds that are closely related to the one that was originally of interest, or represent data obtained by similar techniques under similar conditions, and in either case provide cross-references that may further illuminate the information already obtained. It should not be inferred that these are complete lists of the compounds about which information is given in the original references, for some of these may have been omitted from this volume as having been studies in insufficient detail or for other reasons.

AA (ANALYTICA CHIMICA ACTA)

Vol. 92 (1977)
AA092-0353 Beckett, A.H., Rahman, N.N., and Smyth, W.F., FA92, FB01, FB61

Vol. 94 (1977)
AA094-0119 Smyth, M.R., Smyth, W.F., and Clifford, J.M., FD59, FE80
AA094-0461 Rogstad, A., and Høgberg, K., FB89

Vol. 96 (1978)
AA096-0143 Jemal, M., and Knevel, A.M., FD39
AA096-0335 Smyth, M.R., and Osteryoung, J.G., FA29, FA42, FB05, FB06
AA096-0345 Jacobsen, E., and Bjørnsen, M.W., FD90
AA096-0353 Lal, S., and Jain, P.S., FA77

Vol. 98 (1978)
AA098-0093 Brunt, K., FB83, FD91

CO (COLLECTION OF CZECHOSLOVAK CHEMICAL COMMUNICATIONS)

Vol. 26 (1961)
CO026-1763 Matrka, M., Navrátil, F., Štěrba, V., and Arient, J., FC06, FF09
CO026-2271 Tvaroha, B., FF46

Vol. 27 (1962)
CO027-0483 Volke, J., FC46
CO027-1861 Holý, A., and Vystrčil, A., FC70
CO027-2447 Hrdý, O., FD92, FD93, FD94, FD95, FE15, FE16, FE29, FE30, FE39, FE40, FE53, FE54, FE55, FE56, FE57, FE58, FE59, FE60, FE61, FE82, FE83, FE84

Vol. 28 (1963)
CO028-1985 Ryvolová, A., FA90, FB00, FB60, FD43, FE08

Vol. 30 (1965)
CO030-4143 Zhdanov, S.I., and Pozdeeva, A.A., FF25, FF27

Vol. 31 (1966)
CO031-1264 Oelschläger, H., Volke, J., and Hoffmann, H., FD22, FD27, FD28

Vol. 32 (1967)
CO032-2140 Petránek, J., Ryba, O., and Doskočilová, D., FD05, FD44, FD45, FD46, FD47, FD61, FD62, FD76, FD77, FD78, FE17

Vol. 34 (1969)
CO034-1615 Matrka, M., Marhold, J., and Ságner, Z., FB50, FB98, FC57, FC58, FC59
CO034-3952 Matrka, M., Marhold, J., Chmátal, V., Štěrba, V., Ságner, Z., and Kroupa, J., FC34, FC36, FC39, FC41, FC49, FC57, FC63, FC91, FD01, FD03

Vol. 35 (1970)
CO035-2944 Matrka, M., Chmátal, V., Pípalová, J., Ságner, Z., Kroupa, J., and Marhold, J., FC33, FC35, FC38, FC40, FC48, FC57, FC62, FC90, FD00, FD02

Vol. 36 (1971)
CO036-0575 Sümmermann, W., and Bäumgartel, H., FE86, FE87, FF08, FF49
CO036-1406 Zacharová-Kalavská, D., and Perjéssy, A., FD08, FD09, FD10, FD11, FD48, FD52, FD54

CZ (CHEMICKÉ ZVESTI)

Vol. 16 (1962)
CZ016-0316 Sohr, H., FD70

EA (ELECTROCHIMICA ACTA)

Vol. 18 (1973)
EA018-0139 Issa, I.M., Issa, R.M., Temerk, Y.M., and Mahmoud, M.R., FB26, FB27, FB31, FB32, FB33, FB34, FB41, FB42, FB45, FB69, FB72, FB73, FB75, FB76, FB79, FB80, FB90, FB91

EA (ELECTROCHIMICA ACTA)(cont.)
Vol. 18 (1973)(cont.)
EA018-0237 Laviron, E., and Roullier, L., FA68
EA018-0241 Tsunaga, M., Iwakura, C., and Tamura, H., FA70, FA86
EA018-0265 Issa, I.M., Issa, R.M., Ghoneim, M.M., and Temerk, Y.M., FD96, FD97
EA018-0327 Saxena, R.S., and Chaturvedi, U.S., FA71
EA018-0331 Coleman, J.P., Fleischmann, M., and Pletcher, D., FA30, FA32, FA33, FA35, FA52, FA87, FB14
EA018-0335 Bannerjee, N.R., and Negi, A.S., FD96
EA018-0373 Imberger, H.E., and Humffray, A.A., FB22, FB23, FB24, FB25
EA018-0433 Kariv, E., Terni, H.A., and Gileadi, E., FA67
EA018-0519 Parker, V.D., FA39, FA45
EA018-0537 Hammerich, O., and Parker, V.D., FB25, FC54, FF14
EA018-0615 Tsunaga, M., Iwakura, C., Yoneyama, H., and Tamura, H., FA85
EA018-0639 Kihara, T., Sukigara, M., and Honda, K., FF85, FF86, FF87, FF89
EA018-0665 Jensen, B.S., and Parker, V.D., FE05
EA018-0691 Moiroux, J., and Fleury, M.B., FA15, FA25, FA26, FA61, FA62, FA99
EA018-0923 Roberson, P.M., Schwager, F., and Ibl, N., FA05
EA018-0933 Saxena, R.S., and Bhatia, S.K., FA16
EA018-0975 Diggle, J.W., and Parker, A.J., FA93, FD18
EA018-1025 Pak, C.M., and Gulick, W.M., Jr., FD85

Vol. 19 (1974)
EA019-0049 Barradas, R.G., Kutowy, O., and Shoesmith, D.W., FA53
EA019-0063 Takamura, T., and Sato, Y., FA01
EA019-0215 Gouverneur, L., Leroy, G., and Zador, I., FB39, FC37, FD37, FD38, FD65, FD68, FD89, FD96, FD97, FD98, FE25, FE38, FE48, FE49, FE50, FE51, FE52, FE88, FE89, FF26, FF36
EA019-0555 Kita, H., Ishikura, S., and Katayama, A., FA09
EA019-0565 Sundholm, F., and Sundholm, G., FA10
EA019-0593 Iwakura, C., Tsunaga, M., and Tamura, H., FA86
EA019-0611 Horner, L., and Degner, D., FA67, FA81, FA97, FB04, FB11
EA019-0629 Lund, H., Doupeux, H., Michel, M.A., Mousset, G., and Simonet, J., FA09, FA22, FA56, FA78, FA80, FA94, FB09, FB36, FB55, FB56, FB77, FB84, FB95, FC10, FC25, FC26, FC32, FC52, FC77, FC84, FC85, FD26, FD82, FD86, FD87
EA019-0681 Landsberg, R., and Müller, S., FA79, FB12
EA019-0865 Jubault, M., Raoult, E., and Peltier, D., FA65
EA019-0951 Troll, T., and Baizer, M.M., FC19, FC21, FC82, FE09, FE97

Vol. 20 (1975)
EA020-0007 Grabner, E.W., FF94
EA020-0021 Savéant, J.M., and Binh, S.K., FD69, FD71, FE71, FF30
EA020-0033 Troll, T., and Baizer, M.M., FB40
EA020-0143 Michel, M.A., Mousset, G., Simonet, J., and Lund, H., FB71, FB77, FB84, FC52, FD86, FE93, FE96, FF13, FF21, FF59
EA020-0323 Skolova, E., FA01, FA04, FA11, FA12, FA18, FA19, FA27, FA28, FA50
EA020-0369 Fleury, D., and Moiroux, J., FA06, FA21, FA24
EA020-0427 Ratajczak, H.M., Pajdowski, L., and Ostern, M., FA03
EA020-0469 Serve, D., FA84, FA86, FB29, FB44, FB86, FC42
EA020-0853 Santiago, E., and Simonet, J., FA95, FB02, FB03, FC50, FC51, FC86, FC98, FE13, FF01
EA020-0951 Fleury, M.B., and Molle, G., FA34, FA40
EA020-0965 Dufresne, J.-C., FA72
EA020-0973 Dufresne, J.-C., FA73, FD12, FD21, FD55
EA020-0981 Fleury, M.B., and Dufresne, J.-C., FA72, FA73, FD63
EA020-1011 Cauquis, G., Cognard, J., and Serve, D., FE74, FF05, FF41

EA (ELECTROCHIMICA ACTA)(cont.)
Vol. 20 (1975)(cont.)
 EA020-1019 Cauquis, G., Delhomme, H.,
          and Serve, D., FE63,
          FE65, FE66, FE69, FE75,
          FE99, FF00, FF03, FF07,
          FF19, FF20, FF23

Vol. 21 (1976)
 EA021-0345 Martigny, P., Lund, H.,
          and Simonet, J., FB54,
          FC01, FD24, FD25, FE12,
          FE79
 EA021-0395 Devaud, M., and
          Le Moullec, Y., FA00,
          FA02, FA13, FA17, FA51
 EA021-0407 Jubault, M., Raoult, E.,
          and Peltier, D., FA65
 EA021-0421 Roullier, L., and
          Laviron, E., FA63
 EA021-0473 Degrand, C., and Belot,
          G., FB62, FB63, FB64,
          FB65, FC43, FC45, FC65,
          FD73
 EA021-0479 Degrand, C., Millet, C.,
          and Belot, G., FB66,
          FC45, FC47
 EA021-0497 Pragst, F., FE42, FF02,
          FF95
 EA021-0557 Cauquis, G., Delhomme, H.,
          and Serve, D., FB28,
          FB30, FB35, FB58, FC00,
          FC09, FC29, FC44, FC55,
          FC56, FC89, FC94, FD36,
          FD41, FE14, FE62, FF10,
          FF12, FF17, FF31, FF33,
          FF39, ГГ64
 EA021-0621 Tabaković, I., Lacán, M.,
          and Damoni, Sh., FE22,
          FE23, FE24, FF65
 EA021-0753 Schluter, D.N., Biegler,
          T., Brown, E.V., and
          Bauer, H.H., FA88
 EA021-0831 Shawali, A.S., Alanadouli,
          B.E., and Sammour, M.H.,
          FC72, FC73, FC74, FC75,
          FC78, FD19, FD29, FD30,
          FD31, FD32
 EA021-0913 Fleury, M.B., FA20, FA72
 EA021-0973 Sharp, M., FA31, FB15,
          FD06, FD07
 EA021-1085 Sharma, L.R., and Kalia,
          R.K., FA58, FA59, FA60
 EA021-1149 Louati, A., Jordan, J.,
          and Gross, M., FF53,
          FF55
 EA021-1171 Serve, D., FB37, FB43,
          FB47, FB48, FB53, FB96,
          FB97, FC30, FC44, FE68,
          FE76, FF04, FF06, FF16

JA (JOURNAL OF THE AMERICAN CHEMICAL
    SOCIETY)

Vol. 92 (1970)
 JA092-4139 Breslow, R., Grubbs, R.,
          and Murahashi, S.I., FC53,
          FE67

Vol. 95 (1973)
 JA095-5482 Santhanam, K.S.V., and
          Elving, P.J., FA44,
          FA57, FB13, FE26, FE27,
          FE28
 JA095-5939 Brown, G.M., Hopf, F.R.,
          Ferguson, J.A., Meyer,
          T.J., and Whitten,
          D.G., FF42, FF44, FF56,
          FF58, FF66, FF67, FF69,
          FF71, FF77
 JA095-6582 Tokel-Takvoryan, N.E.,
          Hemingway, R.E., and
          Bard, A.J., FA89, FB17,
          FB87, FC71, FD64, FF28,
          FF29, FF50, FF51, FF82,
          FF90, FF91, FF92, FF93
 JA095-6688 Breslow, R., Murayama,
          D.R., Murahashi, S.I., and
          Grubbs, R., FE31, FE64,
          FE70, FE91, FE95, FF35
 JA095-7132 Ronlán, A., Hammerich,
          O., and Parker, V.D.,
          FC98, FD40, FD58, FD60,
          FD74, FF48
 JA095-7164 Van Duyne, R.P., FF81,
          FF83, FF84, FF88

JE (JOURNAL OF ELECTROANALYTICAL
    CHEMISTRY AND INTERFACIAL ELECTRO-
    CHEMISTRY)

Vol. 35 (1972)
 JE035-0013 Brown, O.R., and
          Gonzalez, E.R., FA54,
          FC05
 JE035-0369 Holleck, L., and
          Kazemifard, G., FB38,
          FB51, FB82, FB99, FC92
 JE035-0381 Holleck, L., and
          Mahapatra, S., FA23

Vol. 36 (1972)
 JE036-0147 Andruzzi, R., Cardinali,
          M.E., Carelli, I., and
          Trazza, A., FC12, FF01,
          FE03, FE06, FE18, FE19,
          FE36

JE (JOURNAL OF ELECTROANALYTICAL CHEMISTRY AND INTERFACIAL ELECTROCHEMISTRY)(cont.)

Vol. 36 (1972)(cont.)
JE036-0167 Berg, H., Granath, K., Nygård, B., Strassburger, J., and Weist, P., FF24
JE036-0179 Itabashi, E., FD79

Vol. 38 (1972)
JE038-009A Hammerich, O., and Parker, V.D., FE94
JE038-0479 Pedersen, C.T., Hammerich, O., and Parker, V.D., FD56, FD67, FD88, FE41, FE98, FF43, FF47, FF52, FF60, FF70, FF73, FF80

Vol. 39 (1972)
JE039-0385 Bauer, D., and Foucault, A., FC10, FD99, FE94
JE039-0407 Butkiewicz, K., FC14, FC15, FC16, FC18, FC79, FC80, FC81
JE039-0419 Butkiewicz, K., FB78

Vol. 40 (1972)
JE040-0063 Pilloni, G., Valcher, S., and Martelli, M., FF57

Vol. 41 (1973)
JE041-0067 Andruzzi, R., Cardinali, M.E., and Trazza, A., FE02, FE04, FE07, FE20, FE21, FE37
JE041-0405 Avaca, L.A., and Bewick, A., FC10
JE041-0429 Andruzzi, R., Carelli, I., and Trazza, A., FD80, FD81

Vol. 42 (1973)
JE042-005A Valcher, S., Pilloni, G., and Martelli, M., FF78, FF79
JE042-0189 Lamy, E., Nadjo, L., and Savéant, J.M., FA07, FA14, FA23, FA47, FA48, FA49, FA69, FA74, FA83, FA96, FA98, FB08, FB10, FB49, FC76, FD20
JE042-0309 Rieke, R.D., Copenhafer, R.A., Aguiar, A.M., Chattha, M.S., and Williams, J.C., Jr., FE81, FF15, FF18, FF22, FF32, FF40, FF54, FF61, FF62, FF63, FF68

Vol. 43 (1973)
JE043-0397 Farsang, G., Vass, V., Ladányi, L., and Saber, T.M.H., FA46, FB57

Vol. 45 (1973)
JE045-0483 Pilloni, G., Vecchi, E., and Martelli, M., FF75, FF76

Vol. 46 (1973)
JE046-0141 Hazelrigg, M.J., Jr., and Bard, A.J., FE94
JE046-0391 Heyrovský, M., Vavřička, S., and Heyrovská, R., FB46, FD38

Vol. 47 (1973)
JE047-0089 Pilloni, G., and Martelli, M., FF72
JE047-0115 Ammar, F., and Savéant, J.M., FA36, FA37, FA38, FB07, FB19, FB20, FB21, FB70, FB81, FC13, FC24, FC83, FD33, FE92
JE047-0215 Ammar, F., and Savéant, J.M., FB21, FE92

Vol. 48 (1973)
JE048-0081 Ratard, D., Belin, P., and Plichon, V., FE90
JE048-0297 Butkiewicz, K., FB88, FB92, FB93, FB94, FC02, FC31, FC60, FC61, FC93, FC95, FC96, FC97, FD04
JE048-0425 Schiavon, G., Zecchin, S., Cogoni, G., and Bontempelli, G., FD71
JE048-0447 Slifstein, C., and Ariel, M., FF45

Vol. 49 (1974)
JE049-0111 Marcoux, L., and Adams, R.N., FA64, FA75, FA76, FA91, FB16, FB18
JE049-0133 Takamura, K., and Hayakawa, Y., FF34

Vol. 50 (1974)
JE050-0295 Pilloni, G., Zotti, G., and Martelli, M., FF74
JE050-0351 Rusina, A., Volke, J., Černák, J., Kováč, J., and Kollár, V., FD49, FD50, FD53, FD66, FE11, FE73
JE050-0359 Valcher, S., and Casalbore, G., FD83, FD84

JE (JOURNAL OF ELECTROANALYTICAL CHEMISTRY AND INTERFACIAL ELECTROCHEMISTRY)(cont.)

Vol. 51 (1974)
- JE051-0226 Valcher, S., Casalbore, G., and Mastragostino, M., FB52, FC64, FD42, FD75, FD84, FE72
- JE051-0341 Andruzzi, R., Trazza, A., and Bruni, P., FF11

Vol. 52 (1974)
- JE052-0403 Nadjo, L., Savéant, J.M., and Tessier, D., FB71
- JE052-0459 Schiavon, G., Zecchin, S., Cogoni, G., and Bontempelli, G., FD72

Vol. 53 (1974)
- JE053-0407 Ammar, F., Andrieux, C.P., and Savéant, J.M., FD57
- JE053-0439 Katiyar, S.S., Lalithambika, M., and Joshi, G.C., FB67, FB74, FB85, FC17, FC27, FC28, FC87, FC88, FD34, FD35, FD51
- JE053-0449 Katiyar, S.S., Lalithambika, M., and Dhar, D.N., FC68, FC69, FD13, FD14, FD15, FD16

Vol. 54 (1974)
- JE054-0232 Masson, J.P., Devynck, J., and Trémillon, B., FA41, FA43, FA55, FA82, FC10, FC11, FD99
- JE054-0289 Inesi, A., and Rampazzo, L., FC22, FC23
- JE054-0305 Shang, D.T., and Blount, H.N., FE94
- JE054-0411 Malik, W.U., Mahesh, V.K., and Goyal, R.N., FB59, FC03, FC04, FC66, FC67, FE32, FE33, FE34, FE35, FE43, FE44, FE45, FE46, FE47, FE77, FE78
- JE054-0417 Malik, W.U., and Gupta, P.N., FB68, FD17, FE00, FE10

Vol. 55 (1974)
- JE055-0277 Kalinowski, M.K., and Tenderende-Gumińska, B., FC07, FC08
- JE055-0407 Valcher, S., and Ghe, A.M., FD23

Vol. 56 (1974)
- JE056-0259 Margel, S., and Levy, M., FC10
- JE056-0409 Rieke, R.D., and Copenhafer, R.A., FA66, FC20, FF37, FF38

Vol. 57 (1974)
- JE057-0027 Andrieux, C.P., and Savéant, J.M., FA36, FA37, FA38, FB21

JS (JOURNAL OF THE ELECTROCHEMICAL SOCIETY)

Vol. 121 (1974)
- JS121-1555 Dam, H.T. van, and Ponjeé, J.J., FE85

TABLE XI.
AUTHOR INDEX

This is an index to Table X. It lists, in alphabetical order, the names of all the authors cited in Table X and gives, for each author, the number of each page on which a citation of his name appears in Table X. Each reference given on the cited page should be inspected, for there may be two or more references on the page that includes the name sought, and such situations are not identified here.

TABLE XI. Author Index

Adams, R.N., 528
Aguiar, A.M., 528
Alanadouli, B.E., 527
Ammar, F., 528, 529
Andrieux, C.P., 529
Andruzzi, R., 527, 528, 529
Ariel, M., 528
Arient, J., 525
Avaca, L.A., 528

Baizer, M.M., 526
Bannerjee, N.R., 526
Bard, A.J., 527, 528
Barradas, R.G., 526
Bauer, D., 528
Bauer, H.H., 527
Bäumgartel, H., 525
Beckett, A.H., 525
Belin, P., 528
Belot, G., 527
Berg, H., 528
Bewick, A., 528
Bhatia, S.K., 526
Biegler, T., 527
Binh, S.K., 526
Bjørnsen, M.W., 525
Blount, H.N., 529
Bontempelli, G., 528, 529
Breslow, R., 527
Brown, E.V., 527
Brown, G.M., 527
Brown, O.R., 527
Bruni, P., 529
Brunt, K., 525
Butkiewicz, K., 528

Cardinali, M.E., 527, 528
Carelli, I., 527, 528
Casalbore, G., 528, 529
Cauquis, G., 526, 527
Černák, J., 528
Chattha, M.S., 528
Chaturvedi, U.S., 526
Chmátal, V., 525
Clifford, J.M., 525
Cognard, J., 526
Cogoni, G., 528, 529
Coleman, J.P., 526
Copenhafer, R.A., 528, 529

Dam, H.T. van, 523
Damoni, Sh., 527
Degner, D., 526
Degrand, G., 527
Delhomme, H., 527
Devaud, M., 527

Devynck, J., 529
Dhar, D.N., 529
Diggle, J.W., 526
Doskočilová, D., 525
Doupeux, H., 526
Dufresne, J.-C., 526

Elving, P.J., 527

Farsang, G., 528
Ferguson, J.A., 527
Fleischmann, M., 526
Fleury, D., 526
Fleury, M.B., 526, 527
Foucault, A., 528

Ghe, A.M., 529
Ghoneim, M.M., 526
Gileadi, E., 526
Gonzalez, E.R., 527
Gouverneur, L., 526
Goyal, R.N., 529
Grabner, E.W., 526
Granath, K., 528
Gross, M., 527
Grubbs, R., 527
Gulick, W.M., Jr., 526
Gupta, P.N., 529

Hammerich, O., 526, 527, 528
Hayakawa, Y., 528
Hazelrigg, M.J., Jr., 528
Hemingway, R.E., 527
Heyrovská, R., 528
Heyrovský, M., 528
Hoffmann, H., 525
Høgberg, K., 525
Holleck, L., 527
Holý, A., 525
Honda, K., 526
Hopf, F.R., 527
Horner, L., 526
Hrdý, O., 525
Humffray, A.A., 526

Ibl, N., 526
Imberger, H.E., 526
Inesi, A., 529
Ishikura, S., 526
Issa, I.M., 525, 526
Issa, R.M., 525, 526
Itahashi, E., 528
Iwakura, C., 526

Jacobsen, E., 525
Jain, P.S., 525
Jemal, M., 525
Jensen, B.S., 526
Jordan, J., 527
Joshi, G.C., 529
Jubault, M., 526, 527

Kalia, R.K., 527
Kalinowski, M.K., 529
Kariv, E., 526
Katayama, A., 526
Katiyar, S.S., 529
Kazemifard, G., 527
Kihara, T., 526
Kita, H., 526
Knevel, A.M., 525
Kollár, V., 528
Kováč, J., 528
Kroupa, J., 525
Kutowy, O., 526

Lacán, M., 527
Ladányi, L., 528
Lal, S., 525
Lalithambika, M., 529
Lamy, E., 528
Landsberg, R., 526
Laviron, E., 526, 527
Le Moullec, Y., 527
Leroy, G., 526
Levy, M., 529
Louati, A., 527
Lund, H., 526, 527

Mahapatra, S., 527
Mahesh, V.K., 529
Mahmoud, M.R., 525
Malik, W.U., 529
Marcoux, L., 528
Margel, S., 529
Marhold, J., 525
Martelli, M., 528
Martigny, P., 527
Masson, J.P., 529
Mastragostino, M., 529
Matrka, M., 525
Meyer, T.J., 527
Michel, M.A., 526
Millet, C., 527
Moiroux, J., 526
Molle, G., 526
Mousset, G., 526
Müller, S., 526
Murahashi, S.I., 527
Murayama, D.R., 527

Nadjo, L., 528, 529
Navrátil, F., 525
Negi, A.S., 526
Nygård, B., 528

Oelschläger, H., 525
Ostern, M., 526
Osteryoung, J.G., 525

Pajdowski, L., 526
Pak, C.M., 526
Parker, A.J., 526
Parker, V.D., 526, 527, 528
Pedersen, C.T., 528
Peltier, D., 526, 527
Perjéssy, A., 525
Petránek, J., 525
Pilloni, G., 528
Pípalová, J., 525
Pletcher, D., 526
Plichon, V., 528
Ponjeé, J.J., 529
Pozdeeva, A.A., 525
Pragst, F., 527

Rahman, N.N., 525
Rampazzo, L., 529
Ratajczak, M.H., 526
Ratard, D., 528
Rault, E., 526, 527
Rieke, R.D., 528, 529
Robertson, P.M., 526
Rogstad, A., 525
Ronlán, A., 527
Roullier, L, 526, 527
Rusina, A., 528
Ryba, O., 525
Ryvolová, A., 525

Saber, T.M.H., 528
Ságner, Z., 525
Sammour, M.H., 527
Santhanam, K.S.V., 527
Santiago, E., 526
Sato, Y., 526
Savéant, J.M., 526, 528, 529
Saxena, R.S., 526
Schiavon, G., 528, 529
Schluter, D.N., 527
Schwager, F., 526
Serve, D., 526, 527
Shang, D.T., 529
Sharma, L.R., 527

Sharp, M., 527
Shawali, A.S., 527
Shoesmith, D.W., 526
Simonet, J., 526, 527
Školová, E., 526
Slifstein, C., 528
Smyth, M.R., 525
Smyth, W.F., 525
Sohr, H., 525
Štěrba, V., 525
Strassburger, J., 528
Sukigara, M., 526
Sümmermann, W., 525
Sundholm, F., 526
Sundholm, G., 526

Tabaković, I., 527
Takamura, K., 528
Takamura, T., 526
Tamura, H., 526
Temerk, Y.M., 525, 526
Tenderende-Gumińska, B., 529
Terni, H.A., 526
Tessier, D., 529
Tokel-Takvoryan, N.E., 527
Trazza, A., 527, 528, 529
Trémillon, B., 529
Troll, T., 526
Tsunaga, M., 526
Tvaroha, B., 525

Valcher, S., 528, 529
Van Duyne, R.P., 527
Vass, V., 528
Vavřička, S., 528
Vecchi, E., 528
Volke, J., 525, 528
Vystrčil, A., 525

Weist, P., 528
Whitten, D.G., 527
Williams, J.C., Jr., 528

Yoneyama, H., 526

Zacharová-Kalavská, D., 525
Zador, I., 526
Zecchin, S., 528, 529
Zhdanov, S.I., 525
Zotti, G., 528

Corrigenda 537

CORRIGENDA FOR PRECEDING VOLUMES OF THE CRC HANDBOOK SERIES IN ORGANIC ELECTROCHEMISTRY

VOLUME II
Table II

Page	Code Number	Correction
15	AK74	Structure should show Cl^- instead of I^-.

VOLUME III
Table I

Page	Code Number	Correction
200	CG75	The name should read "thianthrene".

VOLUME IV
Table I

Page	Code Number	Correction
201	DF69	First value in col. 16 should read "-0.735".

Table IV

Page	Code Number	Correction
385		"Pyrene, DH44" should be added in alphabetical order.

VOLUME V
Table I

Page	Code Number	Correction
16	EA09	Add: "aa50" to code.
52	EA93	Add: "cb22" to code.
73	EB29	Col. 21, lines 3 and 4 should read "$dE_{\frac{1}{2}}/dlog[N^+]$".
80	EB40	Add: "ag91" to code.
108	EC03	Add: "db62" to code.
	EC04	Add: "db63" to code.
172	ED80	Add: "dd60" to code.

Table IX

Page		Correction
437	Carbon paste	silicone fluid (MS200), EB53 should read EB47

VOLUME V

Table X

441	C0026-2749	The first line should read:	Němečková, A., Maturová,
442	JE038-0245	The first name should read:	Chambers, J.Q.,
443	JE040-0345	The first name should read:	Bréant, M.,
	JE042-0049	The first name should read:	Cardinali, M.E.,
	JE043-0318	The first name should read:	Koch, V.R.,
444	JE050-0073	The second line should read:	and Vianello, É., EC23
445	JE054-0232	The second line should read:	and Trémillon, B., EB09,
	Vol. 57(1974)	The third reference should read:	JE057-0191

Table XI

449   For Breant read Bréant

       For Chambers, J.A., read Chambers, J.Q., 442, 444-45

450   For Hlavatý read Hlávatý

       For Kazemifard read Kazenifard

       For Línek read Linek

451   For Tremillon read Trémillon

ALL OTHER (cont.)

| LUMO | lowest unoccupied molecular orbital |

M

M	merges with final current rise
m	rate of flow of Hg [mg s^{-1}]
MAS, MS	mass spectrometry
MB	McIlvaine
Mc( )	mistake corrected (column)
MeCN	acetonitrile
MG	1,2-dimethoxyethane (monoglyme)
MM	modulation polarography
Mn	minimum
MPT, mp	melting point
MSE	mercury-mercurous sulfate electrode
MWT, mw	molecular weight
Mx	maximum

N

n	number of electrons
n_a	number of electrons transferred through rate-determining step of a cathodic process
n_{app}	an apparent number of electrons transferred
n_b	number of electrons transferred through rate-determining step of an anodic process
N_o	collection efficiency (ring-disc electrode)
NBS	National Bureau of Standards pH scale
NCE	normal calomel electrode
neg	negative
NHE	normal hydrogen electrode
NMR	nuclear magnetic resonance
N-S	αn_a or $k_{s,h}$ calculated from peak separation (Nicholson-Shain equation)
ns	not stated

O

0	no wave observed
o	organic
(oc)	open circuit
oh =	number of hydroxide ions consumed through rate-determining step
oxidn	oxidation

P

P	merges with previous wave
p	preliminary
p=	number of protons consumed through rate-determining step
PA	triangular-wave polarography
PB	higher harmonic ac polarography with phase-sensitive rectification
PC	current-scanning polarography
PD	derivative polarography
PE	pretreated electrode
PF	differential polarography
PG	square-wave polarography
PH	higher harmonic ac polarography
PHEN	phenol
PHOS	phosphate
PHTH	phthalate
PI	alternating-voltage polarography
PK	Kalousek polarography
pK_a	-log(acidic dissociation constant)
pK'	pH at inflection point of polarographic dissociation curve
PL	intermodulation polarography
Pl	plateau
PM	multisweep polarography
PO	oscillopolarography
Po	postwave
pos	positive
PP	pulse polarography
ppt	precipitate, precipitation
PQ	polarographic coulometry
PR	cyclic triangular-wave polarography
Pr	prewave
PrCN	butyronitrile
PROC	propylene carbonate
PRW	Prideaux and Ward
PS	staircase polarography
PT	Tast polarography
PV	total ac polarography
PVC	poly(vinyl chloride)
PVI	in-phase ac polarography
PVQ	quadrature polarography
PW	single-sweep polarography
PX	double-tone polarography
PY	polarography
PYR	pyridine

Q

Q	in characteristic potential column, incision quotient; elsewhere, quantity of charge [μC]
QE	controlled-potential coulometry
QI	quasi-reversible
QP	controlled-current coulometry without a reagent precursor
Qp	integrated charge corresponding to the faradaic current [μC]
QR	chronocoulometry
QT	controlled-current coulometry with a reagent precursor
QU	double potential-step chronocoulometry
QW	electrogravimetry

R

R	reversible
r	(alone) reliable, or (with a subscript) radius
r_d	radius of disc [cm]
$r_{r,i}$	inner radius of ring [cm]
$r_{r,o}$	outer radius of ring [cm]
redn	reduction
rms	root mean square
R_4NCE	Hg|Hg$_2$Cl$_2$($\underline{s}$), R$_4$NCl($\underline{c}$) electrode (followed by identity of R and value of $\underline{c}$)
RP	radio-frequency polarography
rxn	reaction

S

S	merging with succeeding wave
satd	saturated
SCE	saturated calomel electrode (with KCl unless otherwise specified)
Se	energy-sufficient
Sr	surface reaction
St	slope of tangent $E_{\frac{1}{2}}$
sttd	stated
SULF	sulfate
SULN	sulfolane
supp. elect.	supporting electrolyte
sur	surface reaction

T

T	temperature [°C]
t	drop time [s]
$\underline{t}$	time [s]
$\underline{t}_s$	switching (reversal) time
t(c)	controlled drop time [s]
t(oc)	drop time on open circuit [s]
t(sc)	drop time on short circuit [s]
Tafel	Tafel slope (dE/dlogi)
Tc	temperature coefficient of the limiting current [% deg^{-1}]
THF	tetrahydrofuran
TI	thin-layer chronoamperometry
TLC	thin-layer chromatography
TLE	thin-layer electrode
TLQE	thin-layer coulometry
Tomeš	$E_{\frac{s}{4}} - E_{\frac{1}{4}}$ (mV)
TQ	thin-layer chronocoulometry
TRIS	tris(hydroxymethyl)aminomethane
TX100	Triton X-100

U

(u)	arbitrary unit
UB	universal buffer
UVS, UV	ultraviolet spectroscopy

V

V	volts
v	scan rate [mV s^{-1}]
V_f	rate of flow through a tubular electrode [cm^{3}s^{-1}]
VA	triangular-wave voltammetry
VC	current-scanning voltammetry
VD	derivative voltammetry
VF	differential voltammetry
VIS	visible spectroscopy